COMPREHENSIVE SERIES IN PHOTOCHEMISTRY
AND PHOTOBIOLOGY

Series Editors

Donat P. Häder
Professor of Botany

and

Giulio Jori
Professor of Chemistry

European Society for Photobiology

COMPREHENSIVE SERIES IN PHOTOCHEMISTRY AND PHOTOBIOLOGY

Series Editors: Donat P. Häder and Giulio Jori

Titles in this Series:

Volume 1 UV Effects in Aquatic Organisms and Ecosystems
Edited by E.W. Helbling and H. Zagarese

Volume 2 Photodynamic Therapy
Edited by T. Patrice

Volume 3 Photoreceptors and Light Signalling
Edited by A. Batschauer

Volume 4 Lasers and Current Optical Techniques in Biology
Edited by G. Palumbo and R. Pratesi

Volume 5 From DNA Photolesions to Mutations, Skin Cancer and Cell Death
Edited by É. Sage, R. Drouin and M. Rouabhia

Volume 6 Flavins: Photochemistry and Photobiology
Edited by E. Silva and A.M. Edwards

Volume 7 Photodynamic Therapy with ALA: A Clinical Handbook
Edited by R. Pottier, B. Krammer, R. Baumgartner, H. Stepp

Volume 8 Primary Processes of Photosynthesis, Part 1: Principles and Apparatus
Edited by G. Renger

Volume 9 Primary Processes of Photosynthesis, Part 2: Principles and Apparatus
Edited by G. Renger

Volume 10 Biophysical and Physiological Effects of Solar Radiation on Human Skin
Edited by Paolo U. Giacomoni

Volume 11 Photodynamic Inactivation of Microbial Pathogens: Medical and Environmental Applications
Edited by Michael R. Hamblin and Giulio Jori

Visit our website at http://www.rsc.org/Shop/Books/

COMPREHENSIVE SERIES IN PHOTOCHEMISTRY AND PHOTOBIOLOGY–VOLUME 11

Photodynamic Inactivation of Microbial Pathogens: Medical and Environmental Applications

Editors

Michael R. Hamblin
Wellman Center for Photomedicine
Massachusetts General Hospital
55 Fruit Street
Boston, MA 02114
USA

and

Giulio Jori
Department of Biology
University of Padova, Via U. Bassi 58/B
Padova
I-35131 Padova
Italy

RSC Publishing

ISBN: 978-1-84973-144-7
ISSN: 2041-9716

A catalogue record for this book is available from the British Library

Published by The Royal Society of Chemistry,
Thomas Graham House, Science Park, Milton Road,
Cambridge CB4 0WF, UK

Registered Charity Number 207890

For further information see our website at www.rsc.org

Preface for the ESP Series in Photochemical and Photobiological Sciences

"It's not the substance, it's the dose which makes something poisonous!" When Paracelsius, a German physician of the fourteenth century, made this statement he probably did not think about light as one of the most obvious environmental factors. But his statement applies as well to light. While we need light, for example, for vitamin D production, too much light might cause skin cancer. The dose makes the difference. These diverse findings of light effects have attracted the attention of scientists for centuries. The photosciences represent a dynamic multidisciplinary field which includes such diverse subjects as behavioural responses of single cells, cures for certain types of cancer and the protective potential of tanning lotions. It includes photobiology and photochemistry, photomedicine as well as the technology for light production, filtering and measurement. Light is a common theme in all these areas. In recent decades a more molecular centred approach changed both the depth and the quality of the theoretical as well as the experimental foundation of photosciences.

An example of the relationship between global environment and the biosphere is the recent discovery of ozone depletion and the resulting increase in high-energy ultraviolet radiation. The hazardous effects of high-energy ultraviolet radiation on all living systems is now well established. This discovery of the result of ozone depletion put photosciences at the centre of public interest with the result that, in an unparalleled effort, scientists and politicians worked closely together to come to international agreements to stop the pollution of the atmosphere.

The changed recreational behaviour and the correlation with several diseases in which sunlight or artificial light sources play a major role in the causation of clinical conditions (*e.g.* porphyrias, polymorphic photodermatoses, *Xeroderma pigmentosum* and skin cancers) have been well documented. As a result, in some countries (*e.g.* Australia) public services inform people about the potential risk of extended periods of sun exposure every day. The problems are often aggravated by the phototoxic or photoallergic reactions produced by a variety of environmental pollutants, food additives or therapeutic and cosmetic drugs.

Photodynamic Inactivation of Microbial Pathogens: Medical and Environmental Applications
Editors Michael R. Hamblin and Giulio Jori

Published by the Royal Society of Chemistry, www.rsc.org

On the other hand, if properly used, light-stimulated processes can induce important beneficial effects in biological systems, such as the elucidation of several aspects of cell structure and function. Novel developments are centred around photodiagnostic and phototherapeutic modalities for the treatment of cancer, artherosclerosis, several autoimmune diseases, neonatal jaundice and others. In addition, classic research areas such as vision and photosynthesis are still very active. Some of these developments are unique to photobiology, since the peculiar physico-chemical properties of electronically excited biomolecules often lead to the promotion of reactions which are characterized by high levels of selectivity in space and time. Besides the biologically centred areas, technical developments have paved the way for the harnessing of solar energy to produce warm water and electricity or the development of environmentally friendly techniques for addressing problems of large social impact (*e.g.* the decontamination of polluted waters). While also in use in Western countries, these techniques are of great interest for developing countries.

The European Society for Photobiology (ESP) is an organization for developing and coordinating the very different fields of photosciences in terms of public knowledge and scientific interests. Due to the ever-increasing demand for a comprehensive overview of the photosciences the ESP decided to initiate an encyclopaedic series, the *Comprehensive Series in Photochemical and Photobiological Sciences*. This series is intended to give an in-depth coverage over all the very different fields related to light effects. It will allow investigators, physicians, students, industry and laypersons to obtain an updated record of the state-of-the-art in specific fields, including a ready access to the recent literature. Most importantly, such reviews give a critical evaluation of the directions that the field is taking, outline hotly debated or innovative topics and even suggest a redirection if appropriate. It is our intention to produce the monographs at a sufficiently high rate to generate a timely coverage of both well-established and emerging topics. As a rule, the individual volumes are commissioned; however, comments, suggestions or proposals for new subjects are welcome.

Donat-P. Häder and Giulio Jori
Spring 2002

Volume Preface

Photodynamic therapy (PDT)[1] was discovered over 100 years ago by observation of the death of microorganisms upon exposure to dyes and visible light.[2] Shortly afterwards the necessity of the presence of molecular oxygen was discovered and the term "photodynamic action" was coined.[3] Regardless of the origins of PDT in microbiology, the principal medical application of the technology during the last century was in the field of cancer treatment. Prompted by discoveries made by Lipson and Baldes of the ability of porphyrins to localize in tumours,[4] and by pioneering work from Dougherty[5] (among others), a large body of work accumulated on the synthesis of first- and second-generation photosensitizers,[6] their use to kill cancer cells *in vitro*[7] and the ability of many systemically administered photosensitizers based on tetrapyrrole structures to destroy tumours in animal models when they were exposed to red light.[8] Clinical approvals for PDT for various cancers started to be issued in various countries in 1988.[9,10] In 2000 PDT using benzoporphyrin derivative or Visudyne was approved for treatment of choroidal neovascularization secondary to age-related macular degeneration and this approval led to a lot of applications in ophthalmology.[11] The discovery of the ability of 5-aminolevulanic acid (ALA) to stimulate biosynthesis of the photosensitizing protoporphyrin IX[12] when it was topically applied to skin tumours and other lesions led to a large number of applications of PDT in dermatology.

Over the 90-year period from 1900 to 1990 there were scattered papers published on the ability of various combinations of dyes and light to kill a variety of microbial strains.[13] However, many of these reports were concerned with very basic mechanistic studies such as the identity of the various reactive oxygen species produced by PDT (especially singlet oxygen),[14] and whether PDT could damage various cellular constituents such as nucleic acids and proteins.[14] Viral lesions were one of the first actual infections to be clinically treated by PDT. In the 1970s topically applied neutral red and white light were used to treat herpes simplex lesions on the face and genitalia in several clinical studies[15] but this practice ceased when a paper in the *New England Journal of Medicine*[16] showed that not only was the procedure largely ineffective, but that

Photodynamic Inactivation of Microbial Pathogens: Medical and Environmental Applications
Editors Michael R. Hamblin and Giulio Jori

Published by the Royal Society of Chemistry, www.rsc.org

there was a danger of cancer being caused by the technique. However, in recent years PDT mediated by porphyrin photosensitizers has begun to be used again for treating viral lesions, especially human papilloma virus lesions on the genitals and in the airways as discussed in Chapter 11.

Blood product sterilization was one of the first medical applications of antimicrobial PDT. Although the main challenge today is removing viruses, a few years ago bacteria, fungi and parasites could also be dangerous. This application is discussed in Chapter 9.

The discovery of antibiotics in the 1940s revolutionized the treatment of infectious diseases, especially those diseases caused by pathogenic bacteria.[17] Antifungal drugs followed in the 1960s, and antiprotozoal, antihelminthic and antimalarial compounds all became available in due course. The extensive panoply of anti-infective drugs available led to the widespread belief that infectious disease had been conquered, and would never threaten human health as it has done from the beginning of recorded history. Unfortunately the medical profession was in for a rude awakening. Antibiotic resistance began to emerge shortly after the introduction of antibiotics in 1946,[18] although it took some time for it to be appreciated as a serious problem. A combination of factors including widespread inappropriate prescription of antibiotics for minor viral diseases, the use of large amounts of antibiotics in livestock feedstuffs, the tendency of patients who have been correctly prescribed antibiotics not to finish the course of treatment and the rise in global travel have conspired together to make the spreading of antibiotic resistance an exponentially growing global problem.[19]

The rise of antibiotic resistance has led to widespread predictions of the "end of the antibiotic era".[20,21] This alarming prospect has stimulated the search for alternative antimicrobial therapies that could be used against infections that in theory could not respond to any known antibiotic and might therefore be termed untreatable. Many of these alternative antimicrobial approaches are based on discoveries made long before the discovery of antibiotics. For instance, phage therapy was developed in Russia and Georgia before the Second World War and has been clinically used sporadically in these countries ever since.[22] Several research groups in Western countries are now studying phage therapy.[23] Another ancient antimicrobial approach making a comeback is the use of silver.[24] Silver preparations are now used both to treat and to prevent infections usually by topical application.

Antimicrobial PDT is yet another example of a technology making a comeback as a possible approach to treat multi-drug resistant infections. It was known that Gram-positive bacteria were highly susceptible to photoinactivation with a wide range of photosensitizers such as those traditionally used to treat cancer after intravenous injection. Therefore one of the key discoveries that was made in the 1990s was related to ways of making other microorganisms (in particular Gram-negative bacteria) susceptible. It was found that combination approaches that overcame the Gram-negative permeability barrier such as polymyxin nonapeptide,[25] EDTA-mediated removal of divalent cations[26] or induction of competence[27] could allow the same photosensitizers that could easily kill Gram-positives to be effective against Gram-negatives. Furthermore it was soon discovered that simply imparting one (or preferably more than one) cationic charge to the photosensitizer molecule could produce a broad spectrum photosensitizer that could inactivate all classes of micro-organism; Gram-positive, Gram-negative, fungi, viruses, parasites and even highly resistant microbial life-forms such as bacterial spores[28] and parasite cysts[29] could be killed (see Chapters 1–6). Many sophisticated antimicrobial photosensitizers have now been synthesized based on tetrapyrrole backbones and other molecular frameworks.

In recent years applications of PDT to treat bacterial infections both in animal models and in patients have begun to be explored. In Figure 0.1 we show how the number of papers in the broad area of antimicrobial PDT and PDT for infections has increased year-on-year from 1954 to the present day. The small rise in the 1970s was due to the clinical use of PDT for viral lesions referred to above. The second exponential increase started in the 1990s, coinciding with the discovery that Gram-negative bacteria could also be photo-inactivated if cationic photosensitizers were employed.

It is usually accepted that treatment of infections involves topical application of the photosensitizer to the infected tissue followed by delivery of light after a relatively short drug-light interval. One common finding in studies that use PDT to kill bacteria that are infecting tissue is that the total dose of PDT (PS concentration and fluence of light) is much larger (100–1000 times larger) than the dose that is needed to kill microorganisms *in vitro*. A recent publication provided a perspective on why this should be.[30] Since the definition of infection is 10^5 CFU g^{-1} tissue and 10^5 bacteria have a total mass of only 1 microgram, it is evident that there is 1 million times more mass of tissue than there is mass of bacteria and this provides a real challenge for selective binding of the PS to

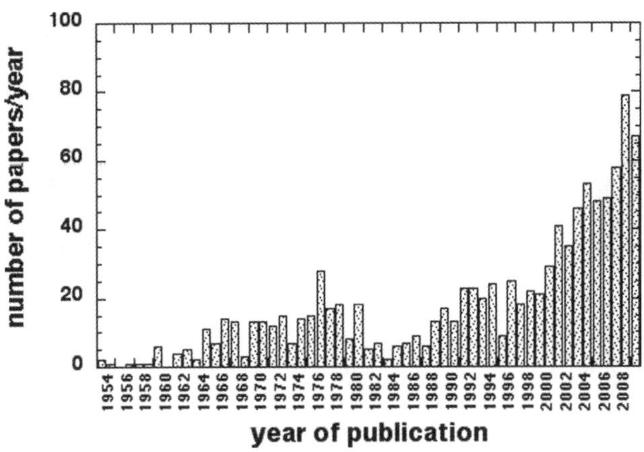

Figure 0.1 The number of papers per year published in the general area of antimicrobial PDT obtained by a search of the MedLine database between 1954 and 2010.

the microbial cells rather than the host cells. Even if the infection is very severe (10^8 CFU g^{-1} tissue) the ratio is still 1000 : 1.

In animal models of infection this is readily easy to accomplish but, as discussed in Chapter 8, the use of bioluminescent pathogens and real-time optical imaging of living infected animals using low-light cameras has simplified the experimental approach dramatically.[31] Clinical applications of antimicrobial PDT have been most widespread amongst dentists and especially among periodontists (Chapter 13). Fifty-thousand clinical procedures using methylene blue and 660-nm light delivered by fibre into the dental pocket have been performed using the Periowave System marketed by Ondine Biopharma.[32] ALA-PDT is clinically used for acne (see Chapter 10). Other clinical trials have taken place for healing of infected leg ulcers (Chapter 15) and using blue light alone for gastric *H. pylori* infection (Chapter 16). Finally, the last chapter (Chapter 17) covers the possible use of antimicrobial PDT in the environment to kill insects and other pests as well as undesirable microbes.

The Editors would like to thank the following for valuable assistance with editing tasks.

Sulbha K. Sharma, PhD
Yingying Huang, MD
Gitika B. Kharkwal, PhD
Tyler G. St. Denis

Michael R. Hamblin
and Giulio Jori

References

1. M.R. Hamblin and P. Mroz, *Advances in Photodynamic Therapy: Basic, Translational and Clinical*, Artech House, Norwood, MA, 2008.
2. O. Raab, Uber die Wirkung fluoreszierender Stoffe auf Infusorien, *Z. Biol.*, 1900, **39**, 524–546.
3. H. Von Tappeiner and A. Jodlbauer, Uber Wirkung der photodynamischen (fluorieszierenden) Stoffe auf Protozoan und Enzyme, *Dtsch. Arch. Klin. Med.*, 1904, **80**, 427–487.
4. R.L. Lipson, E.J. Baldes and A.M. Olsen, The use of a derivative of hematoporphyrin in tumor detection, *J. Natl. Cancer Inst.*, 1961, **26**, 1–11.
5. T.J. Dougherty, J.E. Kaufman, A. Goldfarb, K.R. Weishaupt, D. Boyle and A. Mittleman, Photoradiation therapy for the treatment of malignant tumors, *Cancer Res.*, 1978, **38**, 2628–2635.
6. C.J. Gomer, Preclinical examination of first and second generation photosensitizers used in photodynamic therapy, *Photochem. Photobiol.*, 1991, **54**, 1093–1107.
7. P. Agostinis, E. Buytaert, H. Breyssens and N. Hendrickx, Regulatory pathways in photodynamic therapy induced apoptosis, *Photochem. Photobiol. Sci.*, 2004, **3**, 721–729.
8. T.J. Dougherty, G.B. Grindey, R. Fiel, K.R. Weishaupt and D.G. Boyle, Photoradiation therapy. II. Cure of animal tumors with hematoporphyrin and light, *J. Natl. Cancer Inst.*, 1975, **55**, 115–121.

9. G.R. Prout, Jr, C.W. Lin, R. Benson, Jr, U.O. Nseyo, J.J. Daly, P.P. Griffin, J. Kinsey, M.E. Tian, Y.H. Lao, Y.Z. Mian, X. Chen, F.M. Ren and S.J. Qiao, Photodynamic therapy with hematoporphyrin derivative in the treatment of superficial transitional-cell carcinoma of the bladder, *New Engl. J. Med.*, 1987, **317**, 1251–1255.
10. T.J. Dougherty, An update on photodynamic therapy applications, *J. Clin. Laser Med. Surg.*, 2002, **20**, 3–7.
11. S. Mennel, I. Barbazetto, C.H. Meyer, S. Peter and M. Stur, Ocular photodynamic therapy – standard applications and new indications (part 1). Review of the literature and personal experience, *Ophthalmologica*, 2007, **221**, 216–226.
12. J.C. Kennedy, R.H. Pottier and D.C. Pross, Photodynamic therapy with endogenous protoporphyrin IX: basic principles and present clinical experience, *J. Photochem. Photobiol. B*, 1990, **6**, 143–148.
13. J.D. Macmillan, W.A. Maxwell and C.O. Chichester, Lethal photosensitization of microorganisms with light from a continuous-wave gas laser, *Photochem. Photobiol.*, 1966, **5**, 555–565.
14. H. Singh and J.A. Vadasz, Singlet oxygen quenchers and the photodynamic inactivation of E. coli ribosomes by methylene blue, *Biochem. Biophys. Res. Commun.*, 1976, **76**, 391–397.
15. L.E. Bockstahler, C.D. Lytle and K.B. Hellman, A review of photodynamic therapy for herpes simplex: benefits and potential risks, *NY J. Dent.*, 1975, **45**, 148–157.
16. M.G. Myers, M.N. Oxman, J.E. Clark and K.A. Arndt, Failure of neutral-red photodynamic inactivation in recurrent herpes simplex virus infections, *New Engl. J. Med.*, 1975, **293**, 945–949.
17. J. Clardy, M.A. Fischbach and C.R. Currie, The natural history of antibiotics, *Curr Biol.*, 2009, **19**, R437–R441.
18. E.J. Soulsby, Resistance to antimicrobials in humans and animals, *Br. Med. J.*, 2005, **331**, 1219–1220.
19. N. Khardori, Antibiotics – past, present, and future, *Med. Clin. North Am.*, 2006, **90**, 1049–1076.
20. S.G. Bell, Antibiotic resistance: is the end of an era near?, *Neonatal Netw.*, 2003, **22**, 47–54.
21. M.D. Poole, Are we facing the end of the antibiotic era?, *Ear Nose Throat J.*, 1993, **72**, 433.
22. J.N. Housby and N.H. Mann, Phage therapy, *Drug Discov. Today*, 2009, **14**, 536–540.
23. A. Wright, C.H. Hawkins, E.E. Anggard and D.R. Harper, A controlled clinical trial of a therapeutic bacteriophage preparation in chronic otitis due to antibiotic-resistant Pseudomonas aeruginosa; a preliminary report of efficacy, *Clin. Otolaryngol.*, 2009, **34**, 349–357.
24. M. Spear, Silver: an age-old treatment modality in modern times, *Plast. Surg. Nurs.*, 2010, **30**, 90–93.
25. Y. Nitzan, M. Gutterman, Z. Malik and B. Ehrenberg, Inactivation of gram-negative bacteria by photosensitized porphyrins, *Photochem. Photobiol.*, 1992, **55**, 89–96.
26. G. Valduga, G. Bertoloni, E. Reddi and G. Jori, Effect of extracellularly generated singlet oxygen on gram-positive and gram-negative bacteria, *J. Photochem. Photobiol. B*, 1993, **21**, 81–86.
27. G. Bertoloni, F. Rossi, G. Valduga, G. Jori and J. van Lier, Photosensitizing activity of water- and lipid-soluble phthalocyanines on Escherichia coli, *FEMS Microbiol. Lett.*, 1990, **59**, 149–155.

28. T.N. Demidova and M.R. Hamblin, Photodynamic inactivation of Bacillus spores, mediated by phenothiazinium dyes, *Appl. Environ. Microbiol.*, 2005, **71**, 6918–6925.
29. K. Kassab, T. Ben Amor, G. Jori and O. Coppellotti, Photosensitization of Colpoda inflata cysts by meso-substituted cationic porphyrins, *Photochem. Photobiol. Sci.*, 2002, **1**, 560–564.
30. M.R. Hamblin and T. Dai, Can surgical site infections be treated by photodynamic therapy?, *Photodiagnosis Photodyn. Ther.*, 2010, **7**, 134–136.
31. T.N. Demidova, F. Gad, T. Zahra, K.P. Francis and M.R. Hamblin, Monitoring photodynamic therapy of localized infections by bioluminescence imaging of genetically engineered bacteria, *J. Photochem. Photobiol. B*, 2005, **81**, 15–25.
32. R. Polansky, M. Haas, A. Heschl and G. Wimmer, Clinical effectiveness of photodynamic therapy in the treatment of periodontitis, *J. Clin. Periodontol.*, 2009, **36**, 575–580.

Contents

Photodynamic Inactivation of Microbial Pathogens: Medical and Environmental Applications
Editors Michael R. Hamblin and Giulio Jori

Published by the Royal Society of Chemistry, www.rsc.org

To my dearest wife Angela, without whom this book would not have been possible.

Michael R. Hamblin

Chapter 1

Antimicrobial Photodynamic Therapy: Basic Principles

Giulio Jori, Monica Camerin, Marina Soncin, Laura Guidolin and Olimpia Coppellotti

Department of Biology, University of Padova, Via Ugo Bassi 58B, 35131 Padova, Italy

Table of Contents

Photodynamic Inactivation of Microbial Pathogens: Medical and Environmental Applications
Editors Michael R. Hamblin and Giulio Jori

Published by the Royal Society of Chemistry, www.rsc.org

1.1. Introduction

Photodynamic therapy (PDT) represents a well-established modality for the treatment of a variety of malignant tumours; on the other hand, this technique is presently being extended to several non-cancerous conditions, especially those which are characterized by overgrowth of unwanted or abnormal cells, such as *inter alia* several dermatological diseases, benign prostatic hyperplasia and age-related macular degeneration.[1,2] An increasing amount of attention is also being focused on the application of PDT for the treatment of infectious diseases.[3] In actual fact, the possibility to kill microorganisms by the combined action of visible light and a photosensitizing dye was discovered at the very beginning of the twentieth century, when Raab observed[4] that the presence of some exogenously added visible light-absorbing compounds, such as acridine orange, was necessary for sunlight or artificially produced visible light to promote the death of paramecia. Raab's observations were repeated with a variety of multi- and unicellular organisms, while Jodlbauer and von Tappainer demonstrated[5] that the presence of oxygen is an essential requisite for photosensitization to occur. The combined effect of the three elements, namely light, photosensitizer and oxygen, has been termed photodynamic action.[6] However, the potential of PDT against diseases of microbial origin was not exploited until the beginning of the present century, since it was repeatedly observed that some well-known pathogens, especially Gram-negative bacteria and protozoa in the cystic stage, are poorly responsive to PDT with the most traditional photosensitizing agents, including xanthene or acridine derivatives and those negatively charged porphyrins that have been frequently used in the PDT of tumours (*e.g.* Photofrin, benzoporphyrin derivative, tetrasulfonated porphyrins or phthalocyanines);[7] furthermore, the discovery of antibiotics raised the belief that microbiologically based diseases would have been gradually reduced to a level that no longer had a serious impact on human health. On the contrary, at present the therapy of infectious diseases is increasingly challenged by the rapid evolutionary changes and large number of pathogens encountered, as well as by the continuous onset of multi-drug-resistant pathogens; as a consequence, quite a few types of infections, which, in the past, could be readily treated, now often lead to increased morbidity and may become life-threatening even in a nosocomial environment.[8] The problem is further exacerbated by the truly large variety of mechanisms adopted by microbial cells to increase their resistance to external insults. These include a thickening of their outer wall, encoding of new proteins which prevent the penetration of drugs, onset of mutants deficient in those porin channels allowing the influx of externally added chemicals, *etc.*[9] As a consequence, it has so far proven to be very difficult to identify a comprehensive strategy for overcoming this problem.

A significant step forward in the overall scenario of antimicrobial PDT occurred during the 1990s owing to the discoveries independently made by three research groups showing that visible light-activated cationic photosensitizers belonging to the family of phenothiazines,[10] porphyrins[11] and phthalocyanines[12] induce a fast and extensive killing of typical Gram-negative bacteria, such as *Escherichia coli* and *Pseudomonas aeruginosa*, in addition to the inactivation of fungi and Gram-positive bacteria.

Several lines of evidence indicate[13–15] that PDT is particularly suitable for the treatment of localized microbial infections, including those which have become chronic after prolonged chemotherapy. Favourable features of PDT include:

- the possibility to develop treatment regimes efficiently acting on most classes of microbial pathogens, including Gram-positive and Gram-negative bacteria, yeasts, fungi and parasitic protozoa;
- the high susceptibility to PDT exhibited by antibiotic-resistant microbial strains with no parallel selection of photoresistant species even after multiple treatments;
- the selectivity of microbial cell killing in comparison with the constituents of host tissues;
- the low risk of inducing mutagenic effects;
- the possibility to activate the photodynamic sensitizers by means of inexpensive and safe visible light sources. In particular, while PDT can be easily applied for the therapy of infections in external readily accessible organs, the continuous progresses in optical fibre technology makes even deep-seated infections amenable to the photodynamic approach.

Thus, even though PDT is still in its infancy as regards microbiological applications, it is likely to become a mainstream therapeutic option in the near future, at least for specific indications.

1.2. Photosensitized Inactivation of Microbial Cells: General Aspects

1.2.1. The Target

Microbial cells display a truly large variety of size, sub-cellular architecture and biochemical composition; as a consequence, the susceptibility to photosensitized processes can be significantly different for the various microorganisms.

Thus, Gram-positive bacteria are surrounded by an outer wall, which is separated from the plasma membrane by a periplasmic space. The 20–80 nm thick wall represents a protective mesh mainly constituted by peptidoglycan layers, which are traversed by negatively charged lipoteichoic and teichuronic acids anchored in the membrane. This kind of spatial arrangement does not act as a strict permeability barrier, since several macromolecules with molecular weight up to about 60 000 Da were found to readily diffuse through the wall to reach the inner membrane.[13,14] Therefore, the most common photosensitizing agents, whose molecular weight is generally not greater than 1500 Da, can more or less promptly cross the outer wall and localize in the immediate surroundings of the photosensitive endocellular sites. On the other hand, Gram-negative bacteria are characterized by the presence of an additional 10–15 nm thick structural element, external to the peptidoglycan network, whose constituents (*e.g.* lipoproteins, lipopolysaccharides, teichoic and lipoteichoic acids) provide the outer surface with a quasi-continuum of densely packed negative charges: this highly organized system inhibits the penetration of compounds with molecular weight larger than 600–700 Da.[15] A thick external wall, composed of a mixture of lipoproteins, chitin, glucan and mannan, is also present in yeasts, a group of rather heterogeneous eukaryotic organisms, which are the causative factors of various infectious diseases. The wall is again separated from the cytoplasmic membrane by a periplasmic space. The permeability properties of such a wall are fairly similar to those typical of Gram-positive bacteria.

Pathogenic protozoa also represent a group of structurally rather disparate organisms, which raise a complicated problem for any therapeutic approach. Protozoa include parasitic and opportunistic pathogenic forms, which are characterized by a variety of complex life cycles and are often incapable of living outside the host except as resistant stages.[16] As an example, the general pattern of life cycle of amoebic protozoa, such as *Entamoeba histolytica* and *Acanthamoeba* spp., is a trophic or feeding stage, alternating with a resting cystic stage. The cysts, which develop as environmental conditions become unfavourable (lack of food, waste product accumulation, desiccation), survive in nature until growth conditions improve and can remain viable for as long as 25 years. Indeed, the cysts are enclosed by a wall that may consist of two or three layers, each up to 300 nm thick, which forms a physical barrier. In *Acanthamoeba castellanii* the wall is constituted by fibrous material in which cellulose and an acid-insoluble protein have been identified as the principal components. On the other hand, the trophozoite is poorly equipped to survive for prolonged periods of time and is generally responsible for the lesions in infections of different tissues where it rapidly propagates.[17,18]

On the basis of these considerations, it appears that Gram-negative bacteria and cysts from protozoa represent the most challenging targets for any type of antimicrobial treatment. It is worth emphasizing that in several cases the infections are caused by a heterogeneous microbial flora, hence the protocol adopted for the treatment of such infections cannot be focused on just one type of pathogens, rather it must be characterized by the possibility to efficaciously act on microbial pathogens with very different biological, morphological and physiological characteristics. As discussed below, PDT shows peculiar features which allow this problem to be addressed in an adequate manner.

1.2.2. Photobiological Processes in the Photosensitized Inactivation of Microbial Cells

As it is well known,[19] photodynamic processes proceed by two competitive mechanisms, both of which require the participation of the long-lived photo-excited triplet state of the photosensitizer (^{3}Sens) as the reactive intermediate. The type I pathway involves an electron transfer step between the triplet photosensitizer and a nearby substrate (Sub) with generation of radical species; the latter are then intercepted by ground state molecular oxygen yielding a variety of oxidized products. A typical simplified scheme for a type I photosensitization mechanism can be outlined as follows:

$$^{3}\mathrm{Sens} + \mathrm{Sub} \rightarrow \mathrm{Sens}^{(-)} + \mathrm{Sub}^{(+)} \quad (1.1)$$

$$\mathrm{Sub}^{(+)} + \mathrm{O_2} \rightarrow \mathrm{Sub_{ox}} \quad (1.2)$$

The direction of the electron transfer event between the photosensitizer and the substrate is controlled by the relative redox potentials of the two species.

Alternatively, the type II pathway involves an electronic energy transfer process from the triplet photosensitizer to a suitable acceptor, most frequently oxygen, which is exceptionally a triplet in its ground state. The latter compound is converted to a highly reactive derivative, named singlet oxygen ($^{1}O_2$), which in turns attacks photosensitive targets in its surroundings:

$$^{3}\mathrm{Sens} + {}^{3}\mathrm{O_2} \rightarrow \mathrm{Sens} + {}^{1}\mathrm{O_2} \quad (1.3)$$

$$^{1}\mathrm{O_2} + \mathrm{Sub} \rightarrow \mathrm{Sub_{ox}} \quad (1.4)$$

In both cases, the oxidative attack is of electrophilic nature, hence it is preferentially directed toward specific biomolecules,[20] including

(a) amino acids possessing an aromatic or heterocyclic side chain (*e.g.* tryptophan, tyrosine, histidine), as well as amino acids containing sulfur atoms (*e.g.* cysteine, methionine);
(b) pyrimidine and purine bases of DNA/RNA, with emphasis on guanosine;
(c) unsaturated lipids, such as oleic, linoleic and arachidonic acids, and steroids (especially cholesterol).

It is generally assumed[19,21] that porphyrin- and phenothiazine-photosensitized processes proceed *via* formation of singlet oxygen, which thus represents the main cytotoxic agent responsible for the photodynamic inactivation of microbial cells. Indeed, it has been observed that the cultivation of *Staphylococci* in the presence of singlet oxygen scavengers, such as methionine or histidine, provides a high level of protection for the photosensitized cultures.[22] At the same time, some authors demonstrated that singlet oxygen, generated by a photosensitizer deposited on an inert matrix and physically separated from the microbial cell culture, can diffuse through an air-equilibrated medium and cause an irreversible damage of *Streptococcus faecium* and *E. coli*.[23,24] The decrease in the overall efficiency of the cell photoinactivation process upon increasing the distance between the immobilized photosensitizer and the bacterial cells corresponds very well with the measured lifetime of $^{1}O_2$ in the reaction medium. However, the possibility of an at least partial

Table 1.1 Main functional and morphological effects of porphyrin photosensitized processes in microbial cells.

Type of photodamage	Specific example	Comments
a) Functional damage		
Inhibition of selected enzymic activities	Inactivation of NADH/lactic/succinic dehydrogenase	Drop of activity in membrane enzymes well correlates with drop in survival
Formation of protein-protein cross-links	A number of membrane and cytoplasmic proteins	Cytoplasmic proteins involved upon Sens rediffusion from the initially targeted membrane
Inhibition of metabolic processes	Inhibition of DNA synthesis and glucose transport	RNA and protein synthesis are consequently blocked
b) Morphological damage		
Alteration of the mesosome structure	Increased volume and frequency of appearance	Reflects a disturbed synthesis of membrane and cell wall
Alteration of chromatin	Appearance of electron-transparent areas with highly packed nucleic acid	Late stage in the overall photoprocess

contribution to the overall photoprocess from type I (radical-involving) reaction pathways cannot be ruled out, since a close spatial relationship between a cell-associated photosensitizer and photosensitive cell constituents could favour a direct interaction. In actual fact, the photoinduced formation of a radical anion from an *E. coli*-bound phthalocyanine has been detected.[25]

In general, all the photosensitizers which have been found to be very active as photoantimicrobial agents preferentially localize at the level of the cytoplasmic membrane. One important exception is represented by acridines, such as proflavine or acridine orange, which largely intercalate with DNA bases.[26,27] As a consequence, the main alterations of cell functions and morphology, caused by photodynamic inactivation, are typical of damaged membranous domains (see Table 1.1). No involvement of the genetic material is generally observed until the late stages of the overall photoprocess, which indicates that such photodamage is not correlated with cell death.[21,28] This pattern of photoinduced subcellular damage is in agreement with the lack of mutagenic effects, as it has been repeatedly reported by different investigators,[29,30] as well as with the lack of selection of photoresistant microbial strains even after several photosensitization treatments.[31]

1.2.3. The Photosensitizer

Several photosensitizing agents of photodynamic nature proved to induce an efficient inactivation of at least some classes of microbial pathogens.[19,26,28] On the basis of the experience developed so far and the considerations on the biochemical and morphological features of microbial cells (see paragraph 1.2.1), it is possible to outline which are the optimally desirable properties of an antimicrobial photosensitizer:

(a) Suitable photophysical characteristics, including a high extinction coefficient in the deeply tissue-penetrating wavelength regions (mainly the red and far-red spectral interval) even though for the treatment of superficial infections also the intensely absorbed blue light (400–420 nm) can be very useful; a long-lived electronically excited triplet state; and a high quantum yield for the generation of reactive oxygen species (*in primis* singlet oxygen) when exposed to excitation with visible light.

(b) A large affinity for the broadest possible classes of microbial cells, taking advantage of either the relatively high permeability of the outer wall (*e.g.* Gram-positive bacteria, trophozoites) or the presence of specific biochemical properties in this structural element (*e.g.* the presence of a vast array of negatively charged carboxylate functions in Gram-negative bacteria).

(c) The preferential binding of the photosensitizer with the cytoplasmic membrane, whose photosensitive constituents (*e.g.* unsaturated lipids, proteins) will consequently represent the main targets of the photoinactivation process; as a consequence, cell death will be predominantly due to membrane damage rather than involving the genetic material.

(d) The mechanisms involved in cell photoinactivation should have a minimal risk, if any, to induce the onset of mutagenic processes and to lead to the selection of microbial strains which are resistant to photodynamic action. The multi-target nature of the photodynamic killing of microbial cells can be considered as a strong warranty for reaching this objective.

(e) A broad spectrum of action on bacteria, fungi, yeasts and parasitic protozoa, in order to facilitate the achievement of an efficacious therapeutic effect also in the treatment of those infectious diseases which are characterized by the presence of a heterogeneous flora of pathogens.

(f) The possibility to identify a therapeutic window allowing an extensive mortality of the microbial pathogens with minimal simultaneous damage to the constituents of the host tissue, as well as the re-growth of the pathogens after the treatment. The achievement of the latter goal can especially rely on the lack of mutagenic effects or increased photoresistance in the photodynamically inactivated microbial cells; this would allow the repetition of the PDT approach in case of insufficient response of the infective agents to the primary treatment or the recurrence of the infection after a given time interval from the primary treatment.

While those photosensitizers which fulfil the criteria (a) and (b) are readily inducing an extensive killing of Gram-positive bacteria, fungi and trophozoites from parasitic protozoa, the goals mentioned in point (e) are generally much more difficult to achieve, largely owing to the tight organization of the outer wall and, respectively, the exocyst in Gram-negative bacteria and cysts derived from protozoa. Only photosensitizers which are characterized by the presence of positively charged functional groups in either the axial ligands to the centrally coordinated metal ion or the peripheral substituents located in the pyrrole rings or the meso positions of the macrocycle exhibit a sufficiently high photoactivity against these two classes of microorganisms.[3,30] The substituents imparting a broad and markedly efficient antimicrobial photosensitizing activity are generally represented by moieties, such as pyridine, piperidine, morpholine, *etc*., whose nitrogen atoms can be quaternarized by alkylation reactions.[32,33] In addition, even tertiary amino groups can represent an acceptable alternative provided their basic strength is sufficiently high to allow for protonation of the lone electron pair in the nitrogen atom at neutral pH values.

In general, cationic photosensitizers besides broadening the field of application of PDT by involving essentially all the classes of pathogens in their microbicidal action, also display a substantially larger efficacy as compared with their neutral or anionic counterparts. This conclusion can be drawn through an attentive analysis of the data shown in Table 1.2, namely the comparison between the minimal bactericidal concentration (MBC) for selected photosensitizing agents. For this comparative analysis, MBC is taken as the minimal photosensitizer concentration yielding a 4 log decrease in cell survival under a given set of irradiation conditions.

The values in the table cannot be quantitatively compared since in all cases the irradiations have been performed by using full spectrum visible light and no corrections have been made for differences in the molar extinction coefficient of the individual photosensitizing agents. In any case, it is apparent that the di-anionic haematoporphyrin and the tetra-anionic tetrasulfonated porphine are essentially ineffective against the Gram-negative bacterium *Escherichia coli*, while their photodynamic efficiency against Gram-positive bacteria (namely, both the wild and one antibiotic-resistant strain of *S. aureus*) is significantly lower than that displayed by tetracationic meso-substituted porphyrins. The latter are very active against both Gram-positive and Gram-negative bacteria.

Table 1.2 Minimal bactericidal concentration (MBC, μM) of photosensitizing agents with different chemical structure.

Porphyrin	*S. aureus* wild strain	*S. aureus* methicillin-resistant	*E. coli*
meso-$TPPS_4$	1.0	1.5	>10
Haematoporphyrin	2.2	2.3	>10
meso-TPyP $(N\text{-methyl})_4$	0.8	0.9	1.1
meso-TPyP $(N\text{-methyl})_3$ (N-hexyl)	0.6	0.7	1.2
meso-TPyP $(N\text{-methyl})_3$ (N-dodecyl)	0.2	0.4	0.5
meso-TPyP $(N\text{-methyl})_3$ (N-tetradecyl)	0.8	0.8	1.7
meso-TPyP $(N\text{-methyl})_3$ (N-docosanyl)	5.2	5.0	>10
Toluidine blue	4.0	5.3	6.1

$TPPS_4$ = tetra(4-sulfonatophenyl)porphine.

TPyP = tetra-pyridyl-porphine; the various derivatives are substituted by alkyl chains at the level of the pyridine N atoms.

The **MBC** value is taken as the minimal photosensitizer concentration yielding a 4 log decrease in cell survival under a given set of irradiation conditions.

All irradiations were performed by using photosensitizer concentrations in the micromolar range and visible light at a fluence rate of 50 mW cm^{-2}.

Cationic photosensitizers have also been shown to efficiently promote the inactivation of bacterial spores[34] and protozoa in the cystic stage.[35]

The MBC data in Table 1.2 further suggest that the increase in the length of the alkyl chain bound to one meso-pyridyl ring from 1 to 6 and especially to 12 carbon atoms enhances the antibacterial activity of the porphyrin. Analogous observations were reported by S. Brown (personal communication) upon replacing the four amino-bound methyl groups in methylene blue by four butyl groups. It appears[3,33] that the presence of a hydrophobic hydrocarbon chain brings about two effects: (a) it interferes with the forces stabilizing the three-dimensional architecture of the outer wall thereby increasing its permeability; (b) it anchors the photosensitizer within lipid domains of the cytoplasmic membrane, and as a consequence there is a stronger binding of the photosensitizer itself with the microbial cell.[33,36] The very high photobactericidal activity of the *N*-dodecyl porphyrin derivative (see Table 2.1) is also in agreement with its quantum yield for singlet oxygen generation close to 1.0, that is almost twice the value typical of most meso-substituted tetracationic porphysins.[37] The marked drop in the photosensitizing activity observed for the tetracationic analogue bearing a carbon chain with 20 methylene groups correlates with the drastic decrease in the triplet lifetime and quantum yield for singlet oxygen generation consequent to the formation of micellar-type porphyrin oligomers in aqueous media due to the increased degree of hydrophobicity.[37,38]

The partial loss of the tight organization of the bacterial and fungal outer wall can also be obtained by the covalent attachment of polycationic polypeptide chains to the photosensitizer molecule even when the latter is devoid of positively charged moieties. A significant drop in the survival of both Gram-positive and Gram-negative bacterial cells is achieved by conjugation of the neutral chlorin e_6 molecule with polylysine, provided the length of the polypeptide chain is sufficiently long (*ca.* 30 amino acid residues); shorter chain (*ca.* 10 residues) polylysine conjugates with the photosensitizer do not promote any appreciable phototoxic effect on Gram-negative bacteria.[39] Identical results were obtained upon conjugation of polylysine with porphycenes[40] and porphyrins,[41] as well as by selecting poly-ornithine and poly-ethyleneimine as the polycationic membrane-permeabilizing agents.

Lastly, it is important to underline that the number and the position of the positively charged groups which are present in the photosensitizer molecule often play a major role in modulating the course and efficacy of the photoprocess against microbial cells. Thus, it has been found[33,42] that amphiphilic porphyrin or phthalocyanine derivatives, such as those in which two cationic moieties protrude from two adjacent pyrrole or azaisoindole rings, are endowed with the greatest photoinactivation efficiency on Gram-positive bacteria and yeasts, whereas the more hydrophilic octa-cationic derivatives usually act more efficiently on Gram-negative bacteria.

Some bacteria, including *Helicobacter pylori*[43] and *Propionibacterium acnes*,[44] can also produce a number of endogenous porphyrins (predominantly, protoporphyrin IX) through a variant of the normal haem biosynthetic pathway. An analogous situation has been observed for some black-pigmented bacteria, which normally colonize the oral cavity, *e.g. Prevotella nigriscens*.[45] As a consequence, these bacteria are sensitive to irradiation with visible light, and in particular with blue light (400–420 nm), which is most efficiently absorbed by porphyrins. For such cases, direct irradiation of the areas affected by the bacterial infections is possible and has so far given promising results in preliminary clinical applications, *e.g.* in the treatment of some forms of acne.[46] For those bacteria which do not naturally produce significant amounts of endogenous porphyrins, the possibility exists to stimulate an active biosynthesis of protoporphyrin IX through the addition of 5-amino-levulinic acid (ALA),[47] even though this possibility appears to be successful only in some bacterial cells and the detailed mechanism of the processes leading to this modification of the natural biosynthesis of haem has not been elucidated.

1.3. Kinetics and Mechanism of the Photosensitized Inactivation of Microbial Cells

The photosensitized inactivation of microbial cells can be schematically subdivided into the following steps:

(a) *The incubation*. This step usually requires a time period of just a few minutes (1–5 min.) in the case of bacteria and vegetative protozoan cells, since the binding is driven by the electrostatic interaction between the positively charged groups on the photosensitizer molecule and the negatively charged groups at the surface of the microbial cells. On the other hand, the cysts of protozoa and yeasts usually require incubation times of the order of at least 30 min. in order to achieve endocellular photosensitizer concentrations which can be photochemically active.[48,49] Such a fast binding process is of prime importance, since the photosensitizer doses which induce a 4–5 log decrease in microbial cell survival are in the millimolar range; under these conditions, no significant accumulation of the cationic photosensitizing agents into mammalian cells takes place.[50] Therefore, irradiation of an infected lesion at a few minutes after topical deposition of millimolar amounts of the photosensitizer normally causes an extensive decrease in the population of the microbial flora with no appreciable damage in correspondence of the host tissue. This conclusion was experimentally demonstrated by *in vitro* investigations showing that fibroblasts and keratinocytes, *i.e.* some out of the most frequent constituents of human tissues, undergo no measurable loss of survival under irradiation conditions which provoke a 5 log decrease in the survival of typical bacterial infective agents, such as meticillin-resistant *S. aureus* and *E. coli*;[51] moreover, during the PDT treatment of *S. aureus* soft tissue infections in mice, the host tissue underwent minimal damage upon application of PDT regimes causing about a 3 log decrease in the bacterial population.[52] Therefore, the antimicrobial PDT technique appears to be endowed with a high degree of selectivity.

(b) *The primary targets of the photosensitized process.* The earliest and most frequent target of microbial cell inactivation photosensitized by phenothiazine, porphyrin or phthalocyanine-type derivatives appears to be represented by the cytoplasmic membrane. A number of direct or indirect experimental observations support this statement

a. Several enzymes, which are associated with the cytoplasmic membrane of bacterial cells, such as NADH, succinic and lactic dehydrogenase, are inactivated by irradiation in the presence of positively charged porphyrin or phthalocyanine photosensitizers at a rate which is faster than that observed for the photoinduced cell death.[51] Several outer membrane and cytoplasmic membrane proteins undergo an extensive cross-linking in the early stages of the phenothiazine-photosensitized killing of *Porphyromonas gingivalis*.[53]

b. Studies on cationic porphyrin-photosensitized *Candida albicans* cells performed by means of freeze-fracture electron microscopy point out that the photoinduced damage progresses from the outer to the inner leaflet of the cytoplasmic membrane.[54]

c. The presence of serum in the incubation medium reduces the amount of photosensitizer bound with the microbial cells owing to the competitive association of the photosensitizer itself with serum albumin and lipoproteins.[55] On the contrary, no appreciable effect on the efficiency of the photosensitized process is induced by changes in the ionic strength or pH of the medium, provided such changes do not affect the chemical structure of the photosensitizing agents.[33,56]

d. The photosensitization process of a number of yeast and bacterial cells is accompanied by a loss of membrane barrier properties, with collapse of the K^+ and ionic balance.[57,58]

e. The photoprocess also results in the massive reduction of the transport capacity for a variety of solutes, with a consequent shortage of essential substrates for anabolic and catabolic pathways.[59]

The overall process is obviously of multi-target nature, thus it is extremely difficult for microbial cells to develop protection strategies which eventually determine the selection of strains showing an enhanced resistance to the photodynamic treatment. This represents one major advantageous feature of antimicrobial PDT.

(c) *The later stages of the photosensitization process.* The examination of bacteria and protozoa which had been incubated in the dark with tetracationic porphyrins or phthalocyanines at the fluorescence microscope showed[33] that in both cases the photosensitizing agent is initially partitioned in the cytoplasmic membrane, which then represents the primary site of attack by the photogenerated reactive oxygen species (see above). However, as irradiation progresses, the photosensitizer gradually diffuses to inner cellular districts; in particular, in the case of trophozoites from *Acanthamoeba palestinensis*, the fluorescence emitted by

> the photosensitizer is very evident in the contractile vacuole for unirradiated cells and becomes appreciably more diffuse in the photosensitized samples.[33]

It is thus reasonable to expect that a variety of non-membranous sites, including DNA, are involved in the photooxidative reactions at later stages of the overall photoprocess. However, several investigators concluded that such additional damage is not directly correlated with cell death, which is already quite extensive when the cytoplasmic membrane has been heavily disrupted (see references 3, 21 and 28 for a detailed description and discussion).

In actual fact, a major contribution by DNA modification to cell death is very unlikely since

(a) wild strains of *E. coli*, as well as *E. coli* strains which are defective for DNA repair mechanisms, exhibit a closely similar sensitivity to photoinactivation promoted by cationic porphyrins;[31]
(b) no mutagenic effects have been reported for a large number of microbial cells which had been extensively photosensitized by phenothiazine, porphyrin or phthalocyanine derivatives of cationic nature;[3,21,28]
(c) the bacterium *Deinococcus radiodurans*, which possesses a very efficient system of DNA repair, undergoes a rapid killing upon exposure to visible light in the presence of photodynamic-type agents.

However, caution must be exerted in drawing general conclusions from the data so far available, since specific situations may be characterized by a different scenario. For example, Salmon-Divon *et al.*[59,60] found that the inactivation of *E. coli* photosensitized by a tetracationic meso-substituted porphyrin is primarily dependent on the photodamage induced at the level of genomic DNA rather than on protein and membrane malfunction. Therefore, more in-depth studies are necessary in order to shed more complete light on this issue.

1.4. Conclusions

The application of PDT for the treatment of microbial infections shows a number of attractive features especially owing to the mild experimental conditions which yield a drastic decrease in the survival of a large number of pathogenic agents with minimal risk of damage to host tissues. The optimal protocol for the application of photodynamic processes against microbial pathogens is outlined in Table 1.3.

The perspectives for the utilization of this technique at a clinical level are particularly stimulating since formulations which can efficiently transfer the photosensitizing agents across the *stratum corneum*, thus allowing for the topical deposition of the infected lesion, are now available.[61] Such formulations are very useful for the use of PDT aimed at the removal of periodontal biofilms[62] and the disinfection of burn wounds.[63] Quite interestingly, virulence

Table 1.3 PDT protocol yielding an efficient photodynamic killing of microbial pathogens.

Parameter	Optimal values
Photosensitizer chemical structure	Cationic phenothiazines, phthalocyanines, porphyrins
Structural factors enhancing the antimicrobial activity	Attachment of hydrocarbon chains, cationic polypeptides, targeting factors (e.g. antibodies)
Cell-photosensitizer incubation time prior to irradiation	5–10 min. (Gram-positive bacteria, protozoa trophozoites); 30 min. (yeasts, Gram-negative bacteria, protozoa cysts)
Photosensitizer concentration	0.1–5 μM
Ionic strength	Efficiency is independent of the salt concentration
Fluence rate	Lower than 50 mW cm^{-2}
Total light fluence (irradiation time)	Lower than 10 J cm^{-2} (30 min.)

factors, such as V8 protease, sphingomyelinase and α-haemolysin, which cause cellular damage following a wound, can also be treated using PDT.[64] At present, the tetrabutyl analogue of methylene blue is in phase II clinical trials for the sterilization of infected wounds.[65]

Pharmacokinetic studies on different animal models showed no detectable recovery of the *in situ* administered porphyrins or phthalocyanines from either serum or liver, which is the main site involved in porphyrin clearance from the organism *via* the bile-gut pathway;[2,66] thus, no appreciable leakage of the photosensitizer to the dermis should occur, which guarantees against the development of persistent cutaneous photosensitivity. This is the main undesired side effect of tumour PDT.[1]

In summary, antimicrobial PDT is presently emerging from its infancy era and its scope and potential are likely to undergo an important expansion in the near future, giving this therapeutic modality a specific role at least for the treatment of localized infections.

References

1. T.J. Dougherty, C.J. Gomer, B.W. Henderson, G. Jori, D. Kessel, M. Korbelik, J. Moan and Q. Peng, Photodynamic therapy: a review, *J. Natl. Cancer Inst.*, 1998, **80**, 889–902.
2. J.G. Levy and M. Obochi, New applications of photodynamic therapy. Introduction, *Photochem. Photobiol.*, 1996, **64**, 737–739.
3. M.R. Hamblin and T. Hasan, Photodynamic therapy: a new antimicrobial approach to infectious diseases?, *Photochem. Photobiol. Sci.*, 2004, **3**, 436–450.
4. O. Raab, Über die Wirkung fluoresceinder Stoffe and Infusorien, *Zeit für Biol.*, 1900, **39**, 524–546.
5. A. Jodlbauer and H. von Tappainer, Über die Beteiligung der Sauerstoffes bei der photodynamischem Wirkung fluorezierender Stoffe, *Münch. Med. Wochenschr.*, 1904, **26**, 1139–1141.
6. J.D. Spikes and R. Livingston, The molecular biology of photodynamic action: sensitised photoautoxidations in biological systems, in *Advances in Radiation*

Biology, ed. L.G. Augenstein, R. Mason and M. Zelle, Academic Press, New York, 1969, vol. 3, pp. 30–121.

7. W. Wainwright, Photoantimicrobials – a PACT against resistance and infection, *Drug Future*, 2004, **29**, 85–93.
8. R.A. Gleckman and F. Borrego, Adverse reactions to antibiotics: clues for recognizing, understanding and avoiding them, *Postgrad. Med.*, 1997, **101**, 97–108.
9. T.L. Smith, M.L. Pearson and K.R. Wilcox, Emergence of vancomycin resistance in *Staphylococcus aureus:* glycopeptide-intermediate. *Staphylococcus aureus* working group, *New England J. Med.*, 1999, **340**, 493–501.
10. B. Wilson, T. Burns, J. Pratten and G.J. Pearson, Bacteria in supragingival plaque samples can be killed by low power light in the presence of a photosensitiser, *J. Appl. Bacteriol.*, 1995, **78**, 569–574.
11. M. Merchat, G. Bertoloni, P. Giacomoni, A. Villanueva and G. Jori, Meso-substituted cationic porphyrins as efficient photosensitisers of Gram-positive and Gram-negative bacteria, *J. Photochem. Photobiol., B: Biol.*, 1996, **32**, 153–157.
12. A. Minnock, D.I. Vernon, J. Schofield, J. Griffiths, J.H. Parrish and S.B. Brown, Photoinactivation of bacteria: use of a cationic water-soluble zinc phthalocyanine to photoinactivate both Gram-negative and Gram-positive bacteria, *J. Photochem. Photobiol., B: Biol.*, 1996, **32**, 159–164.
13. C.L. Friedrich, D. Moyles, T.J. Beveridge and R.E. Hancock, Antibacterial action of structurally diverse cationic peptides on Gram-positive bacteria, *Antimicrob. Agents Chemother.*, 2000, **44**, 2086–2092.
14. H. Nikaido, Prevention of drug access to bacterial targets: permeability barrier and active efflux, *Science*, 1994, **264**, 362–368.
15. L. Leive, The barrier function of the Gram-negative envelope, *Ann. NY Acad. Sci.*, 1994, **235**, 109–129.
16. F.L. Schuster and G.S. Visvesvara, Amoebae and ciliated protozoa as causal agents of water-borne zoonotic diseases, *Vet. Parasitol.*, 2004, **126**, 91–120.
17. F.J. Chlapowsky and R.N. Band, Assembly of lipids into membranes in *Acanthamoeba palestinensis*, *J. Cell Biol.*, 1971, **50**, 634–651.
18. D. Lloyd, N.A. Turner, W. Khunkitto, A.C. Hann, J.R. Furr and A.D. Russell, Encystation in *Acanthamoeba castellanii*: development of biocide resistance, *J. Eukaryot. Microbiol.*, 2001, **48**, 11–16.
19. G. Jori and S.B. Brown, Photosensitised inactivation of microorganisms, *Photochem. Photobiol. Sci.*, 2004, **3**, 403–405.
20. M. Ochsner, Photophysical and photobiological processes in the photodynamic therapy of tumours, *J. Photochem. Photobiol., B: Biol.*, 1997, **39**, 1–8.
21. M. Wainwright, Photodynamic antimicrobial chemotherapy, *J. Antimicrob. Chemother.*, 1998, **42**, 13–28.
22. Y. Nitzan, B. Shainberg and Z. Malik, The mechanism of photodynamic inactivation of *Staphylococcus aureus* by deuteropophyrin, *Curr. Microbiol.*, 1989, **19**, 265–269.
23. T.A. Dahl, W.R. Midden and P.E. Hartman, Comparison of killing of Gram-positive and Gram-negative bacteria by pure singlet oxygen, *J. Bacteriol.*, 1989, **171**, 2188–2194.
24. G. Valduga, G. Bertoloni, E. Reddi and G. Jori, Effect of extracellularly generated singlet oxygen on Gram-positive and Gram-negative bacteria, *J. Photochem. Photobiol., B: Biol.*, 1993, **21**, 81–86.
25. J.A. Lacey and D. Phillips, The photobleaching of disulfonated aluminium phthalocyanine in microbial systems, *Photochem. Photobiol. Sci.*, 2002, **1**, 120–125.

26. T. Ito, Photodynamic agents as tools for cell biology, *Photochem. Photobiol. Reviews*, 1983, **7**, 141–186.
27. M. Wainwright, D.A. Poenix, J. Marland, D.R.A. Wareing and F.J. Bolton, *In vitro* photoantibacterial activity of aminoacridines, *J. Antimicrob. Chemother.*, 1997, **40**, 587–589.
28. G. Jori, Photodynamic therapy of microbial infections: state-of-the-art and perspectives, *J. Environ. Pathol. Toxicol. Oncol.*, 2006, **25**, 505–519.
29. Z. Malik, H. Ladan and Y. Nitzan, Photodynamic inactivation of Gram-negative bacteria: problems and possible solutions, *J.Photochem. Photobiol., B: Biol.*, 1992, **14**, 261–266.
30. T. Maisch, R.M. Szeimies, G. Jori and C. Abels, Antibacterial photodynamic therapy in dermatology, *Photochem. Photobiol. Sci.*, 2004, **3**, 907–917.
31. G. Valduga, B. Breda, G.M. Giacometti, G. Jori and E. Reddi, Photosensitisation of wild and mutant strains of *Escherichia coli* by meso-tetra-(N-methyl-4-pyridyl)porphine, *Biochem. Biophys. Res. Commun.*, 1999, **256**, 84–88.
32. D.A. Caminos and E.N. Durantini, Synthesis of asymmetrically meso-substituted porphyrins bearing amino groups as potential cationic photodynamic agents, *J. Porph. Phthaloc.*, 2005, **9**, 334–342.
33. G. Jori, C. Fabris, M. Soncin, S. Ferro, O. Coppellotti, D. Dei, L. Fantetti and G. Roncucci, Photodynamic therapy in the treatment of microbial infections: basic principles and perspective applications, *Lasers Surg. Med.*, 2006, **38**, 468–481.
34. T.N. Demidova and M.R. Hamblin, Photodynamic inactivation of *Bacillus* spores mediated by phenothiazinium dyes, *Appl. Environ. Microbiol.*, 2005, **71**, 6918–6925.
35. S. Ferro, O. Coppellotti, G. Roncucci, T. Ben Amor and G. Jori, Photosensitised inactivation of *Acanthamoeba palestinensis* in the cystic stage, *J. Appl. Microbiol.*, 2006, **101**, 206–211.
36. G. Bertoloni, F.M. Lauro, G. Cortella and M. Merchat, Photosensitising activity of haematoporphyrin on *Staphylococcus aureus* cells, *Biochim. Biophys. Acta*, 2000, **1475**, 169–174.
37. E. Reddi, M. Ceccon, G. Valduga, G. Jori, J.C. Bommer, F. Elisei, L. Latterini and U. Mazzucato, Photophysical properties and antibacterial activity of meso-substituted cationic porphyrins, *Photochem. Photobiol.*, 2002, **75**, 462–470.
38. M. Merchat, J.D. Spikes, G. Bertoloni and G. Jori, Studies on the mechanism of bacteria photosensitisation by meso-substituted cationic porphyrins, *J. Photochem. Photobiol., B: Biol.*, 1996, **35**, 149–157.
39. M.R. Hamblin, D.A. O'Donnell, N. Murthy, K. Rajagopalan, N. Michaud, M.E. Sherwood and T. Hasan, Polycationic photosensitiser conjugates: effects of chain length and Gram classification on the photodynamic inactivation of bacteria, *J. Antimicrob. Chemother.*, 2002, **49**, 941–951.
40. L. Polo, A. Segalla, G. Bertoloni, G. Jori, K. Schaffer and E. Reddi, Polylysine-porphycene conjugates as efficient photosensitisers for the inactivation of microbial pathogens, *J. Photochem. Photobiol., B: Biol.*, 2000, **59**, 152–158.
41. J. Tomé, M.G.P.M. Neves, A.C. Tomé, J.A.S. Cavaleiro, M. Soncin, M. Magaraggia, S. Ferro and G. Jori, Synthesis and anti-bacterial activity of new poly-L-lysine – porphyrin conjugates, *J. Med. Chem.*, 2004, **47**, 6649–6652.
42. G. Jori and O. Coppellotti, Inactivation of pathogenic micro-organisms by photodynamic techniques: mechanistic aspects and perspective applications, *Anti-infective Agents in Medicinal Chemistry*, 2007, **6**, 119–131.

43. M.R. Hamblin, J. Viveiros, C. Yang, A. Ahmadi, R.A. Ganz and M. Tolkoff, *Helicobacter pylori* accumulates photoactive porphyrins and is killed by visible light, *Antimicrob. Agents Chemother.*, 2005, **49**, 2822–2827.
44. W.L. Lee, A.R. Shalita and M.B. Poh-Fitzpatrick, Comparative studies of porphyrin production in *Propionibacterium acnes* and *Propionibacterium granulosum*, *J. Bacteriol.*, 1994, **133**, 811–815.
45. N.S. Soukos, S. Som, A.D. Abernethy, K. Ruggiero, J. Dunham, C. Lee, A. G. Doukas and J.M. Goodson, Phototargeting of oral black-pigmented bacteria, *Antimicrob. Agents Chemother.*, 2005, **49**, 1391–1396.
46. R.A. Ganz, J. Viveiros, A. Ahmadi, A. Khalil, M.J. Tolkoff, N.S. Nishioka and M.R. Hamblin, *Helicobacter pylori* in patients can be killed by visible light, *Lasers Surg. Med.*, 2005, **36**, 260–265.
47. Y. Nitzan, M. Salmon-Divon, E. Shporen and Z. Malik, ALA-induced photodynamic effects on Gram-positive and Gram-negative bacteria, *Photochem. Photobiol. Sci.*, 2004, **3**, 430–435.
48. D. A. Caminos, M.B. Spesia and E.N. Durantini, Photodynamic inactivation of *Escherichia coli* by novel meso-substituted porphyrins by 4-(3-N,N,N-trimethylammoniumpropoxy)-phenyl and 4-(trifluoromethyl)phenyl groups, *Photochem. Photobiol. Sci.*, 2006, **5**, 56–65.
49. G. Bertoloni, F. Rossi, G. Valduga, G. Jori, H. Ali and J.E. van Lier, Photosensitising activity of water- and lipid-soluble phthalocyanines on prokaryotic and eukaryotic microbial cells, *Microbios*, 1999, **71**, 33–46.
50. N. Oleinick, R.L. Morris and I. Belichenko, The role of apoptosis in response to photodynamic therapy: what, where, why and how, *Photochem. Photobiol. Sci.*, 2002, **1**, 1–21.
51. M. Soncin, C. Fabris, A. Busetti, D. Dei, D. Nistri and G. Roncucci, Approaches to selectivity in the Zn(II)-phthalocyanine-photosensitised inactivation of wild-type and antibiotic-resistant *Staphyloccus aureus*, *Photochem. Photobiol. Sci.*, 2002, **1**, 815–819.
52. F. Gad, T. Zahra, K.P. Francis, T. Hasan and M.R. Hamblin, Targeted photodynamic therapy of established soft tissue infections in mice, *Photochem. Photobiol. Sci.*, 2004, **3**, 451–458.
53. M. Bhatti, S.P. Nair, A.J. McRobert, B. Henderson, P. Shepherd, J. Cridland and M. Wilson, Identification of photolabile outer membrane proteins of *Porphyromonas gingivalis*, *Curr. Microbiol.*, 2001, **43**, 96–99.
54. S.A.G. Lambrechts, M.C.G. Aalders and J. van Arle, Mechanistic study of the photodynamic inactivation of *Candida albicans* by a cationic porphyrin, *Antimicrob. Agents Chemother.*, 2005, **49**, 2026–2034.
55. S.A.G. Lambrechts, M.C.G. Aalders, F.D. Verbraak, J.W.M. Lagerberg, J. Dankert and J.J. Schuitmaker, Effect of albumin on the photoinactivation of microorganisms by a cationic porphyrin, *J. Photochem. Photobiol., B: Biol.*, 2005, **79**, 51–57.
56. M. Paardekopper, A.W. deBrujine, J. van Steveninck and J.A. van den Broek, Inhibition of transport systems in yeast by photodynamic treatment with toluidine blue, *Biochim. Biophys. Acta*, 1992, **1151**, 143–148.
57. M. Paardekopper, P.J. A van den Broek, A.W. de Brujine, J.G.R. Elferink, T.M.A.R. Dubbelman and J. van Steveninck, Photodynamic treatment of yeast cells with the dye toluidine blue: all-or-none loss of plasma membrane barrier proteins, *Biochim. Biophys. Acta*, 1992, **1108**, 86–90.

58. Z. Malik, T. Babushkin, S. Sher, J. Hanania, H. Ladan, Y. Nitzan and S. Salzberg, Collapse of K^+ and ionic balance during photodynamic inactivation of leukemic cells, erythrocytes and Staphylococcus aureus, *Int. J. Biochem.*, 1993, **25**, 1399–1406.
59. M. Salmon-Divon, Y. Nitzan and Z. Malik, Mechanistic aspects of *Escherichia coli* photodynamic inactivation by cationic tetra-meso(N-methyl-pyridyl)porphine, *Photochem. Photobiol. Sciences*, 2004, **3**, 423–429.
60. M. Schafer, C. Schmitz and G. Horneck, High sensitivity of *Deinococcus radiodurans* to photodynamically produced singlet oxygen, *Int. J. Radiat. Biol.*, 1998, **74**, 249–253.
61. R.F. Donnelly, C.M. Cassidy, R.G. Loughlin, A. Brown, M.M. Tunney, M.G. Jenkins and P.A. McCarron, Delivery of methylene blue and meso-tetra (N-methyl-pyridyl)porphine tetratosylate from cross-linked poly(vinyl alcohol) hydrogels: a potyential means of photodynamic therapy of infected wounds, *J. Photochem. Photobiol., B: Biol.*, 2009, **26**, 223–231.
62. K. Konopka and T. Goslinski, Photodynamic therapy in dentistry, *J. Dent. Res.*, 2007, **86**, 694–707.
63. D. Dai, G.P. Tegos, Z. Lu, L. Huang, T. Zhiyentayev, M.J. Franklin, D. Baer and M.R. Hamblin, Photodynamic therapy for *Acinectobacter baumanii* burn infections in mice, *Antimicrob. Agents Chemother.*, 2009, **53**, 3929–3934.
64. S. Tubby, M. Wilson and S.P. Nair, Inactivation of the staphylococci virulence factors using a light-activated antimicrobial agent, *BMC Microbiol.*, 2009, **9**, 211–215.
65. S. B. Brown, personal communication.
66. Z. Huang, A review of the progress in clinical photodynamic therapy, *Technol. Cancer Res. Treatment*, 2005, **4**, 283–293.

Chapter 2

Phenothiazinium Salts as Antimicrobial Photosensitizing Agents

Mark Wainwright

School of Pharmacy & Biomolecular Science, Liverpool John Moores University, Liverpool L3 3AF, UK

Table of Contents

Photodynamic Inactivation of Microbial Pathogens: Medical and Environmental Applications
Editors Michael R. Hamblin and Giulio Jori

Published by the Royal Society of Chemistry, www.rsc.org

2.1. Introduction

Photoantimicrobial action represents the combination of light-absorbing chemicals, light and oxygen leading to the production of reactive oxygen species (ROS) intimately with non-economic microbial cells, leading to cell destruction/inactivation. Plainly this can be demonstrated more simply *in vitro* than in more complicated, clinical situations, but the photosensitizer properties allowing such potential are well established: microbial selectivity, ROS production, long wavelength light absorption, compound purity and ease of formulation. Among photosensitizer classes, there can be few examples better fitted to the antimicrobial application, which haa a strong historical pedigree and provides a better rationale for use in this field than any other, in combination with synthetic production methodologies which are well understood and allow relatively straightforward and wide-ranging analogue production.

2.1.1. History

In the broad sweep of dyes used in medicine, the synthetic dyes derived from anilines must surely hold pride of place, having been involved since the middle of the nineteenth century. Initially, the concept of textile coloration encouraged the production of synthetic aniline dyes, but the link between textile dyeing and differential staining in the contemporarily developing science of microscopy allowed the quantum jump to the use of dyes as therapeutics. Famously, Christian Gram developed his system for the broad staining classification of microscopic bacteria, using the basic dye crystal violet. Contemporary workers, including Ehrlich, Koch, Romanovsky and Leishman, followed similar principles in the study of tuberculosis and tropical disease.[1]

There are three fundamental sub-divisions of aniline dyes: azoics, azines and cyanines, each of which was well established by the 1920s. However, the azine class has been the most productive in terms of biologically active compounds, providing ubiquitously used stains for bacteria, fungi, viruses, protozoa and malignancy/pre-malignancy.[2] In addition, Ehrlich's realization that such selective stains might be used in the treatment of disease provided the foundation for modern chemotherapy, standard dyes such as methylene blue, acriflavine and proflavine being employed as front-line therapeutics by the early part of the twentieth century.[3,4] Following this, it is also possible to make a strong link between such dye-drugs and later conventional drugs, such as chloroquine (Figure 2.1).

The early demonstration of the antimicrobial capabilities of dyes such as methylene blue and acriflavine led to their testing in many areas of infectious disease, the associated tissue staining being acceptable in a period when there were very few alternatives – highly toxic arsenicals and mercurials constituting the main opposition until the arrival of the sulfonamides in the late 1930s.

The acridines acriflavine and proflavine had been introduced into clinical wound antisepsis during the First World War,[4] and proflavine was widely used during the Second World War, both as a single agent,[5] and also in combination

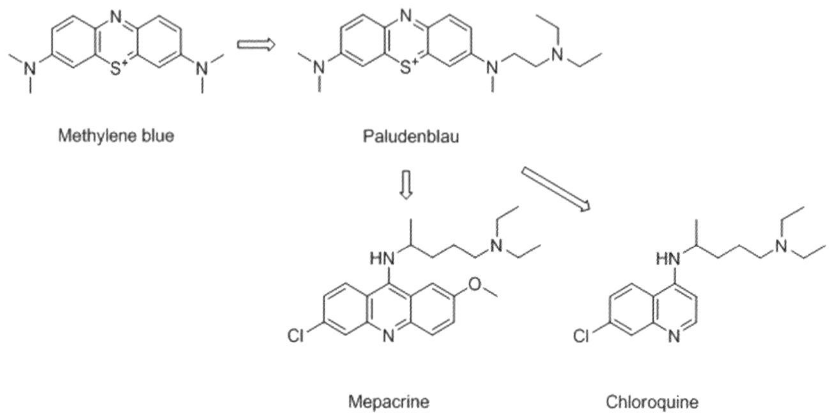

Figure 2.1 Antimalarial development from methylene blue.

with sulfonamides.[6] Conversely, little clinical use was made of methylene blue, and dye-based therapy was dwarfed by the introduction of the penicillins and later antibiotics. However, by the middle of the twentieth century methylene blue was known to be a powerful *photo*antimicrobial.

The cytotoxic photodynamic effect was first reported by Raab in 1900, using acridine and eosin against paramecia,[7] and the approach was also used against tumours by several other workers in the early part of the twentieth century.[8] However, the specific photoantimicrobial action of methylene blue was demonstrated in 1928 against a bacteriophage,[9] followed shortly by a similar demonstration for the related phenothiazinium toluidine blue,[10] and extended to photobactericidal activity by 1935.[11] It should be remembered that this demonstration of photoantimicrobial activity occurred between 1928 and 1935, the period bookended by Fleming's and Domagk's breakthrough publications on penicillin and prontosil (sulfonamides), respectively. One of the principal authors during this early realization of photoantimicrobial action, T'ung, had extended the work to include tropical pathogens by 1938.[12] Thus, allowing for very simple systems of illumination, much of the basis for the modern use of photoantimicrobials had been provided before the start of the Second World War.

Clearly the combination of technologies required to facilitate a proper phototherapeutic approach to infection control was not achievable during wartime. Indeed, the compact, user-friendly light sources routinely taken for granted today could not have been dreamed of by workers in the field 70 years ago, even if development funding had been available. Consequently, interest in the field languished.

However, as noted above, much of the basis for photoantimicrobial action lies in the staining of cells and tissue. Dyes such as the phenothiazinium methylene blue and the related benzo[*a*]phenoxazinium Nile blue A were in constant use in pathology throughout the photodynamic hiatus during the mid-twentieth century, and often used as lead compounds in drug research programmes. Notably, a large number of Nile blue derivatives were tested both as anti-cancer and anti-tuberculosis candidates by Crossley and coworkers,[13] following the useful vital staining of animal tumours *in vivo* by the parent compound.[14] Contemporary further, fuller reporting of photoantiviral action supported the suggestion of methylene blue as a photodisinfectant for blood products in 1955,[15] although this was contained in a US military report and thus not open to the wider scientific community until the 1980s. However, a considerable number of studies were carried out during the 1960s on the inactivation of animal viruses,[16] and this laid the foundation for the first proper clinical application of photoantimicrobials in the treatment of herpes lesions – usually herpes genitalis – in the United States during the early part of the following decade.[17] Notably, a photosensitizer commonly used in much of this antiviral work was the acridine proflavine, the precise interaction of which with nucleic acids had been published in 1961,[18] thus providing a biomolecular rationale for viral inactivation. By the time that the phototherapy of genital herpes was discontinued in the late 1970s, a significant body of work concerning more general photoantimicrobial action was accruing, and this has more recently been given added impetus by global viral pandemics such as HIV-AIDS and serious increases in conventional, typically antibacterial, drug resistance in healthcare. The broad spectrum photoantimicrobial activity of the lead compounds methylene blue and toluidine blue is demonstrated in Table 2.1.

Table 2.1 Antimicrobial examples for methylene blue and toluidine blue.

Photosensitizer	**Organism**	**Ref.**
	Bacteria	
	Gram-positive bacteria	
TBO	*Staphylococcus aureus*	61
	Methicillin-resistant *S. aureus*	
MB	*Staphylococcus aureus*	
	Staphylococcus epidermidis	62
MB/TBO	*Listeria monocytogenes*	63
MB/TBO	*Bacillus cereus* (+ spores)	64
MB	*Streptococcus mutans*	65
TBO	*Streptococcus mutans*	66
MB	*Enterococcus faecalis*	67
MB	*Streptococcus pyogenes*	68
MB/TBO	*Streptococcus pneumoniae*	69
MB	Vancomycin resistant *Enterococcus faecium*	70
	Gram-negative bacteria	
MB/TBO	*Pseudomonas aeruginosa*	71
	Escherichia coli	
TBO	*Porphyromonas gingivalis*	72
MB	*Salmonella typhimurium*	73
TBO	*Vibrio vulnificus*	74
MB/TBO	*Helicobacter pylori*	75

Table 2.1 (*Continued*)

Photosensitizer	Organism	Ref.
MB/TBO	*Haemophilus influenzae*	69
MB	*Yersinia enterocolitica*	76
MB	*Propionibacterium acnes*	68
	Corynebacterium minitissimum	
MB	*Prevotella intermedia*	77
	Fusobacterium nucleatum	
	Mycobacteria	
MB	*Mycobacterium smegmatis*	78
	Fungi	
MB	*Trichophyton mentagrophytes*	79
	Microsporum gypseum	
MB	*Saccharomyces cerevisae*	80
MB	*Candida albicans*	81
	Candida dublinensis	
	Candida krusei	
	Candida tropicalis	
MB	*Cryptococcus neoformans*	82
MB/TBO	*Aspergillus nidulans*	83
	Viruses	
	Phages	
MB	Bacteriophage T2	84
MB/TBO	Bacteriophage T5	85
MB	Qβ-bacteriophage	86
	Enveloped	
MB	Hepatitis C	87
MB	Dengue	88
MB	Yellow Fever	89
MB/TBO	Foot-and-mouth	90
MB	Herpes simplex	91
MB	Human Immunodeficiency-1	92
MB	West Nile	93
	Non-enveloped	
MB	Adenovirus	94
TBO	Polio	95
	Coxsackie	
	Echo	
	Protozoa	
MB	*Plasmodium falciparum*	96
MB	*Trypanosoma brucei*	12
MB	*Trypanosoma cruzi*	97

The preceding history demonstrates the patchy nature of photoantimicrobial endeavour until almost the end of the last century. However, the presence throughout of cationic photosensitizers is indicative of their utility. That compounds such as proflavine and methylene blue are simple water soluble structures reflects their antiquity – the organic chemical industry was in its early stages when they were first produced – and also a lack of chemical input into biological screening programmes. As will be apparent from the following discussion, the attitude of the majority of researchers to structural variation in phenothiazinium photosensitizer discovery has been conservative, to say the least. However, there is a sound rationale to the use of this particular class of agents in photoantimicrobial research, stretching back to Ehrlich and his peers, which can not be claimed for other classes.

2.2. Phenothiazine/Phenothiazinium Chemistry

As mentioned already, the nascent field of organic chemistry – and more particularly that of heterocyclic chemistry – grew rapidly as a result of textile dye research, much of which was based on the oxidation of crude coal tar extracts. The presence of natural sulfur as part of the crude anilines *etc.*, led to its structural inclusion in the oxidized products, which were usually considerable mixtures. Thus, while 10*H*-phenothiazine (Figure 2.2) is considered to be the "parent" chromophore, this was not isolated until 1883, seven years after Caro's reporting of methylene blue from aniline oxidation.[19]

Drugs based on the reduced 10*H*-phenothiazine system are widely prescribed in disorders of the central nervous system, particularly in schizophrenia, and chlorpromazine (Figure 2.2) remains one of the more widely investigated synthetic therapeutics, probably second only to chloroquine (Figure 2.1). However, the reduced (10*H*-) and oxidized chromophores furnish vastly different physico-chemical properties, not least that of intense visible light absorption by the latter, which is, of course, of great importance in the use of methylene blue and its analogues in photodynamic processes. Conversely, drugs such as chlorpromazine are strong absorbers of ultraviolet light, which can lead to patient photosensitivity, but this is a deleterious clinical side effect, rather than a desirable outcome.[20] Similarly, the chromophore planarity ensured by the oxidized, aromatic state of the central 1,4-thiazinium ring in methylene blue and its congeners is absent in the reduced form, resulting molecules often having a dihedral angle of around 120°. This can have a drastic effect on biomolecular targeting. Another major difference between the two

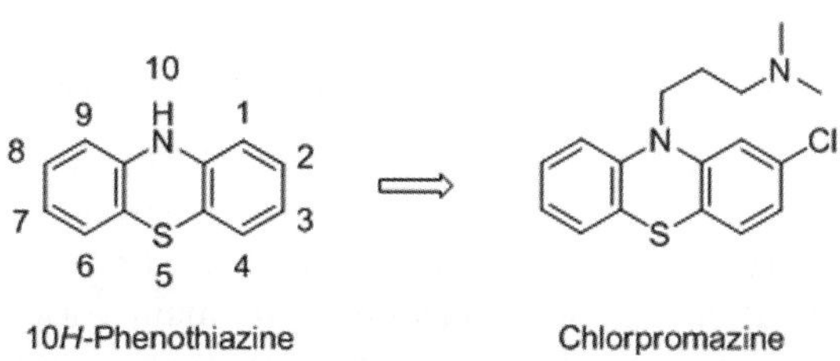

Figure 2.2 Parent 10H-phenothiazine, numbering and chlorpromazine.

systems is the delocalized cationic charge associated with phenothiaziniums. Again, this affects physico-chemical properties such as the hydrophilic/lipophilic balance, compounds closely resembling methylene blue generally being hydrophilic, whereas 10*H*-phenothiazine derivatives are lipophilic, drugs such as chlorpromazine having LogP values in the region of $+5$.[21] The relevance of such comparison may be understood in the biological reduction of the phenothiazinium to the 10*H*-phenothiazine chromophore, since this provides a potential route of distribution not normally associated with an aromatic cation (see below).

2.2.1. *Synthesis*

Since the biological use of phenothiazinium photosensitizers is fundamentally based on biological stains such as methylene blue and toluidine blue, it is not surprising that, initially at least, recourse was made to the much earlier scientific literature for information on synthesis. As noted above, early experimentation utilized standard oxidizing agents, including chromium(VI) salts, widely employed in the textile dye industry. A typical synthesis of methylene blue itself is given in Scheme 2.1.

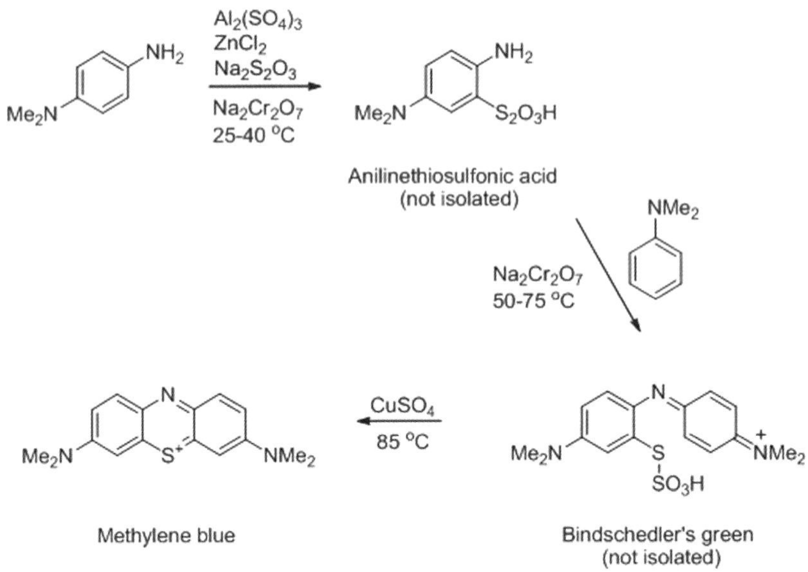

Scheme 2.1 Traditional textile dye synthesis of methylene blue.

The synthesis shown in Scheme 2.1 is suitable for the preparation of methylene blue for dyeing purposes, in that it produces a mixture of derivatives which are not separable *via* standard "wet chemistry" methods, such as dissolution/precipitation or recrystallization. The main reason for this is the strength of the oxidant, in combination with the temperatures and low pH employed – in addition to driving the oxidation of the initial phenylenediamine

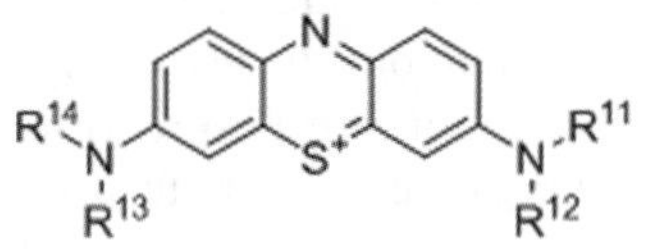

R^{11}	R^{12}	R^{13}	R^{14}	Derivative
Me	Me	Me	Me	Methylene blue
Me	Me	Me	H	Azure B
Me	Me	H	H	Azure A
Me	H	H	H	Azure C
H	H	H	H	Thionin

Figure 2.3 Demethylated derivatives of methylene blue.

to the thiosulfonic acid/Bindschedler's green and effecting the thiazinyl ring closure, other oxidative steps are possible, such as the dealkylation of the auxochromic dimethylamino groups. This explains the presence in the reaction product of the azure stains (Figure 2.3).

The presence of such mixtures is plainly unsuitable in medical applications. In addition, there are other disadvantages in the general synthetic approach, since *p*-phenylenediamines are toxic and not readily available, the synthetic methodology uses an aqueous system which leads to solubility problems with more lipophilic aniline derivatives, and ring functionality in the precursor anilines is a potential site for side-reaction oxidation, as above. Plainly, other methodologies were required, which would allow a more "organic-friendly" approach – almost from the commencement of phenothiazinium photosensitizer discovery projects in the late 1980s.

Such methodologies were supplied, at least in part, by the use of iodine or bromine as oxidants, in conjunction with 10*H*-phenothiazine as a starting material. The resulting phenothiazinium cations could then be attacked at positions 3- and 7- by amine nucleophiles, thus furnishing a range of 3,7-disubstituted derivatives (Scheme 2.2). While both reagents allow the production of symmetrical or asymmetrical derivatives (with respect to the auxochromes), the iodine route allows simpler purification at each step and thus better yields of asymmetrical material, while the bromine oxidation route features 3,7-dibromophenothiazinium bromide as an intermediate, which is ideal for symmetrical replacement.[22]

While the routes given in Scheme 2.2 have provided a range of novel derivatives, these were all of the methylene blue-type, *i.e.* having variation only in the

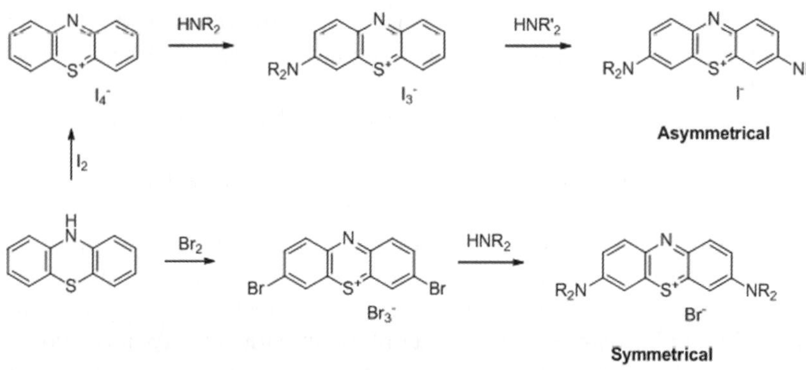

Scheme 2.2 Synthetic routes to symmetrical/asymmetrical methylene blue derivatives.

auxochromic positions. In order to explore substitution in the ring periphery it is still necessary to use the oxidative ring synthesis approach, but the use of milder oxidizing agents, such as silver(I) salts, is usually sufficient for ring closure without damage to other groups.[23] Modern syntheses, employing silver carbonate, usually also isolate the anilinethiosulfonic derivative which thus improves the purity of the product. The use of methanol, rather than water, as a reaction solvent for the ring synthesis step has also increased the range of structural possibilities.

Thus there are three main routes for the provision of novel phenothiazinium salts. While some examples have been examined in relation to the photodynamic anti-cancer application, the lead compound most employed as a photoantimicrobial is methylene blue, thus it is not surprising that this arena has seen a greater examination of novel phenothiazinium salts.

2.3. Methylene Blue as a Photoantimicrobial

The early demonstrations of the photoantimicrobial action of methylene blue, it has been noted, mainly concerned viruses. Much of the modern basis was supplied by Mohr's publication concerning the use of methylene blue as a photodisinfectant for blood plasma,[24] reflecting both the appreciation of its antimicrobial pedigree and its general acceptance as a safe chemical entity.

While it was, and remains, important to have procedures in place to inactivate human immunodeficiency virus (HIV) in the blood supply, this was not the sole reason since there are other possible contaminants. Logically, the approach must be antimicrobial, rather than merely antiviral (see *Pathogen Inactivation* section below). The use of methylene blue and its congeners in such research has thus supported and dovetailed with other infection control applications since the early 1990s.

2.3.1. Pathogen Inactivation

Human blood is a precious commodity which may be used whole in replacement, or as one of its constituent parts – plasma, platelets, red cells or derivatives of these. However, since it is sourced by donation directly from humans it is subject to any microbial contamination carried by the donor, and it is paramount that such contamination is not passed on to a recipient – a lesson painfully learned by the HIV infection of haemophiliacs treated with clotting factors derived from tainted blood.[25]

It has long been established that methylene blue interacts with nucleic acids *via* intercalation – *i.e.* stacking of the phenothiazinium chromophore between consecutive base pairs.[26] Given that viruses represent simple cellular structures based around a nucleic acid core, the use of a known photosensitizer/intercalator was a logical concept. Indeed, each of the simple, direct methylene blue derivatives shown in Figure 2.3 demonstrated useful photoantiviral activity in fractionated blood.[27] In addition, the majority of commercially available examples of the phenothiazinium class exhibit long wavelength (red) absorption in the visible spectrum, thus not suffering from competition with endogenous absorbers such as the haem pigments present in red blood cells. This is not a necessary consideration for the non-pigmented plasma and platelet fractions.

Methylene blue has been licensed by various European blood collecting agencies for the photodecontamination of plasma since the early 1990s.[28] However, its use has not been extended to platelet and red cell fractions, mainly due to collateral damage suffered by what are cellular rather than protein concentrates. Maintaining the integrity of the cells is clearly essential in transfusion, and this has encouraged considerable research into alternative photosensitizer structures, including novel phenothiaziniums.

For example, although known entities, methylene violet and Taylor's blue (1,9-dimethyl methylene blue, both Figure 2.4) were both investigated as potential agents for red cell decontamination, mainly on the basis of improved cellular uptake, due to either electronic neutrality or increased lipophilicity, respectively.[23,29] Neither derivative progressed to clinical trial due to the inhibition of action of methylene violet by plasma and the increased mammalian cell toxicity of dimethyl methylene blue (compared to that of methylene blue).[28] However, it should be noted that the dimethylated derivative proved significantly more effective as a photoantiviral than the parent compound.[23] Consequently, the search for improved phenothiazinium derivatives for use in this field has continued, several series of compounds having been produced recently which demonstrate increased *in vitro* DNA binding and antimicrobial efficacies. Examples of these new derivatives are dealt with in sections below.

2.3.2. *Clinical Infection Control*

Given the considerable amount of research carried out concerning the use of phenothiaziniums in blood product pathogen inactivation, it is perhaps surprising that there has been less progress towards the direct clinical use of these agents in infection control protocols. Again, here, methylene blue has been the lead compound, and thanks to its long record of safe use in humans – for example in the treatment of methaemoglobinaemia[2] – there are now licensed procedures and equipment which may be used, for example in oral disinfection,[30] while it has also been shown to be highly effective in smaller trials against onychomycosis and herpes labialis.[31]

Although these examples all represent forward steps in greater clinical recognition, the fact remains that methylene blue is, itself, a relatively weak

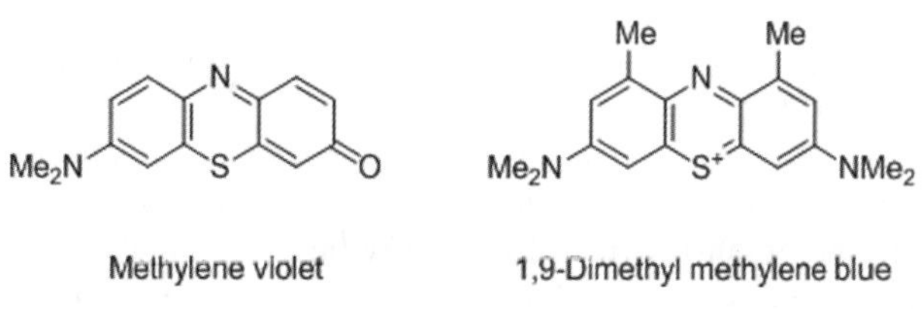

Figure 2.4 Commercially available alternative photoantimicrobials to methylene blue.

photoantimicrobial and this has further encouraged the synthesis of improved analogues. However, in terms of progressing such compounds towards the clinic, the major obstacle remains that of toxicity screening. The very low mammalian toxicity of the parent compound is usually sufficient to dissuade interested parties from pursuing new entities. Without significant investment in clinical trials, it is impossible to move new chemical entities to the clinic, and such investment is not generally encouraged when there is a ready-made, non-toxic agent available. The fact remains, however, that some of the recent work concerned with improved methylene blue derivatives has produced compounds which are so much more active than the parent compound that further investigation has been justifiable.

2.4. Improved Derivatives

As noted above, the use of milder oxidizing agents has allowed an enormous expansion in the number of new derivatives based on methylene blue. For the most part, these derivatives are covered by the patent literature, having been produced mainly by private companies or academic-industrial collaborations.

Examination of the available literature concerning derivatives produced by the dye industry before the Second World War does not provide many imaginative diversions from methylene blue itself, but this is undoubtedly due, at least in part, to the contemporary synthetic methodology. It is somewhat disappointing, then, to recognize a similar lack of enterprise in modern phenothiazinium discovery, with most groups venturing only as far as auxochrome replacement. The fact remains that even for simple derivatives of the phenothiazinium chromophore, there are eight possible positions available for substitution at the outset (ten in benzannelated chromophores). Why use only two? Such absence of adventure is indicative of a scarcity of input by chemists among drug discovery groups. Novelty does not guarantee success, of course, but a greater diversity of compounds would aid in gaining improved derivatives and in a greater understanding of structure-activity relationships. Various approaches to novel phenothiazinium photoantimicrobials are covered below.

2.4.1. *Simple Symmetricals*

Given the simple, symmetrical form of methylene blue, coupled with a straightforward synthetic route, it is perhaps surprising that the homologous series (methyl – *n*-hexyl, Figure 2.5) based on the parent compound took such a long time to appear in the literature,[32] although the tetraethyl and tetrapropyl derivatives had been published by Taylor in 1961.[33] The series reported by Mellish was originally synthesized for the purposes of anti-cancer PDT testing, but it has since been screened against bacteria, the longer-chain examples exhibiting the greater photobactericidal activities.[34]

During the original work by Mellish *et al.*, it was apparent from mammalian cell studies that the lengthening of the auxochromic alkyl chains also led to

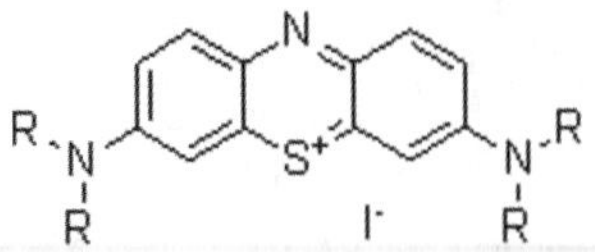

R	MBC (μM) S. aureus	MBC (μM) E. coli	λ_{max} (nm, MeOH)
Me	25	25	657
Et	6	6	664
n-Pr	3	6	669
n-Bu	3	3	671
n-Pe	3	3	672
n-Hx	3	3	673

Figure 2.5 Minimum photobactericidal concentrations (MBe) and λ_{max} data for simple symmetrical homologues of methylene blue.[34]

altered cellular distribution, again the heavier analogues being associated with lysosomes rather than the nucleus, as with the parent compound.[32] Such non-nuclear localization is appealing from a regulatory viewpoint, given the DNA-intercalative properties of planar cations and the obvious link to mutagenic/oncogenic potential.

It should be noted that symmetrical phenothiaziniums of this type are best prepared *via* the bromine oxidation/dibromination pathway covered previously, particularly from the point of view of product yield but also that of aqueous solubility, since the iodide salts tend to much greater lipophilicity. Alternative anions may, of course, be provided by salting out or ion-exchange methods.

2.4.2. Asymmetricals

Simple asymmetrical analogues of the original series have been provided using the iodine oxidation method. Such an approach, *i.e.* isolation of a 3-dialkylamino-phenothiazinium triiodide intermediate followed by reaction with a different dialkylamine derivative, allows the synthesis of derivatives having both lipophilic and hydrophilic "ends",[35] for example combining a dihexylamino residue at C-3- with a diethanolamino- or piperazino at C-7. Such substitution patterns would also allow the inclusion of functionality aimed at chemical linkage of the resulting photosensitizer to polymers, biomolecules *etc.*, for targeting or drug-delivery purposes.

Use of the asymmetric approach also offers the opportunity to examine the effects of substitution patterns in auxochromic groups. To this end, 3-dialkylaminophenothiazinium salts were reacted with a series of benzylamines.[36] Again, the more lipophilic examples were the most active (Figure 2.6), although a larger compound range would be required to determine absolute structure-activity patterns.

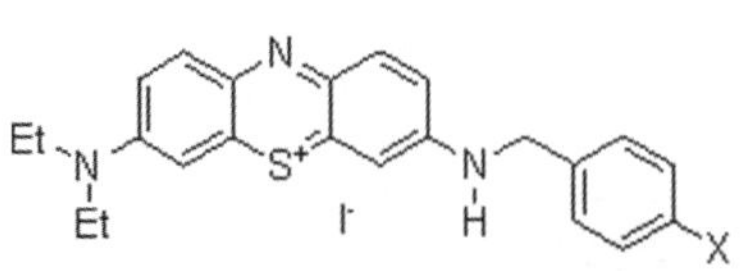

X	MBC (μM) S. aureus	MBC (μM) E. coli	λ_{max} (nm, MeOH)
H	6	6	644
Me	6	6	648
MeO	3	3	646
F	3	3	644
Cl	3	3	642

Figure 2.6 Photobactericidal activities of 3-(substituted benzylamino) derivatives (MBC of MB = 25 μM in each case).

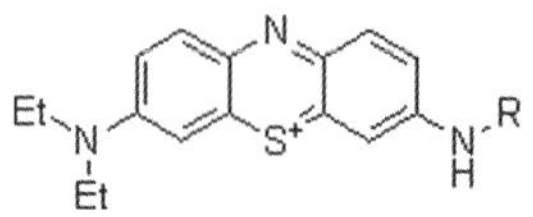

R	λ_{max} (nm, MeOH)
Me	649
CH_2Ph	644
Ph	658

Figure 2.7 Effect of aromatic attachment to auxochrome on λ_{max}.

Interestingly, whereas the benzylamino series exhibited singlet oxygen efficiencies similar to that of methylene blue *in vitro* (isobenzofuran photoxidation assay), this was not the case with either the corresponding 3-anilino or 3,7-bis(anilino) analogues. However, these did demonstrate photobactericidal activity, admittedly to a lesser extent compared to the benzylamino derivatives, in the *in vitro* antibacterial screen.[36] This provides a useful example of the importance of realistic testing in drug development. A measure of the effect of the aromatic moiety attached directly to the amino group may be seen in Figure 2.7, the *N*-phenyl derivative having a considerably longer λ_{max} than the corresponding unsubstituted or benzyl-substituted analogues.

Variations in activity linked to lipophilicity were also reported by Crossley *et al.* in relation to the conventional screening of benzo[*a*]phenoxazinium (Nile blue) derivatives having combinations of dialkylamino and either benzylamino- or anilino-substitution, against murine *Mycobacterium bovis*.[13] In this work, derivatives having either 5-di-*n*-propylamino- or di-*n*-butylamino- were associated with the greatest activities against *M. bovis* in the mouse model.

2.4.3. Benzo[a]phenothiazinium Derivatives

As noted above, the research carried out by Crossley, Clapp *et al.*, aimed at both anti-tumour and anti-mycobacterial activity, followed *in vivo* tumour staining work reported in the late 1940s by Lewis, concerning the use of Nile blue in an animal tumour model.[14] Cincotta, Foley and coworkers built on this, initially aiming to produce selective photodynamic anti-tumour agents based on the benzo[*a*]phenoxazinium framework. The lack of photodynamic activity exhibited by the original Nile blue examples was overcome by the use of the heavy atom effect, utilizing lower group atoms in the parent molecular structure. Thus, brominated and (especially) iodinated versions of NBA were shown to produce significantly higher yields of singlet oxygen in the *in vitro* assay.[37] This work also included examples having the ring oxygen (O-7) replaced by sulfur or selenium, the resulting derivatives again exhibiting greater photosensitizing potential (Figure 2.8).

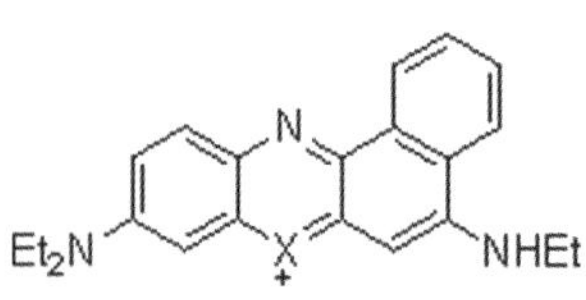

X	$^1O_2$ yield	λ_{max} (nm, MeOH)
O	0.005	632
S	0.025	652
Se	0.650	659

Figure 2.8 Heavy atom inclusion, singlet oxygen yield and bathochromicity.

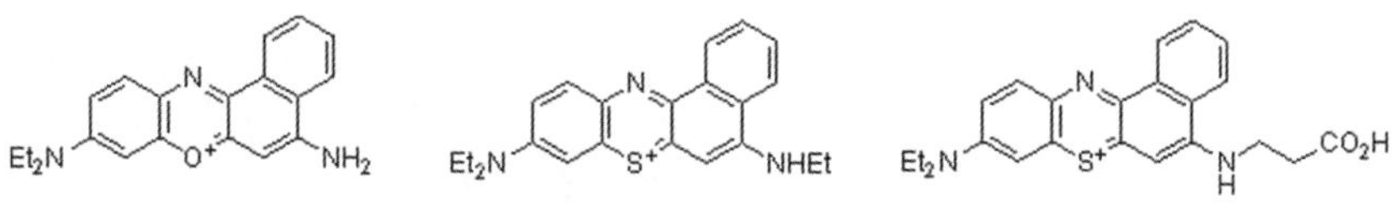

NBA EtNBS EtNBS-CO_2H

Figure 2.9 Nile blue derivatlves.[39]

Most recently, the optimal compound obtained from this research, EtNBS (Figure 2.9), was also shown to be an excellent photoantimicrobial agent *in vitro*, including significant activity against the parasites responsible for cutaneous leishmaniasis.[38] Interestingly, the influence of photosensitizer molecular charge on Gram-positive/Gram-negative photobactericidal effect has been demonstrated with two close analogues of EtNBS, one containing a carboxylic acid in the side chain (Figure 2.9). This anionic group-containing derivative was active against *S. aureus*, but not against the Gram-negative *E. coli*.[39]

Given that EtNBS had previously been thoroughly screened in animals as a PDT agent, a considerable weight of evidential proof exists regarding its toxicity levels in mammals and this should supply a sound basis for its progression into clinical trials.

2.4.4. Chromophore-substituted Derivatives

Given over 150 years of phenothiazinium synthesis in the laboratory, the preponderance of auxochrome-functionalized analogues of methylene blue among these is surprising. Derivatives do exist, of course, having substituents at positions other than 3- and 7-, including highly effective photoantimicrobial agents such as toluidine blue (2-methyl), new methylene blue (2,8-dimethyl) and Taylor's blue (1,9-dimethyl), each of which was shown to be effective in killing clinical isolates of MRSA, for example.[40]

In order to investigate non-auxochromic substitution, a series of chromophore-mono-, di- and trimethylated derivatives were synthesized, as analogues of toluidine blue – *i.e.* having different methylation patterns.[41] This work benefited from the use of a milder oxidant in the ring synthesis, since it allowed maintenance of the methyl substituents. The resulting series exhibited high activity against both Gram-positive and Gram-negative bacteria, considerably higher than that of the lead compound, toluidine blue. However, it was noticeable that di- and trimethylation caused hypsochromic shifts in the associated absorption spectra. In addition, the presence of a dialkylamino auxochrome effectively blocked methylation of either adjacent position. Thus, for long wavelength absorption – a key criterion in photosensitizer design – chromophore substitution is available for methylene blue derivatization only at positions 1- and 9-, as with Taylor's blue. Taylor's blue itself exhibits a hypsochromic shift of approx. 10 nm compared to the parent compound in water.

Work aimed at the production of improved photoantivirals for use in blood decontamination included the synthesis of a number of "Taylor's blue derivatives" (Figure 2.10)[23].

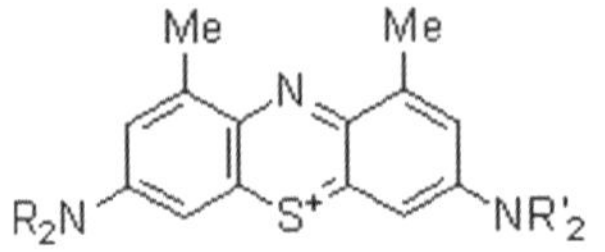

R	R'	LogP	λ_{max} (nm, MeOH)
Me	Me	+0.61	649
Et	Me	+1.26	651
n-Bu	Et	+3.55	657

Figure 2.10 "Taylor's blue derivatives".[23]

In terms of antiviral activity, Taylor's blue itself represents the optimum compound out of those tested (including methylene blue), exhibiting the greatest efficacy against both enveloped and non-enveloped viruses *in vitro*. Clearly, in targeting viral particles inside red blood cells, it is important to be able to enter the cell – viruses may be intra- or extracellular – but target binding also logically raises the efficiency of inactivation. Where methylene blue is too hydrophilic to enter erythrocytes (logP = –0.95), Taylor's blue is sufficiently amphiphilic to do so. In addition, Taylor's blue was the most efficient DNA binder of the series examined, although it was ultimately too toxic in mammalian studies for ultimate use in blood product pathogen inactivation.

2.4.5. Tetra- and Pentacyclic Derivatives

In order to produce improved versions of the standard photosensitizer methylene blue, with more effective photoantimicrobial properties and longer absorption wavelengths, it is necessary to alter the given structure. As has been covered above, either or both of the chromophore and the auxochromes may be altered, but bathochromic shifting is more easily achievable *via* auxochrome extension (Figure 2.5). However, such extension does not provide large shifts, and increasing alkyl group size is obviously associated with decreasing aqueous solubility which may lead to problems in formulation.

Extension of the aromatic system of the chromophore is one way to increase the absorption wavelength of the system. This can be seen in the series commencing with the diethylamino derivative of azure A, followed by benzannelation, as seen in Figure 2.11. Again, increasing the size of the aromatic chromophore usually leads to lowered solubility.

A solution which appears to offer the advantages mentioned above without the resulting compounds being insoluble is to extend the auxochromic moiety by joining it to the chromophore *via* the distal terminus, *i.e.* forming a further

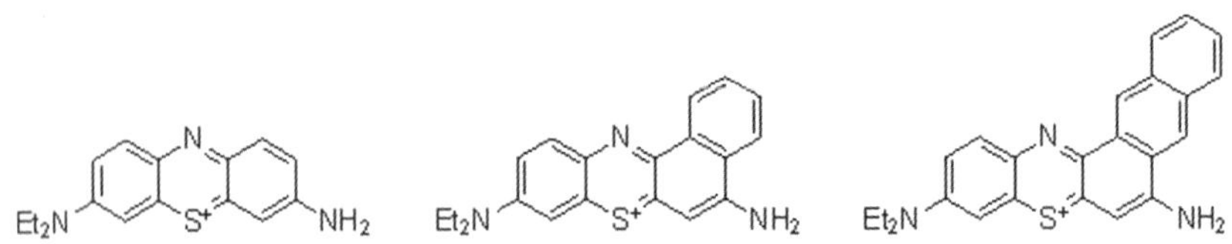

628 nm 645 nm 660 nm

Figure 2.11 Effect of benzannelation on λ_{max} (in MeOH) for simple phenothiazinium derivatives.

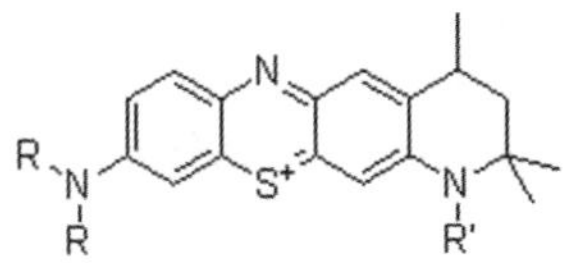

Tetracycles

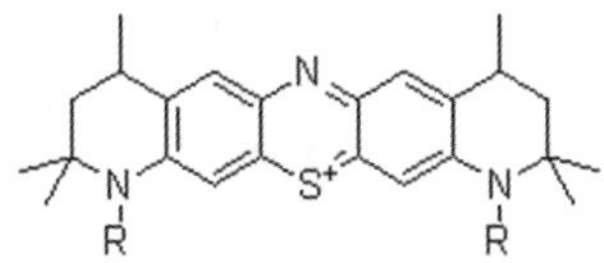

Pentacycles

R	R'	MBC (μM) *S. aureus*	MBC (μM) *E. coli*	λ_{max} (nm, MeOH)
Et	Et	3	3	664
n-Bu	Me	1	1	685

R	MBC (μM) *S. aureus*	MBC (μM) *E. coli*	λ_{max} (nm, MeOH)
Me	3	3	685
Et	0.7	3	693

Figure 2.12 Tetra- and pentacyclic phenothiazinium derivatives.[42]

annelated, but non-benzenoid, ring. In addition to providing the advantages of lengthened auxochromic pendants, the ring formation rigidifies the auxochrome itself, the inhibition of amine group rotation ensuring permanent coplanarity of the amine lone pair with the remainder of the chromophore. This has the effect of both promoting bathochromic shifting and increasing photosensitizing activity in derivatives.

The lead compound in this research, first synthesized as an anti-cancer PDT candidate, was derived from an alkylated tetrahydroquinoline *via* the dichromate oxidation route, yielding a pentacyclic derivative (R = Et, Figure 2.12). Owing to the very high, broad-spectrum activity of this lead in microbiological screening, a series of related penta- and tetracycles has been synthesized, a representative sample of which may be seen in Figure 2.12. Notably, this series of compounds constitutes the closest approach in absorption so far reported for phenothiazinium derivatives to the 700 nm mark. In addition, several examples of the series exhibited singlet oxygen yields greater than that of methylene blue in the chemical assay.

Given such improved properties it should be of no surprise that many of the series mentioned exhibit considerably higher antimicrobial activities than the lead compound, methylene blue. This includes activity against Gram-positive and Gram-negative bacteria, yeasts, moulds, viruses and protozoa,[42] and represents significant potential in terms of local anti-infection.

2.5. Toxicity Issues

Chemotherapy – taken literally – infers the use of chemicals to treat disease. Paul Ehrlich's definition of more than a century ago was based on the principle of *selective toxicity*, *i.e.* damaging the target cells to a significantly greater degree than those of the host. Ehrlich's Holy Grail of a magic bullet, totally selective for the target cells, has never been realized, and much of the work entailed in drug development remains devoted to minimizing side-effects.

Taking toxicity as a general term, at the cellular level this can mean biomolecular damage leading to gross structural breakdown and necrosis or to organellar damage and the initiation of death-signalling pathways, or apoptosis.[43] Similarly, toxicity to DNA can be merely and grossly destructive, or it can cause damage and alteration to the genetic material but still allow this to be copied, providing an incorrect or *mutated* template and leading to incorrect daughter molecules.

2.5.1. Toxicity of Phenothiazinium Derivatives

Both methylene blue and toluidine blue are in daily clinical use, normally at a concentration of 1% w/v, equivalent to 31.3 mM and 32.7 mM, respectively, for the chloride salts. Both 100 mg and 500 mg orally administered doses of methylene blue are in use in anti-malarial protocols.[44] Toxicity data for related compounds relies on cellular studies, usually from anti-cancer PDT screens, a sample of which is shown in Table 2.2.

As can be seen from the table, methylene blue is by far the least phototoxic to the cell lines employed in the study. However, this lack of toxicity must be balanced against the far greater comparative activity of more active derivatives, *e.g.* dimethyl methylene blue (Taylor's blue) against bacteria and viruses.

Table 2.2 90% Phototoxicity (IC_{90}) levels for methylene blue derivatives.

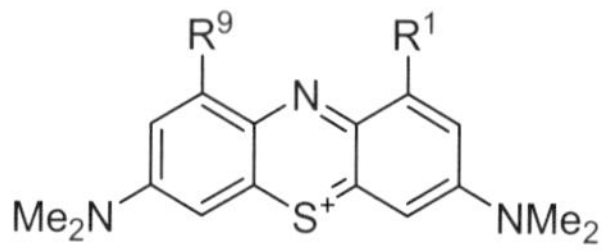

	R^1	R^9	IC_{90} (iM) EMT-6[a]	SK-23[b]	SK-MEL-28[c]
Methylene blue	H	H	38	40	39
Methyl methylene blue	Me	H	5	5	10
Dimethyl methylene blue	Me	Me	0.3	0.4	0.1

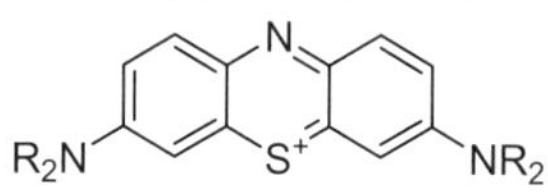

R	IC_{90} (iM) RIF-1[d]
Methyl	200
Ethyl	20
n-Propyl	2
n-Butyl	5
n-Pentyl	3
n-Hexyl	5

Cell lines:
[a]mouse mammary carcinoma
[b]mouse melanoma
[c]human melanoma, white light, 7 J cm^{-2}.[45]
[d]mouse fibrosarcoma cell line, white light, 3 J cm^{-2}.[32]

In addition, it should be remembered that photoantimicrobials are mainly intended for topical and local administration.

2.5.2. Redox Properties

While it is largely accepted that the phenothiazinium cation is responsible for cellular phototoxicity, the propensity for reduction-oxidation (redox) cycling by the phenothiazinium/phenothiazine system is well known. Methylene blue itself was used for many years as an indicator of bacterial activity in milk testing,[46] and may be reduced to the leuco-form by ascorbic acid or other standard reductants.[47] From a photoantimicrobial viewpoint, such reduction may be useful, as the neutral leuco-form is more easily taken up by cells than the hydrophilic cation. However, effective action requires the re-oxidation of the leuco-form to the cation once inside the cell (Scheme 2.3). The difference in photobactericidal activity between methylene blue and the more effective toluidine blue has been reported to be due to higher concentrations of the latter cation inside bacterial cells.[48] Similarly, when methylene blue exhibited a lack of activity of against tuberculosis in a rabbit model, removed grey tubercles sectioned and left open to the air developed the blue colour of the re-oxidized cation.[49]

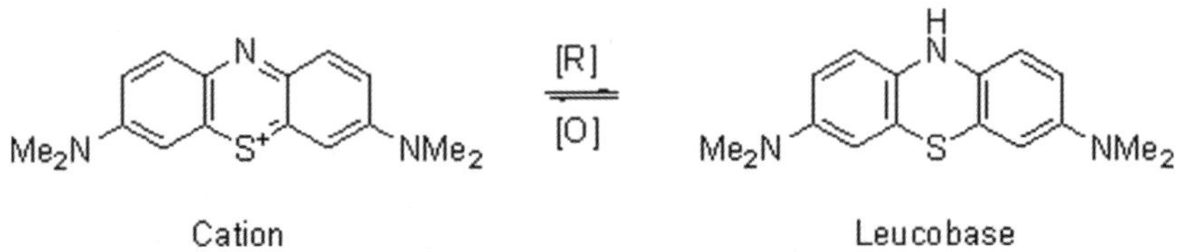

Scheme 2.3 Redox equilibrium for methylene blue.

The molecular features required either to inhibit or to promote phenothiazinium reduction are not yet well established. However, the two methylated derivatives of methylene blue shown in Table 2.2 were found to be more difficult to reduce using the NADH-diaphorase system, and this may be due to steric protection of the ring nitrogen.[50]

2.5.3. Mutagenicity

Lerman's report of the interaction of the acridine drug proflavine with DNA was an elegant piece of research, aiding efforts in nucleic acid research on a wide front.[18] However, it has been highly deleterious to the potential of molecules having structural similarities to the acridine derivative in that a broad-brush attitude has since been taken by medical and regulatory authorities in assuming that such molecules must damage DNA and thus lead to mutagenicity.

It should be stated at the outset that Lerman (and others following) investigated the proflavine-DNA interactions in isolation, *i.e.* the two interacting components were not hindered by potential sequestration or inhibition by other cellular or metabolic processes. Such studies should not, therefore, be cited as proof of mutagenic activity *in vivo*. Similarly, the standard screen for assaying mutagenicity *in vitro* is the Ames test, which relies on DNA damage in particular strains of *Salmonella typhimurium* or *Escherichia coli*. Once again, this should not be taken as proof-positive that any compound is a mutagen in humans. Plainly, there are many differences between the behaviour of any compound in a single prokaryotic cell and a massively complex collection of eukaryotic ones. However, it should be stated that both lead compounds, methylene blue and toluidine blue, have been reported to be mutagenic in the Ames test.[51,52]

Considering other compounds based on the phenothiazinium chromophore, work reported by Mellish *et al.* showed that the cellular localization of simple methylene blue derivatives varied with structure.[32] For example, while the lead compound did exhibit nuclear localization, more lipophilic analogues (by benefit of longer alkyl chain substitution in the auxochrome) were concentrated in the lysosome, thus being isolated from nuclear genetic material. Using simple spectrophotometric solution measurement – *i.e.* DNA and compound in buffer – each of the analogues exhibited DNA binding in the same way as methylene blue.[34] Taken in isolation, this piece of research would therefore – erroneously – suggest a potential for mutagenicity throughout the series.

In addition, and importantly, quite a range of antibacterial agents in everyday use has been shown to affect DNA deleteriously *in vitro* and *in vivo*. This is seldom mentioned in the literature, but the activities exhibited are no different from those of simple phenothiazinium compounds in the same tests. Considering DNA-active antibacterials, ciprofloxacin has been shown to exert genotoxic effects *in vitro*,[53] and, indeed, a considerable number of quinolone antibacterials has exhibited positive Ames tests.[54] The anti-tubercular agent rifampicin has also been shown to produce chromosomal aberration in mice.[55] Among other, general antibacterial agents, both trimethoprim[56] and metronidazole[57] exhibit mutagenic potential, as do two widely used anti-mycobacterials, isoniazid[58] and the anti-leprosy drug clofazimine.[59] Given its basic side chain and fused heteroaromatic chromophore, it is perhaps not surprising that the anti-malarial chloroquine also falls into this category.[60] Each of these drugs is well accepted and normally administered systemically. A topically or locally applied photosensitizer must, by definition, offer far lower risk.

Given the few chemists involved in photosensitizing drug discovery worldwide, and the fraction of these concerned with phenothiazinium synthesis, the progress made in producing candidates which exhibit photoantimicrobial activities of such a high magnitude should not be under-rated, particularly in view of the paucity of financial support for research in this area. In terms of the passage of such compounds to the clinic, the funding required in trialling is more difficult still to obtain, perhaps understandable given the likelihood of compound failure on transferring from cellular studies to animal testing. However, antimicrobial drug resistance, particularly in respect of bacteria, is constantly increasing, and effective solutions to this, even if unconventional and only applicable topically or locally, must still be deserving of consideration and trial.

References

1. M. Wainwright, The use of dyes in modern biomedicine, *Biotech. Histochem.*, 2003, **78**, 147.
2. M. Wainwright and K.B. Crossley, Photosensitising agents – circumventing resistance and breaking down biofilms, *Int. Biodeter. Biodeg.*, 2004, **53**, 119.
3. P. Guttmann and P. Ehrlich, Ueber die wirkung des methylenblau bei malaria, *Berlin Klin. Wochschr.*, 1891, **39**, 953.
4. C.H. Browning, R. Gulbransen, E.L. Kennaway and L.H.D. Thornton, Flavine and brilliant green, powerful antiseptics with low toxicity to the tissues: their use in the treatment of infected wounds, *B.M.J.*, 1917, **i**, 73.
5. G.A.G. Mitchell and G.A.H. Buttle, Proflavine in closed wounds, *Lancet*, 1943, **242**, 749.
6. J. McIntosh and F.R. Selbie, Proflavine-sulphathiazole mixture, *Lancet*, 1944, **244**, 66.
7. O.Z. Raab, Uber die wirking fluoreszierender stoffe auf infusorien, *Z.Biol.*, 1900, **39**, 524.
8. H. Von Tappeiner and A. Jesionek, Therapeutische Versuche mit fluoreszierenden Stoffen, *Munch. Med. Wochenschr.*, 1903, **47**, 2042.
9. E.W. Schultz and A.P. Krueger, Inactivation of Staphylococcus bacteriophage by methylene blue, *Proc. Soc. Exper. Biol. Med.*, 1928, **26**, 100.
10. C.E. Clifton and T.G. Lawler, Inactivation of *Staphylococcus* bacteriophage by toluidine blue, *Proc. Soc. Exper. Biol. Med.*, 1930, **27**, 1041.
11. T. T'ung, Photodynamic action of methylene blue on bacteria, *Proc. Soc. Exp. Biol. Med.*, 1935, **33**, 328.
12. T. T'ung, *In vitro* photodynamic action of methylene blue on *Trypanosoma brucei*, *Proc. Soc. Exp. Biol. Med.*, 1938, **38**, 29.
13. M.L. Crossley, R.J. Turner, C.M. Hofmann, P.F. Dreisbach and R.P. Parker, Chemotherapeutic Dyes. II. 5-Arylamino-9-dialkylaminobenzo[a]phenoxazines, *J.A.C.S.*, 1952, **74**, 578.
14. M.R. Lewis, P.P. Goland and H.A. Sloviter, The action of oxazine dyes on tumors in mice, *Cancer Res.*, 1949, **9**, 736.
15. F. Heinmets, J.R. Kingston and C.W. Hiatt, Inactivation of viruses in plasma by photosensitized oxidation, *Walter Reed Army Inst. Res. Rep.*, 1955, **53**.
16. C. Wallis and J.L. Melnick, Photodynamic inactivation of animal viruses: a review, *Photochem. Photobiol.*, 1965, **4**, 159.
17. T.D. Felber, E.B. Smith, J.M. Knox, C. Wallis and J.L. Melnick, Photodynamic inactivation of herpes simplex: report of a clinical trial, *J.A.M.A.*, 1973, **223**, 289.
18. L.S. Lerman, Structural considerations in the interaction of deoxyribonucleic acid and acridines, *J. Mol. Biol.*, 1961, **3**, 18.
19. H. Caro, *Ber. Deutsch Chem.*, 1876, **xi**, 360.
20. J.A.A. Hunter, L.K. Bhutani and I.A. Magnus, Chlorpromazine photosensitivity in mice: its action spectrum and the effect of anti-inflammatory agents, *Br. J. Dermatol.*, 1970, **82**, 157.

21. M.H. Bickel, B.E. Graber and M. Moor, Distribution of chlorpromazine and imipramine in adipose and other tissues of rats, *Life Sci.*, 1983, **33**, 2025.
22. M. Wainwright and R.M. Giddens, Phenothiazinium photosensitisers: choices in synthesis and application, *Dyes Pigments*, 2003, **57**, 245.
23. S.J. Wagner, A. Skripchenko, D. Robinette, J.W. Foley and L. Cincotta, Factors affecting virus photoinactivation by a series of phenothiazine dyes, *Photochem. Photobiol.*, 1998, **67**, 343.
24. B. Lambrecht, H. Mohr, J. Knüver-Hopf and H. Schmitt, Photoinactivation of viruses in human fresh plasma by phenothiazine dyes in combination with visible light, *Vox Sang.*, 1991, **60**, 207.
25. J. Rovner, U.S. bill aids HIV-1-infected haemophiliacs, *Lancet*, 1998, **352**, 1610.
26. E.M. Tuite and J.M. Kelly, Photochemical interactions of methylene blue and analogues with DNA and other biological substrates, *J. Photochem. Photobiol. B: Biol.*, 1993, **21**, 103.
27. A. Klein-Struckmeier and H. Mohr, Photodynamic virus inactivation of thrombocyte concentrates by phenothiazine dyes, *Transfusionsmed.*, 1997, **34**, 43.
28. M. Wainwright, H. Mohr and W. Walker, Phenothiazinium derivatives for pathogen inactivation in blood products, *J. Photochem. Photobiol. B: Biol.*, 2007, **86**, 45.
29. H. Morrison, T. Mohammad and R. Kurukulasuriya, Photobiological properties of methylene violet, *Photochem. Photobiol.*, 1997, **66**, 245.
30. R. Andersen, N. Loebel, D. Hammond and M. Wilson, Treatment of periodontal disease by photodisinfection compared to scaling and root planing, *J. Clin. Dent.*, 2007, **18**, 34.
31. J.P. Tardivo and M.S. Baptista, Treatment of osteomyelitis in the feet of diabetic patients by photodynamic antimicrobial chemotherapy, *Photomed. Laser Surg.*, 2009, **27**, 145.
32. K.J. Mellish, R.D. Cox, D.I. Vernon, J. Griffiths and S.B. Brown, *In vitro* photodynamic activity of a series of methylene blue analogues, *Photochem. Photobiol.*, 2002, **75**, 392.
33. K.B. Taylor, The influence of molecular structure of thiazine and oxazine dyes on their metachromic properties, *Stain Technol.*, 1961, **36**, 73.
34. M. Wainwright, A. Shah, K. Meegan, C. Loughran, A. Smith and N. Valli, Phenothiazinium-fluoroquinolone drug conjugates, *Int. J. Antimicrob. Agents*, 2010, **35**, 405.
35. S.A. Gorman, A.L. Bell, J. Griffiths, D. Roberts and S.B. Brown, The synthesis and properties of unsymmetrical 3,7-diaminophenothiazin-5-ium iodide salts: potential photosensitisers for photodynamic therapy, *Dyes Pigments*, 2006, **71**, 153.
36. M. Wainwright, S. Brandt, A. Smith, A. Styles, K. Meegan and C. Loughran, Phenothiazinium photosensitisers VII. Novel substituted asymmetric *N*-benzylphenothiaziniums as photoantimicrobial agents, *J. Photochem. Photobiol. B: Biol.*, 2010, **99**, 74.
37. L. Cincotta, J.W. Foley and A.H. Cincotta, Novel red absorbing benzo[*a*] phenoxazinium and benzo[*a*]phenothiazinium photosensitizers: *in vitro* evaluation, *Photochem. Photobiol.*, 1987, **46**, 751.
38. O.E. Akilov, S. Kosaka, K. O'Riordan, X. Song, M. Sherwood, T.J. Flotte, J.W. Foley and T. Hasan, The role of photosensitizer molecular charge and structure on the efficacy of photodynamic therapy against Leishmania parasites, *Chem. Biol.*, 2006, **13**, 839.
39. S. Verma, U.W. Sallum, H. Athar, L. Rosenblum, J.W. Foley and T. Hasan, Antimicrobial photodynamic efficacy of side-chain functionalized benzo[*a*] phenothiazinium dyes, *Photochem. Photobiol.*, 2009, **85**, 111.

40. M. Wainwright, D.A. Phoenix, S.L. Laycock, D.R.A. Wareing and P.A. Wright, Photobactericidal activity of phenothiazinium dyes against methicillin-resistant strains of *Staphylococcus aureus*, *FEMS Microbiol. Lett.*, 1998, **160**, 177.
41. M. Wainwright, Phenothiazinium photosensitisers V. Photobactericidal activities of chromophore-methylated phenothiazinium salts, *Dyes Pigments*, 2006, **73**, 7.
42. M. Wainwright, Ring-fused derivatives of 37-diaminophenothiazinium and phenoselenazinium salts for use as antimicrobial agents, *GB Pat.* 2 453 564, 2009.
43. N.L. Oleinick, R.L. Morris and I. Belichenko, The role of apoptosis in response to photodynamic therapy: what, where and how?, *Photochem. Photobiol. Sci.*, 2002, **1**, 1.
44. J. Burhenne, K.D. Riedel, J. Rengelshausen, P. Meissner, O. Müller, G. Mikus, W.E. Haefeli and I. Walter-Sack, Quantification of cationic anti-malaria agent methylene blue in different human biological matrices using cation exchange chromatography coupled to tandem mass spectrometry, *J. Chromatog. B*, 2008, **863**, 273.
45. L. Rice, M. Wainwright and D.A. Phoenix, Phenothiazine photosensitisers III. Activity of methylene blue derivatives against pigmented melanoma cell lines, *J. Chemother.*, 2000, **12**, 94.
46. H.R. Whitehead, The reduction of methylene blue in milk, *Biochem. J.*, 1930, **24**, 579.
47. S. Mowry and P.J. Ogren, Kinetics of methylene blue reduction by ascorbic acid, *J. Chem. Ed.*, 1999, **76**, 970.
48. A. Blázquez-Castro, J.C. Stockert, F. Sanz-Rodríguez, A. Zamarrón and A. Juarranz, Differential photodynamic response of cultured cells to methylene blue and toluidine blue: role of dark redox properties, *Photochem. Photobiol. Sci.*, 2009, **8**, 371.
49. L.M. DeWitt, Preliminary report of experiments in the vital staining of tubercles, *J. Infect. Dis.*, 1913, **12**, 68.
50. M. Wainwright, D.A. Phoenix, L. Rice, S.M. Burrow and J.J. Waring, Increased cytotoxicity and phototoxicity in the methylene blue series *via* chromophore methylation, *J. Photochem. Photobiol., B: Biol.*, 1997, **40**, 233.
51. National Toxicology Program, *Tech. Rep. Ser.*, 2008, **540**, 1.
52. A.J. Dunipace, R. Beaven, T. Noblitt, Y. Li, S. Zunt and G. Stookey, Mutagenic potential of toluidine blue evaluated in the Ames test, *Mut. Res.*, 1992, **279**, 255.
53. B.A. Herbold, S.Y. Brendler-Schwaab and H. J. Ahr, Ciprofloxacin: *in vivo* genotoxicity studies, *Mut. Res.*, 2001, **498**, 193.
54. Y. Hayasaki, S. Itoh, M. Kato and K. Furuhama, Mutagenesis induced by 12 quinolone antibacterial agents in *Escherichia coli*, *Toxicol. in vitro*, 2006, **20**, 342.
55. F.A.E. Aly and S.M. Donya, *In vivo* antimutagenic effect of vitamins C and E against rifampicin-induced chromosome aberrations in mouse bone-marrow cells, *Mut. Res.*, 2002, **518**, 1.
56. A. Abou-Eisha, Evaluation of cytogenic and DNA damage induced by the antibacterial drug, *trimethoprim, Toxicol. in vitro*, 2006, **20**, 601.
57. M.D. Mudry, M. Carballo, M. Labal de Venuesa, M. Gonzalez Cid and I. Larripa, Mutagenic bioassay of certain pharmacological drugs: III. Metronidazole (MTZ), *Mut. Res.*, 1994, **305**, 127.
58. G. Rohrborn, P. Propping and W. Buselmaier, Mutagenic activity of isoniazid and hydrazine in mammalian test systems, *Mut. Res.*, 1972, **16**, 189.
59. R.K. Das and B. Roy, Evaluation of genotoxicity of clofazimine, an antileprosy drug, in mice *in vivo*. I. Chromosome analysis in bone marrow and spermatocytes, *Mut. Res.*, 1990, **241**, 161.
60. E.E. Obaseiki-Ebor and E.E. Obasi, Aspects of chloroquine mutagenicity, *Mut. Res. Lett.*, 1986, **175**, 51.

61. M. Sharma, L. Visai, F. Bragheri, I. Cristiani, P.K. Gupta and P. Speziale, Toluidine blue-mediated photodynamic effects on staphylococcal biofilms, *Antimicrob. Agents Chemother.*, 2008, **52**, 299.
62. F. Gad, T. Zahra, T. Hasan and M.R. Hamblin, Effects of growth phase and extracellular slime on photodynamic inactivation of Gram-positive pathogenic bacteria, *Antimicrob. Agents Chemother.*, 2004, **48**, 2173.
63. N.A. Romanova, L.Y. Brovko, L. Moore, E. Pometum, A.P. Savitsky, N.N. Ugarova and M.W. Griffiths, Assessment of photodynamic destruction of *Escherichia coli* O157: H7 and *Listeria monocytogenes* by using ATP bioluminescence, *Appl. Env. Microbiol.*, 2003, **69**, 6393.
64. T.N. Demedova and M.R. Hamblin, Photodynamic inactivation of *Bacillus* spores, mediated by mediated phenothiazinium dyes, *Appl. Env. Microbiol.*, 2005, **71**, 6918.
65. S. Wood, D. Metcalf, D. Devine and C. Robinson, Erythrosine is a potential photosensitizer for the photodynamic therapy of oral plaque biofilms, *J. Antimicrob. Chemother.*, 2006, **57**, 680.
66. I.C.J. Zanin, R.B. Goncalves, A.B. Junior, C.K. Hope and J. Pratten, Susceptibility of *Streptococcusmutans* biofilms to photodynamic therapy: an *in vitro* study, *J. Antimicrob. Chemother.*, 2005, **56**, 324.
67. N.S. Soukos, P.S.Y. Chen, J.T. Morris, K. Ruggiero, A.D. Abernethy, S. Som, F. Foschi, S. Doucette, L.L. Bammann, C.R. Fontana, A.G. Doukos and P.P. Stashenko, Photodynamic therapy for endodontic disinfection, *J. Endodont.*, 2006, **32**, 979.
68. B. Zeina, J. Greenman, W.M. Purcell and B. Das, Killing of cutaneous microbial species by photodynamic therapy, *Br. J. Dermatol.*, 2001, **144**, 274.
69. M.N. Usacheva, M.C. Teichert and M.A. Biel, The role of methylene blue and toluidine blue monomers and dimers in the photoinactivation of bacteria, *J. Photochem. Photobiol. B: Biol.*, 2003, **71**, 87.
70. M. Wainwright, D.A. Phoenix, M. Gaskell and B. Marshall, Photobactericidal activity of methylene blue derivatives against vancomycin-resistant *Enterococcus* spp, *J. Antimicrob. Chemother.*, 1999, **44**, 823.
71. G.P. Tegos and M.R. Hamblin, Phenothiazinium antimicrobial photosensitizers are substrates of bacterial multidrug resistance pumps, *Antimicrob. Agents Chemother.*, 2006, **50**, 196.
72. N. Komerik, H. Nakanishi, A.J. MacRobert, B. Henderson, P. Speight and M. Wilson, *In vivo* killing of *Porphyromonas gingivalis* by toluidine blue-mediated photosensitization in an animal model. *Antimicrob, Agents Chemother.*, 2003, **47**, 932.
73. B.L. Taylor and D.E. Koshland, Intrinsic and extrinsic light responses of *Salmonella typhimurium* and *Escherichia coli*, *J. Bacteriol.*, 1975, **123**, 557.
74. T.W. Wong, Y.Y. Wang, H.M. Sheu and Y.C. Chuang, Bactericidal effects of toluidine blue-mediated photodynamic action on *Vibrio vulnificusAntimicrob, Agents Chemother.*, 2005, **49**, 895.
75. C.E. Milson, M. Wilson, A.J. MacRobert and S.G. Bown, Ex-vivo treatment of gastric *Helicobacter* infection by photodynamic therapy, *J. Photochem. Photobiol, B: Biol.*, 1996, **32**, 59.
76. M. Wainwright, D.A. Phoenix, D.R.A. Wareing and T.E. Smillie, Photobactericidal activity of phenothiaziniums against *Yersinia enterocolitica*, *J. Chemother.*, 2001, **13**, 503.
77. J.L. Fimple, C.R. Fontana, F. Foschi, K. Ruggiero, X. Song, T.C. Pagonis, A.C.R. Tanner, R. Kent, A.G. Doukas, P.P. Stashenko and N.S. Soukos,

Photodynamic treatment of endodontic polymicrobial infection *in vitro*, *J. Endodont.*, 2008, **34**, 728.
78. E. Feese and R.A. Ghiladi, Highly efficient *in vitro* photodynamic inactivation of *Mycobacteriumsmegmatis*, *J. Antimicrob. Chemother.*, 2009, **64**, 782.
79. C. Propst and L. Lubin, *In vitro* and *in vivo* photosensitized inactivation of dermatophyte fungi by heterocyclic dyes, *Infect. Immun.*, 1978, **20**, 136.
80. W.H. Koch and G.E. Bass, Sodium azide affects methylene blue concentration in *Salmonella typhimurium* and *Saccharomyces cerevisiae*, *Photochem. Photobiol.*, 1984, **39**, 841.
81. S.C. de Souza, J.C. Junquiera, I. Balducci, C.Y. Koga-Ito, E. Munin and A.O.C. Jorge, Photosensitization of different *Candida* species by low power laser light, *J. Photochem. Photobiol. B: Biol.*, 2006, **83**, 34.
82. R.A. Prattes, E.G. Da Silva, P.F. Chaves, A.J.S. Santos, C.R. Paula and M.S. Ribeiro, Photodynamic therapy can kill Cryptococcus neoformans in *in vitro* and *in vivo* models, *Proc. SPIE*, 2009, **7165**, 71650H1.
83. F.P. Gonzales, S.H. da Silva, D.W. Roberts and G.U. Braga, Photodynamic inactivation of conidia of the fungi *Metarhizium anisopliae* and *Aspergillus nidulans* with methylene blue and toluidine blue, *Photochem. Photobiol.*, 2010, **86**, 653.
84. J.N. Welsh and M.H. Adams, Photodynamic inactivation of bacteriophage, *J. Bacteriol.*, 1954, **68**, 122.
85. N. Yamamoto, Photodynamic inactivation of bacteriophage and its inhibition, *J. Bacteriol.*, 1958, **75**, 443.
86. D. Lee, M. Foux and E.F. Leonard, The effects of methylene blue and oxygen concentration on the photoinactivation of Qâ bacteriophage, *Photochem. Photobiol.*, 1997, **65**, 161.
87. K. Muller-Breitkreutz and H. Mohr, Hepatitis C and human immunodeficiency virus RNA degradation by methylene blue/light treatment of human plasma, *J. Med. Virol.*, 1998, **56**, 239.
88. Q. Huang and W.L. Fu, Inactivation of dengue virus by methylene blue/narrow bandwidth light system, *J. Photochem. Photobiol. B: Biol.*, 2004, **77**, 39.
89. G.M. Findlay, Immunization against yellow fever, *Trans. Roy. Soc. Trop. Med. Hyg.*, 1934, **27**, 437.
90. O.N. Fellowes, Inactivation of foot-and-mouth disease virus by interaction of dye and visible light, *Appl. Env. Microbiol.*, 1966, **14**, 86.
91. T.W. Chang and L. Weinstein, Photodynamic inactivation of herpes virus hominis by methylene blue, *Proc. Soc. Exp. Biol. Med.*, 1975, **148**, 291.
92. B. Bachmann, J. Knuyver-Hopf, B. Lambrecht and H. Mohr, Target structures for HIV-1 inactivation by methylene blue and light, *J. Med. Virol.*, 1995, **47**, 172.
93. R.A. Floyd and J.E. Schneider, D. P. Dittmer. Methylene blue photoinactivation of RNA viruses, *Antiviral Res.*, 2004, **61**, 141.
94. B. Heemskerk, L.A. Veltrop-Duits, T. Van Vreeswijk, M.M. ten Dam, S. Heidt, R.E.M. Toes, M.J.D. van Tol and M.W. Schilham, Extensive cross-reactivity of CD4+ adenovirus-specific T cells: implications for immunotherapy and gene therapy, *J. Virol.*, 2003, **77**, 6562.
95. C. Wallis and J.L. Melnick, Photodynamic inactivation of enteroviruses, *J. Bacteriol.*, 1965, **89**, 41.
96. D.E. Rounds, W. Opel, R.S. Olson and I.W. Sherman, The potential use of laser energy in the management of malaria, *Biochem. Biophys. Res. Comm.*, 1968, **32**, 616.
97. N. Gironés, B.L. Bueno, J. Carrión, M. Fresno and E. Castro, The efficacy of photochemical treatment with methylene blue and light for the reduction of *Trypanosoma cruzi* in infected plasma, *Vox sanguinis*, 2006, **91**, 285.

Chapter 3

Approaches to Kill Gram-negative Bacteria by Photosensitized Processes

Yeshayahu Nitzan and Izabella Pechatnikov

The Mina and Everard Goodman Faculty of Life Sciences, Bar-Ilan University, Ramat-Gan 52900, Israel

Table of Contents

Photodynamic Inactivation of Microbial Pathogens: Medical and Environmental Applications
Editors Michael R. Hamblin and Giulio Jori

Published by the Royal Society of Chemistry, www.rsc.org

3.1. Introduction

Antibiotic resistance of pathogenic bacteria is a major concern and presents a special challenge for the development of alternative antibacterial modalities.[1] Research was therefore directed towards photodynamic therapy (PDT) as an alternative approach for killing bacteria.[2] The antibacterial activity of porphyrin-induced PDT exhibits unique properties, since it has an efficient microbial killing effect. The efficient and irreversible antimicrobial effects of PDT were found to be independent of the antibiotic susceptibility spectrum of the treated pathogenic bacteria.[3–7] Bacterial photodynamic inactivation is based on a Type II mechanism, which consists of three stages:[2] I) Preferential accumulation of the photosensitizer in the bacterial cell; II) Targeted illumination at an approximate wavelength that induces photochemical activation of O_2 into its excited state – singlet oxygen $^1O_2$, which is a highly reactive oxygen species; and III) $^1O_2$ reacts with various neighbouring macromolecule targets in the bacteria and produces lethal damage to the bacterial cell.[2] Photodynamic inactivation in the absence of exogenous photosensitizers is termed as a Type I mechanism and was found[8] to consist of: I) Excitation of endogenous photosensitizer molecules such as porphyrin, flavins or cytochromes; II) Transfer of electrons to molecular oxygen generating the superoxide radical anion; and III) The superoxide radical converts to hydroxyl free radicals and singlet oxygen in the presence of H_2O_2.[9] It was found that hydroxyl free radicals are also a major cause of bacterial cell death.[10]

Photodynamic inactivation failed to eradicate Gram-negative bacteria that were found to be resistant to this treatment. This phenomenon was seen when using neutral porphyrins or their anionic derivatives such as: haematoporphyrin (HP), haematoporphyrin derivative (HPD) or deuteroporphyrin (DP).[5,11,12] These porphyrins failed to photosensitize *Escherichia coli*,[4,11] *Pseudomonas aeruginosa*,[4,11] *Klebsiella pneumoniae*,[6] *Haemophilus influenzae*[11] and *Bacteroides fragilis*.[11] This resistance to PDT is attributed to differences in the structure and organization of the envelopes of Gram-negative compared to those of Gram-positive bacteria. The envelope of Gram-negative bacteria has a highly complex multilayered structure. It consists of several layers: the cytoplasmic membrane, a planar-thin peptidoglycan layer and an outer membrane. This outer membrane contains phospholipids, lipoproteins, lipopolysaccharides and pore-forming proteins that are present in a high copy number per cell. This structure prevents binding and penetration of porphyrins into Gram-negative bacteria. In contradistinction, the envelope of Gram-positive bacteria consists of only two layers, the cytoplasmic membrane and a peptidoglycan layer that contains teichoic acid. This structure enables the binding and penetration of porphyrin photosensitizers in an amount large enough for eradication. Simple diagrams or a detailed scheme are demonstrated in some works[2,13] and are described in detail by Hamblin and Hasan.[14]

The results of an early study[15] clearly indicated that binding of HPD to the cell is a prerequisite condition for photosensitization of bacteria, regardless of whether they are aerobic[15] or anaerobic.[16] This binding is also pH dependent and the best binding occurs at $pH = 6.5$.[15] Thus, Gram-positive bacteria, which exhibit efficient

HPD binding to the bacterial barriers, are immediately photoinactivated, whereas Gram-negative bacteria bind very little HPD and are resistant to HPD phototreatment. Removal of the cell wall and production of Gram-negative spheroplasts enabled good binding of the porphyrin to the cell membrane. However, this binding did not automatically lead to photoinactivation of the Gram-negative spheroplasts upon illumination. Penetration of porphyrin molecules through the cytoplasmic membrane is apparently a crucial step for the photoinactivation of bacteria after binding to the membrane takes place.[15]

Several approaches have been employed during the last two decades for overcoming the problem of the resistance of Gram-negative bacteria to photoinactivation and to photoinactivate these bacteria as well. Treating Gram-negative microbes has become a high priority, since they are responsible for life-threatening infections in humans, especially in the elderly and in compromised hosts in whom they cause hospital-acquired infections. These bacteria, especially *Pseudomonas aeruginosa* or *Acinetobacter baumannii*, became resistant to most of the commonly used antibiotics and are now the most frequently found bacteria in hospitals.

3.2. Photodynamic Inactivation of Gram-negative Bacteria Using Neutral-uncharged Photosensitizers in the Presence of Membrane Disorganizing Substances

3.2.1. Photodynamic Inactivation of Gram-negative Bacteria in the Presence of Polymyxin Nonapeptide (PMNP)

It is known that the polycationic agent polymyxin B nonapeptide (PMNP) is a non-toxic membrane structure disorganizing agent that acts on growing bacteria in culture medium[17–19] and is very efficient in enabling penetration of antimicrobial drugs into Gram-negative bacteria.

PMNP is prepared by cleavage of the antibiotic polymyxin B in its sulfate salt form by the enzyme ficin following isolation of the pure nonapeptide. It is a polycationic peptide that contains a circular peptide of 7 amino acids and a small tail of 2 amino acids. The positive charges are located on δ-amino groups of the amino acid diaminobutyric acid. The positive charges are responsible for better penetration of neutral-hydrophobic porphyrins through the outer cell membranes and permit better contact and penetration of the porphyrin into the inner membrane of Gram-negative bacteria.

In this laboratory it was found[20] that pre-treatment of *E. coli* and *P. aeruginosa* with PMNP, followed by incubation in Brain Heart Infusion (BHI) with the neutral porphyrin-deuteroporphyrin (DP) and illumination, resulted in complete inhibition of bacterial growth within 30 min. from phototreatment initiation. In parallel, the growth curve of bacteria treated with PMNP alone was not affected and was similar to the untreated control, indicating that PMNP is not toxic to either bacteria. Viability of these bacteria decreased by more than 6 orders of magnitude for *E. coli* and more than 5 orders of magnitude for *Pseudomonas aeruginosa* after 3 h. of illumination.

The effective concentration of DP on both bacteria is 10 μg ml^{-1}, whereas the effective amount of PMNP varies. After screening the optimal concentration of PMNP needed for an effective action of the porphyrin, it was found that 500 μg ml^{-1} are needed for *E. coli*, while only 50 μg ml^{-1} are needed for *P. aeruginosa* cultures.

An effective photodynamic effect was also obtained[7] on *A. baumannii* grown in BHI medium after treatment of the bacteria with 17 μM DP or Cd-texaphyrin in the presence of 200 μM PMNP. Under these conditions, the culture was found to be sterile after 1 h. of treatment, with a decrease of more than 8 orders of magnitude in viability. The light source in all these experiments was two unfiltered tungsten lamps with an irradiance of 140 W m^{-2}.

Protein content of the medium in which photoinactivation takes place is a very important factor in the ability of the photosensitizer to bind to Gram-negative bacteria and thus exert inactivation upon illumination.[7] In was shown that DP in the presence of PMNP can cause photoinactivation even in the presence of the high protein concentration found in BHI. This is an important aspect when phototreatment is carried out in body fluids.

Binding of the porphyrin DP in the presence of PMNP at the appropriate concentration was shown[20] by the fluorescence titration technique employing the fluorescence spectrum of DP at increasing amounts of the tested bacterial cells. A clear increase in the intensity of the 622 nm band that resembles the penetration of DP through the cell membrane was observed. It was also calculated that the binding time of DP to *E. coli* in the presence of PMNP was very fast – 91 seconds – while the binding time to *P. aeruginosa* was 20-fold slower (1890 seconds).

3.2.2. Photodynamic Inactivation of Gram-negative Bacteria by a Natural Photosensitizer – Protochlorophyllide in the Presence of PMNP

A special approach to eradication of bacteria was the use of a neutral intermediate of chlorophyll biosynthesis. The photosensitizer is a green pigment that is accumulated in *Rhodobacterium capsulatum* and has a mutation in the chlorophyll pathway.[21] The photosensitizer protochlorophyllide (Pchlide) can be obtained as a pure product. It can photoinactivate *Y. pseudotuberculosis* or *E. coli* at a concentration of 10 μg ml^{-1} upon addition PMNP at 50 and 20 μg ml^{-1}, respectively. Pchlide did not exhibit any dark toxicity against these bacteria. Bacterial inactivation only occurs in the presence of light. It has a strong absorbance with a high extinction coefficient in the range of 600–800 nm, where the maximum penetration of light is observed. An 8 orders of magnitude reduction in the viable count was observed for *E. coli* after 1.5 h., and the culture was rendered sterile. Furthermore, the control in the presence of 20 μg ml^{-1} PMNP without the photosensitizer but with illumination did not result in any reduction of the *E. coli* cultures' viability. Growth inhibition of *Y. pseudotuberculosis* was observed immediately after the addition of the photosensitizer in the presence of PMNP and illumination. A reduction of 7 orders of magnitude in the viable count was observed after 2 h.

Pchlide-mediated inactivation can be achieved using Pchlide concentrations of 10 or 20 μg ml^{-1}. These values are similar to the concentrations used for photoinactivation by deuteroporphyrin, haematoporphyrin, or photophrin.[6,22,23] Illumination was generated from a 60 W Tungsten lamp placed at a distance of 30 cm. It should be mentioned that the light intensity needed for photoinactivation by Pchlide is very low. An intensity of 0.1 mW cm^{-2} is sufficient for eradication of the tested bacteria, whereas the intensities used for the corresponding photosensitizers are from 1 to 100 mW cm^{-2}.[23–25] Thus, Pchlide generates a larger quantity of reactive oxygen species than other photosensitizers.

3.2.3. *Photodynamic Inactivation of Gram-negative Bacteria by Antioxidant-substituted Porphyrin Photosensitizers in the Presence of PMNP*

Another similar approach was used with two synthesized compounds: 3,7-bis (2-carboxyethyl)-2,8,12,17-tetramethyl-13,18-bis [1-(3,5-di-tert-butyl-4-hydroxybenzyloxy)ethyl] porphyrin (**ACS-2**) and 3,7-bis(2-carboxyethyl)-2,8,12,17-tetramethyl-13,18-bis[1-(5-propoxycarbonyl-3-4-dihydroxy-phenyloxy)ethyl] porphyrin (**ACS-3**). These molecules, named antioxidant carrier sensitizers (ACSs), have the structure of haematoporphyrin that carries two residues of butyl hydroxy toluene (ACS-2) or two residues of propyl gallate (ACS-3).[26] The logic of working with these porphyrins, which possess dual activity as photosensitizers and radical inhibitors, is as follows.

It was assumed that photosensitizers substituted by radical inhibitors will affect the efficiency of the photodynamic treatment. This was based on the fact that when inhibitor molecules react with free radicals and decrease the reactivity of the latter, the efficacy of ground state oxygen in competition for the triplet sensitizer is increased, thus increasing the contribution of the singlet oxygen-mediated pathway.

The antibacterial activities of the photosensitizers ACS-2 and ACS-3 were tested after illumination by a light system that utilizes high-intensity narrowband blue light in the range of 407–420 nm with total UV cutoff on bacterial strains. The Gram-negative photoinactivated strains were *E. coli* and *A. baumannii*. Treatment of these bacteria with ACS-2 at a concentration of 38.6 μM and illumination with intense blue light at an intensity of 56 J cm^{-2} resulted in total eradication of *A. baumannii*. The viability dropped by 8 orders of magnitude, to a sterile culture. With *E. coli*, a decrease of only 3 orders of magnitude in the viability of the culture was obtained by 75 J cm^{-2} blue light. In this case, photoinactivation of the Gram-negative bacteria could only be achieved in the presence of PMNP (500 μM), which disorganizes bacterial membranes. Treatment with the photosensitizer ACS-3 revealed that results can be obtained with smaller amounts of the photosensitizer. Treatment of *A. baumannii* with 10 μM ACS-3 and 500 μM PMNP resulted in complete eradication of the visible bacteria with a light fluence of only 7.5 J cm^{-2}, and a decrease of more than 7 orders of magnitude was achieved. The same decrease (7 orders of magnitude) can be achieved using 20 μM ACS-3 and a light fluence of only 5 J cm^{-2}. A smaller amount of the photosensitizer, 1 μM, also leads to a decrease in viability of 7 orders of magnitude, but requires a light fluence of about 40 J cm^{-2}. A 7-fold higher light fluence was necessary for total eradication of *E. coli*. Total inactivation was achieved with 37.5 J cm^{-2} and 20 μM ACS-3 (in the presence of 500 μM PMNP). Lower amounts of the photosensitizer ACS-3, such as 10 and 1 μM, caused a decrease of only 4 to 5 orders of magnitude in the viability of *E. coli* B.

ACS-3 was found to be a specific and efficient photoinactivation agent against Gram-negative bacteria when illuminated with blue light. ACS-3 is the first photosensitizer to show better photoinactivation selectivity for Gram-negative than for Gram-positive bacteria. All other photosensitizers tested to date act selectively better on Gram-positive bacteria. A decrease of only 2 orders of magnitude in the viable count of *S. aureus* was achieved by photosensitization with ACS-3 under the same conditions in which *E. coli* were totally eradicated.

3.2.4. *Photodynamic Inactivation of Gram-negative Bacteria by Chlorine e_6 Conjugated to Polycationic Molecules*

Another poly-cationic peptide was suggested based on using cationic peptides as agents for permeabilizing neutral porphyrins across the membranes of Gram-negative bacteria. This peptide was a polylysine consisting of 20 lysine residues. This peptide failed to photoinactivate Gram-negative bacteria with DP as photosensitizer.[27] The group of Soukos, Hamblin and Hasan[28,29] proposed a new approach to photodynamic inactivation of Gram-negative bacteria. The approach was to covalently bind the neutral photosensitizer molecule chlorine e_6 to the cationic polylysine peptide, which will alter the organization of the outer membrane of Gram-negative bacteria and result in efficient inactivation of these bacteria. It was demonstrated that the conjugate between chlorine e_6 and a chain of 20 lysine residues resulted in more than 99% photoinactivation of the oral Gram-negative *Porphyromonas gingivalis* after 1 min. illumination with red light. In this experiment, photoinactivation of the bacteria was carried out by the conjugate in the presence of epithelial cells. Illumination by red light preferentially killed the bacterium since the conjugate was taken up by the bacteria and not by the epithelial cells. Polo *et al.*[30] showed that the conjugation of polylysine to porphycene (an electronic isomer of porphyrin) does not appear to interfere with the photobiological activity of the porphycene itself. About 95% of the treated *E. coli* strain lost its viability after 15 min. of visible light irradiation at a concentration of 10 mM.

The same group[31] compared the effectiveness of polylys–porphycene conjugates with chain lengths of either 8 or 37 lysines and found that the 37-lysine conjugate was the best for photodestruction of Gram-negative bacteria. These findings indicate that the rate and extent of bacterial photoinactivation can be modulated through suitable selection of the length of the polylysine polymers as well as the concentration of the polylysine–porphycene conjugate.

Tang *et al.*[32] showed that the conjugate chlorine e_6–polylysine polymer, with an average length of 164 lysine residues, can conjugate to an average of six chlorine (e_6) molecules and in spite of its size can be photoactive against Gram-negative bacteria. This conjugate attained a 5 orders of magnitude decrease in the viability of *E. coli* at a concentration of 8 μM and a light intensity of 30 J cm^{-2}.

Helander *et al.*[33,34] reported that polyethyleneimine is a polycationic polymer that consists of primary, secondary and tertiary amino groups. This polymer acts as a permeabilizer of the outer membranes of Gram-negative *E. coli*, *P. aeruginosa* and *S. typhimurium*. It is an efficient permeabilizer and concentrations of only 20 μg ml^{-1} or less are needed for its action. Tegos *et al.*[35] showed that polycationic conjugates between chlorine (e_6) and large cross-linked polyethyleneimine molecules had the ability to photoinactivate Gram-negative *E. coli* and *P. aeruginosa* after 10 min. illumination. Suspensions containing 10^8 ml^{-1} bacterial cells and treated with 10 μM of the conjugate resulted in a reduction of almost 5 orders of magnitude in both bacteria (after illumination with red light by diode laser at 665 nm). The advantage of this new approach is that, contrary to polylysine conjugates, the polyethyleneimine conjugates are resistant to degradation by proteases such as trypsin, which can hydrolyze lysine–lysine peptide bonds.

3.3. Photodynamic Inactivation of Gram-negative Bacteria by Positively Charged (Cationic) Photosensitizers

3.3.1. Photodynamic Inactivation of Gram-negative Bacteria by Positively Charged (Cationic) Porphyrins

One of the most effective antibacterial photodynamic inactivation approaches against Gram-negative bacteria is the use of cationic porphyrins as photosensitizers.

Porphyrins can be transformed into cationic entities through the insertion of positively charged substituents in the peripheral positions of the tetrapyrrole macrocycle. This substitution affects the kinetics and extent of binding to microbial cells. The degree of the porphyrins' hydrophobicity can be modulated by the number of cationic moieties (up to four in *meso*-substituted porphyrins) or by the introduction of hydrocarbon chains of different lengths on the amino nitrogens.[36]

Merchat *et al.*[37] reported that the antibacterial photodynamic effects of cationic *meso*-substituted porphyrins, namely tetra(4*N*-methyl-pyridyl) porphine tetraiodide (T_4MPyP) and tetra(4*N*,*N*,*N*-trimethyl-anilinium) porphine tetraiodide (T_4MAP), on Gram-negative bacteria can be demonstrated in the absence of any addition of a membrane disorganizing peptide for increasing membrane permeability. These porphyrins can efficiently photoinactivate *V. anguillarum* at a concentration of 10 μg ml^{-1} after 5 minutes of irradiation. For *E. coli*, a 5 order of magnitude reduction in viable count was obtained after 30 min. of irradiation with visible light at 6 mW cm^{-2}.

The same cationic photosensitizer T_4MPyP, at a concentration of 32 μM, inactivated the Gram-negative *Acinetobacter baumannii* when illuminated with visible light during its growth in nutrient broth.[7] Photoinactivation of *A. baumannii* by TMPyP in a protein-rich medium such as BHI failed, and it can take place only in poorer nutritional media. Total eradication of *A. baumannii* and *E. coli* cultures was determined immediately after initiation of illumination by blue (400–450 nm), green (480–550 nm) or red (600–700 nm) light when these bacteria were treated with TMPyP at a concentration of 29.4 μM. Light intensity for eradication varied from 2 J cm^{-2} with blue light, 15 J cm^{-2} with green light and between 24 and 32 J cm^{-2} with red light. The same effect was obtained for *A. baumannii* and *E. coli* after treatment of bacterial cultures with only 3.7 μM TMPyP and illumination with intense narrowband blue light (407–420 nm) at 7 and 9 J cm^{-2}, respectively.[38]

Some studies[39,40] compared the efficiency of synthetic *meso*-substituted cationic porphyrins with different charge distributions (tetra-, tri-, di- or monocationic porphyrins). However, the various groups obtained different results. Merchat *et al.*[39] reported that some di-and tri-cationic porphyrins were more efficient against Gram-negative bacteria than tetracationic porphyrins. Photoinactivation of *E. coli* using *meso*-substituted cationic porphyrins[41] resulted in higher photo-inactivation of the tricationic porphyrins, causing a 5.5 orders of magnitude decrease in cell survival when treated with 10 μM of sensitizer. Under these conditions, a lower effect of 4 orders of magnitude decrease in viability was found for the dicationic porphyrin.

Banfi *et al.*[42] compared three tetracationic porphyrins which were prepared by *N*-alkylation of 5,10,15,20-tetra-4-pyridylporphyrin with different benzyl chlorides and a dicationic porphyrin obtained in a similar way using a methoxy-benzyl chloride. All of the used porphyrins were active against. *P. aeruginosa*, but the dicationic porphyrin was more efficient.

The photodynamic effect of novel cationic porphyrins with different patterns of *meso*-substitution by 4-(3-*N*,*N*,*N*-trimethylammoniumpropoxy)phenyl (A) and 4-(trifluoromethyl)phenyl (B) groups were studied by other researchers.[43] The cationic centres in these porphyrins are isolated from the tetrapyrrolic macrocycle by a propoxy bridge. Thus, the cationic charges have minimal influence on the photophysical properties of the sensitizers. This chain also provides a higher mobility of the charge, which may facilitate the interaction with the outer membrane of the Gram-negative bacteria. Photoinactivation of *E. coli* suspensions showed a higher photodynamic activity of the tricationic A_3B^{3+}variant than the tetracationic porphyrin. This variant caused about 7 orders of magnitude reduction in viability of *E. coli* at a concentration of 7.3 μM.

Alves *et al.*[40] showed that the number of positive charges, the charge distribution in the structure of the porphyrin and the *meso*-substituted groups have different effects on the photoinactivation of *E. coli*. Cationic porphyrins that have three or four charges are highly efficient photosensitizers. The tricationic photosensitizers, namely tri-Py^+-Me-PF porphyrin and tri-Py^+-Me-CO2-Me, exhibited the highest reduction of cell survival under low fluence illumination of 40 Wm^{-2} white light. Using 5 μM of each photosensitizer at a light dose of 64.8 J cm^{-2} resulted in a 7 orders of magnitude decrease in viability. The tetracationic porphyrin was found to be a good photosensitizer, whereas the dicationic or monocationic porphyrins were the least effective.

These results indicate that the number of positive charges, the charge distribution and an increase in the amphiphilic character of the photosensitizer enhance its affinity for the cell wall of Gram-negative bacteria. They offer a promising molecular architecture of photosensitizing agents with potential applications, especially for inactivation of Gram-negative bacteria by photodynamic treatment.

Although the anionic porphyrin meso-tetra (4-sulfonatophenyl)porphine (TPPS4) can be photosensitized and produce reactive oxygen species, it does not bind to intact bacterial cells[37,39] and does not photoinactivate in either a low protein or a richer medium.[7,44]

3.3.2. Photodynamic Inactivation of Gram-negative Bacteria by Positively Charged Phthalocyanines

An additional approach for efficient photoinactivation of Gram-negative bacteria was by treatment with cationic phthalocyanins. Phthalocyanine zinc(II) complexes with different charges were evaluated as photodynamic agents for bacteria.[24] The diamagnetic metal ions in a closed-shell arrangement such as zinc(II) determine a high fluorescence quantum yield, long triplet lifetimes and high triplet quantum yields which lead to a high probability of energy or electron transfer.

Minnock *et al.*[24] were the first to show that Gram-negative bacteria such as *E. coli* and *P. aeruginosa* can be photoinactivated when illuminated in the presence of a cationic water-soluble zinc pyridinium phthalocyanine.

Photoinactivation of *E. coli* carried out after successive washings of the bacterial cells which were exposed to this phthalocyanine revealed that the photosensitizer is tightly bound. The bound fraction is probably the most important fraction, associated and involved in the photosensitization process. It was also found that the presence of Mg^{2+} in the medium prevented the photoinactivation of *E. coli* cells by phthalocyanin.[45]

Mantareva *et al.*[46] showed that the new cationic phthalocyanine 2,9,16,23[tetrakis-(3-methylpyridyloxy)phthalocyanine zinc(II) [ZnPcMe] is active against the Gram-negative *P. aeruginosa*. When treated with 6 μM ZnPcMe and illuminated with laser at 675 nm at a light intensity of 30 $J cm^{-2}$, this bacterium was photoinactivated leading to a 4.5 orders of magnitude decrease in viability. The viability decrease was even greater at a light dose of 60 $J\ cm^{-2}$ and reached 5.5 orders of magnitude. The rate of photoinactivation is related to the bacterial density in the treatment solution. When $10^8\ ml^{-1}$ *P. aeruginosa* cells are treated the decrease in viability is 4 orders of magnitude, whereas treatment of $10^7\ ml^{-1}$ cells yielded a 6 orders of magnitude decrease in viability. *P. aeruginosa* treated with the anionic tetrakis-(4-sulfophenoxy)-phthalocyanine zinc (II) [ZnPcS] at a light dose of 60 $J\ cm^{-2}$ was not photoinactivated at all. Bertoloni *et al.*[6,47] have already shown that zinc-mono- or di-sulfonated phthalocyanine does not photoinactivate *E. coli* or *Klebsiella pneumoniae*. This again shows that anionic photosensitizers do not photoinactivate Gram-negative bacteria.

3.3.3. Photodynamic Inactivation of Gram-negative Bacteria by Positively Charged Phenothiazines

Another approach to photoinactivation of Gram-negative bacteria is to use photosensitizer molecules with an intrinsic positive charge. The blue dyes known as phenothiazinium salts, such as toluidine blue O and methylene blue, have been studied for many years and are the only antibacterial photosensitizers used clinically for antimicrobial photodynamic therapy.[48] Wilson used toluidine blue O for photodynamic inactivation of bacteria, including Gram-negative bacteria found in the oral cavity.[49,50]

Phenothiazines are amphipathic planar molecules that possess one intrinsic quaternary nitrogen atom and have phototoxic efficiency against a broad range of microorganisms, including Gram-negative bacteria such as *Vibrio vulnificus*,[51] *P. aeruginosa*,[52,53] *E. coli*,[52,53] *Helicobacter pylori*,[54] *Haemophilus influenzae*,[53] oral bacteria such as *P. gingivalis and F. nucleatum*[55] and many other periodontopathogens.[56]

Usacheva *et al.*[53] reported that the complete photodestruction of *H. influenzae, P. aeruginosa* or *E. coli* can be achieved by lower concentrations of toluidine blue O than methylene blue and thus exhibit greater bactericidal activity with toluidine blue O.

After irradiation by HeNe laser red light (632 nm) at a light fluence of 15 J cm^{-2} and a light intensity of 75 mW cm^{-2}, about 8 orders of magnitude decrease in *H. influenzae* survival was obtained (a concentration of about 6 μM of toluidine blue O or 16 μM methylene blue are needed to obtain this eradication rate). The same result was observed when *E. coli* was exposed to a light dose of 40 J cm^{-2} and light intensity of 100 mW cm^{-2} at a concentration of 100 μM toluidine blue O or 200 μM methylene blue. Photoinactivation of *P. aeruginosa* under the same illumination conditions requires 200 μM toluidine blue O in order to achieve a 7 orders of magnitude decrease in viability. Only about 2 orders of magnitude decrease in viability are observed with 200 μM methylene blue.

Tegos *et al.*[55] have recently demonstrated that inhibitors of microbial efflux pumps can have a dramatic effect on increasing microbial killing upon illumination when phenothiazinium salts are employed as antibacterial photosensitizers.

3.4. Photodynamic Inactivation of Gram-negative Bacteria by Endogenous Photosensitizers

3.4.1. Photodynamic Inactivation of Gram-negative Bacteria by Endogenous Porphyrins

Photosensitization of Gram-negative bacteria by endogenously produced porphyrins is another approach for eradication of Gram-negative bacteria.[57,58]

Endogenous porphyrins can be produced in high quantities after incubation with δ-aminolevulinic acid.[59] The latter compound is a precursor of the protoporphyrin synthesis pathway in bacterial cells for the production of haemin, the iron-containing protoporphyrin. It has previously been shown that induction of porphyrin synthesis is a consequence of the incubation of the cells with δ-ALA.[60] Another more hydrophobic derivative, δ-ALA methyl ester, can also induce porphyrin synthesis.[57,58]

Most Gram-negative bacteria do not contain large amounts of endogenous porphyrins. Enhanced porphyrin production by δ-ALA in Gram-negative bacteria has been observed in *E. coli*,[57,60] *Aeromonas hydrophila*, *P. aeruginosa*, *Acinetobacter baumannii* and *Serratia marcescens*.[58]

The phenomenon of porphyrin induction by δ-ALA has been demonstrated by fluorescence measurements of extracted porphyrins from the bacteria.[60] It was possible to inactivate cultures of *E. coli* by photosensitization after the cultures were grown with δ-ALA or δ-ALA methyl ester. Photosensitization of the bacteria was carried out under controlled illumination parameters of wavelength and light dose. The viability of *E. coli* cultures after induction of porphyrin synthesis by δ-ALA or δ-ALA methyl ester decreased very significantly but was dependent on the illumination method. Illumination of *E. coli* with blue light (400–500 nm) after incubation with δ-ALA revealed a 4 orders of magnitudes decrease in viability by a light dose of 50 J cm^{-2}. With red light (600–700 nm), a 10-fold higher light dose than with blue light was needed for a 2 orders of magnitude decrease in the viability of *E. coli*. This decrease was obtained after incubation with δ-ALA methyl ester. Illumination with a HeNe laser beam at 632.8 nm revealed that a very high dose of light will be needed for the same decrease in *E. coli* viability. It was not possible to obtain a decrease of viability beyond one order of magnitude with white light and a light dose of 75 J cm^{-2}.

The same phenomenon was observed with *Haemophilus parainfluenzae*,[61] another Gram-negative bacterium, which was incubated with δ-ALA and photosensitized. The killing efficiency was less than that found with exogenously treated porphyrins and light and total eradication of *H. parainfluenzae* by its endogenous porphyrin was not achieved, only a decrease of 2–3 orders of magnitude in viability. It has been suggested[61] that the limited extent of *E. coli* inactivation is a result of photobleaching of the produced endogenous porphyrins. Another suggestion is that this may be a reflection of a situation in which the protoporphyrins produced in *E. coli* cells form aggregates that may effectively shield against photoinactivation.[44] A difference in viability after photoinactivation was also found between the δ-ALA-induced *E. coli* cultures and the δ-ALA methyl ester-induced cultures. A better photokilling rate was always observed in the δ-ALA methyl ester-induced *E. coli* cells than in the δ-ALA-induced cells. This may indicate a less aggregated state of the porphyrins within the cells produced by δ-ALA methyl ester. The different efficacies of photoinactivation by the different light sources correlate well with the extinction coefficient of the endogenous porphyrins produced. For this reason, blue light is more efficient in photokilling.

3.4.2. Photodynamic Inactivation of Gram-negative Bacteria by Endogenous Photosensitizers and Intense Visible Light

The following approach is used for wound healing. Visible light at a high intensity was found to kill bacteria in infected wounds. In a recent study,[62] the phototoxicity of broadband (400–800 nm) 120 J cm^{-2} visible light on the survival of *E. coli* or *S. marcescens* was examined. The illumination caused a 62% reduction in viability of *E. coli* and 56% of *S. marcescens*. The phototoxic effect was found to involve ROS (reactive oxygen species) production by the bacteria. The light source is a device that consists of a halogen lamp equipped with appropriate filters for irradiation in the white region of 400–800 nm. The halogen lamp was coupled to an optic fibre. Heat exerted from the lamp was a maximum of 4°C above room temperature even after 30 min. of irradiation.

The photosensitization of the bacteria is possible since they contain endogenous photosensitizers such as porphyrins, flavins, NADH and cytochromes, which absorb light and can be excited throughout the visible region.

It is also known that *S. marcescens* secretes prodigiosin, a red pigment with a strong absorption band in the visible range, which was found to photosensitize cancerous cells.[63] Based on these results, it is probable that light-induced bacterial killing is not only due to endogenous bacterial photosensitizers, but is also a result of the existence of various photosensitizing metabolites produced and secreted by the bacteria.

No effect on *P. aeruginosa* cultures was observed upon illumination by intense visible light. This result is in agreement with the finding that *P. aeruginosa* has a high resistance to oxidative stress.[64]

E. coli or *S. marcescens* strains were found to form superoxide and hydroxyl free radicals following irradiation. No production of reactive oxygen species was seen by *P. aeruginosa*.

P. aeruginosa also has the ability to produce and secrete pyocyanin, a pigment that absorbs visible light.[65] Pyocyanin itself is a radical upon irradiation. *P. aeruginosa* is resistant to light-induced toxicity, despite the fact that pyocyanin generates ROS upon irradiation. Nevertheless, irradiation of *E. coli* in the presence of pyocyanin further decreased their viability compared to light alone. The extended decrease in the viability of *E. coli* in the presence of pyocyanin and light was significant and reached almost 90%. The high-intensity visible light is used in wound healing to kill bacteria, including Gram-negative bacteria, and to stimulate faster wound healing.

3.5. New Approaches for Photoinactivation of Gram-negative Bacteria

3.5.1. Photodynamic Effects on Gram-negative Bacteria by Encapsulated Water-soluble Photosensitizers

Liposomes are candidate carriers for controlled drug delivery and targeting to cells. They can accommodate hydrophilic molecules in the aqueous spaces and lipophilic molecules in the lipid bilayers. A drug delivery mechanism by liposomes which is mediated by liposome-cell fusion, thus enhancing drug entry into bacterial cells, was examined for Gram-negative bacteria.[66,67]

The action of liposome encapsulated photosensitizers was only recently investigated by Ferro.[68,69] The combination of photodynamic antimicrobial chemotherapy with liposome delivery seems to have wide-ranging possibilities for providing good transport and targeting conditions for photosensitizers acting against bacterial cells. The photoantimicrobial activity of water-soluble photosensitizers encapsulated in liposomes against Gram-negative bacteria is presented in the following approach.[70]

The effect of three water-soluble dyes that are also photosensitizers, methylene blue (MB), neutral red (NR) and Rose Bengal (RB), on Gram-negative bacteria was examined using encapsulated photosensitizers and illumination. The liposomes in this approach are polylamellar. They absorb photosensitizers and are broken into unilamellar liposomes under ultrasonic treatment. Illumination took place at room temperature and was provided by an 18 W white luminescent lamp with an emission range of 400–700 nm. Light intensity of 10.6 klux was measured and the illumination time was between 15 min. and 2 h.

Gram-negative species tested by this approach were *E. coli*, *Salmonella paratyphi B*, *Shigella flexneri*, *P. aeruginosa* and *Klebsiella pneumoniae*. Upon illumination, the liposomes encapsulated with MB were cytotoxic against the *Shigella flexneri*. Liposomes encapsulated with NR inactivated *E. coli* and *Salmonella para B*, and those encapsulated with BR were effective in killing *P. aeruginosa* and *Shigella flexneri*. None of the examined liposome encapsulated photosensitizers showed activity against *Klebsiella pneumoniae*.

In these experiments, the antimicrobial effect after illumination was measured by the decrease in the MIC (minimal inhibitory concentration) of the photosensitizer. Photoinhibition was compared between the encapsulated and the free unencapsulated photosensitizer. Encapsulated MB afforded a stronger antimicrobial effect than the free drug when used against *E. coli* or *Shigella flexneri*. The MIC was 2-fold lower with the encapsulated MB. Encapsulated NR was better than the free drugs against *E. coli* and *Salmonella para B* and the MIC was about 2-fold lower. Encapsulated RB was found to give the same photoinactivation effect as free RB.

Although encapsulation of the photosensitizer can increase the rate of bacterial photoinactivation, this inactivation is dependent on a fit between the bacterial species and the particular encapsulated photosensitizer. The non-inactivation of *Klebsiella pneumoniae* might be a result of the existence of a wide capsule that covers the entire surface of the bacterium and does not enable attachment and fusion of the lipases to the bacterial membranes. It also seems that higher light intensity will increase the effectiveness of the encapsulated photosensitizers.

3.5.2. Photodynamic Inactivation of Gram-negative Bacteria by Chemiluminescence Photodynamic Antimicrobial Therapy (CPAT)

In a new research, it was demonstrated that CPAT is a practical and effective alternative to traditional photodynamic antimicrobial chemotherapy in inhibiting growth and eradicating Gram-negative bacteria. The important advantage of this approach is in bypassing the need for an external light source for activating the photosensitizer and suggests that CPAT might be an effective approach for internal infections where an external light source cannot be used for illumination using traditional photodynamic techniques.

CPAT caused more than a 2 orders of magnitude suppression of *E. coli* when methylene blue was used as a photosensitizer and luminol as the agent that releases light energy. Luminol does not need any external light source or electrical or thermal input. In this approach, luminol[71] was introduced as the light source at non-toxic concentrations and it appears to be a promising approach.[72]

3.6. Damages to Inner-membrane and Ultrastructural Changes in Gram-negative Bacteria upon Photoinactivation

Membrane damage to Gram-negative bacteria was determined in several occasions by X-ray microanalysis. Treatment of *Acinetobacter baumannii* cells with TMPyP and illumination with blue light at 407–420 nm[38] resulted in damage to the sodium-potassium pumps and drastic influx of sodium and chloride and a parallel efflux of potassium from the cells. Treatment of *A. baumannii* with the antioxidant carrier sensitizer 3 (ACS-3) in the presence of PMNP[26] resulted in the loss of 99% of the potassium, a significant loss of phosphate and an increase in sodium, indicating damage to the Na^+/K^+ pump as well as to the ATPase system in the inner membrane.

Treatment of *E. coli* with TMPyP[38] or by ACS-3 in the presence of PMNP[26] also resulted in sodium-potassium pump damage. The damage to the sodium-potassium pump is progressive at increasing light intensities concomitantly to the orders of magnitude of decrease in viability.[73] Furthermore, Valduga *et al.*[74] demonstrated a complete loss of the enzymes lactate dehydrogenase and NADH dehydrogenase. Very little damage to the inner cell membrane of *E. coli* was detected by illumination of the endogenously produced porphyrins induced by δ-ALA. A minute decrease in potassium or increase in sodium was observed.

Electron microscopy of *E. coli* cells treated by DP in the presence of PMNP and exposed to light revealed ultra-structural alterations in the cells' organization.[20] Elongated cells connected to each other without separation of the daughter cells were commonly detected. Furthermore, the chromosomal region was opaque. Irregularities in the cell wall structure included wavy sheets of the wall or tiny vesicular structures stuck on the outer surface of the cells.

P. aeruginosa cells that underwent the same photoinactivation procedure showed a similar alteration pattern as well as chromosomal array modification.

Acinetobacter baumannii treated with TMPyP and illuminated with blue light at 407–420 nm exhibited[38] multilamellar membrane structures that look like mesosomes. These mesosome-like structures are directly connected to the inner membrane. It seems that the newly synthesized membrane is folded into a multilamellar form instead of resulting in septation between daughter cells. Large electron-dense black dots are detected in the cytoplasm of the treated *A. baumannii*. These dots may represent aggregation of polyphosphate inclusions.

When *A. baumannii* is photosensitized by the photosensitizer ACS-3 (antioxidant carrier senstizers-3) in the presence of PMNP and illuminated with blue light (407–420 nm),[26] it exhibits mesosome-like folded membrane structures, which are the result of membrane disruption that causes leakage of cytoplasm content, leading to the inability of the cell to multiply and produce viable bacteria.

The ultra-structural changes observed after incubation of *E. coli* with δ-aminolevulinic acid that led to the induction of endogenous porphyrins and illumination with blue light (400–450 nm) revealed that the changes are not different from the changes observed when these bacteria were photosensitized with exogenous porphyrins.[57]

3.7. Mode of Photoinactivation of Gram-negative Cells

In early studies on the photodynamic effects of porphyrins on bacteria, it was assumed that DNA damage initiated cell death. This assumption was based on the fact that DNA synthesis was the first to be decreased upon interaction with photosensitized porphyrins.[4] Furthermore, the disappearance of plasmidal supercoiled fractions and changes in chromosomal DNA patterns were also observed in bacteria after photosensitization.[75] It was also found that the essential components in the bacterial cell that can be damaged by these active oxygen species might be sodium-potassium pumps, inner-membrane structures[38] and loss of cytoplasmic enzymatic systems such as lactate dehydrogenase and NADH dehydrogenase.[74] These injuries strongly affect bacterial cell survival because these systems cannot be repaired. X-ray microanalysis demonstrated that leakage of substances from the bacterial cells may cause rapid cell death.[26,38] This leads to the conclusion that phototreated bacterial cell death is unrelated to DNA damage. This conclusion is in agreement with the conclusion of a previous work[74] stating that DNA does not represent a critical target of TMPyP photosensitization. In a recent study,[73] the investigators again reached the conclusion that with TMPyP, the primary act of photokilling is related to alterations caused to the bacterial cell membranes, and that damage to DNA appears only later. All of these conclusions are based on the fact that damage to the bacterial cell membrane could also be shown at a relatively low intensity of the appropriate illumination, whereas a high light intensity is necessary for demonstrating DNA breakage.[38] With the latter phenomenon, it was proven[76] that low light intensity (4 J cm^{-2}) resulted in the death of *Acinetobacter calcoaceticus* phototreated with TMPyP. The DNA of the bacteria remained intact, despite their photoinactivation. Significant DNA breakage was visualized after illumination with an intensity of at least 30 J cm^{-2}. It seems that low illumination and photosensitizer cause breakages in the DNA, which cause enhancement of DNA repair by the enzyme Reca. Meanwhile, the damage in the inner cell membrane causes linkage of the cytoplasm's content, including ATP, which is necessary for the DNA repair. All processes that repair the damage to the chromosomal DNA consequently stop. Thus, the major cause of bacterial death by photosensitization is the damage of essential systems in the cell membrane. The damage to DNA appears when the cell is already dead.

3.8. Conclusions

Photoinactivation of Gram-negative bacteria was investigated by several groups in the last two decades. The resistance of these bacteria to the photodynamic inactivation effects of neutral porphyrin photosensitizers has challenged the search for other approaches in order to overcome this problem.

These neutral photosensitizers did not act on Gram-negative bacteria due to their failure to attach to the bacteria's inner membrane after penetration through the outer membrane. Although the neutral photosensitizers induce the production of reactive oxygen species (ROS) upon illumination with the appropriate light intensity at the proper range of the visible light range, the ROS cannot affect Gram-negative bacteria because of their distance from the inner membrane.

The approaches for overcoming the described resistance employed several changes: a) adding an agent that will open channels for the neutral photosensitizer to approach and cross the inner membrane; b) changing the charge of the photosensitizer; and c) photosensitizing endogenous porphyrins and other inner photosensitizers of the bacteria either by direct illumination or by illumination after enhancement of the porphyrin content in the bacterial cytoplasm.

Gram-negative bacteria were susceptible to photosensitizers that possess positive (cationic) charges. These charges can be an integral part of the photosensitizer's molecule, as with meso-substituted cationic porphyrins, cationic phthalocyanines or positively charged phenothiazines. The positive charge can also be on a cationic molecule such as polymyxin nonapeptide, polylysine or polyethyleneimine, which can be free or conjugated to a neutral photosensitizer. These cationic peptides must be present during photosensitization with neutral porphyrins in order to cause disorganization of the outer and inner membranes of the Gram-negative bacteria, allowing the photosensitizers to come within close vicinity of the inner membrane. When the positive charge is located in the photosensitizer molecule, the same action is involved by the molecule itself. Upon photosensitization, ROS are produced. These damage the inner cell membrane by damaging active transport systems such as sodium-potassium pumps and other enzymatic systems that are located on the inner membrane and the consequence is leakage of solutes from the cytoplasm. Penetrated ROS also cause chromosomal breakages that cannot be repaired by the RecA system due to the loss of essential substrates in the cytoplasm.

Porphyrins are produced naturally in the regular biosynthesis pathway or their production is enhanced by the precursor δ-ALA. During photoinactivation of Gram-negative bacteria by their endogenous porphyrins, the ROS are produced in the cytoplasm. The efficiency of photoinactivation is relatively lower than with exogenous photosensitizers and needs a much greater photosensitization intensity. The proposed reason is that the available level of the photosensitizer is low, as in the case of regular production of porphyrins. When the amount of porphyrins is high, in the case of induction by δ-ALA, most of the porphyrins are in the aggregated state and only a small portion in the monomeric state. In these cases, ROS are formed in relatively small quantities. These ROS encounter first molecules within the cytoplasm and only later cause damage to the inner-membrane systems.

All approaches described in this chapter are more effective in eradication of Gram-positive bacteria than Gram-negative bacteria, except for the antioxidant-substituted porphyrin ACS-3, which is more specific for Gram-negative bacteria.

Although the clinical use of photosensitizers is limited to oral and superficial skin infections in which Gram-negative bacteria participate as a cause of infection, further investigation of the introduction of these approaches into clinical practice must continue.

Acknowledgements

The authors thank Ms Sharon Victor for assistance in typing and formatting this chapter. Technical assistance for this chapter was supported by funds from the Mina and Everard Goodman Faculty of Life Sciences of Bar-Ilan University.

References

1. I.K. Hosein, D.W. Hill, L.E. Jenkins and J.T. Magee, Clinical significance of the emergence of bacterial resistance in the hospital environment, *J. Appl. Microbiol.*, 2002, **92**, 90S–97S.
2. Z. Malik, J. Hanania and Y. Nitzan, Bactericidal effects of photoactivated porphyrins – an alternative approach to antimicrobial drugs, *J. Photochem. Photobiol. B*, 1990, **5**, 281–293.
3. Z. Malik, S. Gozhansky and Y. Nitzan, Effect of photoactivated hematoporphyrin derivative on bacteria and antibiotic resistance, *Microbios. Lett.*, 1982, **21**, 103–112.
4. Y. Nitzan, S. Gozhansky and Z. Malik, Effect of photoactivated hematoporphyrin derivative on the viability of Staphylococcus aureus, *Curr. Microbiol.*, 1983, **8**, 279–284.
5. G. Bertoloni, M. Dall'Acqua, M. Vazzoler, B. Salvato and G. Jori, Bacterial and yeast cells as models for studying hematoporphyrin photosensitization, *in Porphyrin in Tumor Phototherapy, ed. A. Andreoni and R. Cubeddu, Plenum Press, New York*, 1983, pp. 177–183.

6. G. Bertoloni, B. Salvato, M. Dall'Acqua, M. Vazzoler and G. Jori, Hematoporphyrin-sensitized photoinactivation of *Streptococcus faecalis*, *Photochem. Photobiol.*, 1984, **39**, 811–816.
7. Y. Nitzan, A. Balzam-Sudakevitz and H. Ashkenazi, Eradication of *Acinetobacter baumannii* by photosensitized agents *in vitro*, *J. Photochem. Photobiol. B*, 1998, **42**, 211–218.
8. A. Lipovsky, Y. Nitzan, H. Friedmann and R. Lubart, Sensitivity of *Staphylococcus aureus* strains to broadband visible light, *Photochem. Photobiol.*, 2009, **85**, 255–260.
9. O. Feuerstein, D. Moreinos and D. Steinberg, Synergic antibacterial effect between visible light and hydrogen peroxide on Streptococcus mutans, *J. Antimicrob. Chemother.*, 2006, **57**, 872–876.
10. Y. Nitzan, B. Shainberg and Z. Malik, The mechanism of photodynamic inactivation of Staphylococcus aureus by deuteroporphyrin, *Curr. Microbiol.*, 1989, **19**, 265–269.
11. Z. Malik, S. Gozhansky and Y. Nitzan, Effects of photoactivated HPD on bacteria and antibiotic resistance, *Microbios Letters*, 1982, **21**, 103–112.
12. Z. Malik, H. Ladan, J. Hanania and Y. Nitzan, Mesosomal structures and antimicrobial activity induced by hemin oxidation or porphyrin photodynamic sensitization, *Curr. Microbiol.*, 1988, **16**, 321–328.
13. G. Jori and S.B. Brown, Photosensitized inactivation of microorganisms, *Photochem. Photobiol. Sci.*, 2004, **3**, 403–405.
14. M.R. Hamblin and T. Hasan, Photodynamic therapy: a new antimicrobial approach to infectious disease?, *Photochem. Photobiol. Sci.*, 2004, **3**, 436–450.
15. B. Ehrenberg, Z. Malik and Y. Nitzan, Fluorescence spectral changes of hematoporphyrin derivative upon binding to lipid vesicles, *Staphylococcus aureus* and *Escherichia coli*, *Photochem. Photobiol.*, 1985, **41**, 429–435.
16. Y. Nitzan, H.M. Wexler and S.M. Finegold, Inactivation of anaerobic bacteria by porphyrins or by hemin, *Curr. Microbiol.*, 1994, **29**, 125–131.
17. M. Vaara and T. Vaara, Polycations sensitize enteric bacteria to antibiotics, *Antimicrob. Agents Chemother.*, 1983, **24**, 107–113.
18. M. Vaara and T. Vaara, Polycations as outer membrane-disorganizing agents, *Antimicrob. Agents Chemother.*, 1983, **24**, 114–122.
19. J. Lam, J. Hildebrandt, E. Schutze and A.F. Wenzel, Membrane-disorganizing property of Polymyxin B nonapeptide, *J. Antimicrob. Chemother.*, 1986, **18**, 9–15.
20. Y. Nitzan, M. Gutterman, Z. Malik and B. Ehrenberg, Inactivation of Gram negative bacteria by photosensitized porphyrins, *Photochem. Photobiol.*, 1992, **55**, 89–96.
21. J. Walther, M.J. Brocker, D. Watzlich, M. Nimtz, M. Rohde, D. Jahn and J. Moser, Protochlorophyllide: a new photosensitizer for the photodynamic inactivation of Gram-positive and Gram-negative bacteria, *FEMS Microbiol. Lett.*, 2009, **290**, 156–163.
22. Y. Nitzan, B. Shainberg and Z. Malik, Photodynamic effects of deuteroporphyrin on Gram positive bacteria, *Curr. Microbiol.*, 1987, **15**, 251–258.
23. T. Maisch, Anti-microbial photodynamic therapy: useful in the future?, *Laser. Med. Sci.*, 2007, **22**, 83–91.
24. A. Minnock, D.I. Vernon, J. Schofield, J. Griffiths, J.H. Parish and S.B. Brown, Photoinactivation of bacteria. Use of a cationic water-soluble zinc phthalocyanine to photoinactivate both Gram-negative and Gram-positive bacteria, *J. Photochem. Photobiol. B Biol.*, 1996, **32**, 159–164.

25. D. Matevski, R. Weersink, H.C. Tenenbaum, B. Wilson, R.P. Ellen and G. Lepine, Lethal photosensitization of periodontal pathogens by a red-filtered Xenon lamp in vitro, *J. Periodontal Res.*, 2003, **38**, 428–435.
26. H. Ashkenazi, Y. Nitzan and D. Gal, Photodynamic effects of antioxidant substituted porphyrin photosensitizers on gram-positive and -negative bacteria, *Photochem. Photobiol.*, 2003, **77**, 186–191.
27. Z. Malik, H. Ladan and Y. Nitzan, Photodynamic inactivation of Gram-negative bacteria: Problems and possible solutions, *J. Photochem. Photobiol. B Biol.*, 1992, **14**, 262–266.
28. N.S. Soukos, M.R. Hamblin and T. Hasan, The effect of charge on cellular uptake and phototoxicity of polylysine chlorin(e6) conjugates, *Photochem. Photobiol.*, 1997, **65**, 723–729.
29. N.S. Soukos, L.A. Ximenez-Fyvie, M.R. Hamblin, S.S. Socransky and T. Hasan, Targeted antimicrobial photochemotherapy, *Antimicrob. Agents Chemother.*, 1998, **42**, 2595–2601.
30. L. Polo, A. Segalla, G. Bertoloni, G. Jori, K. Schaffner and E. Reddi, Polylysine-porphycene conjugates as efficient photosensitizers for the inactivation of microbial pathogens, *J. Photochem. Photobiol. B Biol.*, 2000, **59**, 152–158.
31. M.R. Hamblin, D.A. O'Donnell, N. Murthy, K. Rajagopalan, N. Michaud, M.E. Sherwood and T. Hasan, Polycationic photosensitizer conjugates: effects of chain length and Gram classification on the photodynamic inactivation of bacteria, *J. Antimicrob. Chemother.*, 2002, **49**, 941–951.
32. H.M. Tang, M.R. Hamblin and C.M. Yow, A comparative in vitro photo-inactivation study of clinical isolates of multidrug-resistant pathogens, *J. Infect. Chemother.*, 2007, **13**, 87–91.
33. I.M. Helander, H.L. Alakomi, K. LatvaKala and P. Koski, Polyethyleneimine is an effective permeabilizer of Gram-negative bacteria, *Microbiology UK*, 1997, **143**, 3193–3199.
34. I.M. Helander, K. Latva-Kala and K. Lounatmaa, Permeabilizing action of polyethyleneimine on Salmonella typhimurium involves disruption of the outer membrane and interactions with lipopolysaccharide, *Microbiology UK*, 1998, **144**, 385–390.
35. G.P. Tegos, M. Anbe, C. Yang, T.N. Demidova, M. Satti, P. Mroz, S. Janjua, F. Gad and M.R. Hamblin, Protease stable polycationic photosensitizer conjugates between polyethyleneimine and chlorin(e6) for broad spectrum antimicrobial photoinactivation, *Antimicrob. Agents Chemother.*, 2006, **50**, 1402–1410.
36. G. Jori, C. Fabris, M. Soncin, S. Ferro, O. Coppellotti, D. Dei, L. Fantetti, G. Chiti and G. Roneucci, Photodynamic therapy in the treatment of microbial infections: Basic principles and perspective applications, *Laser. Surg. Med.*, 2006, **38**, 468–481.
37. M. Merchat, G. Bertoloni, P. Giacomini, A. Villanueva and G. Jori, Meso-substituted cationic porphyrins as efficient photosensitizers of Gram-positive and Gram-negative bacteria, *J. Photochem. Photobiol. B Biol.*, 1996, **32**, 153–157.
38. Y. Nitzan and H. Ashkenazi, Photoinactivation of *Acinetobacter baumannii* and *Escherichia coli* B by a cationic hydrophilic porphyrin at various light wavelengths, *Curr. Microbiol.*, 2001, **42**, 408–414.
39. M. Merchat, J.D. Spikes, G. Bertoloni and G. Jori, Studies on the mechanism of bacteria photosensitization by meso-substituted cationic porphyrins, *J. Photochem. Photobiol. B Biol.*, 1996, **35**, 149–157.
40. E. Alves, L. Costa, C.M.B. Carvalho, J.P.C. Tome, M.A. Faustino, M.G.P.M.S. Neves, A.C. Tome, J.A.S. Cavaleiro, A. Cunha and A. Almeida, Charge effect on

the photoinactivation of Gram-negative and Gram-positive bacteria by cationic meso-substituted porphyrins, *BMC Microbiol.*, 2009, **9**, 70.
41. D. Lazzeri, M. Rovera, L. Pascual and E.N. Durantini, Photodynamic studies and photoinactivation of Escherichia coli using meso-substituted cationic porphyrin derivatives with asymmetric charge distribution, *Photochem. Photobiol.*, 2004, **80**, 286–293.
42. S. Banfi, E. Caruso, L. Buccafurni, V. Battini, S. Zazzaron, P. Barbieri and V. Orlandi, Antibacterial activity of tetraaryl-porphyrin photosensitizers: An in vitro study on Gram negative and Gram positive bacteria, *J. Photochem. Photobiol. B Biol.*, 2006, **85**, 28–38.
43. D.A. Caminos, M.B. Spesia and E.N. Durantini, Photodynamic inactivation of Escherichia coli by novel meso-substituted porphyrins by 4-(3-N,N,N-trimethylammoniumpropoxy)phenyl and 4-(trifluoromethyl)phenyl groups, *Photochem. Photobiol. Sci.*, 2006, **5**, 56–65.
44. Y. Nitzan, R. Dror, H. Ladan, Z. Malik, S. Kimel and V. Gottfried, Structure-activity relationship of porphyrins for photoinactivation of bacteria, *Photochem. Photobiol.*, 1995, **62**, 342–347.
45. A. Minnock, D.I. Vernon, J. Schofield, J. Griffiths, J.H. Parish and S.B. Brown, Mechanism of uptake of a cationic water-soluble pyridinium zinc phthalocyanine across the outer membrane of Escherichia coli, *Antimicrob. Agents Chemother.*, 2000, **44**, 522–527.
46. V. Mantareva, V. Kussovski, I. Angelov, E. Borisova, L. Avramov, G. Schnurpfeil and D. Wohrle, Photodynamic activity of water-soluble phthalocyanine zinc(II) complexes against pathogenic microorganisms, *Bioorg. Med. Chem.*, 2007, **15**, 4829–4835.
47. G. Bertoloni, F. Rossi, G. Valduga, G. Jori and J. Van Lier, Photosensitizing activity of water and lipid soluble phthalocyanines on *Escherichia coli*, *FEMS Microbiol. Lett.*, 1990, **71**, 149–155.
48. G.P. Tegos and M.R. Hamblin, Phenothiazinium antimicrobial photosensitizers are substrates of bacterial multidrug resistance pumps, *Antimicrob. Agents Chemother.*, 2006, **50**, 196–203.
49. M. Wilson, Photolysis of oral bacteria and its potential use in the treatment of caries and periodontal disease, *J. Appl. Bacteriol.*, 1993, **75**, 299–306.
50. M. Wilson, J. Dobson and W. Harvey, Sensitization of oral bacteria to killing by low-power laser radiation, *Curr. Microbiol.*, 1992, **25**, 77–81.
51. T.W. Wong, Y.Y. Wang, H.M. Sheu and Y.C. Chuang, Bactericidal effects of toluidine blue-mediated photodynamic action on Vibrio vulnificus, *Antimicrob. Agents Chemother.*, 2005, **49**, 895–902.
52. M. Wainwright, D.A. Phoenix, J. Marland, D.R.A. Wareing and F.J. Bolton, A study of photobactericidal activity in the phenothiazinium series, *FEMS Immunol. Med. Microbiol.*, 1997, **19**, 75–80.
53. M.N. Usacheva, M.C. Teichert and M.A. Biel, Comparison of the methylene blue and toluidine blue photobactericidal efficacy against gram-positive and gram-negative microorganisms, *Laser Surg. Med.*, 2001, **29**, 165–173.
54. C.E. Millson, M. Wilson, A.J. Macrobert, J. Bedwell and S.G. Bown, The killing of *Helicobacter pylori* by low-power laser light in the presence of a photosensitiser, *J. Med. Microbiol.*, 1996, **44**, 245–252.
55. G.P. Tegos, K. Masago, F. Aziz, A. Higginbotham, F.R. Stermitz and M.R. Hamblin, Inhibitors of bacterial multidrug efflux pumps potentiate antimicrobial photoinactivation, *Antimicrob. Agents Chemother.*, 2008, **52**, 3202–3209.

56. M. Wilson, Lethal photosensitisation of oral bacteria and its potential application in the photodynamic therapy of oral infections, *Photochem. Photobiol. Sci.*, 2004, **3**, 412–418.
57. Y. Nitzan and M. Kauffman, Endogenous porphyrin production in bacteria by δ-aminolevulinic acid and subsequent bacterial photoeradication, *Laser Med. Sci.*, 1999, **14**, 269–277.
58. Y. Nitzan, M. Salmon-Divon, E. Shporen and Z. Malik, ALA induced photodynamic effects on gram positive and negative bacteria, *Photochem. Photobiol. Sci.*, 2004, **3**, 430–435.
59. J.C. Kennedy and R.H. Pottier, Endogenous protoporphyrin IX, a clinically useful photosensitizer for photodynamic therapy, *J. Photochem. Photobiol. B Biol.*, 1992, **14**, 275–292.
60. Y. Nitzan, Z. Malik, M. Kauffman, B. Ehrenberg, K. Berg and J. Moan, Induction of endogenic porphyrin production in bacteria and subsequent photoinactivation by various light sources, *Photochemotherapy: Photodynamic Therapy and Other Modalities III*, 1997, 89–94.
61. F.W. Van der Meulen, K. Ibrahim, H.J.C.M. Sterenborg, L.V. Alphen, A. Maikoe and J. Dankert, Photodynamic destruction of *Haemophilus parainfluenzae* by endogenously produced porphyrins, *J. Photochem. Photobiol. B Biol.*, 1997, **40**, 204–208.
62. A. Lipovsky, Y. Nitzan and R. Lubart, A possible mechanism for visible light-induced wound healing, *Laser. Surg. Med.*, 2008, **40**, 509–514.
63. G. Park, J.T. Tomlinson, J.A. Misenheimer, G.L. Kucera and R.A. Manderville, Photoinduced cytotoxicity of prodigiosin analogues, *Bull. Korean Chem. Soc.*, 2007, **28**, 49–52.
64. W. Sabra, E.J. Kim and A.P. Zeng, Physiological responses of Pseudomonas aeruginosa PAO1 to oxidative stress in controlled microaerobic and aerobic cultures, *Microbiology*, 2002, **148**(10), 3195–3202.
65. K.J. Reszka, Y. O'Malley, M.L. McCormick, G.M. Denning and B.E. Britigan, Oxidation of pyocyanin, a cytotoxic product from Pseudomonas aeruginosa, by microperoxidase 11 and hydrogen peroxide, *Free Radic. Biol. Med.*, 2004, **36**, 1448–1459.
66. S. Sachetelli, H. Khalil, T. Chen, C. Beaulac, S. Senechal and J. Lagace, Demonstration of a fusion mechanism between a fluid bactericidal liposomal formulation and bacterial cells, *Biochim. Biophys. Acta*, 2000, **1463**, 254–266.
67. C. Mugabe, M. Halwani, A.O. Azghani, R.M. Lafrenie and A. Omri, Mechanism of enhanced activity of liposome-entrapped aminoglycosides against resistant strains of Pseudomonas aeruginosa, *Antimicrob. Agents Chemother.*, 2006, **50**, 2016–2022.
68. S. Ferro, F. Ricchelli, G. Mancini, G. Tognon and G. Jori, Inactivation of methicillin-resistant Staphylococcus aureus (MRSA) by liposome-delivered photosensitising agents, *J. Photochem. Photobiol. B*, 2006, **83**, 98–104.
69. S. Ferro, F. Ricchelli, D. Monti, G. Mancini and G. Jori, Efficient photoinactivation of methicillin-resistant Staphylococcus aureus by a novel porphyrin incorporated into a poly-cationic liposome, *Intl J. Biochem. Cell Biol.*, 2007, **39**, 1026–1034.
70. M. Nisnevitch, F. Nakonechny and Y. Nitzan, Photodynamic antimicrobial chemotherapy by liposome encapsulated water-soluble photosensitizers, *Russ. J. Bioorg. Chem.*, 2010, **36**, 363–369.
71. R. Laptev, M. Nisnevitch, G. Siboni, Z. Malik and M.A. Firer, Intracellular chemiluminescence activates targeted photodynamic destruction of leukaemic cells, *Br. J. Canc.*, 2006, **95**, 189–196.

72. F. Nakonechny, M.A. Firer, Y. Nitzan and M. Nisnevitch, Intracellular antimicrobial photodynamic therapy: A novel technique for efficient eradication of pathogenic bacteria, *Photochem. Photobiol.*, submitted, 2010.
73. M. Salmon-Divon, Y. Nitzan and Z. Malik, Mechanistic aspects of *Escherichia coli* photodynamic inactivation by cationic tetra-meso(N-methylpyridyl)porphine, *Photochem. Photobiol. Sci.*, 2004, **3**, 423–429.
74. G. Valduga, B. Breda, G.M. Giacometti, G. Jori and E. Reddi, Photosensitization of wild and mutant strains of Escherichia coli by meso-tetra (N-methyl-4-pyridyl)porphine, *Biochem. Biophys. Res. Commun.*, 1999, **256**, 84–88.
75. U. Nir, H. Ladan, Z. Malik and Y. Nitzan, In vivo effects of porphyrins on bacterial DNA, *J. Photochem. Photobiol. B*, 1991, **11**, 295–306.
76. H. Ashkenazi, I. Pechatnikov and Y. Nitzan, Low-intensity photosensitization may enhance RecA production, *Curr. Microbiol.*, 2006, **52**, 317–323.

Chapter 4

Innovative Design of Antimicrobial Photosensitizers

Michael R. Hamblin,[a,b,c] Sulbha K. Sharma[a] and Gitika B. Kharkwal[a,b]

[a] Wellman Center for Photomedicine, Massachusetts General Hospital, Boston, MA 02114, USA; [b] Department of Dermatology, Harvard Medical School, Boston, MA 02115, USA; [c] Harvard-MIT Division of Health Sciences and Technology, Cambridge, MA 02139, USA

Table of Contents

Photodynamic Inactivation of Microbial Pathogens: Medical and Environmental Applications
Editors Michael R. Hamblin and Giulio Jori

Published by the Royal Society of Chemistry, www.rsc.org

4.1. Introduction

In recent years the desirable structural features of effective antimicrobial photosensitizers (PS) have become better understood. In addition to the normal requirements of an efficient PS such as a high absorption band in the red/NIR region of the spectrum and a good triplet yield that allows production of reactive oxygen species upon illumination, additional molecular features are needed. The most important structural feature is either one or more constitutive cationic charges provided by, for instance, quaternary nitrogen groups or a set of basic amino groups provided by a cationic polymer. Additional features may include a correct balance between lipophilicity and hydrophilicity and a certain degree of molecular asymmetry.

The first antimicrobial PS studied were the phenothiazinium dyes typified by methylene blue (MB) and toluidine blue O (TBO).[1] Compounds such as these have the additional advantage that MB is clinically approved as an injectable IV therapy for methemoglobinemia and both MB and TBO are generally accepted as safe for topical application to living human tissue.[2] More recently functionalized tetrapyrroles such as porphyrins bearing various quaternized nitrogen functionalities have been studied as antimicrobial PS and there have been several reports of quantitative structure activity relationships (QSAR) being determined on series of analogous compounds.[3,4]

4.2. Conjugates between PS and Cationic Polymers

Several studies have demonstrated that Gram-positive bacteria are particularly susceptible to photodynamic inactivation with most classes of PS, but Gram-negative bacteria require the use of PS with cationic charges.[3,6,7] It was hypothesized in our laboratory that it might be possible to prepare a broad-spectrum antimicrobial PS that can efficiently bind to and penetrate all classes of microbial cells.[4] This could be accomplished by covalently attaching a PS that has an overall anionic charge provided by its carboxyl groups to a polymer with a large number of basic amino groups. These many amino groups would, in turn, cover a sufficiently large range of pKa values that a substantial number of them would be positively charged at physiological pH.

This approach was first tested against oral pathogens *Actinomyces viscosus* (Gram-positive bacteria) and *Porphyromonas gingivalis* (Gram-negative) using a poly-L-lysine (average MWt = 2000) and chlorin(e6) conjugate.[5] To demonstrate the necessity of having basic amino groups we either left the epsilon-lysine amino groups unmodified to give a cationic conjugate pL-ce6 (**1**), or modified them with acetic anhydride to form a neutral conjugate pL-ce6-ac (**2**), or with succinic anhydride to produce an anionic conjugate pL-ce6-succ (**3**) (Figure 4.1A). These conjugates were tested with oral pathogenic bacteria responsible for tooth decay and periodontitis using 1 min. of incubation and exposure to 671-nm light for 10 min. (15 J cm^{-2}). The cationic conjugate **1** was extremely effective in killing *A. viscosus*, with >99.99% killing (Figure 4.1B). Similarly, with

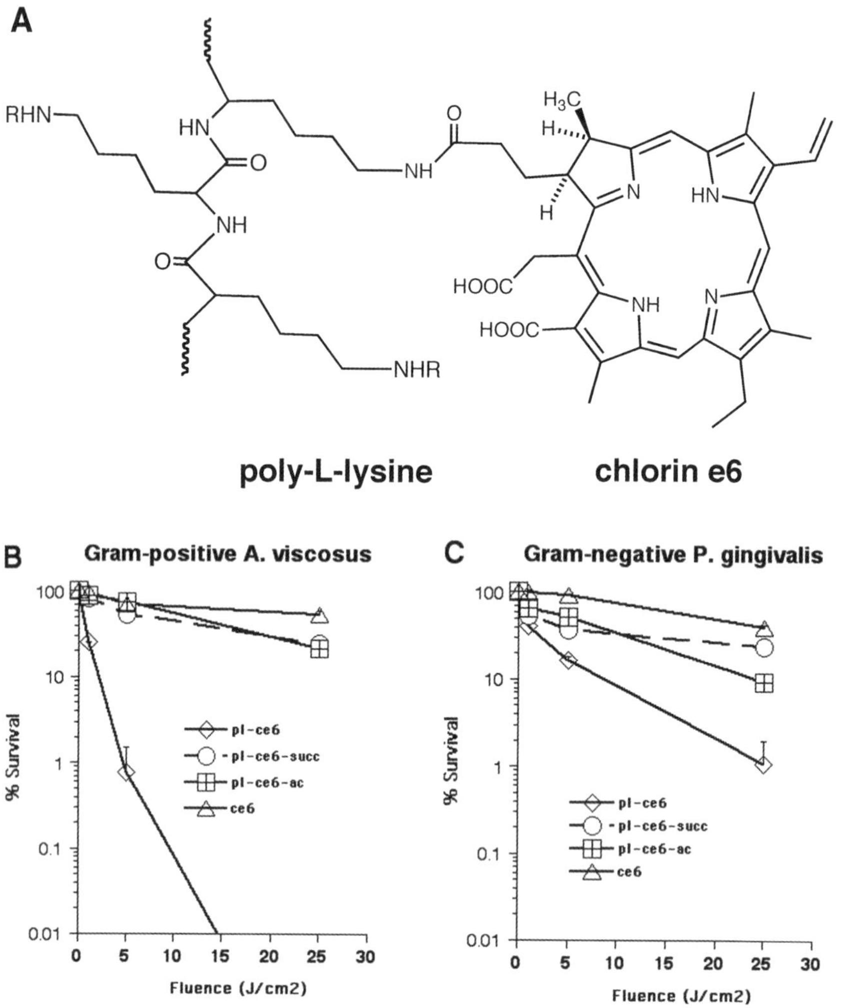

Figure 4.1 (A) Structures of poly-L-lysine chlorin(e6) conjugates with cationic (**1**), neutral (**2**) or anionic (**3**) charges. Fluence-dependent photoinactivation curves mediated by pL-ce6, pL-ce6-ac, pL-ce6-succ and free ce6 of (B) Gram-positive *A. viscosus* and (C) Gram-negative *P. gingivalis*. Adapted from ref. 5.

Gram-negative *P. gingivalis* the cationic conjugate **1** produced 99% killing (Figure 4.1C). Under the same conditions an oral epithelial cell line was unharmed. Furthermore, a mixture – as opposed to a conjugate – of pL with ce6 gave a lesser effect (data not shown).

Previous studies have shown that polycations, such as pL, can bind to the anionic sites of the lipopolysaccharides of the outer membrane in Gram-negative cell wall, thereby disorganizing the structure and rendering it permeable to drugs, anionic detergents or hydrophobic antibiotics.[3–5] Our observation with pL and ce6 mixture on *P. gingivalis* also suggests that positively charged pL altered the outer membrane barrier and helped ce6 to penetrate inside the cell. Moreover the conjugate was not toxic to the oral malignant keratinocytes suggesting a therapeutic window whereby bacteria could be killed without damaging the epithelial cells.

On comparing the activity of a different pL-ce6 conjugate with the traditional PS-toluidine blue O and Rose Bengal against representative members of three classes of pathogens, *E. coli*, *S. aureus* and *C. albicans*, the conjugate was found to be the most powerful sensitizer.[10] Addition of a charged moiety to a PS also increases the spectrum of bacteria against which it is effective as compared to the limited spectrum with the PS alone as reported by Rovaldi *et al.* with ce6 conjugated to a pentalysine moiety.[6]

Increasing the molecular size of polymer added to PS can also broaden the range of bacteria that a particular PS is active against. We showed that PDI with an 8 lysine conjugate **4** (pL-ce6 8) (and with free ce6) was effective against *S. aureus* but not against *E. coli* while a 37-lysine conjugate **5** (pL-ce6 37) was effective against both (see Figure 4.2).[7] These results can be explained by the necessity of a large polycation to penetrate the impermeable outer membrane of the Gram-negative *E. coli*, whereas Gram-positive *S. aureus* is more easily penetrated by small molecules (both cationic and anionic). Polo *et al.* reported similar findings wherein low concentrations of a porphycene conjugate prepared with a smaller sized pL chain effectively killed *S. aureus* but not *E. coli* while a larger sized pL conjugate was effective against *E. coli* even at lower concentrations.[8]

The choice of the polymer itself is also an important variant that can affect PDI. The pL molecule is susceptible to degradation by proteases, such that finding a polymer that can resist protease degradation, and at the same time be

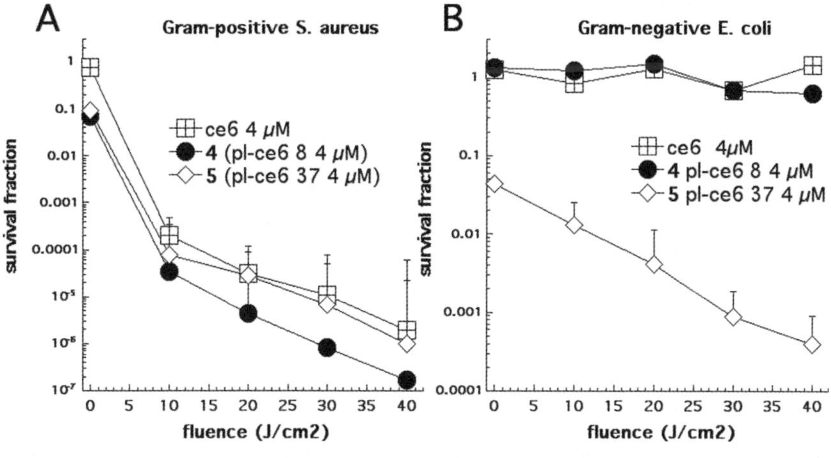

Figure 4.2 Fluence-dependent photoinactivation curves using free ce6, **4** (pL-ce6 8) and **5** (pL-ce6 37) of (A) Gram-positive *S. aureus*; (B) Gram-negative *E. coli*. Adapted from ref. 7.

cost effective, is highly desirable. Ergo, our laboratory prepared a novel set of second-generation polycationic conjugates between ce6 and three molecular forms of polyethylenimine (PEI) – small linear (**6**), small cross-linked (**7**) and a large cross-linked polymer (**8**).[9] As with the pL conjugates, the large cross-linked molecule **8** efficiently killed all organisms (Figure 4.3C), while the small linear conjugate **6** efficiently killed only Gram-positive *S. aureus* and yeast *C. albicans* (Figure 4.3A). The small cross-linked conjugate **7** was inactive at killing any species at these doses (Figure 4.3B).

One important aspect that needs to be addressed in antimicrobial PDT is whether it can lead to the development of bacteria resistant to photo-inactivation or to subsequent antibiotic treatment. A significant finding has been made by Lauro *et al.* to this end.[10] They reported that porphycene-pL conjugates inactivated the bacterial strains responsible for periodontal diseases while repeated photosensitization of surviving cells did not induce selection of resistant bacterial strains nor did it modify their sensitivity to antibiotic treatment.

There are several studies showing the efficacy of cationic conjugates *in vitro* but the question remains whether this approach will be as effective in animal models. Work from our laboratory has shown promising results in this regard[11,12,14,15] Several *in vivo* studies that employed both pL-ce6 and PEI-ce6 are summarized in Chapter 8 (Dai) of this volume.

The field of antimicrobial PDT with conjugates between PS and cationic polymers is still in a nascent stage and several limitations need to be taken into consideration especially in the animal model system. It is a common observation that PDT inactivation of bacteria achieved *in vivo* is less than that seen *in vitro*. This is because first the host tissue competes with the bacteria for binding the PS and second the bacteria tend to penetrate the tissue and form a biofilm. Moreover, the PS can be photobleached or modified by light during PDT, and this fact can influence the therapeutic ratio of PDT positively or negatively.[13] There has not been a conclusive study to compare the best small molecule PS-bearing cationic charges with the best cationic polymer-PS conjugate for

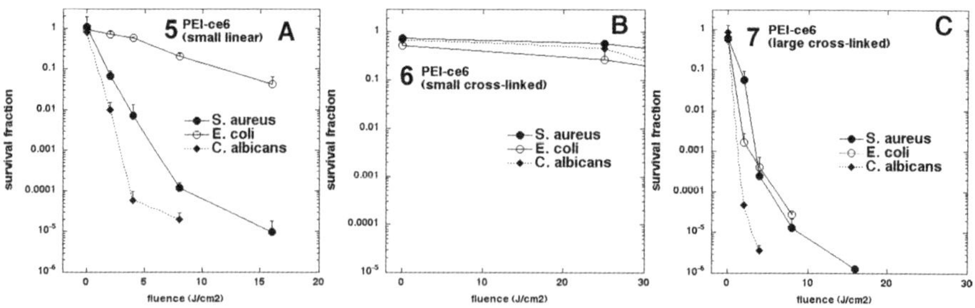

Figure 4.3 Fluence-dependent photoinactivation curves of Gram-positive *S. aureus*; Gram-negative *E. coli*; and yeast *C. albicans* mediated by (A) **5** PEI-ce6 small linear; (B) **6** PEI-ce6 small cross-linked and (C) **7** PEI-ce6 large cross-linked, adapted from ref. 9.

antimicrobial PDI either *in vitro* or *in vivo*. There are other cationic polymers (besides pL and PEI) available for PS conjugation that have yet to be tested. There is therefore a further need to optimize the conjugate structures and PDT procedure so as to make use of cationic-polymer conjugates clinically more relevant.

4.3. Cationic Functionalized Fullerenes

It has been known for several years that fullerenes (buckyballs) have the potential to act as powerful PS. They have high absorption in the visible region of the electromagnetic spectrum, have a high triplet yield and can efficiently generate reactive oxygen species (ROS) upon illumination.[14] Moreover the photochemistry exhibited by light-activated fullerenes tends to preferentially involve Type I species (hydroxyl radicals and superoxide) as compared with the more usual Type 2 ROS, singlet oxygen. Unfortunately, fullerenes present several significant drawbacks in PDT, including their formidable degree of hydrophobicity and their pronounced tendency to aggregate in aqueous media. Furthermore, buckyballs generally have absorption pronounced in the blue region of the visible spectrum, rather than the red. Nonetheless, it is possible to make water-soluble fullerenes, especially by attaching quaternary ammonium groups to the periphery of the spherical cage.

We studied a series of fullerenes that had been functionalized with one (**9**), two (**10**) or three (**11**) methylpyrrolidinium iodide rings (see Figure 4.4A).[15] Using a short incubation time followed by illumination with white light, these compounds had a broad-spectrum antimicrobial activity and the bis- and especially the tris-cationic fullerene **11** were highly active in rapidly killing (4–6 logs) all tested microbes (Gram-positive, Figure 4.4B and Gram-negative Figure 4.4C) and also fungal cells (Figure 4.4D). Although the light absorbance decreases with increased cationic substitution, the increasing positive charge allows the fullerenes to bind to cells and overcome microbial permeability barriers. Cationic fullerenes performed better as antimicrobial photosensitizers than the widely employed antimicrobial photosensitizer toluidine blue O (TBO). Moreover the fullerenes had good selectivity for microbes over mammalian cells (under the same incubation conditions) and the selectivity was also better than that observed for TBO.

Our study was extended when we next compared a further series of innovative cationic fullerenes having the potential to be broad-spectrum light-activated antimicrobials.[16] Their relative effectiveness as broad-spectrum antimicrobial PS against Gram-positive and Gram-negative bacteria, and a fungal yeast was determined by quantitative structure-function relationships (QSAR) The most effective compound against the various classes of microbial cells had a hexacationic structure. The QSAR relationship indicated that increasing the number of cationic charges and lowering the hydrophobicity tended to increase the PDT efficiency against Gram-positive and Gram-negative bacteria.

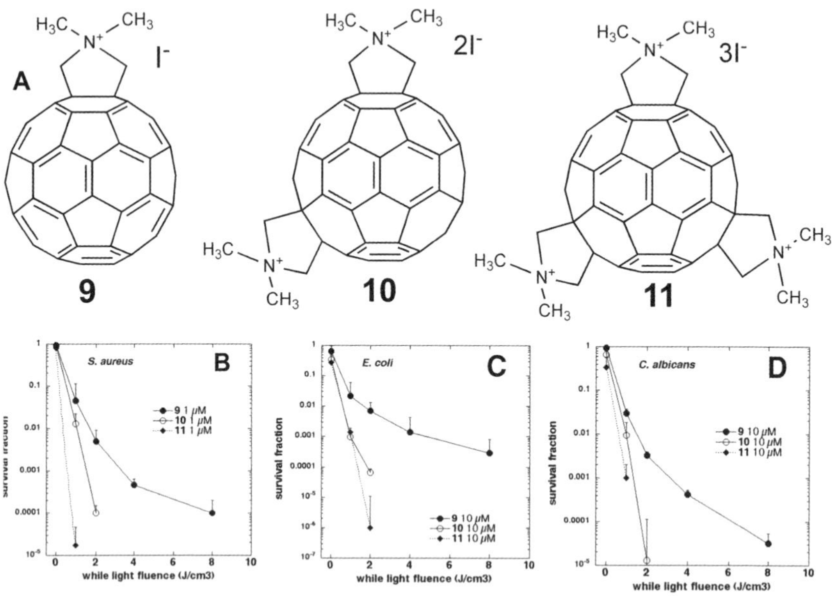

Figure 4.4 (A) Structures of three cationic functionalized fullerenes; **9** mono-*N*-methylpyrrolidinium fullerene, **10** bis-mono-*N*-methylpyrrolidinium fullerene, **11** tris-*N*-methylpyrrolidinium fullerene. Fluence-dependent photoinactivation curves of (B) *S. aureus* at 1 μM; (C) *E. coli* at 10 μM; (D) *C. albicans* at 10 μM, adapted from ref. 15.

The high selectivity and efficacy exhibited by these PS encouraged further testing for ability to treat an infection *in vivo*[17]. Therefore mouse experiments were performed to find out whether this high degree of *in vitro* activity could be translated to an *in vivo* therapeutic effect. For this two potentially lethal mouse models of infection were used for the study. An excisional wound on the back of the mouse was contaminated with one of two bioluminescent Gram-negative microbial strains: *Proteus mirabilis* and *Pseudomonas aeruginosa*. PDT was carried out using tris-cationic **11** (Figure 4.4A) and white light (400–700 nm) at a dose of 180 J cm^{-2}. In case of *P. mirabilis* the fullerene-mediated PDT led to 82% survival of mice compared to only 8% survival in the group without treatment. PDT of mice infected with highly virulent *P. aeruginosa* did not contribute to improved survival (despite killing >90% of bacteria as judged by bioluminescence imaging), but did show a synergistic effect when combined with an antibiotic treatment (tobramycin) with 60% survival, while tobramycin alone showed only 20% survival.

A recent study by Spesia *et al.* reported that a novel *N,N*-dimethyl-2-(4′-*N,N,N*-trimethylaminophenyl)fulleropyrrolidinium iodide (DT-C_{60} (2+)) synthesized by 1,3-dipolar cycloaddition using 4-(*N,N*-dimethylamino) benzaldehyde, *N*-methylglycine and fullerene C_{60} and quaternization with methyl iodide.[18] The photodynamic inactivation produced by these fullerene derivatives was investigated *in vitro* on *E. coli*. Photosensitized inactivation of *E. coli* cellular suspensions by DT-C_{60} (2+) showed approximately 3.5 log decrease of cell survival when the cultures are treated with 1 μM of sensitizer and irradiated for 30 min. This photosensitized inactivation remains high even after one washing step. Also, the photodynamic activity was confirmed by growth delay of *E. coli* cultures. The growth was arrested when *E. coli* was exposed to 2 μM of cationic fullerene and irradiated, whereas a negligible effect was found for the non-charged MA-C_{60}.

Lee *et al.* synthesized four novel hexakis C_{60} derivatives with varying functionalities and their photochemical properties and photodynamic disinfection efficiencies were quantitatively evaluated. Cationic hexakis-aminofullerene, which likely exerted electrostatic attraction, exhibited exceptionally rapid virus inactivation even compared to commercial nano-TiO2 photo catalyst.[19]

Cationic-fullerene-mediated PDT may prove to have a role in the treatment of localized infections in areas such as wounds, burns, skin and mucus membranes, as all of these surfaces are easily accessible by light and do not require deep light penetration.

4.4. Cationic Functionalized Bacteriochlorins

Photosensitizers based on the bacteriochlorin backbone have been studied as potential PDT-agents for cancer and non-oncological applications.[24–27] The large absorption feature in the near-infrared spectral region, which is characteristic of bacteriochlorins, is considered to be ideal for maximizing light penetration through tissue, because both absorption and scattering of light in the 700–800 nm region are minimal. However, in addition to good optical properties, it is necessary for a PS molecule to possess the appropriate structural characteristics that will optimize the binding to and penetration into microbial cells. For antimicrobial applications, the effective molecular features are likely to include the presence of positively charged substituents such as quaternized ammonium groups.

Recently, a *de novo* pathway for the synthesis of bacteriochlorins containing a geminal dimethyl group in each pyrroline ring was developed.[20] This unique structural attribute blocks adventitious dehydrogenation (to form the chlorin) and thereby affords a stable macrocycle. The exciting feature of this discovery is that the synthetic route provided a number of bacteriochlorin building blocks, which provided modular access to bacteriochlorins **12**–**15** (Figure 4.5). The four molecules were designed to allow investigation of the structure-activity relationship among differently charged bacteriochlorins. Bacteriochlorin **12** is a neutral species with two basic amino groups: bacteriochlorins **13**–**15** contain 2, 4 or 6 cationic charges, respectively. The synthesis of bacteriochlorins **12**–**14** was reported, and the synthesis of **15** was described in a recent publication.[21] We tested bacteriochlorins **12**–**15** as antimicrobial PS against a panel of human pathogens of different taxonomic classifications, and determined the selectivity of the four bacteriochlorins for killing microbial cells *versus* mammalian

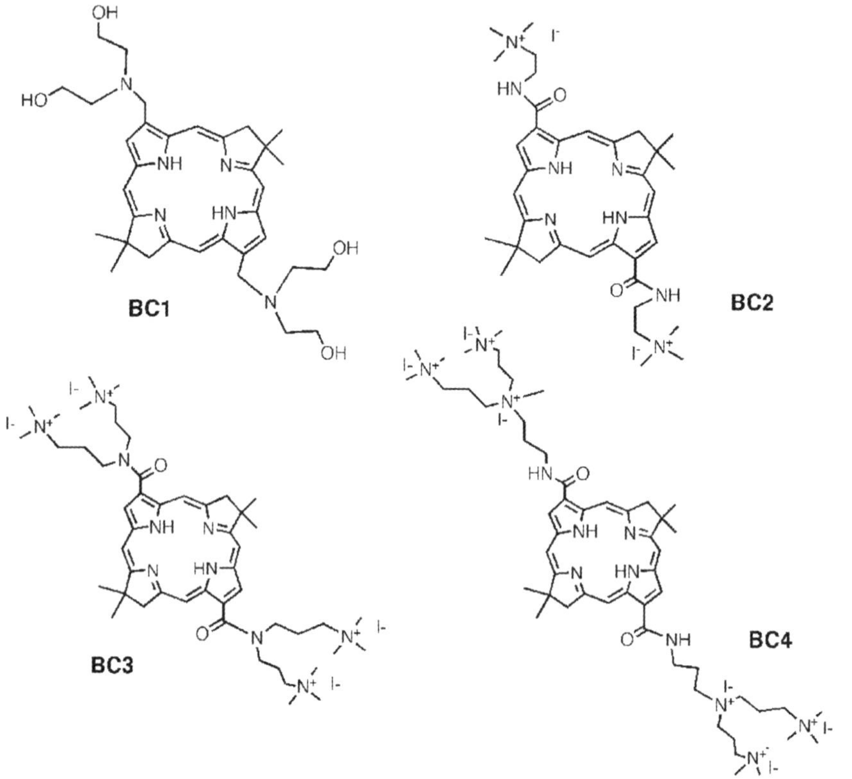

Figure 4.5 Molecular structures of four new synthetic bacteriochlorins bearing two basic groups (**BC1**), or two (**BC2**), four (**BC3**) and six (**BC4**) quaternary ammonium groups, adapted from ref. 25.

(human cancer) cells using the same incubation time and other experimental conditions.

Figure 4.6A displays the survival fraction curves obtained against the Gram-positive bacterium *S. aureus* incubated for 30 minutes using bacteriochlorins **12**–**15** with and without illumination (10 J cm^{-2} 732-nm laser light). The analogous plots are shown for the Gram-negative *E. coli* (Figure 4.6B) and for the fungal yeast in Figure 4.6C. One striking result from this investigation was that the PS structure that gives the maximum PDI effect is different for each class of microbial cell. For the Gram-positive *S. aureus* the bis-quaternized bacteriochlorin **11** was most effective, for the Gram-negative *E. coli* the hexakis cationic **13** was most effective, while for the fungal yeast *C. albicans* only the non-cationic bacteriochlorin **10** had a high PDI killing effect. The only compound that produced any noticeable PDT killing of host mammalian cells (HeLa) was **10** and that was much less than that seen for any microbial cells (data not shown). Therefore choosing the appropriate BC structure can optimize the selectivity and the efficiency of microbial cell killing compared to host cell killing.

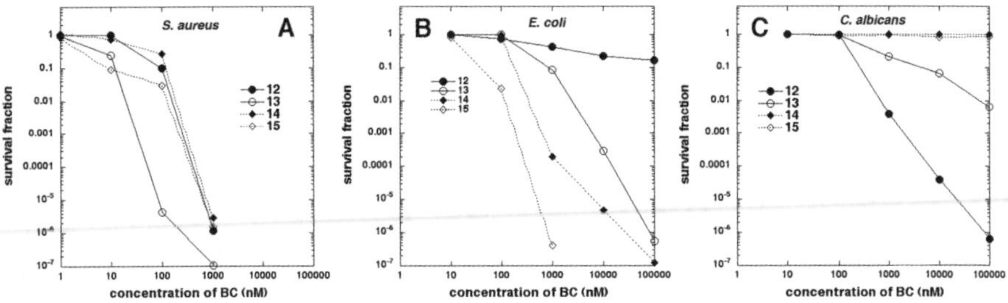

Figure 4.6 Concentration-dependent photoinactivation curves using dilutions of **12**, **13**, **14**, **15** combined with 10 J cm^{-2} of 732-nm light for (A) *S. aureus*; (B) *E. coli*; (C) *C. albicans*, adapted from ref. 25.

4.5. Chalcogen Substituted Benzo[*a*]phenoxazinium Dyes

In another report we studied the benzophenoxazinium family of cationic chromophores as antimicrobial PS.[22] The original observation that the textile dye Nile Blue A (NBA) and a select set of derivatives, when administered orally, strongly stained solid murine tumours suggested that these compounds may act as PS in PDT. Although the Nile Blue derivatives were inefficient PS both *in vitro* and *in vivo*, they nonetheless proved to be a valuable starting point for the rational design of drugs with improved photoactivity. One of the most effective anti-cancer PDT agents discovered early in this program involved the replacement of the oxygen atom of an ethylated-NBA analogue (**16**, EtNBA) with a sulfur atom to give the novel PS 5-(ethylamino)-9-diethylaminobenzo-[*a*]-phenothiazinium chloride (**17**, EtNBS).[23] This drug, when activated with 650-nm light, showed good efficacy against several murine sarcomas (EMT-6, RIF).[23] Consequently, as an extension of this strategy, we were motivated to synthesize EtNBSe (**18**), a chalogen analogue having a still heavier Se atom in place of the centrally located S atom of EtNBS (**17**) (see structures in Figure 4.7A and photophysical properties in Table 4.1).

In an *in vitro* study using the EMT-6 mammary sarcoma cell line, a comparative phototoxicity evaluation of EtNBSe (**18**) with EtNBS (**17**), EtNBA (**16**) and Photofrin (incubated at 0.5 mM concentration) found that the light dose required to kill 50% of the cells was <0.5 J cm^{-2} for EtNBSe (**18**) and 2.0 J cm^{-2} for EtNBS (**17**), whereas the dose to achieve the same response for EtNBA (**16**) and Photofrin could not be determined because of their low phototoxicities.[24] Because these compounds possess a constitutive cationic charge in their heterocyclic ring structure (in a similar fashion to phenothiazinium dyes) we evaluated them as broad-spectrum antimicrobial PS against a panel of prototypical human pathogenic microbes (Gram-positive *Enterococcus faecalis*, Gram-negative *E. coli* and yeast *C. albicans*).

The fluence-dependent killing curves (using wavelengths 630 ± 15 nm for EtNBA (**16**), 652 ± 15 nm for EtNBS (**17**) and 660 ± 15 nm for EtNBSe (**18**)) are shown for the three microbial species in Figure 4.7B–D. We used dyes 2 μM

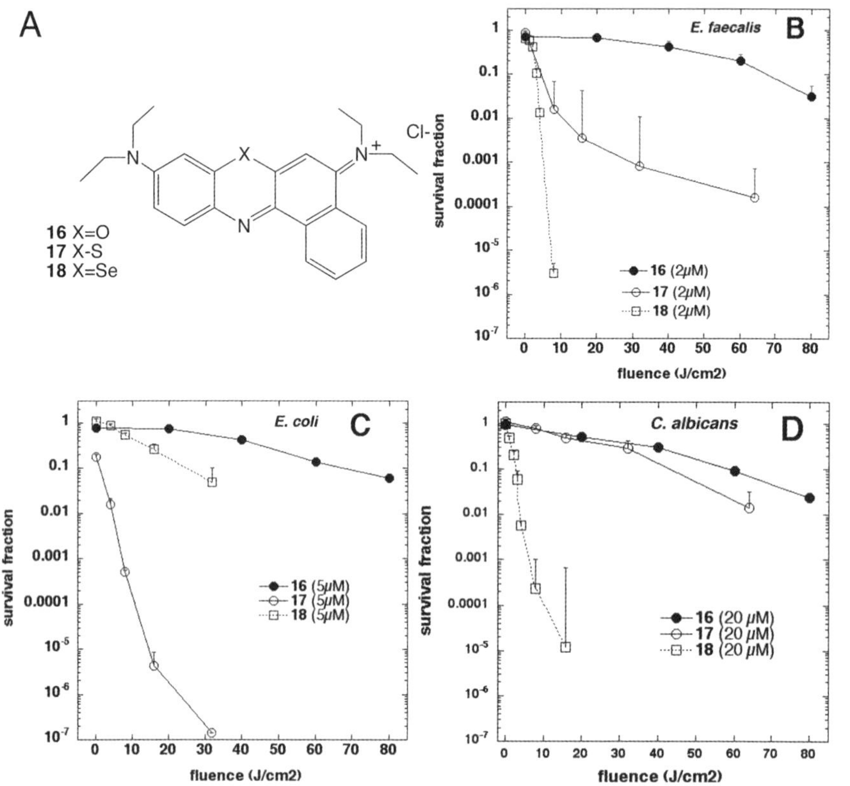

Figure 4.7 (A) Molecular structures of chalcogen analogues of benzo[*a*]phenoxazinium dyes; **16** X=O, **17** X=S, **18** X=Se. Fluence-dependent photoinactivation curves of (B) *E. faecalis* at 2 μM; (C) *E. coli* at 5 μM; (D) *C. albicans* at 20 μM, adapted from ref. 22.

Table 4.1 Physical and photophysical properties of benzophenoxazinium dyes.

Dye	λ_{abs} (nm)[a]	λ_{fl} (nm)[b]	Φ_{fl} [c]	Φ^1O_2 [c]	[d] log P[e]
16	632	660	0.27	0.003	2.69
17	654	693	0.21	0.03	2.76
18	661	703	0.03	0.78	2.08

concentration for *S. aureus*, 5 μM for *E. coli* and 20 μM for *C. albicans* (at 10(7) cells per mL) and performed a wash before illumination. EtNBSe (**18**) was markedly better at killing all three classes of microbial cells than EtNBS (**17**), which in turn was better than ETNBA (**16**). Since these dyes have similar charge and hydrophobicity due to their structural similarities it is assumed that the differences in phototoxic efficiency are primarily due to the heavy atom effect increasing the triplet yield and lifetime and therefore the quantum yield of singlet oxygen and other ROS.

4.6. Conclusion and Future Directions

It is now possible to design highly effective antimicrobial PS to destroy Gram-positive and Gram-negative bacteria. However, the precise determinants of molecule structure necessary to optimize PDI killing of yeast and fungal cells remain incompletely understood. Greater or lesser selectivity for microbial cells over host mammalian cells can be obtained depending on the exact PS structure by employing a short incubation time (minutes). The cationic PS mostly used in antimicrobial PDI bind and penetrate microbial cells rapidly while they are taken up by mammalian cells using the relatively slow process of endocytosis. Further synthetic chemistry efforts will no doubt continue to provide even more effective and selective antimicrobial PS.

Acknowledgements

Research in the Hamblin laboratory is supported by the NIH (grant RO1AI050875), US Air Force MFEL program (contract FA9550-04-1-0079), Center for Integration of Medicine and Innovative Technology (DAMD17-02-2-0006), Congressionally Directed Medical Research Program (W81XWH-09-1-0514).

References

1. M. Wainwright, Photodynamic antimicrobial chemotherapy (PACT), *J. Antimicrob. Chemother.*, 1998, **42**, 13–28.
2. M. Wainwright and K.B. Crossley, Methylene Blue – a therapeutic dye for all seasons?, *J. Chemother.*, 2002, **14**, 431–443.
3. Z. Malik, H. Ladan and Y. Nitzan, Photodynamic inactivation of Gram-negative bacteria: problems and possible solutions, *J. Photochem. Photobiol. B*, 1992, **14**, 262–266.
4. M.R. Hamblin and T. Hasan, Photodynamic therapy: a new antimicrobial approach to infectious disease?, *Photochem. Photobiol. Sci.*, 2004, **3**, 436–450.
5. N.S. Soukos, L.A. Ximenez-Fyvie, M.R. Hamblin, S.S. Socransky and T. Hasan, Targeted antimicrobial photochemotherapy, *Antimicrob. Agents Chemother.*, 1998, **42**, 2595–2601.
6. C.R. Rovaldi, A. Pievsky, N.A. Sole, P.M. Friden, D.M. Rothstein and P. Spacciapoli, Photoactive porphyrin derivative with broad-spectrum activity against oral pathogens in vitro, *Antimicrob. Agents Chemother.*, 2000, **44**, 3364–3367.
7. M.R. Hamblin, D.A. O'Donnell, N. Murthy, K. Rajagopalan, N. Michaud, M.E. Sherwood and T. Hasan, Polycationic photosensitizer conjugates: effects of chain length and Gram classification on the photodynamic inactivation of bacteria, *J. Antimicrob. Chemother.*, 2002, **49**, 941–951.
8. L. Polo, A. Segalla, G. Bertoloni, G. Jori, K. Schaffner and E. Reddi, Polylysine-porphycene conjugates as efficient photosensitizers for the inactivation of microbial pathogens, *J. Photochem. Photobiol. B*, 2000, **59**, 152–158.
9. G.P. Tegos, M. Anbe, C. Yang, T.N. Demidova, M. Satti, P. Mroz, S. Janjua, F. Gad and M.R. Hamblin, Protease-stable polycationic photosensitizer conjugates

between polyethyleneimine and chlorin(e6) for broad-spectrum antimicrobial photoinactivation, *Antimicrob. Agents Chemother.*, 2006, **50**, 1402–1410.
10. F.M. Lauro, P. Pretto, L. Covolo, G. Jori and G. Bertoloni, Photoinactivation of bacterial strains involved in periodontal diseases sensitized by porphycene-polylysine conjugates, *Photochem. Photobiol. Sci.*, 2002, **1**, 468–470.
11. T.N. Demidova, F. Gad, T. Zahra, K.P. Francis and M.R. Hamblin, Monitoring photodynamic therapy of localized infections by bioluminescence imaging of genetically engineered bacteria, *J. Photochem. Photobiol. B*, 2005, **81**, 15–25.
12. T. Dai, Y.Y. Huang and M.R. Hamblin, Photodynamic therapy for localized infections – state of the art, *Photodiagnosis Photodyn. Ther.*, 2009, **6**, 170–188.
13. T. Dai, G.P. Tegos, T. Zhiyentayev, E. Mylonakis and M.R. Hamblin, Photodynamic therapy for methicillin-resistant Staphylococcus aureus infection in a mouse skin abrasion model, *Laser Surg Med.*, 2010, **42**, 38–44.
14. P. Mroz, G.P. Tegos, H. Gali, T. Wharton, T. Sarna and M.R. Hamblin, Photodynamic therapy with fullerenes, *Photochem. Photobiol. Sci.*, 2007, **6**, 1139–1149.
15. G.P. Tegos, T.N. Demidova, D. Arcila-Lopez, H. Lee, T. Wharton, H. Gali and M.R. Hamblin, Cationic fullerenes are effective and selective antimicrobial photosensitizers, *Chem. Biol.*, 2005, **12**, 1127–1135.
16. L. Huang, M. Terakawa, T. Zhiyentayev, Y.Y. Huang, Y. Sawayama, A. Jahnke, G.P. Tegos, T. Wharton and M.R. Hamblin, Innovative cationic fullerenes as broad-spectrum light-activated antimicrobials, *Nanomedicine*, 2009.
17. Z. Lu, T. Dai, L. Huang, D.B. Kurup, G.P. Tegos, A. Jahnke, T. Wharton and M.R. Hamblin, Photodynamic therapy with a cationic functionalized fullerene rescues mice from fatal wound infections, *Nanomedicine UK*, 2010, **5**, 1525–1533.
18. M.B. Spesia, M.E. Milanesio and E.N. Durantini, Synthesis, properties and photodynamic inactivation of Escherichia coli by novel cationic fullerene C60 derivatives, *Eur. J. Med. Chem.*, 2008, **43**, 853–861.
19. I. Lee, Y. Mackeyev, M. Cho, D. Li, J.H. Kim, L.J. Wilson and P.J. Alvarez, Photochemical and antimicrobial properties of novel C60 derivatives in aqueous systems, *Environ. Sci. Technol.*, 2009, **43**, 6604–6610.
20. H.J. Kim and J.S. Lindsey, De novo synthesis of stable tetrahydroporphyrinic macrocycles: bacteriochlorins and a tetradehydrocorrin, *J. Org. Chem.*, 2005, **70**, 5475–5486.
21. C. Ruzie, M. Krayer, T. Balasubramanian and J.S. Lindsey, Tailoring a bacteriochlorin building block with cationic, amphipathic, or lipophilic substituents, *J. Org. Chem.*, 2008, **73**, 5806–5820.
22. J.W. Foley, X. Song, T.N. Demidova, F. Jalil and M.R. Hamblin, Synthesis and properties of benzo[a]phenoxazinium chalcogen analogues as novel broad-spectrum antimicrobial photosensitizers, *J. Med. Chem.*, 2006, **49**, 5291–5299.
23. L. Cincotta, J.W. Foley and A.H. Cincotta, Novel red absorbing benzo[a] phenoxazinium and benzo[a]phenothiazinium photosensitizers: in vitro evaluation, *Photochem. Photobiol.*, 1987, **46**, 751–758.
24. L. Cincotta, J.W. Foley and A.H. Cincotta, Phototoxicity, redox behavior, and pharmacokinetics of benzophenoxazine analogues in EMT-6 murine sarcoma cells, *Cancer Res.*, 1993, **53**, 2571–2580.
25. L. Huang, Y.Y. Huang, P. Mroz, G.P. Tegos, T. Zhiyentayev, S.K. Sharma, Z. Lu, T. Balasubramian, M. Krayer, C. Ruzie, E. Yang, H.L. Kee, C. Kirmaier, J.R. Diers, D.F. Bocian, D. Holten, J.S. Lindsey and M.R. Hamblin, Stable synthetic cationic bacteriochlorins as selective antimicrobial photosensitizers, *Antimicrob. Agents Chemother.*, 2010, **54**, 3834–3841.

Chapter 5

Porphyrins as Antimicrobial Photosensitizing Agents

A. Almeida,[a] A. Cunha,[a] M. A. F. Faustino,[b] A. C. Tomé[b] and M. G. P. M. S. Neves[b]

[a] CESAM & Department of Biology, University of Aveiro, 3810-193 Aveiro, Portugal; [b] QOPNA & Department of Chemistry, University of Aveiro, 3810-193 Aveiro, Portugal

Table of Contents

Photodynamic Inactivation of Microbial Pathogens: Medical and Environmental Applications
Editors Michael R. Hamblin and Giulio Jori

Published by the Royal Society of Chemistry, www.rsc.org

List of Acronyms and Abbreviations

ALA	*delta*-aminolevulinic acid
aPDT	Antimicrobial photodynamic therapy
AZT	Azidothymidine
BPD	Benzoporphyrin derivative
CBSC	Human cord blood stem cell
DHE	Dihaematoporphyrin ether
DP	Deuteroporphyrin
EDTA	Ethylenediaminetetraacetic acid
ESBL	Extended-spectrum β-lactamase
Gram (–)	Gram-negative
Gram (+)	Gram-positive
HBV	Hepatitis B virus
HCV	Hepatitis C virus
HIV	Human immunodeficiency virus
HpD	Haematoporphyrin derivative
MB	Methylene blue
MRSA	Methicillin-resistant *Staphylococcus aureus*
MSSA	Methicillin-sensitive *Staphylococcus aureus*
NOD/SCID	Non-obese diabetic/severe combined immunodeficiency
PDT	Photodynamic therapy
PI	Photoinactivation
PS	Photosensitizer
RB	Rose Bengal
ROS	Reactive oxygen species
TBO	Toluidine blue O

5.1. Introduction

Chemotherapy has shown to be a rapid and effective method to treat or prevent microbial infections, but the regular use of antimicrobials has resulted in the development of drug resistance in common pathogenic microbial strains. The increasing problems with resistance against antibiotics and other antimicrobials, the concern about spreading of antibiotics in the environment due to the inappropriate or excessive prescription of these chemicals and the more and more frequent transmission of microorganisms due to the global travelling and the truly large variety of mechanisms adopted by microbial cells to increase their resistance to external aggression bring the need for the development of novel, convenient and inexpensive methods for combating microbial diseases.

Antimicrobial photodynamic therapy (aPDT) is a non-antibiotic approach to inactivate microorganisms.[1–5] This new methodology is a potential alternative to conventional antibiotics and has already proved to be effective *in vitro* against bacteria, viruses, fungi and protozoa.[4,6–12,13]

Antimicrobial photodynamic therapy combines a non-toxic photosensitizer (PS), with visible light to generate singlet oxygen and free radicals that kill microbial cells.[14,15] The main advantages of aPDT are the non-target specificity and the potential lack of development of resistance mechanisms due to the mode of action and type of biochemical targets (multi-target process).[6,16] The oxidative mechanisms responsible for the photoinactivation of microorganisms, both type I and type II pathways,[4,17] lead to highly reactive oxygen species (ROS) such as singlet oxygen, hydrogen peroxide, superoxide and hydroxyl radical, which are able to irreversibly oxidize microorganisms' vital constituents resulting in lethal damage.[18,19] Although antioxidant enzymes such as superoxide dismutase, catalase and peroxidase give protection against some ROS, they do not protect against singlet oxygen,[20] which, according to the literature, is the main ROS through which the PS exert their photodynamic action.[21–25] Moreover, singlet oxygen has been shown to inactivate these enzymes.[26]

The main targets of the antibacterial and antiviral photodynamic activity are the external microbial structures, as cell walls, cell membranes, protein capsids, lipid envelopes and nucleic acids.[14,27–33] The damages to the external microbial structures can involve leakage of cellular contents or inactivation of membrane transport systems and enzymes.[34,35] Some damages produced in the nucleic acid chain can be repaired by the action of DNA repairing systems.[36] It has been concluded that although nucleic acids damage occurs, it cannot be the principal cause of microbial photodynamic inactivation.[32,37,38] Confirming that, it was shown that bacteria *Deinococcus radiodurans*, which is known to have a very efficient DNA repair mechanism, is easily killed by aPDT.[39]

Fungi present much more complex targets than bacteria and photoinactivation seems to be less dependent on cell-bound PS. Free PS induces some initial functional alterations of the cytoplasmic membrane that allow the penetration of the PS with a more extensive photoinactivation of sub-cellular targets such as the mitochondria[40] or the nucleus.[41]

As in most microorganisms the main targets of aPDT are the external structures, the photosensitizer (PS) does not need to enter in the microorganism; specific and proper adhesion to those structures is sufficient for its light-activated destruction. So, target microorganisms have no chance to develop resistance by stopping uptake, increasing metabolic detoxification or increasing export of the drug.[16]

All studies that tested the possible development of microbial resistance to aPDT[42–44] proved that, as first hypothesized, microorganisms are not able to develop resistance.[45–51]

Moreover, all studies concerning the inactivation of antimicrobial resistant microorganisms by aPDT have found them to be equally as susceptible as their native counterparts.[4,52–58]

Recent studies have demonstrated that aPDT can be effective in the selective inactivation of microorganisms and it can become a potential alternative for the treatment and eradication of microbial infections,[32,59] disinfection of contaminated blood[30,60–70] and also of environmental waters.[8–10,51,71–74]

5.2. Porphyrins Used in aPDT

Porphyrins are a class of aromatic heterocyclic compounds that are ubiquitous in nature. They participate in vital biochemical processes, namely the oxygen transport (haem) and the photosynthesis (chlorophylls). Due to their unique physico-chemical properties, porphyrins are currently being used in a myriad of applications ranging from artificial photosynthesis, oxidation catalysis, sensors, nonlinear optics and nanomaterials to the photodynamic therapy (PDT) of cancer or, more recently, to the photoinactivation of microorganisms.[75] The use of porphyrins in all these applications is challenging the synthetic organic chemists to design porphyrin derivatives with chemical, physical and electronic properties adequate for each application. As shown in Tables 5.1 and 5.2, the porphyrin derivatives used as photosensitizers in the photoinactivation of microorganisms can be divided into two groups. The first group includes natural porphyrins and chemically modified natural porphyrins. The second group is constituted only by synthetic porphyrins, which can be neutral, cationic or anionic.

Considering the macrocycles with natural features, the derivatives based on chlorophyll pattern (**2–8**) and haem pattern (**9–19**) have received special attention due to their proved efficacy in PDT applications and to their relatively easy availability for further derivatization. It is worth referring to the fact that *delta*-aminolevulinic acid (ALA), or the corresponding esters, are prodrugs that when administrated are able to bypass the haem feedback control mechanism affording mainly protoporphyrin IX, the effective photosensitizer. Other endogenous porphyrins can also be produced.

Among the chlorophyll derivatives, chlorin e_6 (**3**) (already approved for clinical use in photodynamic therapy – PDT) and pheophorbide *a* (**2**, **Pheopho *a***) have been the most frequently used; they are commercially available or can be easily isolated from natural sources. The degradative approaches to give access to these derivatives are well established in the literature.[76–78] For instance, the metal-free pheophytins *a* and *b* (**2**, **Pheophy *a*** and **Pheophy *b***) can be obtained by treatment of the corresponding chlorophylls (**2**, **Chl *a*** and **Chl *b***) with mild acid conditions. The hydrolysis of the phytyl residue in chlorophylls or pheophytins affords the corresponding pheophorbides.

Pheophorbide *a* can also be further degraded with alkali affording chlorin e6 (**3**). This derivative was used as an important platform for further simple derivatizations namely with amines and polylysine (**4**).

In the derivatives with a haem pattern we can include, for instance, the haematoporphyrin monomer (**10a**), dihaematoporphyrin ether dimer (**11**) and Photofrin (**12**). It is worth referring to the fact that HpD, a mixture of haematoporphyrin monomer, dimer and oligomers, was the first PS used in PDT and the enormous research involving this mixture led to its approval for clinical use firstly in Canada in 1993 and then in other countries. Purified versions of HpD led to the well-known PS Photofrin, Photogem and Photosan.

The haem group (**9a**, M = Fe), other complexes (**9a**, M = Zn, Mg, Co, Ga) and the free-base counterpart, protoporphyrin IX (**9a**, M = 2H), were also

Table 5.1 Porphyrin derivatives of natural origin used as antimicrobial photosensitizers.

Photosensitizer	Microorganism
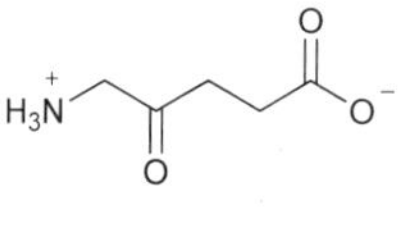 **ALA**	*Acinetobacter baumannii*[85] *Aeromonas hydrophila*[85] *Bacillus cereus*[85,81] *Candida albicans*[89,82] *Escherichia coli*[29,83–86] *Enterococcus hirae*[29] *Haemophilus parainfluenzae*[87] Herpes simplex virus[88] Human immunodeficiency virus[88] Onychomycosis[89] *Osteomyelitis*[90] *Pseudomonas aeruginosa*[86,91,92] *Propionibacterium acnes*[93] *Salmonella enterica*[94] *Staphylococci*[85,95] *Staphylococcus aureus*[84,85] *Streptococcus epidermis*[85] *Streptococcus faecalis*[85] *Staphylococcus aureus* (methicillin-resistant)[86]
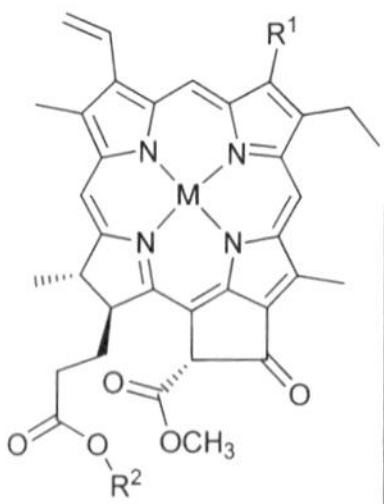 **2**	*Bacillus subtilis*[96] *Escherichia coli*[96] *Listeria monocytogenes*[96] *Staphylococcus aureus*[96] *Staphylococcus aureus* (methicillin-resistant)[146,97] *Yersinia pseudotuberculosis*[96]
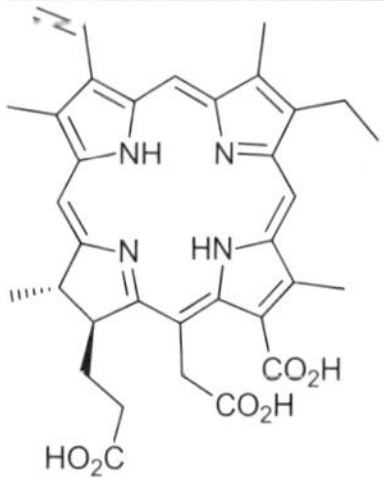 **3** (Ce6)	20 Gram (+) and 30 Gram (–)[98] *Bacillus subtilis*[99] *Capnocytophaga gingivalis*[103] *Escherichia coli*[99–101] *Fusobacterium nucleatum*[103,104] *Mycobacterium smegmatis*[106] Papilloma warts[102] *Pseudomonas aeruginosa*[101] *Proprionobacterium gingivalis*[103,104] Periodontal pathogens[105] *Actinobacillus actinomycetemcomitans* *Actinomyces viscosus* *Bacteroides forsythus* *Campylobacter rectus*

Substituents of compound 2:

	M	R^1	R^2
Protochl	Mg	CH_3	H
Chl *a*	Mg	CH_3	$C_{20}H_{39}$
Chl *b*	Mg	CHO	$C_{20}H_{39}$
Pheophy *a*	2H	CH_3	$C_{20}H_{39}$
Pheophy *b*	2H	CHO	$C_{20}H_{39}$
Pheopho *a*	2H	CH_3	H
Pheopho *b*	2H	CHO	H

Table 5.1 (*Continued*)

Photosensitizer	Microorganism
	Eikenella corrodens *Fusobacterium nucleatum* subsp. *polymorphum* *Porphyromonas gingivalis* *Streptococcus sobrinus* *Streptococcus mitis* *Streptococcus oralis* *Streptococcus mutans* *Staphylococcus aureus*[100–108] *Staphylococcus aureus* (bioluminescent)[109] *Staphylococcus aureus* (methicillin-resistant)[101] *Streptococcus pyrogens*[110] *Yersinia enterocolitica*[106]
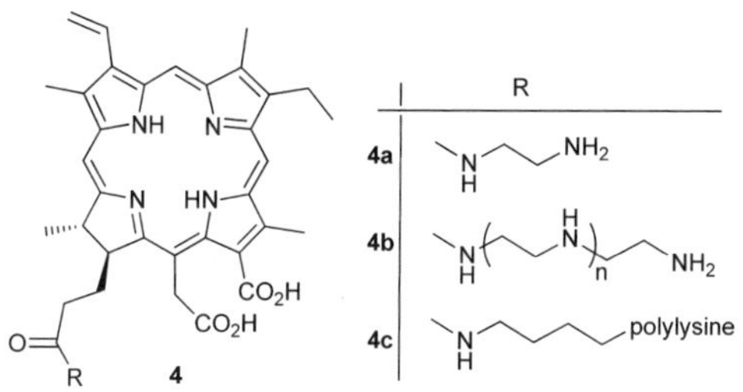	*Acinetobacter baumanii* (bioluminescent)[111] *Actinomyces viscosus*[112] *Bacillus* endospores[113] *Bacillus cereus* *Bacillus subtilis* *Candida albicans*[117,120] *Cryptococcus neoformans*[114] *Escherichia coli*[117–120] *Escherichia coli* (bioluminescent)[115] *Escherichia coli* (resistant)[118] *Pseudomonas aeruginosa* (bioluminescent)[115,116,120] *Proteus mirabilis* (bioluminescent)[116] *Porphyromonas gingivalis*[112] *Staphylococcus aureus*[117–120] *Staphylococcus aureus* (bioluminescent)[109,121] *Staphylococcus aureus* (methicillin-resistant)[107,108,118] *Streptococcus pyogenes*[120]
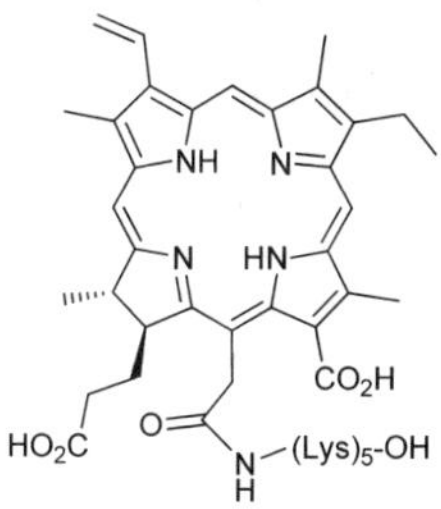 4d	Periodontal pathogens[105] *Actinobacillus actinomycetemcomitans* *Actinomyces viscosus* *Bacteroides forsythus* *Campylobacter rectus* *Eikenella corrodens* *Fusobacterium nucleatum* subsp. *polymorphum* *Porphyromonas gingivalis* *Streptococcus sobrinus* *Streptococcus mitis* *Streptococcus oralis* *Streptococcus mutans*

Table 5.1 (*Continued*)

Photosensitizer	Microorganism
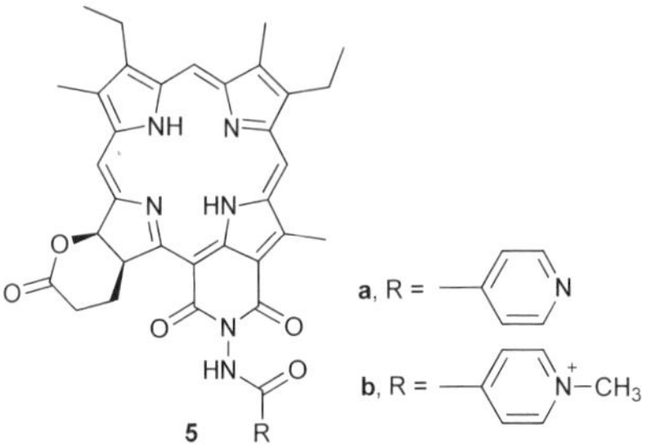 5	*Escherichia coli*[122] *Micrococcus luteus*[122]
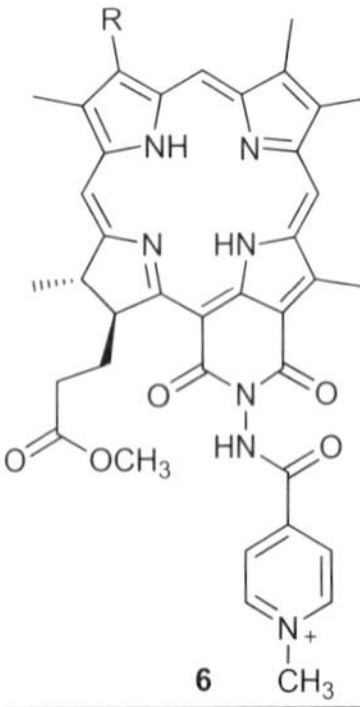 6 **a**, R = CH_2CH_3 **b**, R = $COCH_3$ **c**, R = $C(OH)CH_3$	
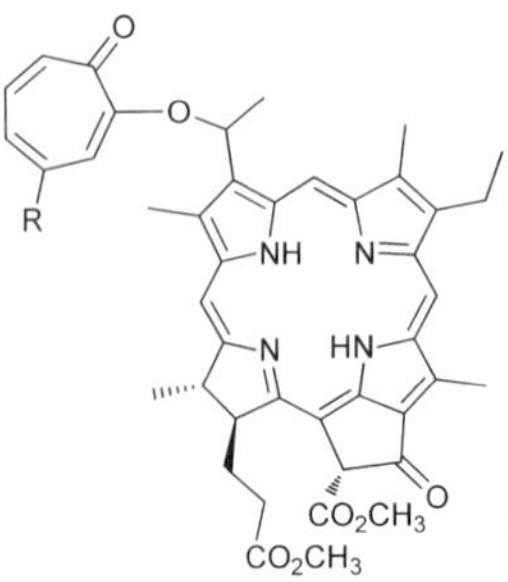 7 **a,** R = H **b,** R = $C(CH_3)_3$ 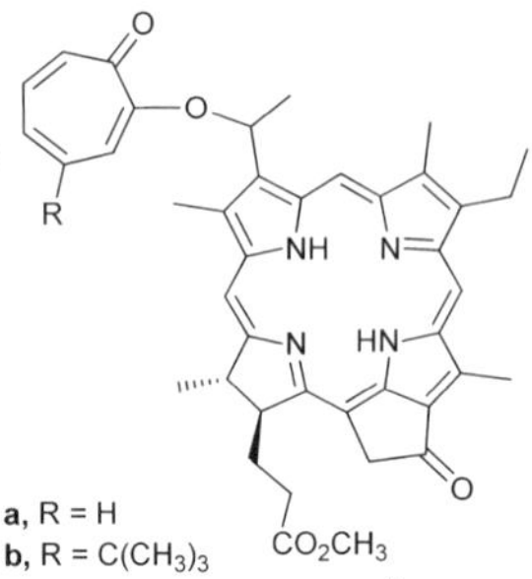 8	*Escherichia coli*[123] *Staphylococcus aureus*[123] *Streptococcus epidermidis*[123]
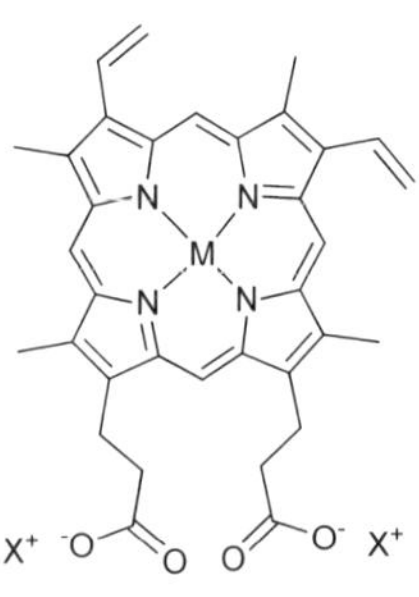 9 9 | X^+ a | H^+ b | Na+ c | L-arginine-H^+	*Actinomyces odontolyticus*[132] *Candida guilermondii*[124] *Escherichia coli*[125] Gram (–) anaerobes[126] *Bacteroides fragilis* Other *Bacteroides* spp. *Bilophila wadsworthia* *Fusobacterium* spp. *Porphyromonas* spp. *Prevotella* spp. Gram (+) anaerobes[126] *Clostridium* spp. Gram (+) cocci Gram (+) rods *Helicobacter pylori*,[127,128]

Table 5.1 (*Continued*)

Photosensitizer	**Microorganism**
	Hepatitis B virus[231] Human immunodeficiency virus[30,129] *Mycobacterium smegmatis*[106] Murine leukaemia virus[130] *Porphyromonas gingivalis*[131,132] *Propionibacterium acnes*[118] *Saccharomyces cerevisiae*[124] *Staphylococcus aureus*[106,125,133,134] Yersinia enterocolitica[106]
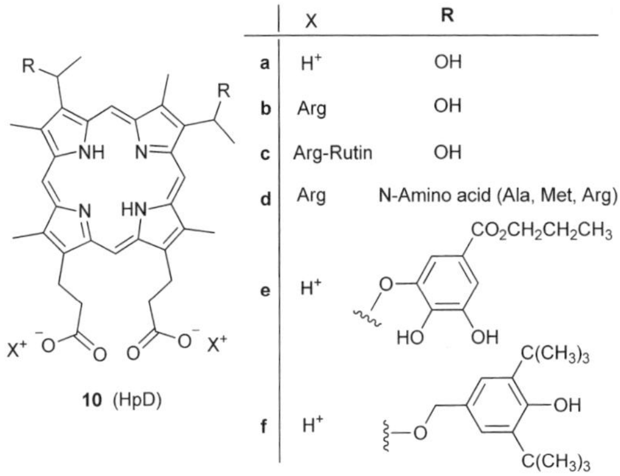10 (HpD)	*Acinetobacter baumannii*[142] *Bacillus subtilis*[141,143,145] *Bacillus cereus*[143] *Cytomegalovirus*[65] Enterobacter cloacae[141] *Escherichia coli*[141,142] *Helicobacter pylori*[136] *Herpes simplex virus* 1[30,65,68,137,138] *Influenza virus*[137] *Klebsiella pneumoniae*[141] *Klebsiella javanica*[143] *Micrococcus luteus*[141] *Pseudomonas aeruginosa*[140,141,144] *Proteus vulgaris*[141] *Rhodotorula mucilaginosa*[143] *Saccharomyces cerevisiae*[143,139] *Staphylococcus aureus*[140–145] *Staphylococcus aureus* (methicillin-resistant)[146] *Streptococcus faecalis*[135] Simian immunodeficiency virus[65]
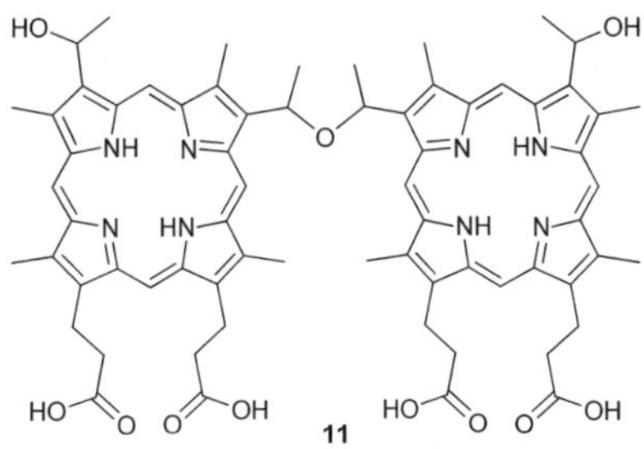11	*Cytomegalovirus*[65] Herpes simplex virus 1[138] Papilloma virus[147] Simian immunodeficiency virus[65,148]

Table 5.1 (*Continued*)

<table>
<tr><th>Photosensitizer</th><th>Microorganism</th></tr>
<tr><td>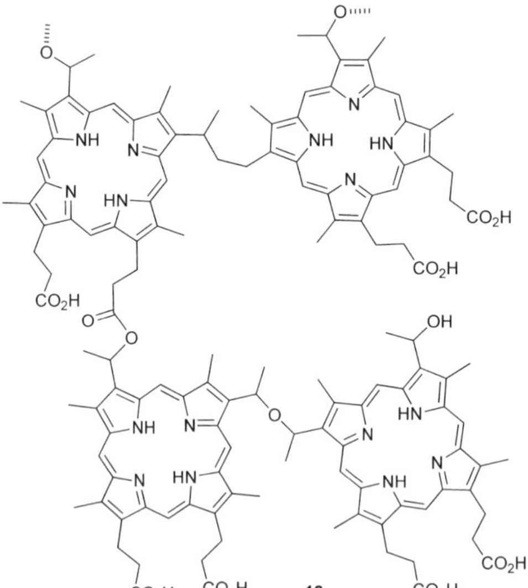
12
(representative bonds in Photofrin)</td><td>Candida albicans[149,150]
Candida dubliniensis[149]
Candida tropicalis[149]
Candida krusei[149,150]
Candida glabrata[150]
Human simplex viru[151]
Lactobacillus acidophilus[152]
Staphylococcus aureus[95]
Streptococcus epidermidis[95]
Streptococcus mutans[152,153]</td></tr>
<tr><td>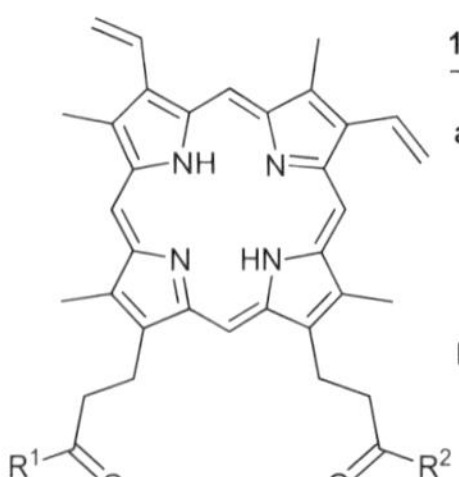
13
<table><tr><th>13</th><th>R¹</th><th>R²</th></tr><tr><td>a</td><td>OCH3</td><td>OCH3</td></tr><tr><td>b</td><td>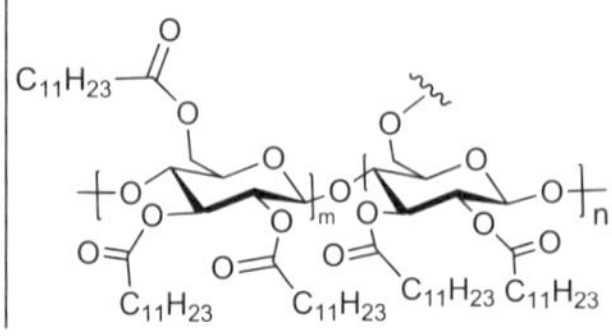</td><td>OH</td></tr></table></td><td>Staphylococcus aureus[154,190,191]</td></tr>
<tr><td>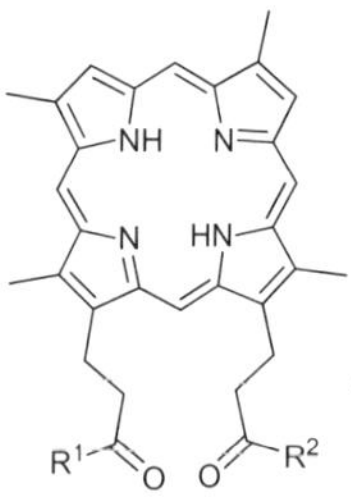
14
<table><tr><th>14</th><th>R¹</th><th>R²</th></tr><tr><td>a</td><td>OH</td><td>OH</td></tr><tr><td>b</td><td>OH or OCH3</td><td>OH or OCH3</td></tr><tr><td>c</td><td>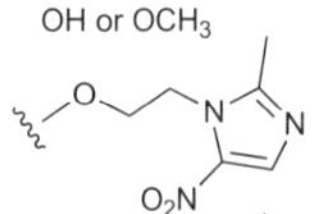 or OH</td><td>OH or 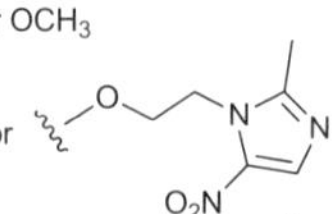</td></tr><tr><td>d</td><td>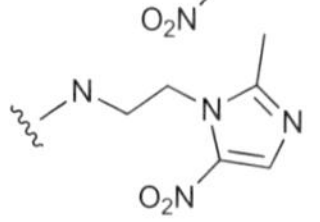 or OH</td><td>OH or 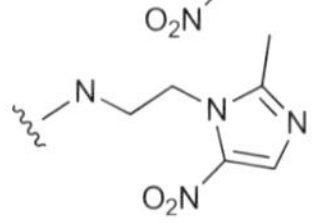</td></tr></table></td><td>Acinetobacter baumanii[155]
Deinococcus radiodurans[156]
Escherichia coli[157]
Gram (–) anaerobes[126]
Bacteroides fragilis,
B. gracilis, B. distasonis,
B. thetaiotaomicron and other Bacteroides spp.
Bilophila wadsworthia
Fusobacterium mortiferum
F. variurn and other Fusobacterium spp.
Porphyromonas spp.
Prevotella spp.
Gram (+) anaerobes[126]
Clostridium spp.
Gram (+) cocci
Gram (+) rods</td></tr>
</table>

Table 5.1 (*Continued*)

Photosensitizer	**Microorganism**
	Human immunodeficiency virus[30] *Pseudomonas aeruginosa*[157] *Porphyrinomonas gingivalis*[158] *Staphylococcus aureus* (methicillin-resistant)[159] *Trichophyton rubrum*[160]
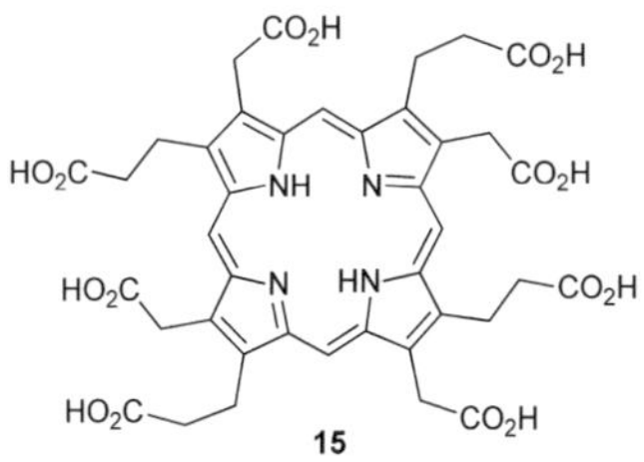 **15**	Recombinant adenovirus (AdCMVLuc)[64]
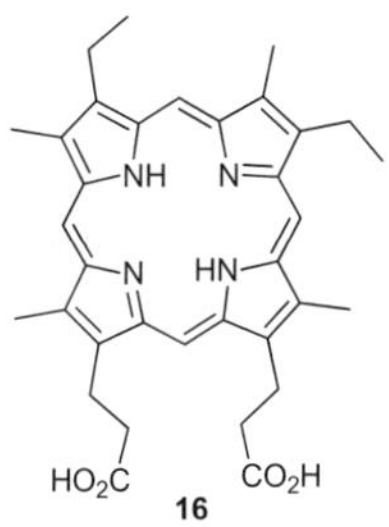 **16**	*Mycobacterium smegmatis*[106] *Staphylococcus aureus*[106] *Yersinia enterocolitica*[106]
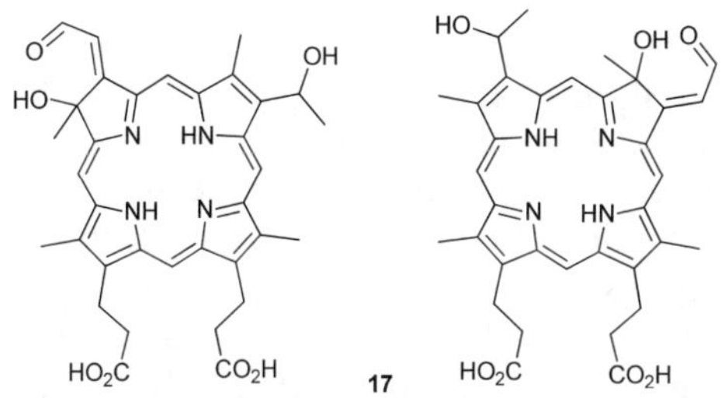 **17**	Human immunodeficiency virus[79]
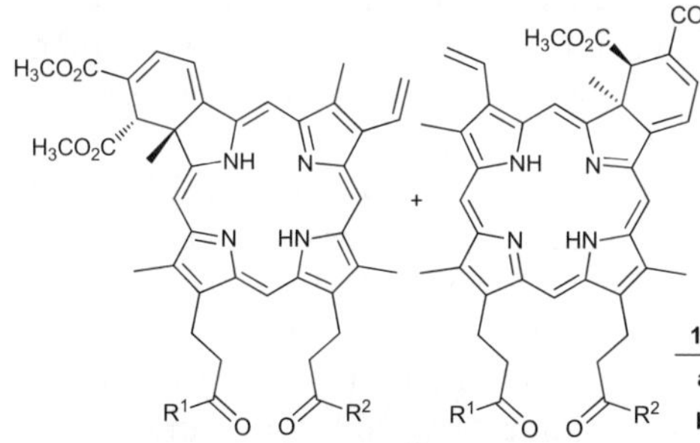 **18** 18: R[1] / R[2] a: OCH_3 / OH b: OH / OCH_3 c: OH / OH	*Bacillus* endospores[113] Feline leukaemia virus[66,162] Gram (+) bacteria and yeasts[143] Human immunodeficiency virus[52] Simian immunodeficiency virus[138] Vesicular stomatitis virus,[161,162]

Table 5.1 (*Continued*)

Photosensitizer	Microorganism
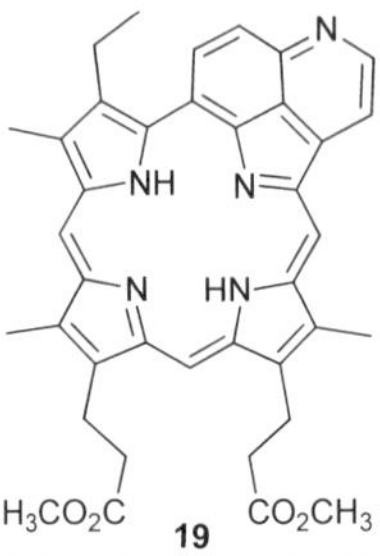 **19**	*Trichophyton rubrum*[160]

considered in some studies. These compounds are available commercially but the degradative approaches to give access to protoporphyrin IX and derivatives such as deuteroporphyrin (**14a**) are well established in the literature.[76–78] Compounds **17**, obtained photochemically from the reaction of partially dehydrated haematoporphyrin with $^1O_2$ were also considered in some studies against HIV strains by Brault.[79]

The potentiality of the monoacid ring A benzoporphyrin derivative **18a,b** (also approved to treat age-related macular degeneration or under clinical trials for the treatment of several cancer diseases) was also evaluated in aPDT. Isomeric compounds **18a,b** are efficiently obtained from protoporphyrin IX (**9a**, M = 2H) *via* Diels–Alder reaction with dimethyl acetylenedicarboxylate.

The synthetic PS studied in aPDT are based mainly in *meso*-tetra-arylporphyrins. The popularity of this type of porphyrins results from their easy synthesis and potentiality toward further elaboration. In fact, the synthetic approaches usually involve the condensation of pyrrole with adequate aldehydes in one-flask or two-step one-flask procedures.[80] The wide range of available aldehydes provides porphyrins with different aryl or hetaryl substituents at the *meso*-positions. Further manipulations of those substituents can give access to a high number of porphyrins with chemical, physical and electronic properties adequate for the desired application. For instance, the 5,10,15,20-tetra(4-pyridyl)porphyrin, the neutral counterpart of the tetracationic porphyrins **20**, **23** and **24**, can be easily obtained from the reaction of pyrrole with pyridine-4-carbaldehyde in acidic conditions.[80] Quaternization of the pyridyl groups with adequate alkyl halides gives access to a series of cationic porphyrins with already proved efficacy in aPDT.

Porphyrins bearing two different *meso*-substituents (porphyrins **27–44**, for instance) are usually prepared by a mixed-aldehyde condensation where higher ratios of the desired porphyrins can be achieved by using the adequate ratio of the two aldehydes. The corresponding neutral porphyrins are obtained from the condensation of pyrrole with a mixture of pyridine-4-carbaldehyde and benzaldehyde or a mixture of pyridine-4-carbaldehyde and pentafluorobenzaldehyde (to obtain **36**). Quaternization of the pyridyl groups gives access to the corresponding cationic porphyrins **35**, **43a–c**.

Table 5.2 Porphyrin derivatives of synthetic origin used as antimicrobial photosensitizers.

Photosensitizer	Microorganism
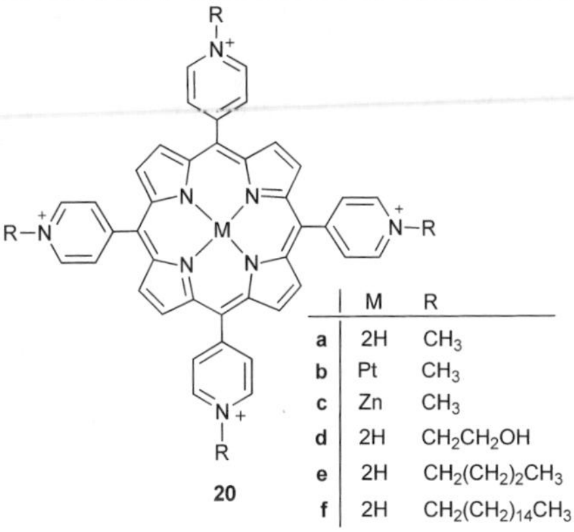 20	*Acinetobacter baumannii*[155,165] *Bacillus cereus* endospores[163] *Candida albicans*[164] *Deinococcus radiodurans*[156] *Escherichia coli*[12,37,165–168] *Escherichia coli* (bioluminescent)[9] *Escherichia coli*[169] *Enterococcus fecalis*[12] Faecal coliforms[8,73] Faecal *Strepococci*[8] Hepatitis A virus[172] Helminth eggs[170] Lambda phage[171] *Listeria monocytogenes*[169] MS2 phages[172] *Mycobacterium smegmatis*[173] *Pseudomonas aerugiosa*[174,166] *Saprolegnia*[72] *Staphylococcus aureus*[72,166,167,174] T7 phages[33] T4 phages[10] Vesicular stomatitis virus[175]
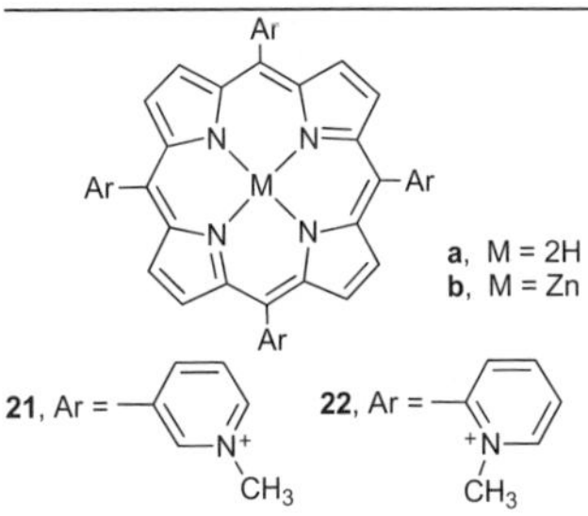 21, 22	*Escherichia coli*[168]
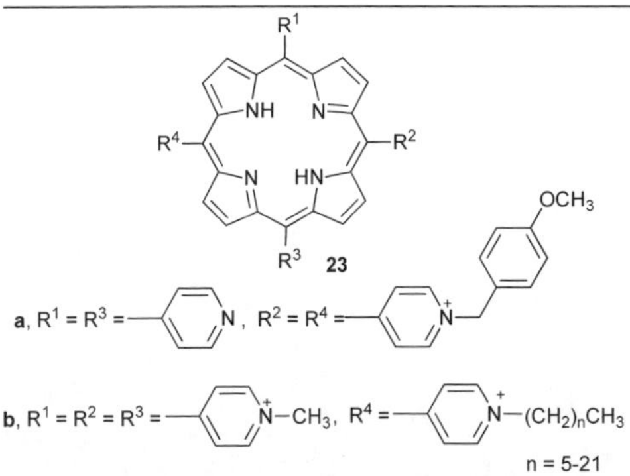 23	*Escherichia coli*[166,176] *Pseudomonas aeruginosa*[166] *Saprolegnia*[72] *Staphylococcus aureus*[72,166,176] *Staphylococcus aureus* (methicillin-resistant)[177]

Table 5.2 (*Continued*)

Photosensitizer	Microorganism
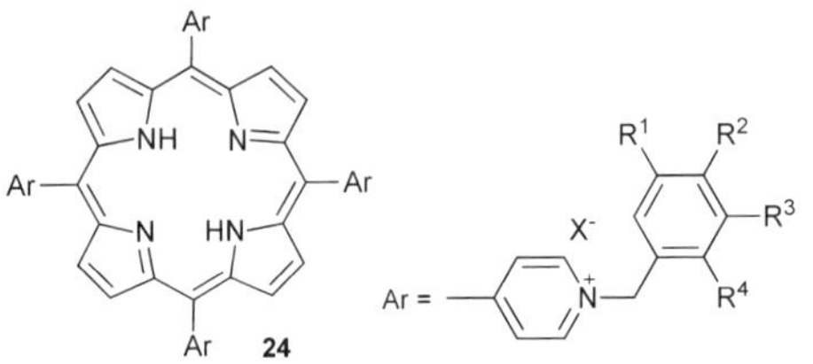 24 **a**: R^1 = H, R^2 = H, R^3 = H, R^4 = H, X = Cl **b**: R^1 = H, R^2 = OCH_3, R^3 = H, R^4 = H, X = Cl **c**: R^1 = OCH_3, R^2 = OCH_3, R^3 = OCH_3, R^4 = H, X = Cl **d**: R^1 = H, R^2 = Cl, R^3 = H, R^4 = Cl, X = Cl **e**: R^1 = H, R^2 = Cl, R^3 = Cl, R^4 = H, X = Cl **f**: R^1 = H, R^2 = F, R^3 = H, R^4 = F, X = Br **g**: R^1 = H, R^2 = NO_2, R^3 = H, R^4 = H, X = Br	*Candida albicans*[178] *Escherichia coli*[166,178] *Pseudomonas aeruginosa*[166,178] *Staphylococcus aureus*[166,178]
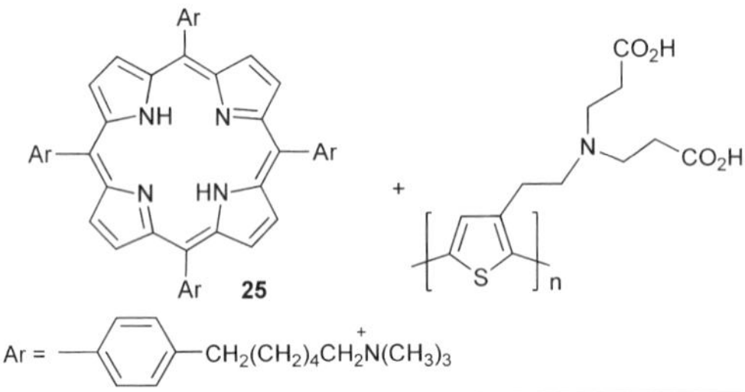 25	*Escherichia coli*[179] *Bacillus subtilis*[179]
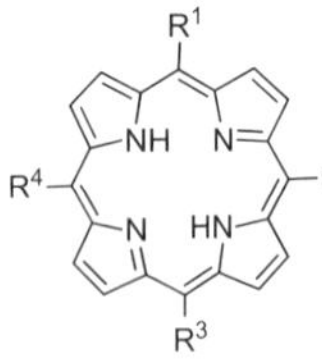 26-34 **26** 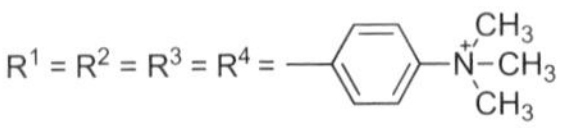 **27** 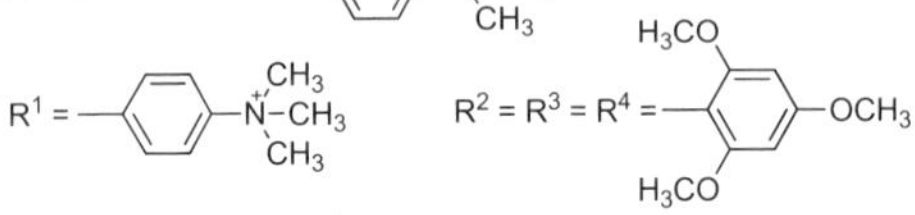 **28** 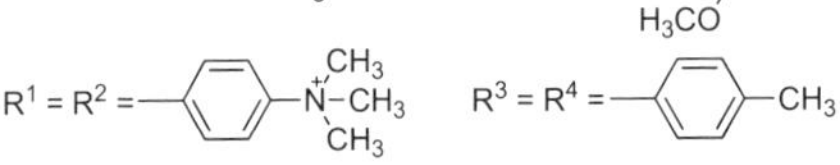 **29** 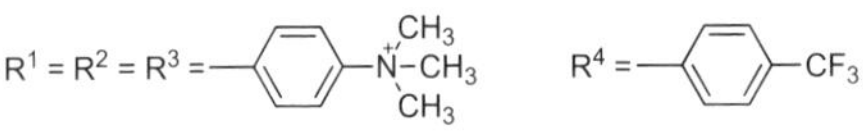 **30** 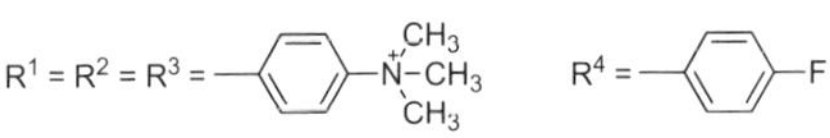 **31** 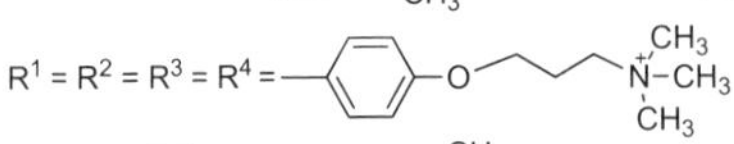 **32** 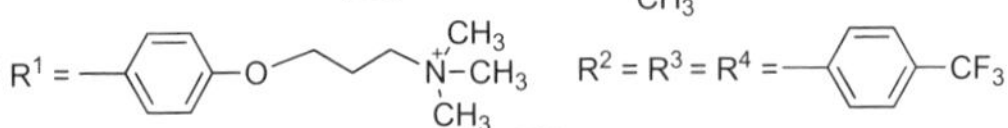 **33** 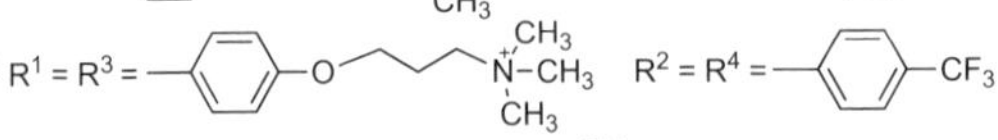 **34** 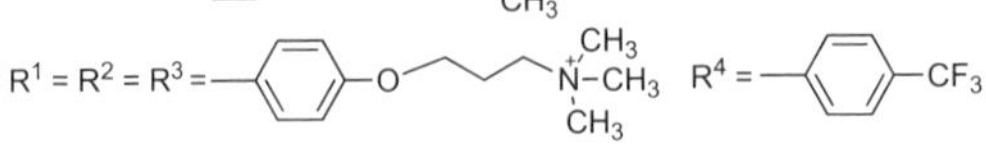	*Candida albicans*[164] *Escherichia coli*[3,166,180] *Mycobacterium smegmatis*[173] *Pseudomonas aeruginosa*[166] *Staphylococcus aureus*[166] Vesicular stomatitis virus[175]

Table 5.2 (*Continued*)

Photosensitizer	Microorganism
	Bacillus endospores[113,163] *Candida albicans*[186,183] *Escherichia coli*[12,187] *Escherichia coli* (bioluminescent)[9] *Enterococcus fecalis*[12] Faecal coliforms[73] Human immunodeficiency virus[183] Herpes simplex virus 1[184] *Pseudomonas aerugiosa*[186,185] *Staphylococcus aureus*[34,174,186,187] *Staphylococcus aureus* in cord blood stem cells products[188] *Staphylococcus aureus* (methicillin-resistant)[187] *Staphylococcus aureus* (bioluminescent)[189] *Streptococcus epidermidis*[34] T4 phages[10] T7 phages[33]
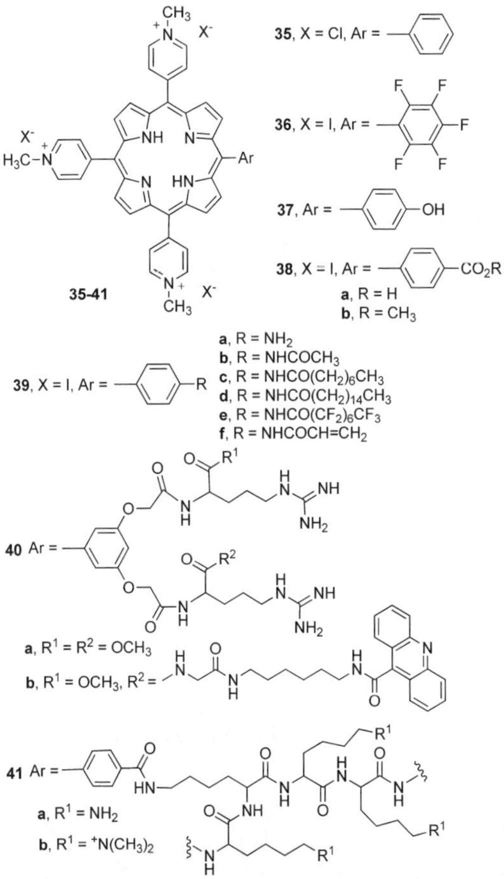	*Trichophyton rubrum*[160] Vesicular stomatitis virus[55,175]

Table 5.2 (*Continued*)

Photosensitizer	Microorganism
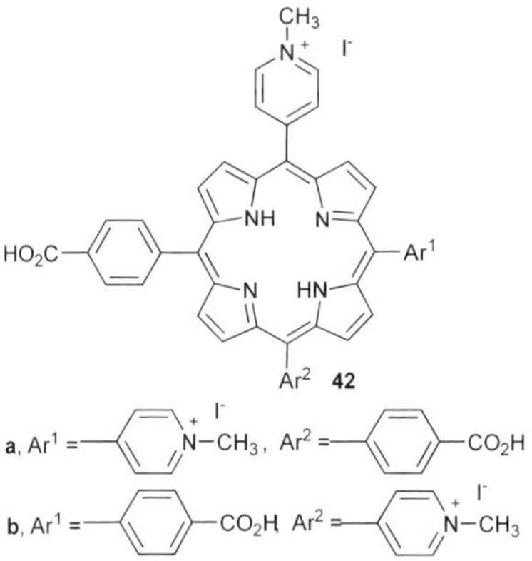 **42**	*Bacillus cereus* endospores[163] *Escherichia coli*[12] *Enterococcus fecalis*[12] T4 phages[10]
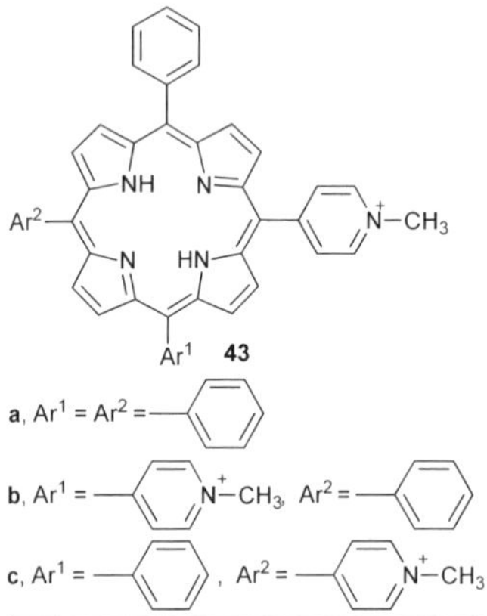 **43**	Vesicular stomatitis virus[175]
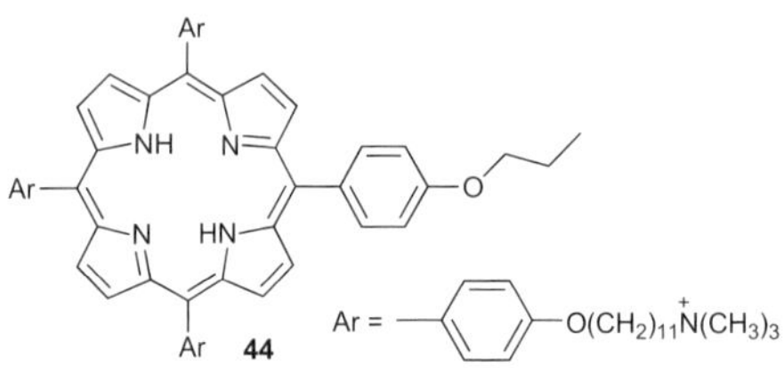 **44**	*Escherichia coli*[54] *Staphylococcus aureus*[54] *Streptococcus epidermidis*[54]
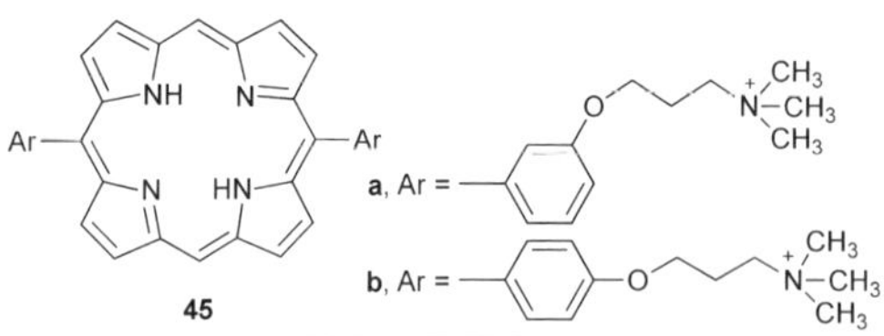 **45**	*Escherichia coli*[54] *Staphylococcus aureus*[54] *Streptococcus epidermidis*[54]
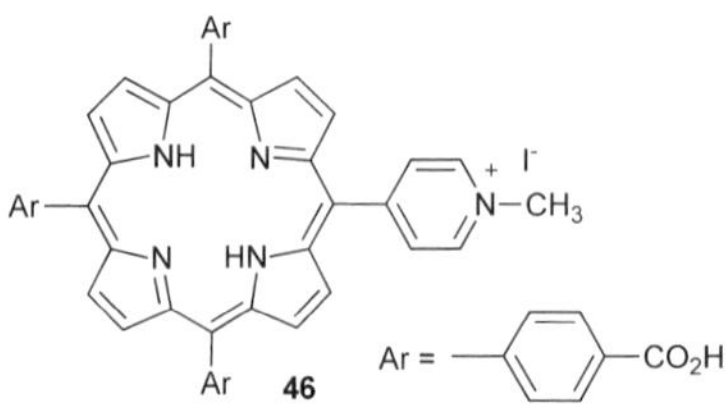 **46**	*Bacillus cereus* endospores[163] *Escherichia coli*[12] *Enterococcus fecalis*[12]

Table 5.2 (*Continued*)

Photosensitizer	Microorganism
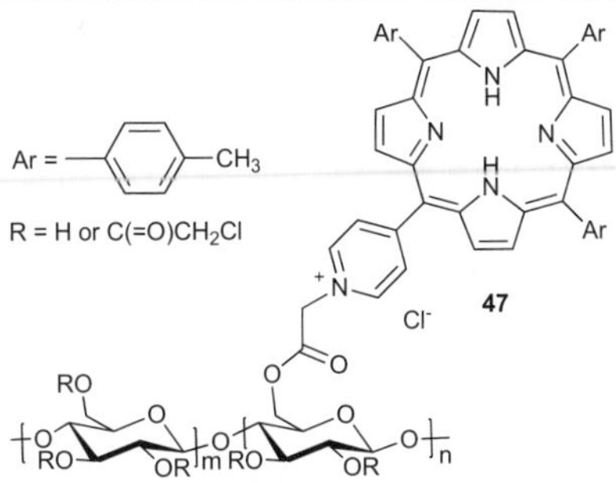 **47**	*Escherichia coli*,[190,191] *Staphylococcus aureus*[190,191]
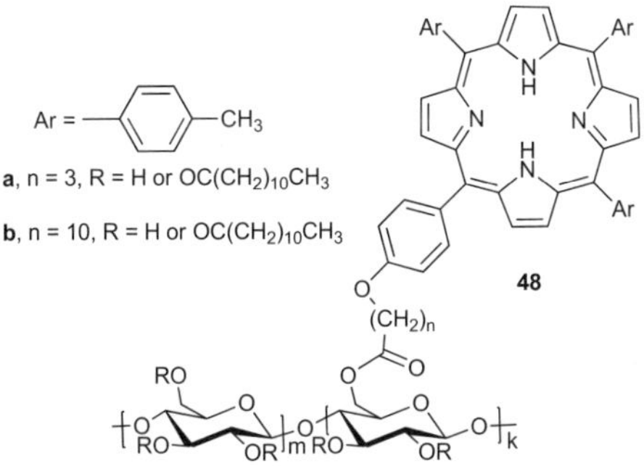 **48**	
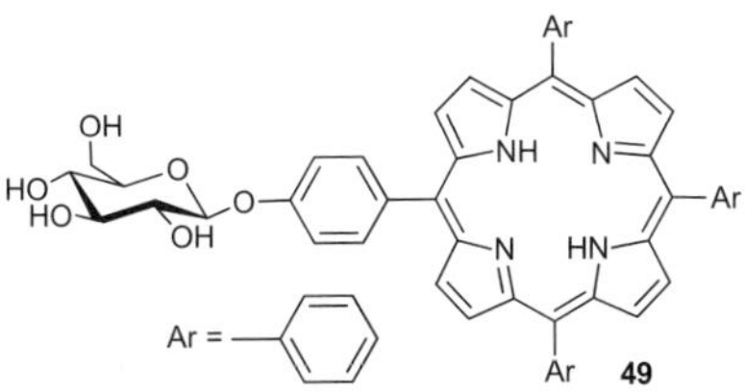 **49**	Herpes simplex virus 1[192] Herpes simplex virus 2[192]
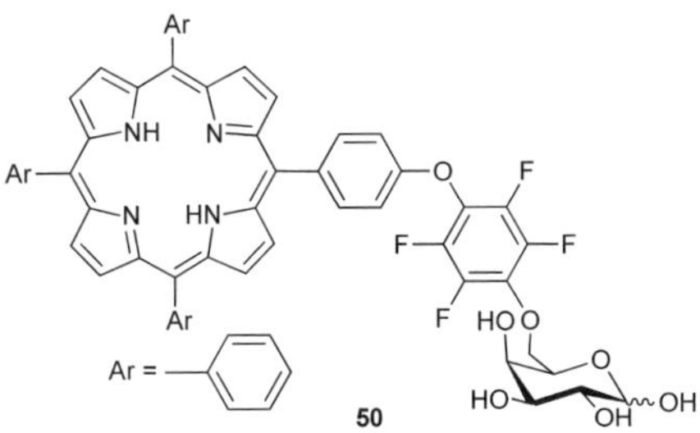 **50**	Herpes simplex virus 1[192] Herpes simplex virus 2[192]
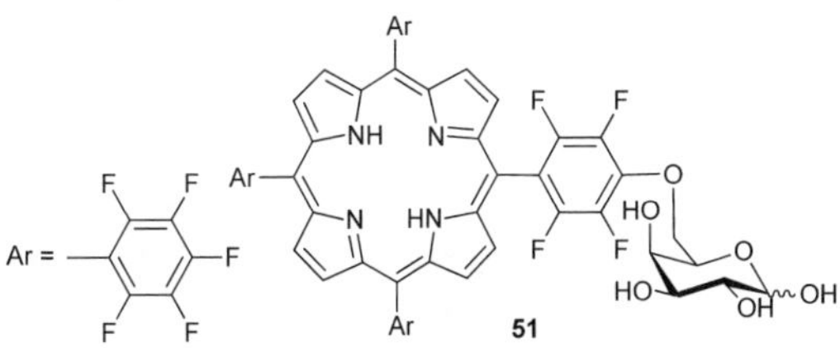 **51**	Herpes simplex virus 1[192] Herpes simplex virus 2[192]

Table 5.2 (*Continued*)

Photosensitizer	Microorganism
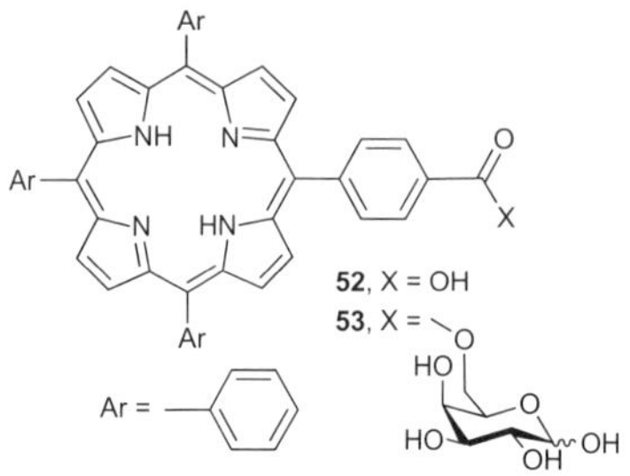 52, X = OH; 53	Herpes simplex virus 1[192] Herpes simplex virus 2[192]
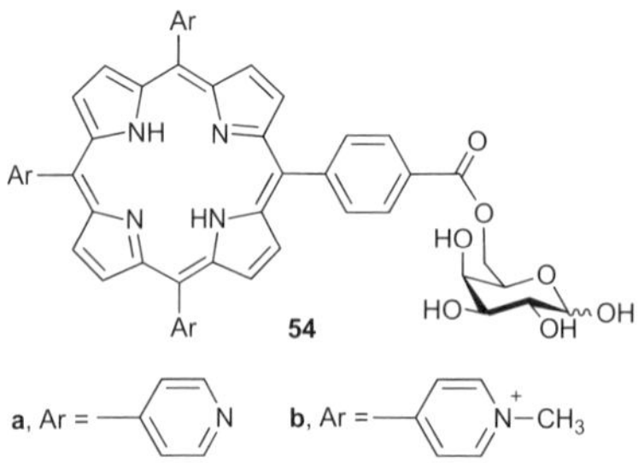 54	Herpes simplex virus 1[184]
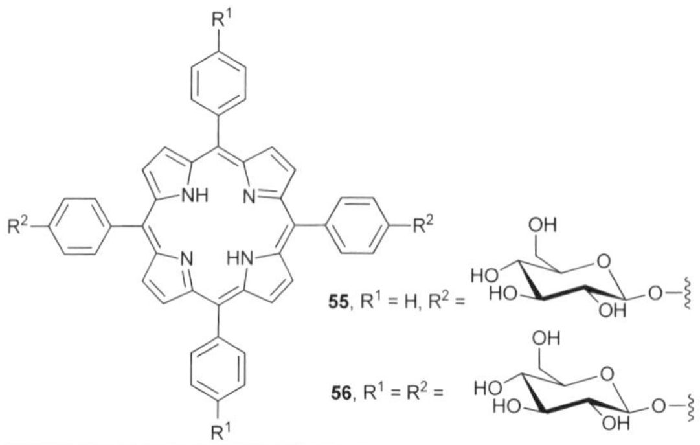 55, 56	*Escherichia coli*[29] *Enterococcus hirae*[29] T7 phages[193]
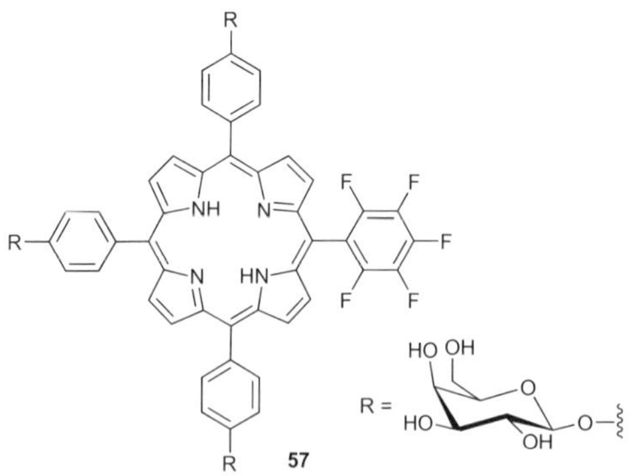 57	T7 phages[31]

Table 5.2 (*Continued*)

Photosensitizer	Microorganism
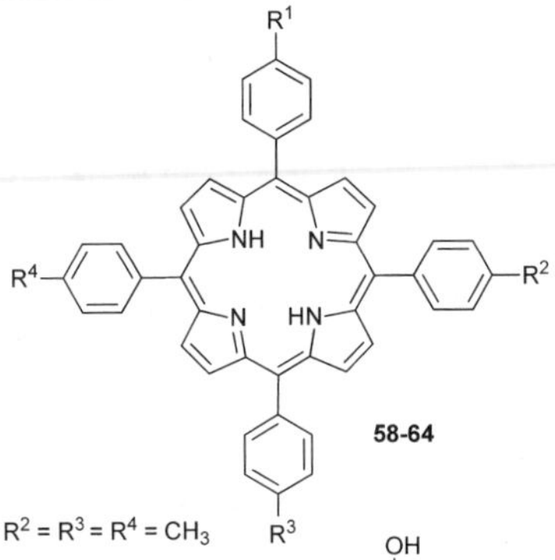 **58-64** **58**, $R^1 = R^2 = R^3 = R^4 = CH_3$ **59**, $R^1 = R^2 = R^3 = CH_3$ R^4 = 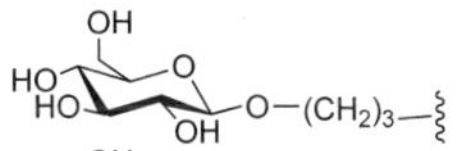 **60**, $R^1 = R^2 = CH_3$ $R^3 = R^4$ = 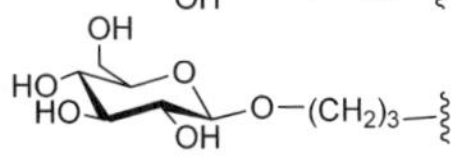 **61**, $R^1 = R^3 = CH_3$ $R^2 = R^4$ = 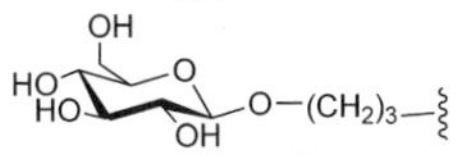 **62**, $R^1 = CH_3$ $R^2 = R^3 = R^4$ = 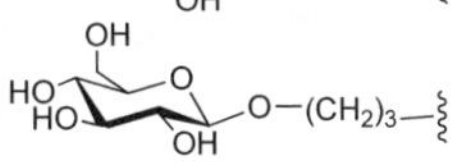 **63**, $R^1 = R^2 = R^3 = R^4$ = 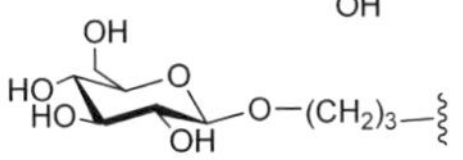 **64**, $R^1 = R^2 = R^3 = R^4$ = 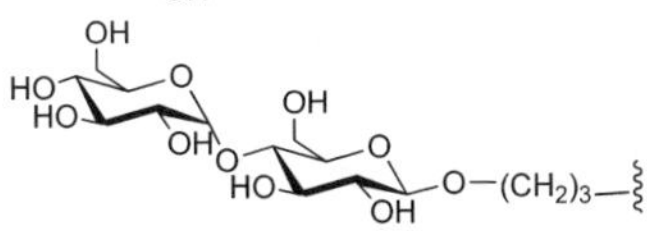	*Saccharomyces cerevisiae*[49]
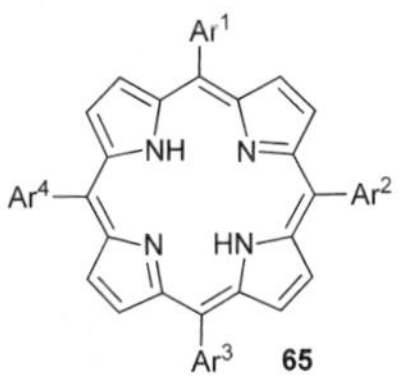 **65** **a**, Ar^1 = 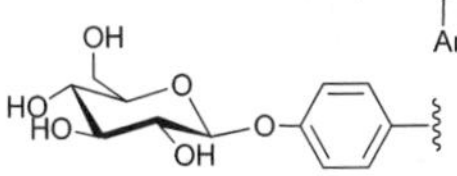 $Ar^2 = Ar^3 = Ar^4$ = 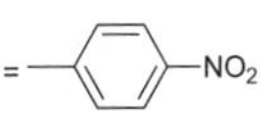 **b**, $Ar^1 = Ar^2$ = 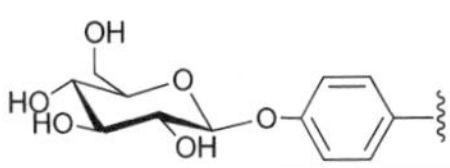 $Ar^3 = Ar^4$ = 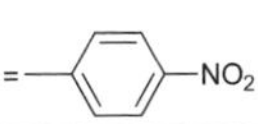	*Escherichia coli*[203] *Staphylococcus aureus*[203]
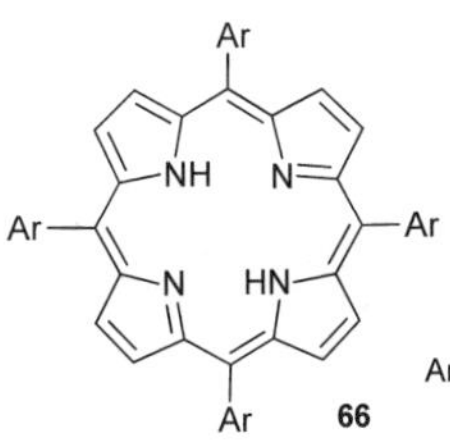 **66** Ar = 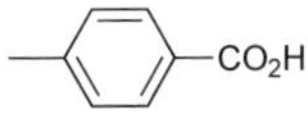	Hepatitis C virus[70] Hepatitis B virus[70] Human immunodeficiency virus[30,70] *Mycobacterium smegmatis*[173]

Table 5.2 (*Continued*)

Photosensitizer	Microorganism
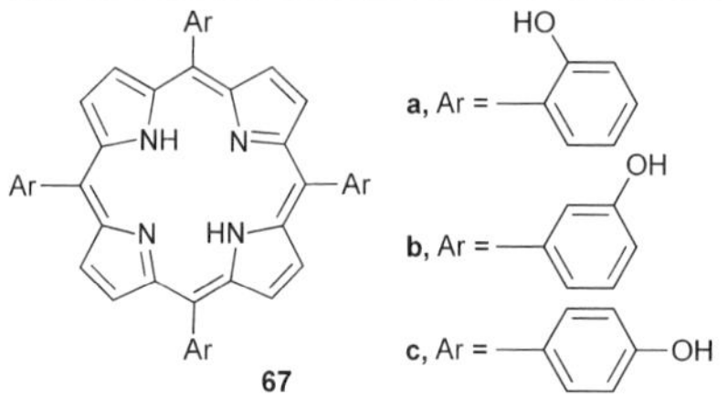 **67**	*Escherichia coli*[167] *Pseudomonas aeruginosa*[194,166] *Staphylococcus aureus*[167,194]
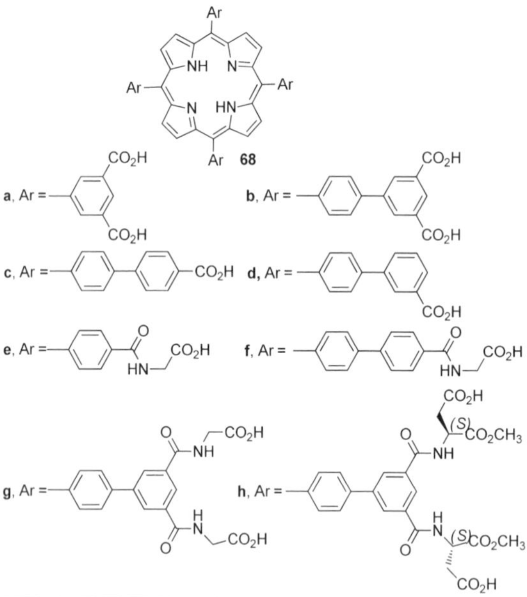 **68**	Hepatitis C virus[70] Hepatitis B virus[70] Human immunodeficiency virus[70]
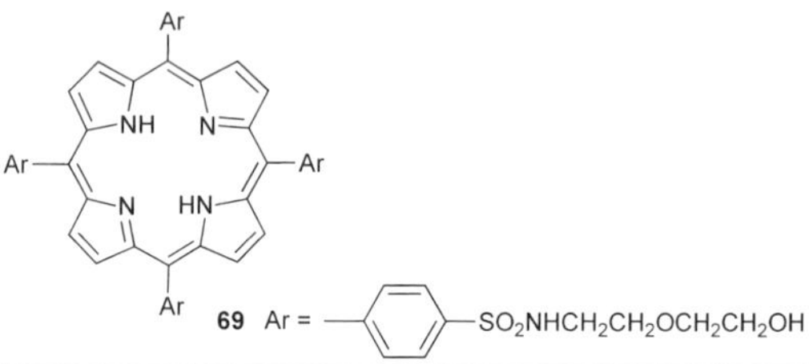 **69**	*Escherichia coli*[166] *Pseudomonas aeruginosa*[166] *Staphylococcus aureus*[166]
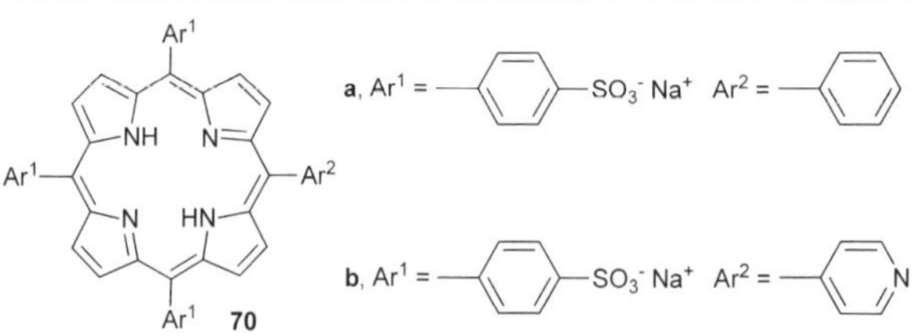 **70**	Human immunodeficiency virus[30]

Table 5.2 (*Continued*)

Photosensitizer	Microorganism
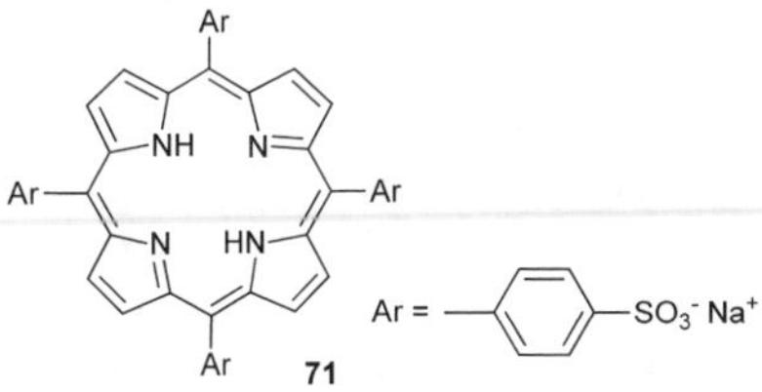 **71**	*Acinetobacter baumanii*[155] *Candida albicans*[164] *Escherichia coli*[167,195] Hepatitis A virus[172] HIV[30,196,197] Herpes simplex virus 1[196] Lambda phage[171] MS2 phages[172] *Mycobacterium smegmatis*[173] *Staphylococcus aureus*[167]
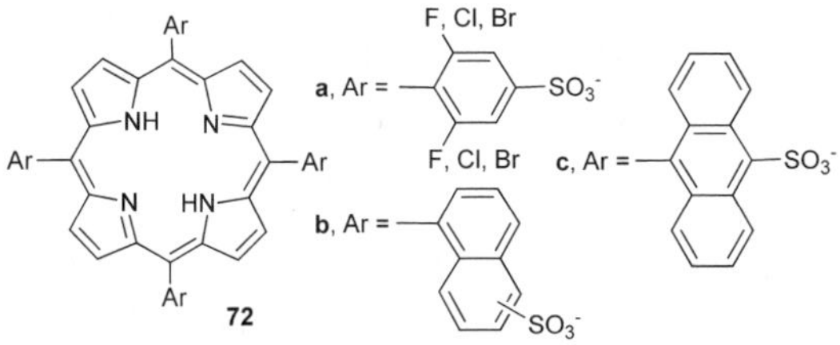 **72**	Human immunodeficiency virus[30]
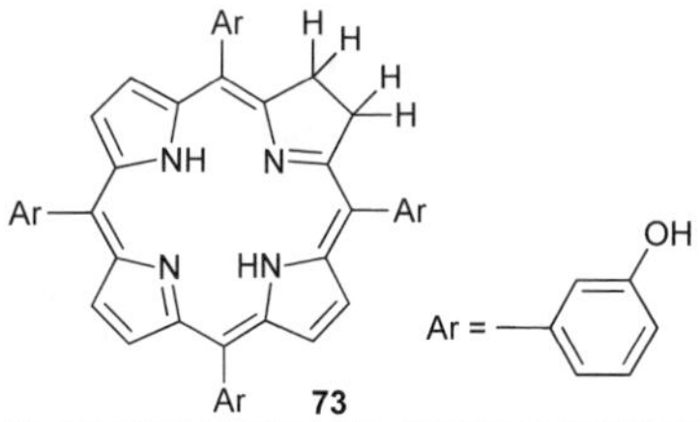 **73**	*Staphylococcus aureus*[57,198]
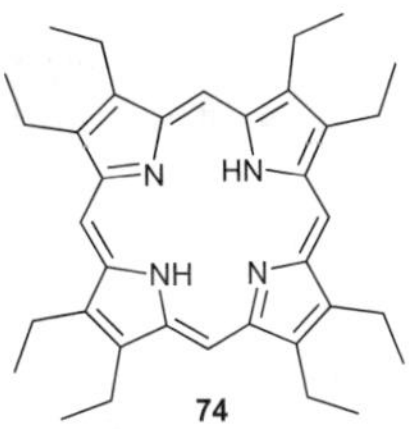 **74**	*Mycobacterium smegmatis*[106] *Staphylococcus aureus*[106] *Yersinia enterocolitica*[106]
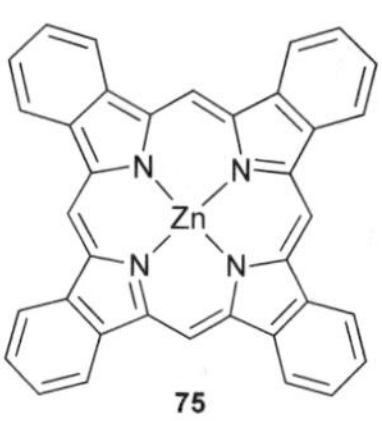 **75**	*Escherichia coli*[48] *Staphylococcus aureus*[48]

5.3. Porphyrins as Antimicrobial Agents in the Clinic

5.3.1. Practical Aspects

The use of light in the treatment of diseases goes back to antiquity[199] but its role in medicine has gained importance only in the last few decades, namely in the treatment of tumours (PDT). Although the main use of PDT has been focused on cancer treatment, the worldwide emergence of antimicrobial resistance amongst pathogenic microorganisms has led to a major research effort to find alternative antimicrobial therapeutics.[200] The aPDT may be the alternative to conventional antimicrobial therapeutics.

The aPDT relies on the accumulated experience in the treatment of malignant tumours by PDT. In that way, the studies on the photosensitizer distribution, local light exposure, fibre-optics and endoscopic equipment previously carried out for PDT can be extended to the aPDT.

In the last years, many porphyrin derivatives (Tables 5.1 and 5.2) have been tested against numerous infectious agents (Tables 5.3 to 5.11). *In vitro* and *in vivo* studies pointed out the great potentialities of porphyrins in aPDT. However, as the delivery of light is a localized process, this biomedical application is restricted to localized infections and not to systemic infections, such as sepsis.[32] Contrarily to PDT for cancer, where the PS is usually injected into the bloodstream and accumulates preferentially in the tumour cells, aPDT for localized infections is mostly carried out by local delivery of the PS to the infected area by topical application, instillation, interstitial injection or aerosol delivery.[32]

With the exception of sporadic use of haematoporphyrin derivative (HpD), dihaematoporphyrin ether (**11**) and ALA in the treatment of local viral infections, presently there is no routine application of aPDT for treatment of microbial infections. Actually, one of the main potential applications of porphyrin derivatives in the clinical area consists in the sterilization of blood and blood products, preventing viral contamination.[30,67–70]

Considering the results of laboratorial assays, one could envisage other applications of porphyrins in aPDT such as the disinfection of infected skin and cornea surface, treatment of wounds, the oral cavity and soft-tissue infections and the cure of abscesses.[2,24,31,32,56] However, there are only a few reports about the use of porphyrins in aPDT in animal models and clinical trials (Tables 5.3 to 5.11). Moreover, although most of the currently approved clinical photosensitizers, namely Photofrin (a purified porphyrin mixture), Foscan or Temoporfin (a chlorin), Visudyne (a chlorin), Levulan and derivatives (*e.g.* Metvix) for cancer treatment belong to the porphyrin family, from those, only ALA-based PS have been tested *in vivo* for aPDT, namely to treat bacterial skin infections by *Propionibacterium acnes*, viral infections by human papilloma viruses and cutaneous leishmaniosis (Tables 5.6, 5.7 and 5.11).

There are some aspects that have to be considered when aPDT is used to inactivate microorganisms in clinic and that must be taken into account when porphyrin derivatives are used:[32] 1) the PS must be effective in destroying the disease-causing pathogens; 2) the PS must have selectivity for the microbes,

Table 5.3 *In vitro* assays of photoinactivation of bacteria.

Photosensitizer	Microorganism	Reference
Haematoporphyrin derivative	*Bacillus subtilis* *Streptococcus faecalis*	Shawer and Cooper, 1990[135]
Chlorin e6 derivatives	*Escherichia coli* *Bacillus subtilis*	Fomichev *et al.*, 1991[99]
Deuteroporphyrin (DP)	*Escherichia coli* *Pseudomonas aeruginosa*	Nitzan *et al.*, 1992[157]
Mg and Zn-tetrabenzoporphyrin	*Staphylococcus aureus* *Escherichia coli*	Ehrenberg *et al.*, 1993[48]
Fe-protoporphyrin (hemin); Co-protoporphyrin; Mg-protoporphyrin; Zn-protoporphyrin	*Staphylococcus aureus*	Ladan *et al.*, 1993[133]
Deuteroporphyrin (DP); haematoporphyrin (HP); haematoporphyrin derivative (HpD); protoporphyrin	350 strains of anaerobic microorganisms	Nitzan *et al.*, 1994[126]
Neutral, cationic and anionic *meso*-tetraarylporphyrins	*Escherichia coli* *Staphylococcus aureus*	Nitzan *et al.*, 1995[167]
Cationic porphyrins: *meso*-tetrakis (1-methylpyridinium-4-yl)porphyrin tetraiodide and *meso*-tetrakis(4-trimethylammoniumphenyl)porphyrin Anionic porphyrin: *meso*-tetrakis (4-sulfonatophenyl)porphyrin	*Vibrio anguillarum* *Escherichia coli* *Enterococcus seriolicida*	Merchat *et al.*, 1996[6]
Bacteriochlorophyll	*Staphylococcus aureus*	Gross *et al.*, 1997[201]
Deuteroporphyrin (DP) and hemin	*MRSA*	Orenstein *et al.*, 1997[159]
ALA	*Pseudomonas aeruginosa*	Sailer *et al.*, 1997[92]
Haematoporphyrin derivative (HpD) Arginine-haematoporphyrin derivative (HpD-Arg(2))	*Pseudomonas aeruginosa* *Staphylococcus aureus* *Bacillus subtilis* *Micrococcus luteus* *Escherichia coli* *Proteus vulgaris* *Klebsiella pneumoniae* *Enterobacter cloacae* *Serratia marcescens* *Haemophilus parainfluenzae*	Szpakowska *et al.*, 1997[141]
ALA		van der Meulen *et al.*, 1997[202]
Deuteroporphyrin (DP), *meso*-tetrakis(1-methylpyridinium-4-yl)porphyrin, *meso*-tetrakis(4-sulfonatophenyl)porphyrin	*Acinetobacter baumannii*	Nitzan *et al.*, 1998[155]
Nitroglycosylated *meso*-arylporphyrins	*Escherichia coli and Staphylococcus aureus*	Sol *et al.*, 1998[203]
ALA, Photofrin	*Staphylococcus aureus* *Staphylococcus epidermidis*	Karrer *et al.*, 1999[95]
meso-tetrakis(3-hydroxyphenyl)chlorin (mTHPC) and Photofrin	*Staphylococcus aureus*	Kubin *et al.*, 1999[198]
Plant-derived chlorin	*Staphylococcus aureus* *Escherichia coli*	Lapotko *et al.*, 1999[204]

Table 5.3 (*Continued*)

Photosensitizer	Microorganism	Reference
Haematoporphyrin diarginate derivative (HpD-Arg(2))	*Staphylococcus aureus*	Lasocki *et al.*, 1999[144]
ALA	*Staphylococcus aureus* *Escherichia coli*	Nitzan and Kauffman, 1999[84]
meso-Tetrakis(1-methylpyridinium-4-yl)porphyrin and deuteroporphyrin	*Deinococcus radiodurans*	Nitzan *et al.*, 1999[156]
Non-iron metalloporphyrins	*Yersinia enterocolitica* *MRSA* *Mycobacterium smegmatis*	Stojiljkovic *et al.*, 1999[106]
ALA	*Escherichia coli*	Szocs *et al.*, 1999[83]
meso-Tetrakis(1-methylpyridinium-4-yl)porphyrin	*Escherichia coli* *Escherichia coli*	Valduga *et al.*, 1999[37]
meso-Tetrakis(3-hydroxyphenyl)chlorin (mTHPC) and haematoporphyrin derivative	*Staphylococcus aureus*	Wierrani, 1999[205]
Haematoporphyrin	*Staphylococcus aureus*	Bertoloni *et al.*, 2000[145]
Endogenous photosensitizing porphyrins	*Propionibacterium acnes* *Actinomyces odontolyticus* *Porphyromonas gingivalis*	Konig *et al.*, 2000[132]
Tetraphenylporphyrin derivatives, ALA	*Escherichia coli* *Enterococcus hirae*	Gábor *et al.*, 2001[29]
Haematoporphyrin (HP) and sodium chlorophyllin	*Staphylococcus aureus* *Bacillus subtili* *Bacillus cereus*	Kreitner *et al.*, 2001[143]
meso-Tetrakis(1-methylpyridinium-4-yl)porphyrin	*Acinetobacter baumannii* *Escherichia coli*	Nitzan and Ashkenazi, 2001[165]
Haematoporphyrin derivative with rutin and arginine substituents (HpD-rut(2)-arg(2))	*MSSA* *MRSA* *Pseudomonas aeruginosa*	Szpakowska *et al.*, 2001[140]
ortho, *meta* and *para* isomers of cationic *meso*-(*N*-alkylpyridinium)porphyrins and their Zn(II) complexes	*Escherichia coli*	Benov *et al.*, 2002[168]
IgG-Tin(IV) chlorin e6 conjugate (SnCe6-IgG)	*MRSA*	Embleton *et al.*, 2002[107]
Chlorin e6-poly(epsilon-benzyloxycarbonyl)lysine conjugates	*Staphylococcus aureus* *Escherichia coli*	Hamblin *et al.*, 2002[119]
meso-Tetrakis(1-methylpyridinium-4-yl)porphyrin	*Faecal coliforms*	Jemli *et al.*, 2002[8]
ALA	*Pseudomonas aeruginosa* (antibiotic resistant)	Sharma *et al.*, 2002[91]
Propyl gallate substituted haematoporphyrins	*Escherichia coli* *Acinetobacter baumannii* *Staphylococcus aureus*	Ashkenazi *et al.*, 2003[142]
Endogenous porphyrins and ALA	*Propionibacterium acnes*	Ashkenazi *et al.*, 2003[93]
meso-Tetrakis(*N*-methyl-6-quinolinyl)-substituted porphyrins and *meso*-tetrakis(*N*-methyl-6-quinolinyl)-substituted chlorins	*Escherichia coli* *Pseudomonas aeruginosa* *Staphylococcus aureus*	Philippova *et al.*, 2003[206]
meso-Tetrakis(1-methylpyridinium-4-yl)porphyrin tetratosylate	*Listeria monocytogenes* *Escherichia coli O157:H7*	Romanova *et al.*, 2003[169]
Aminoporphyrin derivatives (24 compounds)	*Escherichia coli* *Staphylococcus aureus*	Sol *et al.*, 2004[207]

Table 5.3 (*Continued*)

Photosensitizer	Microorganism	Reference
SnCe6 conjugated to MRSA antibody	*MRSA*	Embleton *et al.*, 2004[208]
Chlorin e6-polylysine conjugate	*Staphylococcus aureus*	Gad *et al.*, 2004[109]
meso-Tetrakis(1-methylpyridinium-4-yl)porphyrin and 5-phenyl-10,15,20-tris(1-methylpyridinium-4-yl)porphyrin	*Pseudomonas aeruginosa* *Staphylococcus aureus*	Lambrechts *et al.*, 2004[174]
5,10-bis(4-methylphenyl)-15,20-bis (4-trimethylammoniumphenyl) porphyrin iodide, 5-(4-trifluorophenyl)-10,15,20-tris (4-trimethylammoniumphenyl) porphyrin iodide and 5-(4-carboxyphenyl)-10,15,20-tris (4-methylphenyl)porphyrin	*Escherichia coli*	Lazzeri *et al.*, 2004[182]
ALA	*Staphylococcal strains* *Bacillus cereus*	Nitzan *et al.*, 2004[85]
ALA	*Staphylococcus aureus*	Bisland *et al.*, 2006[90]
Polylysine-chlorin e6 conjugate, 5-phenyl-10,15,20-tris(1-methylpyridinium-4-yl)porphyrin chloride and benzoporphyrin derivative (BPD)	*Spores of* *Bacillus atrophaeus* *Bacillus cereus* *Bacillus megaterium* *Bacillus subtilis* *Bacillus thuringiensis*	Demidova and Hamblin, 2005[113]
poly-L-lysine-chlorin(e6) conjugate	*Escherichia coli* *Staphylococcus aureus*	Demidova and Hamblin, 2005[117]
5-phenyl-10,15,20-tris(1-methylpyridinium-4-yl)porphyrin chloride	*Staphylococcus aureus*, *Pseudomonas aeruginosa*	Lambrechts *et al.*, 2005[186]
Porphyrin-based photosensitizers (structures **44**, **45a** and **45b** in Table 5.2)	*MRSA* *MSSA* *methicillin-resistant* *Staphylococcus epidermidis* *Escherichia coli*	Maisch *et al.*, 2005[24]
5-(4-trimethylammoniumphenyl)-10,15,20-tris(2,4,6-trimethoxyphenyl)porphyrin iodide, 5,10-di(4-methylphenyl)-15,20-di (4-trimethylammoniumphenyl) porphyrin iodide and 5-(4-trifluorophenyl)-10,15,20-tris (4-trimethylammoniumphenyl) porphyrin iodide	*Escherichia coli*	Spesia *et al.*, 2005[181]
Polycationic conjugates between chlorin(e6) and three molecular forms of polyethyleneimine	*Staphylococcus aureus* *Streptococcus pyogenes* *Escherichia coli* *Pseudomonas aeruginosa*	Tegos *et al.*, 2006[120]
cationic and neutral porphyrin derivatives	*Escherichia coli* *Pseudomonas aeruginosa* *Staphylococcus aureus*	Banfi *et al.*, 2006[166]
5-[4-(3-Trimethylammoniumpropoxy)phenyl]-10,15,20-tris (4-trifluoromethylphenyl)porphyrin,	*Escherichia coli*	Caminos *et al.*, 2006[209]

Table 5.3 (*Continued*)

Photosensitizer	Microorganism	Reference
5,15-di[4-(3-trimethylammoniumpropoxy)phenyl]-10,20-di(4-trifluoromethylphenyl)porphyrin iodide, 5,10,15-tris[4-(3-trimethylammoniumpropoxy)phenyl]-20-(4-trifluoromethylphenyl)porphyrin iodide and 5,10,15,20-tetrakis[4-(3-trimethylammoniumpropoxy)phenyl]porphyrin iodide		
meso-Tetraphenylporphyrin (*meso*-tetrakis(4- trimethylammoniumphenyl)porphyrin *p*-tosylate and *meso*-tetrakis(4-sulfonatophenyl)porphyrin sodium salt		
Haematoporphyrin	*MRSA*	Ferro *et al.*, 2006[146]
Films composed of porphyrinated lipophilic cellulose esters (protoporphyrin IX)	*Staphylococcus aureus* *Escherichia coli*	Krouit *et al.*, 2006[154]
Benzoporphyrin	*Mycobacterium bovis*	O'Riordan *et al.*, 2006[210]
Photolon and Protoporphyrin IX	*Staphylococcus aureus* *Escherichia coli*	Podbielska *et al.*, 2006[125]
5-Phenyl-10,15,20-tris(1-methylpyridinium-4-yl)porphyrin chloride	*Staphylococcus aureus*, *Pseudomonas aeruginosa* *Yersinia enterolitica*	Trannoy *et al.*, 2006[55]
5,10,15,20-tetrakis(4-trimethylammoniumphenyl)porphyrin and 5,10,15,20-tetrakis(4-sulfonatophenyl)porphyrin	*Escherichia coli*	Caminos and Durantini, 2007[211]
Neutral and cationic porphyrin derivatives	*Faecal coliforms* *Faecal enterococci*	Carvalho *et al.*, 2007[73]
Tetra sulfonated porphyrin	*Cyanobacteria*	Drábková *et al.*, 2007[212]
5-(1-dodecanylpyridinium-4-yl)-10,15,20-triphenylporphyrin incorporated into a poly-cationic liposome	*MRSA*	Ferro *et al.*, 2007[177]
Protoporphyrin IX, haematoporphyrin diarginate derivative and ALA	*MRSA*	Grinholc *et al.*, 2007[58]
Poly-L-lysine-chlorin(e6) conjugate	*MRSA* *Escherichia coli* *E. coli* (ESBL producing)	Tang *et al.*, 2007[118]
Cationic *meso*-substituted porphyrins	*Escherichia coli*	Alves *et al.*, 2008[9]
meso-tetrakis(3-hydroxyphenyl)chlorin in cationic liposomes	MRSA	Bombelli *et al.*, 2008[57]
5,10,15-tris[4-(3-trimethylammonium propoxy)phenyl]-20-(4-trifluoromethylphenyl)porphyrin iodide and 5,10,15,20-tetrakis(4-trimethylammoniumphenyl)porphyrin tosylate	*Escherichia coli*	Caminos *et al.*, 2008[3]
Haematoporphyrin	*Staphylococcus aureus* *Escherichia coli* *Pseudomonas aeruginosa*	El-Adly, 2008[213]
meso-Tetrakis(1-methylpyridinium-4-yl)porphyrin	*Enteroccoccus hirae* *Escherichia coli*	Ergaieg *et al.*, 2008[19]
ALA alkyl esters	*Escherichia coli*	Fotinos *et al.*, 2008[86]

Table 5.3 (*Continued*)

Photosensitizer	Microorganism	Reference
	Pseudomonas aeruginosa	
	MRSA	
Protoporphyrin diarginate	*MRSA*	Grinholc *et al.*, 2008[134]
	MSSA	
	Staphylococcus aureus	
Pyridinium porphyrinic chloroacetyl cellulose ester chlorides covalently bonded	*Staphylococcus aureus* *Escherichia coli*	Krouit *et al.*, 2008[190]
Endogenous porphyrins	*Staphylococcus aureus*	Maclean *et al.*, 2008[25]
Tetracationic *meso*-substituted tetrahydroporphyrin tetratosylate and dicationic water-soluble chlorin e6 Photolon	*Staphylococcus aureus* – *MRSA* *Escherichia coli* *Pseudomonas aeruginosa*	Schastak *et al.*, 2008[101]
5-phenyl-10,15,20-tris(1-methylpyridinium-4-yl)porphyrin	Gram (+) and Gram (–) bacteria	Trannoy *et al.*, 2008[188]
5,10,15,20-tetrakis(1-methylpiridinium-4-yl)porphyrin tetra-iodide; 5-(pentafluorophenyl)-10,15,20-tris(1-methylpiridinium-4-yl)porphyrin tri-iodide; 5-(4-methoxicarbonylphenyl)-10,15,20-tris(1-methylpiridinium-4-yl)porphyrin tri-iodide; 5-(4-carboxyphenyl)-10,15,20-tris(1-methylpiridinium-4-yl)porphyrin tri-iodide; 5,10-bis(4-carboxyphenyl)-15,20-bis(1-methylpiridinium-4-yl)porphyrin di-iodide; 5,15-bis(4-carboxyphenyl)-10,20-bis(1-methylpiridinium-4-yl)porphyrin di-iodide and 5-(1-methylpiridinium-4-yl)-10,15,20-tris(4-carboxyphenyl)porphyrin iodide	*Escherichia coli* *Enterococcus faecalis*	Alves *et al.*, 2009[12]
ALA	*Salmonella enterica*	Buchovec *et al.*, 2009[94]
meso-Tetrakis(1-methylpyridinium-4-yl)porphyrin tetratosylate, *meso*-tetrakis(4-trimethylammoniumphenyl)porphyrin tetrachloride, *meso*-tetrakis(4-sulfonatophenyl)porphyrin, *meso*-tetrakis(4-carboxyphenyl)porphyrin-Pd(II)	*Mycobacterium smegmatis*	Feese and Ghiladi, 2009[173]
meso-tetrakis(1-methylpyridinium-4-yl)porphyrin tetratosylate cross-linked to poly(vinyl alcohol) hydrogels	*MRSA*	Donnelly *et al.*, 2009[214]
meso-Tetrakis(4-diphenylaminophenyl)porphyrin and its complex with Pd(II) on optically transparent indium tin oxide (ITO) electrodes	*Escherichia coli*	Funes *et al.*, 2009[215]
Porphyrin derivatives grafted to alkylated cellulose films	*Staphylococcus aureus* *Escherichia coli*	Krouit *et al.*, 2009[191]
ALA	*Bacillus cereus*	Luksiene *et al.*, 2009[81]
5,10,15,20-tetrakis(1-methylpiridinium-4-yl)porphyrin tetra-iodide; 5-(pentafluorophenyl)-10,15,20-tris(1-methylpiridinium-4-yl)porphyrin tri-iodide; 5-(4-	*Bacillus cereus* endospores and vegetative cells	Oliveira *et al.*, 2009[163]

Table 5.3 (*Continued*)

Photosensitizer	Microorganism	Reference
carboxyphenyl)-10,15,20-tris(1-methylpiridinium-4-yl)porphyrin tri-iodide; 5,10-bis(4-carboxyphenyl)-15,20-bis (1-methylpiridinium-4-yl)porphyrin di-iodide; 5-(1-methylpiridinium-4-yl)-10,15,20-tris(4-carboxyphenyl)porphyrin iodide; 5-(4-carboxyphenyl)-10,15,20-tri(4-pyridil)porphyrin tri-iodide; 10,15,20-tris(1-methylpyridinium-4-yl)-5-phenylporphyrin triiodide		
dicationic porphyrin	*Staphylococcus aureus*	Ooi *et al.*, 2009[216]
5,10,15,20-Tetrakis(3-hydroxyphenyl)porphyrin incorporated in calcium phosphate nanoparticles	*Staphylococcus aureus* *Pseudomonas aeruginosa*	Schwiertz *et al.*, 2009[194]
Protochlorophyllide	*Staphylococcus aureus* *Listeria monocytogenes* *Yersinia pseudotuberculosis* *Bacillus subtilis* *Escherichia coli*	Walther *et al.*, 2009[96]
Complex of a new anionic water-soluble polythiophene and a cationic porphyrin	*Escherichia coli* *Bacillus subtilis*	Xing *et al.*, 2009[179]
Cationic derivatives based on *meso*-tetrapyridylporphyrin	*Staphylococcus aureus* *Escherichia coli* *Pseudomonas aeruginosa*	Yu *et al.*, 2009[178]
Mono(acridyl)bis(arginyl)porphyrin Mono(acridyl)bis(arginyl)porphyrin complexed to [poly(dG-dC)]2	*Pseudomonas aeruginosa*	Steenkeste *et al.*, 2010[185]
5,10,15-tris(1-methylpyridinium-4-yl)-20-(pentafluorophenyl)-porphyrin triiodide	*Vibrio fischeri* *Escherichia coli*	Tavares *et al.*, 2010[217]

MRSA – Methicillin resistant *Staphylococcus aureus*
MSSA – Methicillin sensitive *Staphylococcus aureus*
ESBL – Extended spectrum ß lactamase

avoiding an unacceptable degree of host tissue damage in the area of infection; 3) the re-growth of the pathogens must be avoided; 4) the efficacy of porphyrin derivatives in aPDT of clinical multi-resistant microbial strains must be similar to that of sensible strains; 5) the PS must be effective in the presence of biological material and 6) to a better evaluation of the potential of porphyrin derivatives as PS in aPDT, the comparison of porphyrin derivatives efficacy with that of other PS in clinical aPDT is also needed.

5.3.1.1. Effectiveness of aPDT

Several studies have already confirmed that both natural and synthetic porphyrin derivatives are able to inactivate high concentrations of different microorganisms either *in vitro* or *in vivo* (Tables 5.3–5.11). For instance, Ashkenazi and coworkers[142] reported the complete eradication *in vitro* of

Table 5.4 *In vitro* assays of photoinactivation of viruses.

Photosensitizer	Microorganism	Reference
Haematoporphyrin	Influenza viruses Herpes viruses	Perlin *et al.*, 1987[137]
Hemin	Murine leukaemia virus	Tsutsui and Mueller, 1987[130]
Haematoporphyrin derivative	HIV	Matthews *et al.*, 1988[138]
Protoporphyrin	HIV	Asanaka *et al.*, 1989[129]
Dihaematoporphyrin ether	Simian immunodeficiency virus	Chanh *et al..*, 1989[148]
Sulfonated derivatives of tetraphenylporphyrin tetraarylporphyrins	HIV	Dixon *et al.*, 1990[218]
Benzoporphyrin derivative	Vesicular stomatitis virus Feline leukaemia virus	Neyndorff *et al.*, 1990[219]
Haem administered together with 3′-azido-3′-deoxythymidine (AZT)	HIV	Levere *et al.*, 1991[220]
Boronated porphyrins	HIV	DeCamp *et al.*, 1992[62]
Amino- and hydroxytetraphenylporphyrins	Human immunodeficiency virus	Dixon *et al.*, 1992[221]
meso-Tetrakis(1-methylpyridinium -4-yl)porphyrin	Lambda phage	Kasturi and Platz, 1992[171]
Dihaematoporphyrin ether (DHE) Haematoporphyrin derivative (HpD) Benzoporphyrin derivatives (BPD)	Herpes simplex virus 1 Cytomegalovirus Simian immunodeficiency virus	Matthews *et al.*, 1992[65]
Benzoporphyrin	Vesicular stomatitis virus Feline leukaemia virus	North *et al.*, 1992[162]
Benzoporphyrin derivative monoacid ring A (BPD-MA)	HIV	North *et al.*, 1993[222]
Porphyrin derivatives	HIV Drug-resistant HIV	Debnath *et al.*, 1994[60]
Benzoporphyrin derivative	HIV	North *et al.*, 1994[52]
Chlorin-type compounds (structures **17** in Table 5.1)	HIV Drug-resistant HIV	Grandadam *et al.*, 1995[79]
Haem and the protoporphyrin complexes	HIV	Staudinger *et al.*, 1996[223]
Dihaematoporphyrin ether (DHE)	Papilloma viruses	Shikowitz *et al.*, 1998[147]
Haem	HIV	Argyris *et al.*, 1999[224]
Carborane porphyrins derivatives	HIV	Debnath *et al.*, 1999[225]
Protoporphyrin Uroporphyrin	Adenovirus	Schagen *et al.*, 1999[70]
Glycoconjugated *meso*-tetraarylporphyrins	T7 phage	Gábor *et al.*, 2001[29]
Chlorophyll derivatives from silkworm excreta	Vesicular stomatitis virus	Lim *et al.*, 2002[226]
Natural porphyrins and sulfonated tetraarylporphyrins	HIV	Vzorov *et al.*, 2002[30]
Zinc protoporphyrin, copper haematoporphyrin, *meso*-tetrakis(2,6-dihydroxyphenyl)porphyrin, sulfonated tetra-1-naphthyl and tetra-1-anthracenylporphyrins, sulfonated derivatives of halogenated tetraarylporphyrins	Poxvirus	Chen-Collins *et al.*, 2003[227]

Table 5.4 (*Continued*)

Photosensitizer	Microorganism	Reference
5,10,15-(4-β-D-galactosylphenyl)-20-(pentafluorophenyl)porphyrin	T7 phage	Egyeki *et al.*, 2003[31]
Haematoporphyrin derivatives, *meso*-tetraphenylporphyrins and a chlorin	HIV	Vanyúr *et al.*, 2003[68]
Cationic, anionic and amphiphilic porphyrins	Hepatitis A virus Bacteriophage MS2	Casteel *et al.*, 2004[172]
Anionic sulfonated porphyrins	HIV	Dairou *et al.*, 2004[69]

5,10-Diphenyl-15,20-bis (1-methylpyridinium-4-yl)porphyrin dichloride
5,15-diphenyl-10,20-bis (1-methylpyridinium-4-yl)porphyrin dichloride
5,10,15-triphenyl-20-(1-methylpyridinium-4-yl)porphyrin chloride
meso-Tetrakis(1-methylpyridinium-4-yl)porphyrin chloride;
meso-tetrakis(4-trimethylammonium-phenyl)porphyrin chloride;
meso-tetrakis[1-(2-hydroxyethyl)pyridinium-4-yl]porphyrin chloride;
5,10,15-tris(1-methylpyridinium-4-yl)-20-phenylporphyrin
	Vesicular stomatitis virus	Trannoy *et al.*, 2004[175]

meso-Tetrakis(1-methylpyridinium-4-yl)porphyrin	T7 bacteriophage	Zupán *et al.*, 2004[228]
meso-Tetrakis(4-trimethylammonium-phenyl)porphyrin and its metal complexes [M = Zn(II), Cu(II), Fe(III), Co(III), Ni(II), Mn(III)]	Influenza virus type A (H3N2) and B Herpes simplex virus, type 1 and 2	Liu *et al.*, 2005[229]
Carbohydrate-substituted *meso*-tetraarylporphyrins (structures **50**–**53**)	Herpes simplex virus, type 1 and 2	Tomé *et al.*, 2005[192]
5,10,15-tris(1-methylpyridinium-4-yl)-20-phenylporphyrin	Vesicular stomatitis virus Bovine viral diarrhoea virus HIV Pseudorabies virus	Trannoy *et al.*, 2006[55]
meso-Tetrakis(sulfonatophenyl)porphyrin	Herpes simplex virus type 1	Ion *et al.*, 2007[196]
Neutral and cationic tripyridylporphyrin-D-galactose conjugates (structures **38a**, **54a** and **54b** in Table 5.2)	Herpes simplex virus type 1	Tomé *et al.*, 2007[73]
Sulfonated porphyrin derivatives	HIV	Vzorov *et al.*, 2007[230]
10,15,20-Tetrakis(1-methylpiridinium-4-yl)porphyrin tetra-iodide; 5-(pentafluorophenyl)-10,15,20-tris(1-methylpiridinium-4-yl)porphyrin tri-iodide; 5-(4-methoxicarbonylphenyl)-10,15,20-tris(1-methylpiridinium-4-yl)porphyrin tri-	T4-like phage	Costa *et al.*, 2008[10]

Table 5.4 (*Continued*)

Photosensitizer	Microorganism	Reference
iodide; 5-(4-carboxyphenyl)-10,15,20-tris(1-methylpiridinium-4-yl)porphyrin tri-iodide; 5,10-bis (4-carboxyphenyl)-15,20-bis (1-methylpiridinium-4-yl)porphyrin di-iodide and 5,15-bis(4-carboxyphenyl)-10,20-bis(1-methylpiridinium-4-yl)porphyrin di-iodide		
Protoporphyrin IX	Hepatitis B virus (HBV)	Lin and Hu, 2008[231]
Sulfonated *meso*-tetraphenylporphyrins complexed with metal ions: Cu(II), Zn(II), Pd(II), Mn(II), Fe(III), Ni(II) and Co(II)	HIV	Vargas *et al.*, 2008[197]
meso-Tetrakis(1-methylpyridinium-4-yl)porphyrin and 5,10,15-tris(1-methylpyridinium-4-yl)-20-phenylporphyrin	T7 bacteriophage	Zupán *et al.*, 2008[33]
meso-Tetrakis(1-methylpyridinium-4-yl)porphyrin	T7 phage	Csík *et al.*, 2009[232]
Immobilized chlorin 3-phorbinepropanol into anionic unilamellar liposomes	Bovine viral diarrhoea virus Encephalomyocarditis virus	Sagristá *et al.*, 2009[233]
Zanamivir-porphyrin conjugates	Influenza virus	Wen *et al.*, 2009[234]
Tetraarylporphyrin derivatives	Hepatitis C virus	Cheng *et al.*, 2010[70]

HIV – Human immunodeficiency virus

Acinetobacter baumannii at low fluence (7.5 J cm^{-2}) and of *E. coli* at higher fluences (37.5 J cm^{-2}) under the same conditions with the antioxidant carrier sensitizer propyl gallate substituted haematoporphyrin **10e**. A much poorer performance was referred for the analogue **10f**. *Propionibacterium acnes* was also inactivated by its endogenic porphyrins by 4–5 orders of magnitude.[93] When ALA-triggered *Propionibacterium acnes* cultures were used the viability decreased by 7 orders of magnitude.[93]

Synthetic cationic porphyrins inactivated *in vitro Escherichia coli*, *Pseudomonas aeruginosa*, *Yersinia enterolitica*, *Deinococcus radiodurans*, *Mycobacterium smegmatis* and *Staphylococcus aureus* by 5–7 log[55,84,166,173,181,182] and also inactivated methicillin-resistant *S. aureus* and methicillin-resistant *S. epidermidis* by 8 log.[54] Poly-L-lysine chlorin e6 conjugate (pL-ce6) (**4c**) also inactivated clinical methicillin resistant *Staphylococcus aureus* (MRSA) and extended-spectrum β-lactamase (ESBL) producing *E. coli*, as well as their corresponding culture collection strains, by 3–5 log[118] and also *Actinomyces viscosus* and *Porphyromonas gingivalis* by 2–3 log.[112]

The *in vivo* inactivation of high concentrations of disease-causing pathogens with natural and synthetic porphyrins has also been observed. Smetana *et al.*[88] drastically reduced the herpes simplex virus titre by $\geq$5 log with the administration of ALA in guinea pigs. ALA also proved to be effective *in vivo* against *Trichophyton rubrum* confirming photodynamic inactivation as an alternative non-invasive approach for the treatment of onychomycosis.[249] Topical use of ALA was also demonstrated to be effective and safe in photodynamic treatment of cutaneous leishmaniosis.[257] A conjugate between polyethyleneimine and chlorin e6 led to a reduction of *Proteus mirabilis* and *Pseudomonas aeruginosa* in biofilms of root canals by 2 log.[116] This conjugate was used in the treatment of *Acinetobacter baumannii*, being applied topically on the surface of 10 burns made on the dorsal surface of shaved female mice. PDT carried out on day 0 (soon after application of the bacteria) gave over 3 log units of bacterial loss while PDT carried out on day 1 and day 2 gave approximately 1.7 log reductions.[111]

5.3.1.2. Toxicity

In situ and *in vivo* aPDT assays show that, in general, negative effects of porphyrin derivatives on host cells do not occur when the PS is used at concentrations that are effective to microbial inactivation.

Several studies carried out *in vitro* indicate that porphyrins have no effect on blood cell viability[65,162,188,197,204] and on other mammalian cells.[112,172,192] The positively charged poly-L-lysine chlorin e6 conjugate (**4c**) proved to be effective in killing bacteria without affecting the viability of the epithelial cell.[112] Casteel *et al.*[172] showed that among cationic, anionic or amphiphilic porphyrins none were mutagenic in the Ames assay and only the congener with the longest chain-length, *meso*-tetrakis(1-hexadecylpyridinium-4-yl)porphyrin (**20f**), was appreciably toxic to mammalian cells under ambient lighting conditions.

Photodynamic inhibition of microorganisms occurs at porphyrin concentrations lower than those found to be cytotoxic for the host mammalian cells. The cytotoxicity of the water soluble porphyrin **26** and a series of its metal complexes with Zn(II), Cu(II), Ni(II), Fe(III), Co(III), Mn(III) was determined for Madin–Darby canine kidney cells and Vero cells.[301] Among the complexes evaluated it was found that the inhibition of influenza virus types A (H3N2) and B and herpes simplex virus type 1 and type 2 occurred at lower concentrations than those found cytotoxic for the model eukaryotic cells.[298] Strong inhibition yields of herpes simplex virus type 1 and type 2 were also observed with glycoporphyrins **49**–**51** and **53** at concentrations much lower than their maximum non-cytotoxic concentrations.[192]

Studies with basic blood cells and neutrophils showed that their viability is not affected by aPDT.[65,162,197,204] The *ex vivo* expansion potential of the progenitor cells, and their capacity to differentiate various haematopoietic cell lineages was not affected.[188] However, PDT reduced the engraftment of human cord blood stem cell (CBSC) in non-obese diabetic/severe combined immunodeficiency mice (NOD/SCID) mice, particularly affecting the B-cell lineage engraftment[188] and platelets were also modified over time.[65]

The *in vivo* studies with porphyrin derivatives indicate also that aPDT has no effect on the host cells. For instance, one day later after the injection of poly-L-lysine chlorin e6 conjugate (**4c**) into a muscle of mice infected with *S. aureus*, no muscle cells damage was observed.[109] However, infections treated with free chlorin e6 (**3**) did not respond equally well and the infection subsequently increased over the succeeding days, probably due to PDT-mediated damage.

Haematoporphyrin (**10a**) did not produce any damaging effect on the mucous membrane of the stomach but it was effective to inactivate *Helicobacter pylori* cells.[136] Dihaematoporphyrin ether (**11**) was also used to treat respiratory papillomatosis and there was no evidence of photosensitizer-induced carcinoma in a universe of 41 patients.[147] Bauman and Smith[283] tested PS **11** and did not observe any evidence of photosensitizer-induced virus-related elevation of disease or of carcinoma in a universe of 81 patients. The intravenous administration of dihaematoporphyrin ether was also tested to treat 26 patients with recurrent laryngeal papillomatosis and no long-term debilitating sequelae were noted.[282]

5.3.1.3. Re-growth of the Pathogens after aPDT

Photofrin II and the pL-ce6 conjugate were used to treat papilloma warts in the cutaneous tissue of rabbits and total wart regression was achieved without recurrence.[102] The topical use of ALA to inactivate *Propionibacterium acnes* in 13 patients with facial acne demonstrated good clinical improvement of the condition of the patients and reduction of acne lesions for up to 6 months following aPDT treatment.[262] The reduction of *E. coli*[119] and of *P. aeruginosa*[269] cells was observed in four sterile excisional wounds on the backs of several mice when treated with pL-ce6 conjugate (**4c**) and light. A different situation occurred with all the control mice that died while almost all aPDT treated mice survived.[119,269] The *in situ* application of this conjugate on the burned dorsal surface of 10 shaved female mice surface to inactivate *Acinetobacter baumannii* showed that some bacterial re-growth in the treated burn was observed, but was generally modest.[111] Chlorin e6 (**3**) or pL-ce6 conjugate (**4c**) were also used to treat artificial infections with *S. aureus* in thigh muscles of mice.[109] With pL-ce6 there was a light dose-dependent inactivation of bacteria that was not observed in the control groups. In addition, no infection was observed in the succeeding days.

The *meso*-tetrakis(3-hydroxyphenyl)chlorin **73** was used to treat recurrent respiratory papillomatosis[292] but, although PDT reduces the severity of laryngeal papillomas, failure to maintain remission with time was observed. The cationic *meso*-substituted porphyrin **43a** was also used to treat third-degree burn wounds in mice infected with bioluminescent *S. aureus*.[189] More than 98% of the bacteria were eradicated, however bacterial re-growth was observed.

Recently, it was found that cells of *Vibrio fischeri* and of *Escherichia coli*, and also T4-like phages, that were photoinactivated *in vitro* to the detection limit in the presence of 5,10,15-tris(1-methylpyridinium-4-yl)-20-(pentafluorophenyl) porphyrin triiodide (**36**) do not recover their viability in laboratory conditions after one week.[217]

5.3.1.4. Efficacy on Multi-resistant Microbial Strains

The efficacy of porphyrin derivatives against a range of clinical multi-resistant microbial strains has been demonstrated (Tables 5.1 and 5.2). It is worth referring to the fact that the photosensitizing activity is independent of the antibiotic-resistance spectrum of the given bacterial pathogen. Reliable killing of antibiotic resistant strains, important in hospital infections such as

Table 5.5 *In vitro* and *in vivo* assays of photoinactivation of fungi.

Photosensitizer	Microorganism	Reference
***In vitro* assays**		
Haematoporphyrin	*Candida albicans*	Bertoloni *et al.*, 1987[40]
Endogenous porphyrin-type compound	*Candida guilliermondii* *Saccharomyces cerevisiae*	Fraikin *et al.*, 1996[124]
Haematoporphyrin derivatives	*Saccharomyces cerevisiae*	Zofadek *et al.*, 1997[139]
Meso-arylglycosylporphyrins	*Saccharomyces cerevisiae*	Carré *et al.*, 1999[49]
Palladium tetraphenyltetra-benzoporphyrin sulfate	*Aspergillus fumigatus*	Friedberg *et al.*, 2001[235]
Haematoporphyrin Sodium chlorophyllin	*Saccharomyces cerevisiae* *Kloeckera javanica* *Rhodotorula mucilaginosa*	Kreitner *et al.*, 2001[143]
meso-Tetrakis(*N*-methyl-6-quinolinyl)-substituted porphyrins and chlorins	*Candida albicans* *Rhodotorula bogoriensis*	Philippova *et al.*, 2004[236]
Photofrin	*Candida albicans* *Candida glabrata* *Candida krusei*	Bliss *et al.*, 2004[237]
Haematoporphyrin	*Rhizopus oryzae* *Aspergillus flavus* *Trichothecium roseum* *Fusarium avenaceum*	Luksiené *et al.*, 2004[238]
Quinolino^{4,5,6,7}-7-demethyl-8-deethyl-*meso*-porphyrin dimethyl ester (structure **19** in Table 5.1)	*Trichophyton rubrum*	Smijs *et al.*, 2004[160]
Photofrin	*Candida albicans*	Chabrier-Roselló *et al.*, 2005[239]
Poly-L-lysine chlorin(e6) conjugate	*Candida albicans*	Demidova and Hamblin, 2005[117]
ALA	*Candida albicans* *Trichophyton interdigitale*	Donnelly *et al.*, 2005[89]
5-Phenyl-10,15,20-tris(1-methylpyr-idinium-4-yl)porphyrin chloride	*Candida albicans*	Lambrechts *et al.*, 2005[183,186]
Haematoporphyrin	*Alternaria alternat* *Fusarium avenaceu* Acremonium strictum *Rhizopus oryzae*	Luksiené *et al.*, 2005[240]
ALA	*Candida albicans*	Monfrecola *et al.*, 2005[241]
meso-Tetrakis(1-methylpyridinium-4-yl)porphyrin 5,10,15-tris(1-Methylpyridinium-4-yl)-20-(1-tetradecylpyridinium-4-yl)porphyrin	*Saprolegnia* spp. zoospores *Saprolegnia* spp.	Magaraggia *et al.*, 2006[72]
5,10,15-tris(1-Methylpyridinium-4-yl)-20-phenylporphyrin Deuteroporphyrin monomethyl ester	*Trichophyton rubrum*	Smijs *et al.*, 2007[242]
5,10,15,20-Tetrakis[4-N-(2-oxyethylpyr-idyl)]porphinatoAg(II) tetranitrate	*Candida albicans*	Ghazaryan *et al.*, 2008[243]
Haematoporphyrin	*Yeast cells*	Ito, 2008[244]
Protoporphyrin IX	*Fusarium poae* *Fusarium culmorum*	Vorobey and Pinchuk, 2008[245]
5-(4-Trifluorophenyl)-10,15,20-tris(4-tri-methylammoniumphenyl)porphyrin	*Candida albicans*	Cormick *et al.*, 2009[164]

Table 5.5 (*Continued*)

Photosensitizer	Microorganism	Reference
iodide and 5,10,15,20-tetra(4-trimethylammoniumphenyl)porphyrin tosylate		
Quaternary ammonium derivatives of *meso*-tetrapyridylporphyrin	*Candida albicans*	Yu *et al.*, 2009[178]
5,10,15-tris(1-methylpyridinium-4-yl)-20-phenylporphyrin	*Trichophyton rubrum*	Smijs *et al.*, 2009[246]
Porphyrin derivative bearing a fluconazole unit	*Candida albicans*	Mora *et al.*, 2010[247]
ALA and ALA methyl ester	*Candida albicans*	Oriel and Nitzan, 2010[82]
***In vivo* assays**		
ALA	*Candida albicans* *Trichophyton dermagrophytes* *T. rubrum*	Calzavara-Pinton *et al.*, 2004[248]
ALA and ALA methyl ester	*Trichophyton rubrum*	Piraccini *et al.*, 2008[249]

Pseudomonas aeruginosa, *Enterococi*, methicillin-resistant *Staphylococcus aureus* (MRSA) and ESBL-producing Gram (–) bacteria, has been attained by aPDT.[118,134,140,144,155,159]

Photoactivated porphyrin derivatives also revealed antiviral activity against drug-resistant isolates (Table 5.4). The benzoporphyrin derivatives **18a,b** were investigated regarding their ability to destroy azidothymidine (AZT)-resistant strains of HIV-1. The results showed that AZT-resistant strains appear equally

Table 5.6 *In vitro* and *in vivo* assays of photoinactivation of parasites.

Photosensitizer	Microorganism	Reference
***In vitro* assays**		
Pheophorbide derivatives	*Plasmodium falciparum* *Babesia divergens*	Grellier *et al.*, 1997[250]
meso-Tetrakis(1-methylpyridinium-4-yl) porphyrin tetratosylate	*Ascaris lumbricoides* (eggs) *Taenia* sp. (eggs)	Alouini and Jemli, 2001[170]
meso-Substituted cationic porphyrins	*Colpoda inflate cysts*	Kassab *et al.*, 2002[251]
ALA Protoporphyrin IX Benzoporphyrin derivative	*Leishmania major*	Akilov *et al.*, 2006[318]
Tetracationic porphyrin derivatives	*Leishmania major*	Bristow *et al.*, 2006[252]
Protoporphyrin IX	*Leishmania* sp.	Akilov *et al.*, 2007[253]
ALA-PpIX	*Leishmania* sp.	Kosaka *et al.*, 2007[13]
Dimethyl and diethyl carbaporphyrin ketals	*Leishmania tarentolae*	Morgenthaler *et al.*, 2008[254]
***In vivo* assays**		
ALA methyl ester	*Leishmania donovani*	Gardlo *et al.*, 2003[255]
ALA methyl ester	*Leishmania* sp.	Gardlo *et al.*, 2004[256]
ALA	*Leishmania major*	Asilian and Davami, 2006[257]
Protoporphyrin IX	*Leishmania* sp.	Akilov *et al.*, 2007[253]
ALA and ALA methyl ester	*Leishmania* sp.	Snoek *et al.*, 2008[258]

Table 5.7 *In vivo* assays of photoinactivation to treat bacterial skin infections.

Photosensitizer	Microorganism	Mode/location of application	Reference
Endogenous porphyrins	*Propionibacterium acnes*	Endogenous production on human skin	Konig *et al.*, 1993[259]
ALA	*Propionibacterium acnes*	Topical use in 22 patients with acne on the back	Hongcharu *et al.*, 2000[260]
ALA	*Propionibacterium acnes*	Topical use in one patient with facial acne	Itoh *et al.*, 2000[261]
ALA	*Propionibacterium acnes*	Topical use in 13 patients with facial acne	Itoh *et al.*, 2001[262]
ALA	*Propionibacterium acnes*	Topical use in one patient with inflammatory acne lesion	Pollock *et al.*, 2004[263]
Endogenous porphyrins	*Propionibacterium acnes*	Topical use in untreated acne patients during standard acne regimens	Borelli *et al.*, 2006[264]
ALA	*Propionibacterium acnes*	Topical use in 30 patients with moderated facial acne	Hörfelt *et al.*, 2006[265]
ALA ALA methyl ester	*Propionibacterium acnes*	Topical use in 15 patients with moderated facial acne	Wiegell *et al.*, 2006[266]
Porphyrin-based photosensitizer XF73	*MRSA* *MSSA*	Topical use in infected *ex vivo* porcine skin	Maisch *et al.*, 2007[267]

MRSA – Methicillin resistant *Staphylococcus aureus*
MSSA – Methicillin sensitive *Staphylococcus aureus*

susceptible to photodynamic inactivation as AZT-sensitive strains of HIV.[52] The chlorin-type photosensitizer **17** (as a mixture of two isomers) was also highly effective for the inactivation of HIV strains including wild-type and drug-resistant mutant isolates.[79] Both strains were totally inactivated using $30\,\mu g\,mL^{-1}$ of chlorins **17** and $0.75\,J\,cm^{-2}$ of 661 nm light.

5.3.1.5. Influence of the Biological Material

The efficacy of porphyrin derivatives in the presence of biological material has been evaluated. It was found that the microbial photoinactivation by aPDT in rich organic matter media is not as effective as in the absence of organic matter. However, a good response can be obtained in organic matter rich environments by increasing moderately the porphyrin concentration. For instance, Nitzan *et al.*[155] found that the cationic photosensitizer *meso*-tetrakis(1-methylpyridinium-4-yl)porphyrin (**20a**) and deuteroporphyrin (**14a**) in the presence of polymyxin nonapeptide effectively inactivated bacterial cells in low protein content environments. At high protein concentration the photoinactivation efficiency depended on the type of protein present in the medium. Lasocki *et al.*[144] studied the antibacterial photodynamic effect of arginine haematoporphyrin derivative **10b** using *P. aeruginosa* and *S. aureus* either as suspension cultures in nutrient broth or as colonies growing on isolated mouse muscles.

Table 5.8 *In vivo* assays of photoinactivation to treat bacterial wound infections.

Photosensitizer	Microorganism	Location of application	Reference
Specific or non-specific tin(IV) chlorin e6-monoclonal antibody conjugates	*Pseudomonas aeruginosa*	Injection on infected wounds of dorsal skin of mice	Berthiaume *et al.*, 1994[268]
Deuteroporphyrin and hemin	*MRSA*	Topical application and injection into infected burn wounds in guinea pigs	Orenstein *et al.*, 1997[159]
Poly-L-lysine and chlorin(e6)	*Escherichia coli*	Topical application to excisional wounds on the backs of mice	Hamblin *et al.*, 2002[119]
Poly-L-lysine-chlorin(e6) conjugate	*Pseudomonas aeruginosa*	Topical application to excisional single wound on the backs of mice	Hamblin *et al.*, 2003[269]
Chlorin e6 or pL-ce6 conjugate	*Fluorescent Staphylococcus aureus*	Topical application or injection into the infected thigh muscles of mice	Gad *et al.*, 2004[109]
5,10,15-tris(1-methylpyridinium-4-yl)-20-phenylporphyrin	*Staphylococcus aureus*	Topical application on third-degree infected burn wounds in mice	Lambrechts *et al.*, 2005[189]
Chlorin(e6)-polyethyleneimine	*Acinetobacter baumannii*	Injection on infected surface of burns on the dorsal surface of mice	Dai *et al.*, 2009[111]
Chlorin(e6)-polyethyleneimine	*MRSA*	Topical application on infected skin abrasion wounds of mice	Dai *et al.*, 2010[121]
Dicationic porphyrin derivative	*Staphylococcus aureus*	Topical application on a burn wound infection of a mouse	Hurtuk *et al.*, 2010[270]

MRSA – methicillin resistant *Staphylococcus aureus*

The antibacterial effect was 1000 times greater in the case of suspension cultures and the concentration of dye necessary for optimal effects was much lower. The tricationic 5,10,15-tris(1-methylpyridinium-4-yl)-20-phenylporphyrin chloride (**35**) was able to photoinactivate successfully bacteria and fungi in phosphate buffered saline, but in the presence of albumin its efficiency for both microorganisms was reduced.[186]

Some *in vitro* studies showed that virus inactivation in whole blood or in plasma is not as effective as in the absence of blood components, but effective inactivation of virus strains in blood or blood derivatives can also be obtained by moderately increasing the porphyrin concentration. Schagen *et al.*[64] showed that, in approximately 90% blood plasma samples tested, the photosensitizers protoporphyrin IX (**9a**) and uroporphyrin (**15**) only cause a 2–3-fold reduction in viral infectivity *versus* a more than 105-fold reduction in PBS. Grandadam *et al.*[79] reported that the efficiency of chlorins **17** against human immunodeficiency virus (HIV) strains, including wild-type and drug-resistant mutant isolates, is maintained in either plasma or whole blood by just increasing moderately the chlorin concentration.

5.3.1.6. Comparison with Other PS

The efficacy of porphyrin derivatives has been compared with that of other photosensitizers used in aPDT of clinical strains. It has been shown that *meso*-substituted cationic porphyrins are equally or even more effective than other PS, such as toluidine blue O (TBO), methylene blue (MB), crystal violet or thionin, in the inactivation of pathogenic microorganisms. For instance, the poly-L-lysine-chlorin e6 conjugate was more effective than TBO for killing ESBL-producing *E. coli*, their corresponding type culture collection strains *E. coli*, MRSA and their corresponding type culture collection strains *S. aureus*.[118] The pL-ce6 conjugate was more effective than Rose Bengal (RB), TBO at killing representatives of three classes of pathogens: *Escherichia coli* (Gram (–) bacteria), *Staphylococcus aureus* (Gram (+) bacteria) and *Candida albicans* (yeast). The pL-ce6 conjugate was equally effective with and without washing.[110] Protoporphyrin IX (**9a**) and haematoporphyrin derivative were also more effective than crystal violet and thionin to inactivate *Helicobacter pylori*.[127] Protoporphyrin IX and haematoporphyrin derivative reduced *Helicobacter pylori* viability as effectively as MB, TBO and disulfonated aluminium phthalocyanine, but at higher PS concentration and/or light dose.[127] Haematoporphyrin derivative (100 pg mL^{-1}) with 160 J cm^{-2}, protoporphyrin IX (**9a**) (150 pg mL^{-1}) with 320 J cm^{-2}, MB (100 pg mL^{-1}) with 21 J cm^{-2}, TBO (50 pg mL^{-1}) with 160 J cm^{-2} and disulfonated aluminium phthalocyanine (100 pg mL^{-1}) with a laser light energy density of 16 J cm^{-2} all reduced bacterial viability by >99%. The effectiveness of *meso*-tetrakis(1-methylpyridinium-4-yl)porphyrin tetratosylate (**20a**) delivered from poly(vinyl alcohol) hydrogels was compared with that of MB to inactivate MRSA growing planktonically and in biofilm culture.[214] With MB, the type of culture grown had a significant effect on the efficacy of photodynamic inactivation. For example, when the isolate was grown planktonically and irradiated at a light dose of 100 J cm^{-2} after incubation with 10 µg ml^{-1} MB, a mean percentage kill of >99.99% was achieved. In contrast, the mean percentage kill for the isolate grown in biofilm culture for the same drug concentration and light dose was significantly lower, 88.19%. This effect was not observed for porphyrin **20a**, where the mean percentage kill of planktonic and biofilm cultures were similar. For example, the mean percentage kill for a planktonic culture exposed to 10 µg mL^{-1} of **20a** and a light dose of 100 J cm^{-2} was >99.99% and the mean percentage kill of the equivalent biofilm culture was 99.71%. However, the effectiveness of porphyrin **20a** was lower than that of TBO and MB for *Listeria monocytogenes* and *E. coli*. Porphyrin **20a** was compared with toluidine blue O (TBO) for treatment of *Pseudomonas aeruginosa* cystic fibrosis pulmonary infection.[214] Porphyrin **20a** diffused more efficiently across artificial cystic fibrosis mucus than TBO, achieving high rates of inactivation (>99%) of *P. aeruginosa* isolates growing both planktonically and in biofilms. Porphyrin **20a** required, however, significantly higher concentrations (2.5 mg mL^{-1}) than TBO to achieve high rates

Table 5.9 *In vitro* and *in vivo* assays of photoinactivation to treat bacterial oral cavity infections.

Photosensitizer	Microorganism	Reference
***In vitro* cell culture**		
Endogenous porphyrins	*Porphyromonas gingivalis* *Prevotella intermedia* *Prevotella denticola* *Porphyromonas endodontalis* Black pigmented bacteria	Henry *et al.*, 1995[271]
Chlorin e6-pentalysine conjugate	*Porphyromonas gingivalis* *Actinobacillus* *Actinomycetemcomitans* *Bacteroides forsythus* *Campylobacter rectus* *Eikenella corrodens* *Fusobacterium nucleatum* *Actinomyces viscosus* *Streptoccocci*	Rovaldi *et al.*, 2000[105]
Endogenous porphyrins	Porphyromonas gingivalis	Izzo and Walsh, 2004[131]
Chlorin e6	Porphyromonas gingivalis Fusobacterium nucleatum Capnocytophaga gingivalis	Pfitzner *et al.*, 2004[103]
Chlorin e6-polyvinylpyrrolidone	20 Gram (+) and 30 Gram (–) clinical anaerobic strains isolated from periodontal diseases	Drulis-Kawa *et al.*, 2005[98]
Porphyrin-linked nitroimidazole antibiotics	*Porphyromonas gingivalis*	Yap *et al.*, 2009[158]
***In vitro* biofilms**		
Haematoporphyrin	*Streptococcus sanguinis* *Porphyromonas gingivalis* *Fusobacterium nucleatum* *Actinomyces actinomycetemcomitans*	Dobson and Wilson, 1992[272]
Cationic, anionic and neutral chlorin e6 conjugated to polylysine	*Actinomyces viscosus* *Porphyromonas gingivalis*	Soukos *et al.*, 1998[112]
Endogenous porphyrins	*Porphyromonas gingivalis* *Actinomyces odontolyticus* *Prevotella intermedia* *Propionibact. acnes* *Streptococcus mutans*	Konig *et al.*, 2000[273]
Polylysine-chlorin e6 conjugate	*Actinomyces naeslundii*	Soukos *et al.*, 2003[274]
ALA	*Pseudomonas aeruginosa*	Lee *et al.*, 2004[275]
Chlorin e6	*Porphyromonas gingivalis* *Fusobacterium nucleatum*	Sigusch *et al.*, 2005[104]
Sn(IV)chlorin e6	*Streptococcus pyogenes*	Hope and Wilson, 2006[110]
Photofrin	*Streptococcus mutans*	Wood *et al.*, 2006[153]
Polyethyleneimine-chlorin e6 conjugate	*Proteus mirabilis* *Pseudomonas aeruginosa*	Garcez *et al.*, 2007[116]
***In vivo* assays**		
Chlorin e6	Dogs infected with *Porphyromonas gingivalis* and *Fusobacterium nucleatum* in all subgingival areas	Sigusch *et al.*, 2005[104]

Table 5.10 *In vitro* and *in vivo* assays of photoinactivation to treat *Helicobacter pylori* infections.

Photosensitizer	Irradiation medium/Location of application	Reference
In vitro studies		
Haematoporphyrin derivative, protoporphyrin IX	Irradiated in Wilkens–Chalgren broth	Millson *et al.*, 1995[276]
Protoporphyrin IX, haematoporphyrin derivative	Irradiated on blood agar surface and in Wilkins–Chalgren broth	Millson *et al.*, 1996[127]
Coproporphyrin and protoporphyrin	Bacteria covered by a layer of gastric mucin before irradiation	Hamblin *et al.*, 2002[277]
Coproporphyrin and protoporphyrin IX	Irradiated in phosphate-buffered saline	Hamblin *et al.*, 2005[128]
In vivo studies		
Haematoporphyrin derivative	Normal swine gastric epithelium	Wolfsen *et al.*, 1993[278]
Haematoporphyrin derivative, protoporphyrin IX	Explanted ferret gastric mucosa infected	Millson *et al.*, 1996[127]
Haematoporphyrin derivative	Gastric mucosa of infected rats	Millson *et al.*, 1996[127]
Haematoporphyrin derivative	Gastric mucosal of rats	Millson *et al.*, 1997[136]
Endogenous haem	Stomachs of infected male ferrets	Hamblin *et al.*, 2002[277]
ALA	Gastric antrum of 13 infected patients	Wilder-Smith *et al.*, 2002[279]
Protoporphyrin IX and coproporphyrin	Gastric antrum of 10 infected patients	Ganz *et al.*, 2005[280]
Endogenous porphyrins	Whole-stomach of 18 infected adults	Lembo *et al.*, 2009[281]

of inactivation (>99%) of *P. aeruginosa* isolates growing planktonically. Likewise, the cationic pL-ce6 and MB gave similar uptakes and killing of *S. aureus* despite a 50-fold difference in incubation concentration being required.[109]

The benzoporphyrin derivatives **18a,b** were tested against HIV and were more selective than MB against intracellular virus in whole blood, giving complete inactivation at a concentration of 0.5 mg mL^{-1} and a light dose of 13 J cm^{-2}.[222]

5.3.2. Clinical Applications

5.3.2.1. Antibacterial aPDT

It has been shown that ALA, as a precursor of endogenous porphyrins, and an ample range of natural and synthetic porphyrins are effective to inactivate bacteria in laboratory conditions (Tables 5.1 and 5.3). In fact, considerable progress with this type of PS has been made in the bacterial photodynamic inactivation *in vitro*, in animal models and in clinical trials to treat bacterial infections of skin, wounds and oral cavity (Tables 5.7–5.9). *In situ* and *in vivo* experiments also give strong suggestions about the application of porphyrins in aPDT of chronic superficial gastritis, such as gastric ulcer, chronic gastritis and gastric cancer infections caused by *Helicobacter pylori* (Table 5.10) because the

Table 5.11 *In vivo* assays of photoinactivation to treat viral infections.

Photosensitizer	Microorganism	Application	Reference
Photofrin II, Chlorin e6	Human papilloma viruses	Induced papilloma warts in the cutaneous tissue of rabbits	Go *et al.*, 1990[102]
Dihaematoporphyrin ether	Human papilloma viruses	Patients with recurrent laryngeal papillomatosis	Mullooly *et al.*, 1990[282]
Dihaematoporphyrin ether	Human papilloma viruses	Patients with recurrent respiratory papillomatosis	Bauman and Smith, 1996[283]
ALA	Human papilloma viruses	Patients with genital warts (condylomata acuminata)	Frank *et al.*, 1996[284]
ALA	Human papilloma viruses	Patients with genital warts of the labia minora and vestibule	Fehr *et al.*, 1996[285]
ALA	Molluscum contagiosum	Patient suffering from molluscum contagiosum	Smetana *et al.*, 1997[88]
Dihaematoporphyrin ether	Human papilloma viruses	Patients with relapsing papillomatosis of the larynx	Shikowitz *et al.*, 1998[147]
ALA	Human papilloma viruses	Patients with viral hand and foot warts verrucae	Stender *et al.*, 1999[286]
ALA	Human papilloma viruses	Patients with viral hand and foot warts verrucae	Stender *et al.*, 2000[287]
ALA	Human papilloma viruses	Patients with plantar warts	Fabbrocini *et al.*, 2001[288]
ALA	Human papilloma viruses	Patients with genital warts (condylomata acuminata)	Fehr *et al.*, 2002[285]
ALA	Human papilloma viruses	Patients with genital warts (condylomata acuminata)	Stefanaki *et al.*, 2003[289]
ALA	Virus-associated epithelial tumours	Five organ recipients with multiple tumours of the face	Schleier *et al.*, 2004[290]
ALA	Human papilloma viruses	Patients with genital warts	Wang *et al.*, 2004[291]
meso-Tetrakis (3-hydroxyphenyl)-chlorin (*m*-THPC)	Human papilloma viruses	Patients with recurrent respiratory papillomatosis	Shikowitz *et al.*, 2005[292]
ALA	Human papilloma viruses	Patients with genital warts	Herzinger *et al.*, 2006[293]
ALA	Human papilloma viruses	Patients with plantar warts	Schroeter *et al.*, 2005[294]
ALA	Human papilloma viruses	Patients with genital warts (condylomata acuminata)	Chen *et al.*, 2007[295]
ALA	Human papilloma viruses	Patients with genital warts (condylomata acuminata)	Kacerovska *et al.*, 2007[296]
ALA	Human papilloma viruses	Patients with genital warts (condylomata acuminata)	Nucci *et al.*, 2009[297]

gastric mucosa is accessible to topical photosensitizer application and to endoscopic light delivery.

5.3.2.2. Antiviral aPDT

Although a great variety of viruses has been effectively inactivated by porphyrins in *in vitro* conditions (Table 5.4), currently the clinical use of porphyrin photosensitizers in this field is limited to the treatment of laryngeal papillomatosis, using haematoporphyrin derivative (HpD) and dihaematoporphyrin ether (**11**)[282,299] and, in a small scale, for the treatment of viral complications in immunocompromised patients, using ALA[88] (Table 5.11). However, considerable progress has been made in the viral photodynamic disinfection of blood products.

The major threat of viral contamination in blood and blood products comes from the immunodeficiency viruses (HIV) and hepatitis viruses B and C (HBV, HCV). Remarkably, the human immunodeficiency virus has been inactivated *in vitro* by aPDT (Table 5.4). Natural porphyrins such as protoporphyrin,[60,129] hemin,[60–61,68,220] other related natural porphyrins[60,65] and metalloporphyrins[30,55,221,300] have been shown to be effective against HIV in the micromolar range. Modified *meso*-tetraphenylporphyrins, namely sulfonated derivatives, and other *meso*-arylporphyrins have also been shown to have antiviral HIV activity.[30,60,61,63,68,69]

The photoinactivation of hepatitis viruses (HBV, HCV) in blood products has also been tested but the studies using porphyrins are scarce. Cheng *et al.*[70] identified a class of carboxylic *meso*-tetraarylporphyrins (**66** and **68**) as potent and specific inhibitors of the HCV replicons. Lin and Hu[231] showed that hemin and related derivatives dramatically suppressed the RT protein-priming reaction of HBV, suggesting that this type of compounds represent a novel class of agents that can block HBV reverse transcriptase functions.

5.3.2.3. *Antifungal aPDT*

Although phenothiazin dyes have traditionally been used more in the aPDT of fungi,[301–303] porphyrin compounds have been successfully tested in the inactivation of yeasts,[49,82,164,178,183,236,239] dermatophytes,[89,242,246,248,249] conidia-forming fungi[235,238,245] and fungal fish pathogens.[72]

5.3.2.4. *Antiparasital aPDT*

Photodynamic therapy based on ALA and analogues, porphyrins of natural and synthetic type and other porphyrin related compounds have been tested for the inactivation of *Leishmania* sp.[254] and for the treatment of cutaneous leishmaniasis[258,304] (Table 5.6). ALA has also been used for the *in vitro* inactivation of *Plasmodium falciparum* haem-cycle intermediates.[305] It was reported that lipophilic pheophorbide derivatives and red light illumination were able to eradicate red blood cells infected with *Plasmodium falciparum*, and with *Babesia divergens* in whole blood indicating that photosensitization with this type of derivatives may be a suitable approach to inactivation of blood-transmissible parasites.[250] Extensive photodamage of *Colpoda inflata* cysts, previously loaded with *meso*-tetrakis(1-methylpyridinium-4-yl)porphyrin tetratosylate (**20a**) and analogues was observed upon visible light irradiation.[251] The same tetracationic porphyrin **20a**, upon irradiation with white light, could successfully inactivate eggs of the helminths *Ascaris lumbricoides* and *Taenia* sp.[170]

5.4. Porphyrins as Antimicrobial Agents – Environmental Applications

5.4.1. Practical Aspects

The aPDT has been applied not only to the clinical area but also to the environmental field.[8,180–182,211] Although only a few studies have been conducted in this area, preliminary results obtained at both laboratory level and pilot station suggest that the photochemical technique, using porphyrin derivatives as PS, has a great potential for environmental application, namely for use in water disinfection in treatment plants[8–10,12,71,73] and in fish-farming plants.[72,74]

To implement this technology in the environment some other aspects less relevant to aPDT in the clinic area need to be taken into consideration, namely: 1) the removal of the sensitizer after photodynamic action to avoid the release of the PS to the water output; 2) the determination of the stability of the PS conjugates under sunlight irradiation conditions; 3) the assessment of the impact of this procedure on the natural non-phatogenic microbial community structure, for instance when this technology is applied in extensive and semi-intensive fish farming plants; 4) the toxicity of the PS on aquatic organisms (*e.g.* fish and shellfish) at doses which induce marked mortality of microbial pathogens; 5) the effect of physical and chemical properties of environmental waters on aPDT; 6) the possibility of using natural sunlight as light source and 7) the efficacy of the porphyrin derivatives relatively to that of other PS in aPDT.

5.4.1.1. Removal of the Sensitizer After Photodynamic Action

The majority of aPDT experiments in laboratory conditions have been carried out with free photosensitizers, but this is far from appropriate for applications to environmental water, where residual traces of photosensitizer in the water output would certainly not be acceptable. Free photosensitizers might not only introduce residual traces of sensitizer but would also make this technology more expensive. An inherent disadvantage of most of the sensitizers is the high solubility in water, which makes their removal from the inactivation medium extremely difficult.[306]

The immobilization of efficient photosensitizers in insoluble supports can be an interesting approach to inactivate pathogenic microorganisms present in environmental waters. In fact, some groups developed PS immobilized on solid supports and tested their efficacy in the photoinactivation (PI) of faecal bacteria[71,180] and viruses[307,308] (see discussion below). The results are similar to those obtained with the free form.[307]

5.4.1.2. Stability of the Porphyrin Conjugates under Sunlight Irradiation

Typically, *meso*-substituted porphyrins do not undergo biodegradation since steric factors prevent the interaction with the active site of enzymes involved in porphyrin catabolism. Jemli *et al.*[8] found that cationic porphyrin **20a** is more photostable than photosensitizers RB and MB in wastewater after irradiation with sunlight. Alves *et al.*[12] and Costa *et al.*[10] also observed that tri- and tetracationic porphyrins are photostable under solar irradiation.

5.4.1.3. *Impact of aPDT on Natural Non-pathogenic Microbial Community*

To implement aPDT to disinfect drinking water, wastewater and fish farming waters of intensive aquaculture plants is potentially easier as all the microorganisms may be destroyed. However, its application to extensive and semi-intensive fish farming plants is not so straightforward because some non-pathogenic microorganisms in these systems have important ecological roles. In extensive and semi-intensive aquaculture systems, non-pathogenic microorganisms, namely bacteria, have a central role in the functioning and productivity of these ecosystems. As aPDT does not present selectivity to microorganisms, inactivating potentially all microorganisms, a careful environmental evaluation must be done before the implementation of a procedure to photoinactivate microorganisms in aquaculture systems.

5.4.1.4. *Toxicity of aPDT to Aquatic Organisms*

The available information about the effect of porphyrin derivatives on aquatic organisms is scarce. However, the toxicity of porphyrin derivatives to eucatiotic cells becomes important only at millimolar concentrations whereas microbial inactivation is effective at micromolar concentration.[8,10,12,50,163,170] Magaraggia *et al.*[72] showed that *meso*-tetrakis(1-methylpyridinium-4-yl)porphyrin **20a** does not display significant toxicity toward adult trouts at photochemically active doses or at doses which induce a marked mortality of microbial pathogens.

5.4.1.5. *Effect of Physical and Chemical Properties of Water*

Some studies have shown that microbial inactivation by aPDT in environmental waters is not as effective as in laboratory conditions in which microorganisms are suspended in buffer solutions. However, an effective inactivation of microbial strains in environmental waters can be obtained by increasing moderately the porphyrin concentration. For instance, Alves[309] verified that changes in temperature, salinity, pH and dissolved oxygen do not affect the inactivation of *Vibrio fischery* by 5,10,15-tris(1-methylpyridinium-4-yl)-20-(pentafluorophenyl)porphyrin triiodide (**36**) in laboratory conditions. The same authors tested the efficacy of the same PS in the treatment of water from aquaculture tanks and concluded that bacterial inactivation is more effective in PBS than in natural water. However, the removal of particulate suspended matter from the aquaculture water increases the success of aPDT.[309] An effective inactivation of bacteria in aquaculture water was obtained by doubling-up the porphyrin concentration.[309] Alouini *et al.*[170] found that the photosensitization of helminth eggs in wastewater by *meso*-tetrakis(1-methylpyridinium -4-yl)porphyrin tetratosylate (**20a**) was influenced by dissolved oxygen concentration, water quality and the type of eggs, but concluded that suspended solids (turbidity) were the most influential solution parameter on the efficiency of the photochemical process. Arrojado[74] showed that for a similar concentration of aquaculture cultivable bacteria, aPDT applied in the same conditions varied during a sampling year period, reaching total inactivation only in 50% of the cases. The different effectiveness of aPDT could be due to differences in bacterial community structure but can be also attributed to differences in suspended matter quantity and quality.[74] These authors showed, however, that increasing moderately the porphyrin concentration the effectiveness of aPDT increases greatly, reaching total inactivation of the cultivable bacteria.

5.4.1.6. The Use of Sunlight as Light Source

The use of sunlight as light source in aPDT should be considered in order to establish an environmentally efficient antimicrobial protocol. Since sunlight penetrates deeply into the water column, a nearly uniform illumination of large volumes is attained; this approach becomes inexpensive since it is based on the use of a low-cost visible light source. Moreover, the Soret absorption band of porphyrins in the 420–430 nm spectral region allows a very efficient interaction with blue light wavelengths, which are endowed with maximum penetration power into natural waters out of the wavelengths which are present in the solar spectrum[310] and which are the most effective of the visible range in killing microorganisms by cationic porphyrins.[165]

5.4.1.7. Efficacy of Porphyrin Derivatives vs. Other PS

The efficacy of porphyrin derivatives in environmental aPDT has been compared with that of other sensitizers. It has been shown that *meso*-substituted cationic porphyrins are more effective in the photoinactivation of environmental strains than other PS such as RB or MB. For instance, Jemli *et al.*[8] compared the photoinactivation of faecal coliforms by the tetracationic porphyrin **20a**, RB and MB. The results showed that porphyrin **20a** is more effective against faecal bacteria in wastewater than MB.

5.4.2. Water Disinfection

5.4.2.1. Antibacterial Photoinactivation

Jemli *et al.*[8] applied the aPDT methodology to disinfect wastewater, in a small scale, using cationic porphyrin **20a** and sunlight. After 60 min. of irradiation, 1.5–2 log survival in faecal coliforms counts were observed respectively with 5 and 10 mM of **20a**. After 240 min., only 0.47 log survived with both concentrations. In this study the photoinactivation of faecal coliforms with porphyrin **20a** and with RB and MB was also compared. It was concluded that porphyrin **20a** is more efficient than MB and RB in the inactivation of bacteria in wastewater.

Wastewater samples from a secondary-treated sewage plant were also used to study the phototoxicity of neutral and tri- and tetracationic porphyrins.[73] Faecal coliforms and faecal enterococci were quantified in the wastewater samples as indicators of faecal pollution. Two of the cationic porphyrins (**36** and **38a**) revealed to be highly active, even against Gram (–) faecal coliforms, inactivating *ca.* 94–99.8% of the bacteria with white light (9 mW cm^{-2}).

According to Alves *et al.*,[12,311] tri- and tetracationic porphyrins, when irradiated by the appropriate light, can efficiently inactivate Gram (+) and Gram (–) faecal bacteria, whether the PS are free or immobilized on solid matrixes. The authors tested seven synthetic cationic *meso*-substituted porphyrins with two to four charges (**20a**, **36, 38a**, **38b**,**42a**, **42b** and **46**) on the photoinactivation of a faecal Gram (+) bacterium (*Enterococcus faecalis*) and a faecal Gram (–) bacterium (*Escherichia coli*) concluding that the most efficient inactivation of both bacteria occurred in the presence of tri- (**36** and **38b**) and tetracationic photosensitizers (**20a**). The complete inactivation of the bacterial strains at low light fluence (40 W m^{-2}) suggests that photodynamic antimicrobial therapy can be a promising possibility for wastewater disinfection process under natural light conditions. The same authors[9] reported a faster and simpler method to monitor the photoinactivation process of *Escherichia coli*, using a recombinant bioluminescent *Escherichia coli* strain as indicator. The photoinactivation of bioluminescent *E. coli* was effective ($>$4 log bioluminescence decrease) with the three porphyrins used (**20a**, **36** and **38b**), but the tricationic porphyrin **36** was the most efficient compound.

Based on the most effective free porphyrin (**36**) tested by Alves *et al.*,[12] the same group built new inorganic-organic hybrids in order to allow the recovery and recycling of the photosensitizing agent.[307,308,311] Porphyrin **36** was immobilized by covalent coupling on nanomagnetic support materials covered with sili*ca*. Three different magnetic silica nanosphere-porphyrin hybrids were prepared and tested, one of them being quite effective for both Gram types (5 log decrease on cell survival on *E. faecalis* after 90 minutes and on *E. coli* after 180 minutes of irradiation with 20 μM porphyrin).[308]

Few studies have also shown that PS immobilized on solid supports can be applied to the disinfection of drinking waters. Regenerated cellulose impregnated with 5,10,15,20-tetrakis(1-methylpyridinium-4-yl)porphyrin tetratosylate (**20a**) showed photobactericidal activity against *S. aureus*, *E. coli*, *Proteus vulgaris* and *Bacillus subtilis.*[71]

Krouit *et al.*[154,190] showed efficient photoinactivation of Gram (+) and Gram (–) bacterial strains by cellulose films with immobilized porphyrin derivatives. A tricationic porphyrin immobilized on poly(methyl methacrylate) polymer was shown also to have an effective inhibitory effect in the photoinactivation of *Deinococcus radiodurans.*[312]

Arrojado[74] showed that ten bacterial species (*V. anguilarum, V. parahaemolyticus, Photobacterium damselae* subsp. *damselae, Photobacterium damselae* subsp. *piscicida, A. salmonicida, E. coli, Enterobacter* sp., *S. aureus, E. faecalis, Pseudomonas* sp.) isolated from a fish-farming plant waters were effectively inactivated (up to 7 logs) *in vitro* with tricationic porphyrin **36** at micromolar PS doses, with a low light intensity of 4 mW cm^{-2}, suggesting that aPDT can be used to photoinactivate fish bacterial pathogens in fish farm waters even during dark days of winter time.[74]

The cultivable fraction of the heterotrophic bacteria of the same aquaculture plant, including pathogenic and non-pathogenic bacteria, was also inactivated by the photodynamic process, but the efficacy of the inactivation varied during the sampling period.[74] The seasonal variation of photoinactivation observed for the cultivable bacteria can be due to differences in bacterial community structure but also due to variations of the physico-chemical water properties. The results of laboratory experiments in which the effect of temperature, salinity, pH, dissolved oxygen and suspended matter variation was tested showed that photoinactivation was not affected by the first four parameters but suspended matter influenced significantly the aPDT.[309]

5.4.2.2. Antiviral Photoinactivation

Costa *et al.*[10] compared the efficiency of a series of cationic porphyrins **20a**, **36**, **38a**, **38b**, **42a**, **42b** with charges varying between two and four in the photoinactivation of environmental bacteriophages isolated in sewage waters under white light (40 W m^{-2}) at three different photosensitizer concentrations. The studies revealed that the tetra- and tricationic porphyrins **20a**, **36** and **38b** inactivated the T4-like sewage phage to the limits of detection ($>$99.9999% of inactivation, reduction of $\approx$7.2 log). The authors concluded that the complete inactivation of viruses with low light intensity means that this methodology can be used even on cloudy days and during winter, opening the possibility to develop new technologies for wastewater treatment.

The new inorganic-organic hybrids based on the tricationic porphyrin derivative **36a**, and referred to in the previous point, were also tested in the photoinactivation of sewage bacteriophages.[307,313] Preliminary results showed that these hybrid materials present antiviral activity and can be recovered and recycled.[313] One of the tested inorganic-organic hybrids was able to inactivate T4-like bacteriophage to the limits of detection (up to 7 logs of inactivation).[308]

5.4.2.3. Antifungal Photoinactivation

Magaraggia *et al.*[72] showed that a micromolar concentration of an analogue of porphyrin **20a** where one *N*-methyl group has been replaced by an *N*-tetradecyl chain (C14) (**23b**, n = 13) can promote the cure of saprolegniosis in trout-farming pools containing naturally or artificially *Saprolegnia* infected fish (inactivation of 6–7 logs) without perilesional damage of the fish. The onset of the infection in healthy fish was reduced by about 50%. Recurrence of the saprolegniosis in the *Saprolegnia* infected sites or in other sites of the fish was not observed. In the trout set with spontaneous infection by *Saprolegnia* a complete remission of the infection was induced within one week.

5.5. Factors Determining Inactivation Efficiency by Porphyrins

5.5.1. Molecular Structure

It is well recognized that the efficacy of the aPDT approach is largely dependent on the photosensitizer structure. For instance, the PS molecular charge has an important role in the PS antimicrobial activity and positively charged photosensitizers are generally more efficient and can act at lower concentrations than neutral and anionic photosensitizer molecules.[20,117] Porphyrins can be transformed into cationic derivatives through the insertion of positively charged substituents in the peripheral positions which may largely affect the kinetics and extent of binding with microorganisms.[2] The amphiphilic nature of the PS is another important feature that affects the efficiency of the photoinactivation process and can be modulated by the introduction of adequate functionalities in the macrocycle periphery (*e.g.* different number of positive charges, asymmetric charge distribution, cationic charges combined with different patterns of substituents). Although in fungi the adsorption of external substances is positively affected by hydrophilicity and by the presence of charged groups, the response of these organisms to photodynamic processes seems to be less related by structural factors.[302]

5.5.1.1. Meso-substituent Groups

a) Bacteria
The size of the *meso*-substituent groups affects the bacterial photodynamic effect. Reddi *et al.*[176] found a direct correlation between the length of the linear hydrocarbon chain of the tetracationic porphyrins **23b** and the extent of porphyrin accumulation in both Gram (+) and Gram (–) bacteria. Steenkeste *et al.*[185] showed that the bis intercalating compound mono (acridyl)bis(arginyl)porphyrin **40b**, prepared from a bis-arginyl-porphyrin **40a** in which the arginyl arm is connected through a flexible chain to an acridine ring, lead to the excitation of twice as many porphyrin chromophores in **40b** by comparison to **40a**, making the photoinactivation of *Pseudomonas aeruginosa* by **40b** more efficient.

b) Viruses
Casteel *et al.*[172] have observed differences in the photoinactivation rate of hepatitis A virus and MS2 phage when they used tetracationic porphyrins with different alkyl substituent groups. Different photoinactivation results were obtained by Costa *et al.*[10] for tricationic porphyrins differing in a *meso*-aryl group.

c) Fungi
Although the photoinactivation of fungal cells is mostly achieved by porphyrin molecules which are present in the incubation medium rather than attached to the cell,[40] the conjugation of the porphyrin molecule with glycosyl moieties enhances their penetration and access to intracellular targets.[1]

5.5.1.2. Charge Number

a) Bacteria

Some studies have shown that cationic porphyrin derivatives can efficiently photoinactivate both Gram (+) and Gram (–) bacteria types, even without the presence of membrane disrupting agents (*e.g.* $CaCl_2$, EDTA or polymixin B nonapeptide). The presence of these disrupting agents are a fundamental item when neutral and anionic photosensitizers are used to photoinactivate Gram (–) bacteria. In fact, different positively charged porphyrin derivatives have been successfully tested as photoinactivating agents against Gram (+) and Gram (–) bacteria.[4,6,15] However, Gram (–) bacteria are usually resistant to the action of negatively charged or neutral porphyrins which efficiently photoinactivated Gram (+) bacteria.[32,157,173,314] This difference is explained by Gram (+) and Gram (–) distinct three-dimensional architecture.[2] The enhanced susceptibility of Gram (+) cells to PI is due to their cell wall constituted by a peptidoglycan layer, intimately associated with lipoteichoic and negatively charged teichuronic acids, that displays a relatively high degree of porosity. Macromolecules with a molecular weight of 30 000–60 000 Da (glycopeptides, polysaccharides) can readily diffuse into the inner plasma membrane, which does not act as a permeability barrier for the most commonly used PS as porphyrins, whose molecular weight does not generally exceed 1500–1800 Da.[2] In contrast, Gram (–) bacteria contain an additional highly organized outer membrane in the cell wall, located outside the peptidoglycan layer, which shows an asymmetric lipid structure composed by strongly negatively charged lipopolysaccharides, polysaccharides, lipoproteins and proteins with transport function.[50] Such a highly organized system inhibits the penetration of several compounds, even hydrophilic 600–700 Da molecules that can diffuse through the porin channels.[315]

The outer membrane acts as a very effective permeability barrier, making the Gram (–) bacteria resistant against negatively charged or neutral porphyrins, intercepting the photogenerated reactive species.[6,165] The positive charges on the photosensitizer molecule appear to promote a tight electrostatic interaction with negatively charged sites at the outer membrane of Gram (–) bacteria. In a study by Soukos *et al.*,[112] the pronounced photodynamic effect of the cationic conjugate on *Actinomyces viscosus* may be due to the electrostatic attraction between the porphyrin derivatives and the negatively charged membrane of the bacterium. It has been found that the efficacy of a photosensitizer in sensitizing Gram (–) bacteria is related to the charge on the photosensitizer itself.[119] The presence of one or more positively charged groups play an essential role in orientating the photosensitizer toward sites which are critical for the stability of cell organization and/or the cell functions.[6,14,105]

Some studies compared the efficiency of synthetic *meso*-substituted cationic porphyrins with different charge distribution (tetra-, tri-, di- or monocationic) on bacterial inactivation.[6,10,12,14,180–182] However, the results are not identical. Tetracationic porphyrins are efficient PS against both Gram (+) and Gram (–) bacteria on visible light but some di- and tricationic porphyrins were more efficient than tetracationic ones, against both a Gram (+) strain and two Gram (–) strains.[14] Lazzeri *et al.*[182] showed that the tricationic porphyrin **29** having a trifluoromethyl group and its metal complex with Pd(II) were more powerful photosensitizing agents against *Escherichia coli* than both the dicationic porphyrin **28** and also a non-cationic sensitizer. The authors justified the better performance of the tricationic porphyrins due to the presence of the high lipophilic trifluoromethyl group that increases their amphiphilic character. Two dicationic porphyrin-based compounds were more effective than a tricationic porphyrin in killing a range of Gram (+) staphylococcal strains, two methicillin-resistant *Staphylococcus aureus* strains, one methicillin-sensitive *S. aureus* strain, one methicillin-resistant *Staphylococcus epidermidis* strain and an *Escherichia coli* strain.[54] Banfi *et al.*[166] compared the photocytotoxicity of the synthetic porphyrins **23a**, **24a**,**b**,**c** and **70** against *Escherichia coli*, *Pseudomonas aeruginosa* and *Staphylococcus aureus* with that of the commercially available PS **20a** and **26**. The results showed that the dicationic porphyrin **23a** was more efficient than the tetracationic derivatives and the better performance of **24a**,**b**,**c** than **20a** and **26** suggests that the moderate degree of lipophilicity achieved by the introduction of aromatic carbon side chains may improve PS efficiency. Recently, Alves *et al.*[12] showed that the photoinactivation of both Gram (+) and Gram (–) bacteria with seven *meso*-substituted cationic porphyrins [(**20a**), three tricationic (**36**, **38a** and **38b**), two dicationic (**42a**,**b**) and one monocationic (**46**)] varied with porphyrin charge but also with the lipophilic character of the *meso*-substituent groups (pentafluorophenyl, carboxyl and methoxycarbonyl) and charge distribution (adjacent and opposite).

b) Viruses

Positively charged photosensitizers are also generally more efficient than neutral and anionic photosensitizer molecules to inactivate viruses. However, viral photoinactivation is also dependent on the target microorganisms.

It is more likely that positively charged photosensitizers cause nucleic acid damage (oxidation of guanosine residues), whereas anionic photosensitizers act against the viral envelope.[240]

The differences in the photoinactivation rate of a T4-like phage using the cationic porphyrins **20a**, **36**, **38a**, **38b**, **42a** and **42b** showed that the number of the positive charges is a key factor in its PI.[10] In fact the tetracationic and the tricationic PS **20a**, **36** and **38b** were much more efficient than the dicationic ones **42** inactivating the sewage phage below the limit of detection. The ionization of the acid groups in **38a** and **42** can be responsible for the modifications of several physical properties that can justify the lower inactivation rate observed for these PS.

Phage inactivation by neutral porphyrins has been also observed but required higher irradiation periods than cationic porphyrins.[172]

c) Fungi

The effect of the charge of the PS on antifungal aPDT is somewhat different from what happens in antibacterial PDT. Filamentous fungi and yeasts are externally covered by a thick wall of glucan, mannan, chitin and lipoprotein and internally have a relatively loose membrane making them sensitive even to anionic photosensitizers, such as haematoporphyrin.[1]

5.5.1.3. Amphiphilicity

a) Bacteria

The nature and distribution of the functional groups in the molecule affect crucially its amphiphilic character and consequently the PS efficacy.[316] As in the transport system across cell membranes, the membrane lipid bilayer allows the passive transport of hydrophobic molecules: by making the PS hydrophobic, the PS diffusion is facilitated and, in principle, the efficiency of the photodynamic effect should be improved. However, the PS must be in solution, so they must have a hydrophilic character. Therefore, the ideal PS must exhibit both hydrophilic and hydrophobic properties. Actually, the combination of hydrophobic and hydrophilic substituents in the sensitizer structure results in an intramolecular polarity axis, which facilitates membrane penetration, and results in better accumulation of cationic porphyrins in sub-cellular compartments. It enhances the effective photosensitization.[317]

Accordingly to Lazzeri *et al.*,[182] an increase in the amphiphilic character of the photosensitizer **29**, given by the presence of a trifluorophenyl group, appears to enhance its affinity for *E. coli* cells. Sol *et al.*[203] showed that the photocytotoxicity of nitroglycosylated *meso*-arylporphyrins (**65**) on *Escherichia coli* and *Staphylococcus aureus* is markedly dependent on the nature, the number and the linking position of the glycosyl moieties and nitro groups that control the amphiphilic characters of the PS.

A moderate increase just in the hydrophobicity of the photosensitizer molecule enhances on its own its affinity for bacterial cells, hence promoting a more efficient photoinactivation.[166,176] The partial hydrophobicity in porphyrins can be achieved either introducing aromatic hydrocarbon side chains on the pyridyl moieties or modulating the number of positive charges on the photosensitizer.[166]

Alves *et al.*[12] in their study with a series of cationic porphyrins inferred that with those selected PS no direct correlation between the PI pattern and the partition coefficient of the PS could be established and probably the combined effect of the cationic charge and the amphiphilic character is responsible for the photodynamic efficiency.

b) Fungi

Meso-arylglycosylporphyrins **58–64** tested for the photoinactivation of *Saccharomyces cerevisiae* showed that there is a relation between the structure of the molecules and the capacity to permeate the cell wall and membrane thus accessing internal targets and the most active compounds were the ones with the highest amphiphilic characters.[49]

Hydrophilic cationic molecules have proved to be more effective than hydrophobic anionic photosensitizers possibly due to their strong attraction to the negatively charged parasitic surface.[318]

5.5.1.4. Symmetry

a) Bacteria

The distribution of the charges on the sensitizer affects the efficiency of the photodynamic inactivation process. Porphyrins with opposite charged groups are more symmetric than porphyrins with adjacent charged groups. The adjacent positive charges in the porphyrin macrocycle should result in a molecular distortion, due to electrostatic repulsion.

The asymmetric charge distribution at the peripheral position of the porphyrin produces an increase in the amphiphilic character of the structure, which can help a better accumulation in cells.[182] However, in a study by Maisch *et al.*,[50] the presence of a symmetric orientation of positive charges could bring about an additional effect.

b) Viruses
The symmetry of the PS affects the photoinactivation of viral particles. In a study by Gábor *et al.*,[193] the symmetrically substituted glycoconjugated porphyrin **56** was found to be twice as effective as the asymmetrical derivative **55** on the photoinactivation process of T7 phages.

5.5.2. *Effect of Concentration*

It has been shown that increasing the concentration of porphyrinc sensitizers at a fixed light dose leads to increased microbial inactivation. However, the sensitizer saturation can be reached during the photoinactivation process. It has been frequently shown that over a specific concentration, further increases in porphyrin concentration do not lead to higher inactivation.[31,163]

a) Bacteria
Jemli *et al.*[8] showed that an increase in the cationic porphyrin **20a** concentration led to an increase in the photosensitizing activity of faecal coliforms with sunlight in wastewater. A low concentration of 1 mM of a cationic porphyrin **20a** resulted in very little cell death. Higher concentrations of 5 mM or 10 mM enhanced cell death. Nevertheless there was a small difference between 5 and 10 mM concentrations during the fourth hour of treatment. The author concluded that a low concentration of sensitizer can be compensated by increasing the duration of irradiation. The results of Maisch *et al.*[54] also demonstrated concentration-dependent differences in the efficacies in the killing of methicillin-resistant *Staphylococcus aureus* strains, methicillin-sensitive *S. aureus*, methicillin-resistant *Staphylococcus epidermidis* strain and *Escherichia coli* with porphyrin-based photosensitizers **44** and **45**. The effectiveness of ALA methyl and butyl esters was also dependent on drug concentration to inactivate bacteria.[86] According to Alves *et al.*,[12] the rate and extent of the PI with seven *meso*-substituted cationic porphyrins varies with porphyrin concentration. The three tricationic porphyrins **36**, **38a**, **38b** and the tetracationic one **20a** were more efficient against *E. faecalis* and *E. coli* with 5.0 μM after 270 minutes of irradiation. With lower concentrations, 0.5 μM and 1.0 μM, a lower reduction after 270 minutes of irradiation was observed, with the worst inactivation at 0.5 μM. However, the photoinactivation pattern for both dicationic porphyrins **42a** and **42b** was not considered significantly different on *E. faecalis* for 1.0 μM and 5.0 μM. The inactivation of *Bacillus subtilis* spores by tri- (**36** and **38a**) and tetracationic (**20a**) molecules was more effective at high

concentrations but above 10 μM further increases in porphyrin concentration do not lead to higher inactivation.[163]

b) Viruses
Kasturi and Platz[171] verified that increasing the concentration of porphyrin **20a** at a fixed light dose leads to increased viral inactivation. The porphyrin inactivated the T7 phage in a concentration-dependent manner. The same conclusion was obtained by Egyeki *et al.*[31] using 5,10,15-(4-β-D-galactosylphenyl)-20-(pentafluorophenyl)porphyrin (**57**), but over 2 μM concentration the process was saturated. Further increase in porphyrin concentration did not lead to a higher inactivation rate of T7. Aggregation of porphyrin derivatives in polar solvent and/or photobleaching of photosensitizer can be considered as possible reasons for such behaviour. The photocytotoxicity action on a T4-like sewage bacteriophage by the six cationic porphyrins (**20a**, **36**, **38a**, **38b**, **42a** and **42b**) was assessed by exposing the viruses with the PS at different concentrations (0.5, 1.0 and 5.0 μM) to white light during different times (30, 60, 90, 180 and 270 min).[10] The results showed that the viral inactivation was higher with 5.0 μM, followed by 1.0 and 0.5 μM for all the porphyrins.

c) Fungi
Inactivation of yeast cells or fungal conidia usually occurs at similar or slightly higher concentrations of porphyrin PS, relatively to those effective against bacteria and viruses. Concentrations between 0.01 μM and 10 μM of various porphyrin derivatives used as exogenous PS reduced by several orders of magnitude the viability of *Candida albicans* and *Saccharomyces cerevisiae* isolates.[49,164,178,183,186,236] *Saprolegnia* sp. zoospores were completely eradicated (up to 6–7 log) after short incubation under visible light in the presence of 0.5–10 μM of a porphyrin.[72]

Inactivation of conidia-producing fungi is highly dependent on PS accumulation by the fungus.[238] Haematoporphyrin in concentrations between 2.5 and 71 μM completely inhibited the germination of conidia of *Fusarium avenaceum*, *Aspergillus flavus* and *Trichothecium roseum*.[238] Higher concentrations (25–710 μM) of the same PS were also used for the photosensitation of *Alternaria alternata*, *Acremonium strictum* and *Rhizopus oryzae*.[240] The dermatophyte *Tricophyton* sp. can be effectively inactivated with concentrations as high as 160 μM of the tricationic porphyrin **35**.[242,246]

d) Parasites
Micromolar concentrations of porphyrin derivatives were demonstrated to be efficient in the photosensitization of protozoa and helminth eggs under white light irradiation.[170] ALA, ALA methyl ester and Protoporphyrin IX (**9a**) in low concentrations and upon irradiation with red light have been considered as promising PS for the photodynamic treatment of cutaneous leishmaniais.[253,257,319]

5.5.3. Light Conditions

The light parameters such as light source, fluence rate and total light dose play an important role in the effectiveness of the aPDT and should be considered to establish an optimal antimicrobial protocol.

5.5.3.1. Light Sources

A very wide variety of coherent and non-coherent light sources has been tested to inactivate microorganisms, ranging from laser technology to basic tungsten-filament lamps. However, aPDT uses low-power light rather than the lasers used in ablative therapy. Microbial killing is attained with milliwatts rather than tens (or hundreds) of watts.[4]

The wavelength of light necessary for the induction of photoinactivation of microorganisms depends on the structure and electron absorption spectrum of the PS.[320] Common porphyrin derivatives have a strong absorption band between 400 and 430 nm (Soret band) and weaker absorption bands (Q-bands) between 500 and 700 nm.[7,113,321] The emission spectrum of the light source must cover all the PS absorption spectrum or at least some of these absorption bands. The experimental results point out that, for the same PS and microorganism, the photodynamic efficiency depends on the light source used.

Some non-coherent light sources such as metal halide lamp (600–800 nm),[322] short-arc xenon lamp (300–1200 nm) and fluorescent lamp (400–450 nm) (Nitzan *et al.*[85]) were purposely designed for photosensitization in the clinic area. However, the main light sources used in aPDT in the clinic are lasers, especially the red and near-infrared lasers,[127,147] because they produce monochromatic light of a known wavelength, the light dosimetry is easy to calculate and the light can be passed down an optical fibre for localized treatment.[199] Light-emitting diodes (350 nm to 1100 nm), which emit between lasers and conventional lamps and solar light, have also been used as light sources for activation of photosensitizers.[8–9,44,71,323]

In *in vitro* studies of aPDT and in environmental applications of aPDT artificial visible light sources have been the most applied (Tables 5.1–5.6). Relatively to white light, the efficacy of microbial therapy apparently decreases with the increase in illumination wavelength.[156,165] The blue light is more efficient for microbial inactivation than green and red lights. The blue-light of shorter wavelength ($<$400–450 nm) is more phototoxic in comparison to higher wavelength blue lights.[165] Although not as efficient as blue light, red-light irradiation is the most preferable for treatment because it penetrates better than blue light.

Sunlight (100–10^6 nm) has also been used as a light source for activation of photosensitizers to inactivate microorganisms in the environment.[8–9,44,71] The bactericidal effect of sunlight has been known for many years.[324] Sunlight is able to inactivate microorganisms due to the synergistic effect of the ultraviolet and infrared parts of sunlight. The disinfecting effect of solar light is enhanced by addition of a sensitizer. The use of solar energy as an alternative to the other illumination systems could be economically viable for regions with a high degree of sunlight radiation and could be used for the treatment of contaminated wastewaters as well as for the disinfection of drinking water and of aquaculture water.[8–9,44,71,74]

a) Bacteria

A great variety of both coherent and non-coherent artificial lights, with light fluence varying between mW m^{-2} and kW m^{-2}, as well as natural sunlight, has been used for photodynamic inactivation of bacteria by porphyrins.[8,9] As the experimental light conditions described are very different, the results can not be directly compared, but in general the inactivation of bacteria is dependent on light dose delivered, irradiation time, fluence-rate and wavelength of irradiation source.

The efficacies of bacterial photoinactivation by different light sources correlate well with the light source wavelengths, decreasing with the increase in illumination wavelength.[159,165]

Nitzan and Ashkenazi[156] have revealed the effects of illumination by different lights at different wavelengths (blue light, green light and red light) in *D. radiodurans*, using two different porphyrin derivatives, the cationic **20a** and the natural derivative **14a**. The strongest photodynamic effects were obtained for both photosensitizers when *D. radiodurans* cultures were treated with blue light (400–450 nm). Using **20a** as photosensitizer a light dose of only 5 J cm^{-2} was required to obtain reductions of 7 log in the viability of the bacteria. When the treated culture was illuminated with green (480–550 nm) or red (600–700 nm) light, 35 and 85 J cm^{-2}, respectively, were required to obtain similar reduction. With **14a** as photosensitizer the results were similar, although the light dose required was higher.[165] Blue, green and red light were also used to induce photoinactivation of multiple antibiotic-resistant bacteria *Acinetobacter baumannii* and *Escherichia coli* by the cationic photosensitizer **20a** and similar results were obtained.[165] Total eradication of both bacteria was obtained with blue light.[165]

The eradication of *Helicobacter pylori* was studied by Hamblin *et al.*[128] using natural porphyrin derivatives (coproporphyrin and protoporphyrin IX) as photosensitizers and a white light (400 and 700 nm) and a set of six 50-nm band pass filters centred on wavelengths of 400, 450, 500, 550, 600 and 650 nm to isolate specific colours of light. They concluded, once more, that the range between 375 and 425 nm was clearly the most effective combination followed by the range of 425 to 475 nm (blue), with the other four ranges producing only a slight killing effect.

The photoinactivation of *E. coli* with PAR solar radiation (620 W m^{-2}) was also achieved in a more efficient and faster way than with low light fluence rate (40 W m^{-2}) using the cationic porphyrins **20a**, **36** and **38b**.[9] In addition, Gábor *et al.*[193] showed that higher fluence rate (0.25 relative to 0.08 W cm^{-2}) leads to a higher rate of photoinactivation for *Escherichia coli* after the same irradiation time.

b) Viruses

The rate and the extent of viral inactivation by porphyrins are dependent on light source, namely when low fluence rates are used (40 W m^{-2}).[44] Costa *et al.*[44] showed that the efficacy of sewage T4-like bacteriophage inactivation using the same fluence rate is dependent on the light source used (halogen lamp at 600 W m^{-2}, solar irradiation and fluorescent lamps at 40 W m^{-2}) to irradiate the porphyrin **36**, in particular when the light is delivered at low fluence rate (40 W m^{-2}). All the light sources used showed to be effective in the excitation of the selected porphyrinic PS but are dependent on their spectral emission distribution.

5.5.3.2. Fluence Rate and Light Dose

Besides the light source, light dose used during irradiation also influences microbial photoinactivation in a fluence-time dependent manner. The law of reciprocity states that a certain biological effect is directly proportional to the total energy dose in spite of the administered regime. Dose is considered the product of intensity and the duration of exposure and therefore the time required to deliver a certain light dose is influenced by the fluence rate of the source. The survey of the available literature on dose-time relationships in photobiology strengthens the view that, for photochemical reactions, it can be assumed that only the principal of reciprocity is valid, at least within a certain dose range, which has to be individually defined for each reaction.[325] Moreover, responses of cells and tissues to electromagnetic radiation usually involve a sequence of interacting biological reactions making a linear dose-time relationship less likely.[325] Additionally, photosensitizing molecules might induce different cellular and molecular responses than do radiation alone.[325] Consequently, for aPDT applied to human or animal tissues, the knowledge of the validity and the limits of this law are crucial for the understanding of the pathophysiology of photodamage.

To date only few data on the role of the principal of reciprocity in aPDT are available. Data in aPDT studies are often showed using light dose to compare different results on microorganism photoinactivation.[113,326,327] The results of these studies get ahead of the idea that fluence is the most important parameter in aPDT and, consequently, the influence of fluence rate and exposure time is not frequently measured.

The few studies on the influence of light parameters currently available in the literature showed that a high light fluence over a short time period may give different results, in terms of microbial kill, from those of a low fluence rate over a longer time even though the light dose is the same in each case.[4] In general, PI of microorganisms is more extensive at a higher light fluence and longer treatment duration.[170]

a) Bacteria

Light parameters influence bacterial photoinactivation in a rate fluence-time dependent manner. Gábor *et al.*[193] showed a higher rate of *Escherichia coli* and *Enterococcus hirae* inactivation when a similar total light dose of white light (fluence 0.08–0.25 W cm^{-2}) is received over a longer time period. Jemli *et al.*[8] verified also that increasing duration of solar irradiation (1235 mW cm^{-2}) improved the inactivation of faecal coliforms and compensated for a low concentration of sensitizer, a less efficient sensitizer type or a high concentration of microorganisms.[8]

b) Viruses

When the same light dose (using the same light source) was obtained under different fluence rates (150, 300, 600 and 1200 W m^{-2}) the results show that the efficacy of sewage T4-like bacteriophage inactivation was dependent on the fluence rate used to irradiate the cells. At the highest light dose tested (216 J cm^{-2}) a higher rate of photoinactivation was obtained when the lower fluence rates 150,

300 and 600 W m^{-2} were used. This means that a higher rate of T4-like phages inactivation is reached when a similar total light dose is received over a longer time period.

5.6. Conclusion

The possibility of designing an enormous variety of structurally different porphyrin derivatives and the facility of binding these molecules to other compounds with specific properties, such as antibodies, polylysines or solid inert supports, places this class of PS among the most promising for the inactivation of pathogenic microorganisms in the clinic and environmental areas.

The studies carried out up until now bring forward some of the positive aspects of the use of porphyrin derivatives in aPDT: 1) the efficacy of porphyrin derivatives in aPDT has been proved for different microorganisms at nanomolar concentration under different light conditions; 2) porphyrin derivatives have, in general, no effect on host cells; the inhibition of microorganisms occurs, frequently, at concentrations lower that those found to elicit cytotoxic responses in host mammalian cells; 3) the effectiveness of porphyrins depends on the characteristics of the medium where aPDT is intended to occur, therefore, whenever planning aPDT *in vivo* the "sample" type should be specified in order to enable the proper adjustment of concentration and exposure time; 4) porphyrins are effective to inactivate microbial strains resistant to antimicrobial drugs and 5) *meso*-substituted cationic porphyrins are, in general, more effective for microorganism inactivation than other PS as MB and RB, when applied to both clinic and environmental samples.

However, more studies to confirm the effect of the number of charges and of the charge distribution of *meso*-substituted groups in synthetic cationic porphyrins are still required, especially in photodynamic approaches to the inactivation of eukaryotic pathogens, such as fungi and protozoa. Also, more controlled studies are necessary to prove the efficacy of porphyrins in aPDT as compared to the classical methods and their efficacy in therapy-resistant cases. It will be also important to study the synergistic effect between aPDT and antimicrobial classical methods.

Acknowledgements

Thanks are due to the University of Aveiro, Fundação para a Ciência e a Tecnologia and FEDER for funding the Organic Chemistry Research Unit (QOPNA) and to CESAM (Centro de Estudos do Ambiente e do Mar) for funding the Microbiology Research Group.

References

1. G. Jori and S.B. Brown, Photosensitized inactivation of microorganisms, *Photochem. Photobiol. Sci.*, 2004, **3**, 403.

2. G. Jori, C. Fabris, M. Soncin, S. Ferro, O. Coppellotti, D. Dei, L. Fantetti, G. Chiti and G. Roncucci, Photodynamic therapy in the treatment of microbial infections: Basic principles and perspective applications, *Lasers Surg. Med.*, 2006, **38**, 468.
3. D.A. Caminos, M.B. Spesia, P. Pons and E.N. Durantini, Mechanisms of Escherichia coli photodynamic inactivation by an amphiphilic tricationic porphyrin and 5,10,15,20-tetra(4-N,N,N-trimethylammoniumphenyl)porphyrin, *Photochem. Photobiol. Sci.*, 2008, **7**, 1071.
4. M. Wainwright, Photodynamic antimicrobial chemotherapy (PACT), *J. Antimicrob. Chemother.*, 1998, **42**, 13.
5. P.W. Taylor, P.D. Stapleton and J.P. Luzio, New ways to treat bacterial infections, *Drug Discov. Today*, 2002, **7**, 1086.
6. M. Merchat, G. Bertolini, P. Giacomini, A. Villanueva and G. Jori, *Meso*-substituted cationic porphyrins as efficient photosensitizers of gram-positive and gram-negative bacteria, *J. Photochem. Photobiol., B*, 1996, **32**, 153.
7. R. Bonnett, *Chemical aspects of photodynamic therapy*, Gordon and Breach Science Publishers, Amsterdam, 2000.
8. M. Jemli, Z. Alouini, S. Sabbahi and M. Gueddari, Destruction of fecal bacteria in wastewater by three photosensitizers, *J. Environ. Monit.*, 2002, **4**, 511.
9. E. Alves, C.M.B. Carvalho, J.P.C. Tomé, M.A.F. Faustino, M. Neves, A.C. Tomé, J.A.S. Cavaleiro, A. Cunha, S. Mendo and A. Adelaide, Photodynamic inactivation of recombinant bioluminescent Escherichia coli by cationic porphyrins under artificial and solar irradiation, *J. Ind. Microbiol. Biotechnol.*, 2008, **35**, 1447.
10. L. Costa, E. Alves, C.M.B. Carvalho, J.P.C. Tomé, M.A.F. Faustino, M.G.P.M.S. Neves, A.C. Tomé, J.A.S. Cavaleiro, Â. Cunha and A. Almeida, Sewage bacteriophage photoinactivation by cationic porphyrins: a study of charge effect, *Photochem. Photobiol. Sci.*, 2008, **7**, 415.
11. C.M.B. Carvalho, J.P.C. Tomé, M.A.F. Faustino, M.G.P.M.S. Neves, A.C. Tomé, J.A.S. Cavaleiro, L. Costa, E. Alves, A. Oliveira, Â. Cunha and A. Almeida, Antimicrobial photodynamic activity of porphyrin derivatives: potential application on medical and water disinfection, *J. Porphyrins Phthalocyanines*, 2009, **13**, 574.
12. E. Alves, L. Costa, C. Carvalho, J.P.C. Tomé, M.A.F. Faustino, M.G.P.M.S. Neves, A.C. Tomé, J.A.S. Cavaleiro, A. Cunha and A. Almeida, Charge effect on the photoinactivation of Gram-negative and Gram-positive bacteria by cationic meso-substituted porphyrins, *BMC Microbiol.*, 2009, **9**, 70.
13. S. Kosaka, O. Akilov, K. O'Riordan and T. Hasan, A mechanistic study of delta-aminolevulinic acid-based photodynamic therapy for cutaneous leishmaniasis, *J. Invest. Dermatol.*, 2007, **127**, 1546.
14. M. Merchat, J.D. Spikes, G. Bertoloni and G. Jori, Studies on the mechanism of bacteria photosensitization by *meso*-substituted cationic porphyrins, *J. Photochem. Photobiol., B*, 1996, **35**, 149.
15. A. Minnock, D.I. Vernon, J. Schofield, J. Griffiths, J. Howard Parish and S.B. Brown, Photoinactivation of bacteria. Use of a cationic water-soluble zinc phthalocyanine to photoinactivate both gram-negative and gram-positive bacteria, *J. Photochem. Photobiol., B*, 1996, **32**, 159.
16. K.D. Winckler, Special section: Focus on anti-microbial photodynamic therapy (PDT), *J. Photochem. Photobiol. B. Biol.*, 2007, **86**, 43.
17. M.A. Calin and S.V. Parasca, Light sources for photodynamic inactivation of bacteria, *Lasers Med. Sci.*, 2009, **24**, 453.

18. M.C. DeRosa and R.J. Crutchley, Photosensitized singlet oxygen and its applications, *Coord. Chem. Rev.*, 2002, **233**, 351.
19. K. Ergaieg, M. Chevanne, J. Cillard and R. Seux, Involvement of both Type I and Type II mechanisms in Gram-positive and Gram-negative bacteria photosensitization by a meso-substituted cationic porphyrin, *Solar Energy*, 2008, **82**, 1107.
20. M. Wainwright and K.B. Crossley, Photosensitising agents – circumventing resistance and breaking down biofilms: a review, *Int. Biodeter. Biodegr.*, 2004, **53**, 119.
21. K. Müller-Breitkreutz, H. Mohr, K. Briviba and H. Sies, Inactivation of viruses by chemically and photochemically generated singlet molecular oxygen, *J. Photochem. Photobiol. B. Biol.*, 1995, **30**, 63.
22. C. Hadjur, N. Lange, J. Rebstein, P. Monnier, H.v.d. Bergh and G. Wagnières, Spectroscopic studies of photobleaching and photoproduct formation of meta (tetrahydroxyphenyl) chlorin (m-THPC) used in photodynamic therapy. The production of singlet oxygen by m-THPC, *J. Photochem. Photobiol. B*, 1998, **45**, 170.
23. Y. Nitzan, B. Shainberg and Z. Malik, The mechanism of photodynamic inactivation of *Staphylococcus aureus* by deuteroporphyrin, *Curr. Microbiol.*, 1989, **19**, 265.
24. T. Maisch, R.M. Szeimies, N. Lehn and C. Abels, Antibacterial photodynamic therapy. A new treatment for bacterial skin diseases?, *Hautarzt*, 2005, **56**, 1048.
25. M. Maclean, S. MacGregor, J.G. Anderson and G.A. Woolsey, The role of oxygen in the visible-light inactivation of *Staphylococcus aureus*, *J. Photochem. Photobiol., B*, 2008, **92**, 180.
26. S.Y. Kim, O.J. Kwon and J.W. Park, Inactivation of catalase and superoxide dismutase by singlet oxygen derived from photoactivated dye, *Biochimie*, 2001, **83**, 437.
27. C.M. Allen, J.M. Weber and J.E. Vanlier, Sulfophthalocyanines for photodynamic inactivation of viruses in blood products – effect of structural modifications, *Photochem. Photobiol.*, 1995, **62**, 184.
28. F. Käsermann and C. Kempf, Photodynamic inactivation of enveloped viruses by buckminsterfullerene, *Antiviral Res.*, 1997, **34**, 65.
29. F. Gábor, K. Szocs, P. Maillard and G. Csik, Photobiological activity of exogenous and endogenous porphyrin derivatives in *Escherichia coli* and *Enterococcus hirae* cells, *Radiat. Environ. Biophys.*, 2001, **40**, 145.
30. A.N. Vzorov, D.W. Dixon, J.S. Trommel, L.G. Marzilli and R.W. Compans, Inactivation of human immunodeficiency virus type 1 by porphyrins, *Antimicrob. Agents Chemother.*, 2002, **46**, 3917.
31. A. Egyeki, G. Turoczy, Z. Majer, K. Toth, A. Fekete, P. Maillard and G. Csik, Photosensitized inactivation of T7 phage as surrogate of non-enveloped DNA viruses: efficiency and mechanism of action, *Biochim. Biophys. Acta*, 2003, **1624**, 115.
32. M.R. Hamblin and T. Hasan, Photodynamic therapy: a new antimicrobial approach to infectious disease?, *Photochem. Photobiol. Sci.*, 2004, **3**, 436.
33. K. Zupán, M. Egyeki, K. Toth, A. Fekete, L. Herenyi, K. Modos and G. Csik, Comparison of the efficiency and the specificity of DNA-bound and free cationic porphyrin in photodynamic virus inactivation, *J. Photochem. Photobiol., B*, 2008, **90**, 105.
34. H.D. Li, O.S. Fedorova, A.N. Grachev, W.R. Trumble, G.A. Bohach and L. Czuchajowski, A series of *meso*-tris(N-methyl-pyridiniumyl)-(4-alkylamidophenyl)porphyrins: Synthesis, interaction with DNA and antibacterial activity, *Biochim. Biophys. Acta*, 1997, **1354**, 252.

35. S. Mettath, B.R. Munson and R.K. Pandey, DNA interaction and photocleavage properties of porphyrins containing cationic substituents at the peripheral position, *Bioconjug. Chem.*, 1999, **10**, 94.
36. F.P. Imray and D.G. MacPhee, Role of DNA polymerase I and rec system in survival of bacteria and bacteriophages damaged by photodynamic action of acridine-orange, *Mol. Gen. Genet.*, 1973, **123**, 289.
37. G. Valduga, B. Breda, G.M. Giacometti, G. Jori and E. Reddi, Photosensitization of wild and mutant strains of *Escherichia coli* by *meso*-tetra(N-methyl-4-pyridyl)porphine, *Biochem. Biophys. Res. Commun.*, 1999, **256**, 84.
38. E.N. Durantini, Photodynamic inactivation of bacteria, *Curr. Bioact. Compd.*, 2006, **2**, 127.
39. M. Schafer, C. Schmitz and G. Horneck, High sensitivity of Deinococcus radiodurans to photodynamically-produced singlet oxygen, *Int. J. Radiat. Biol.*, 1998, **74**, 249.
40. G. Bertoloni, F. Zambotto, L. Conventi, E. Reddi and G. Jori, Role of specific cellular targets in the hematoporphyrin-sensitized photoinactivation of microbial-cells, *Photochem. Photobiol.*, 1987, **46**, 695.
41. K. Kassab, D. Dei, G. Roncucci, G. Jori and O. Coppellotti, Phthalocyanine-photosensitized inactivation of a pathogenic protozoan, Acanthamoeba palestinensis, *Photochem. Photobiol. Sci.*, 2003, **2**, 668.
42. F.M. Lauro, P. Pretto, L. Covolo, G. Jori and G. Bertoloni, Photoinactivation of bacterial strains involved in periodontal diseases sensitized by porphycene-polylysine conjugates, *Photochem. Photobiol. Sci.*, 2002, **1**, 468.
43. A. Tavares, C.M.B. Carvalho, M.A. Faustino, M.G.P.M.S. Neves, J.P.C. Tomé, A. C. Tomé, J.A.S. Cavaleiro, A. Cunha, N.C.M. Gomes, E. Alves and A. Almeida, Antimicrobial photodynamic therapy: study of bacterial recovery viability and potential development of resistance after treatment, *Mar. Drugs*, 2010, **8**, 91.
44. L. Costa, C. Carvalho, J.P.C. Tomé, M.A.F. Faustino, M.G.P.M.S. Neves, A.C. Tomé, J.A.S. Cavaleiro, A. Cunha and A. Almeida, Study of viral resistance following repeated exposure to aPDT and of viability recovery, *Antiviral Therapy*, submitted.
45. T. Ito and K. Kobayashi, In vivo evidence for photodynamic membrane damage as a determining step of inactivation of yeast-cells sensitized by toluidine blue, *Photochem. Photobiol.*, 1977, **25**, 399.
46. B. Bagchi and B. Sreeradha, Role of dye molecules remaining outside the cell during photodynamic inactivation of *Escherichia coli* in the presence of acriflavine, *Photochem. Photobiol.*, 1989, **29**, 403.
47. M. Bhatti, A. MacRobert, S. Meghji, B. Henderson and M. Wilson, A study of the uptake of toluidine blue O by *Porphyromonas gingivalis* and the mechanism of lethal photosensitisation, *Photochem. Photobiol.*, 1998, **68**, 370.
48. B. Ehrenberg, Z. Malik, Y. Nitzan, H. Ladan, F.M. Johnson, G. Hemmi and J.L. Sessler, The binding and photosensitization effects of tetrabenzoporphyrins and texaphyrin in bacterial cells, *Lasers Med. Sci.*, 1993, **8**, 197.
49. V. Carré, O. Gaud, I. Sylvain, O. Bourdon, M. Spiro, J. Blais, R. Granet, P. Krausz and M. Guilloton, Fungicidal properties of meso-arylglycosylporphyrins: Influence of sugar substituents on photoinduced damage in the yeast *Saccharomyces cerevisiae*, *J. Photochem. Photobiol., B*, 1999, **48**, 57.
50. T. Maisch, R.M. Szeimies, G. Jori and C. Abels, Antibacterial photodynamic therapy in dermatology, *Photochem. Photobiol. Sci.*, 2004, **3**, 907.

51. C.M. Cassidy, M.M. Tunney, P. A. McCarron and R.F. Donnelly, Drug delivery strategies for photodynamic antimicrobial chemotherapy: From benchtop to clinical practice – A review, *J. Photochem. Photobiol., B*, 2009, **95**, 71.
52. J. North, R. Coombs and J. Levy, Photodynamic inactivation of free and cell-associated HIV-1 using the photosensitizer benzoporphyrin derivative, *J. Acq. Immun. Def. Synd.*, 1994, **7**, 891.
53. M. Wilson and C. Yianni, Killing of methicillin-resistant *Staphylococcus aureus* by low-power laser-light, *J. Med. Microbiol.*, 1995, **42**, 62.
54. T. Maisch, C. Bosl, R.M. Szeimies, N. Lehn and C. Abels, Photodynamic effects of novel XF porphyrin derivatives on prokaryotic and eukaryotic cells, *Antimicrob. Agents Chemother.*, 2005, **49**, 1542.
55. L.L. Trannoy, F.G. Terpstra, D. Korte, J.W.M. Lagerberg, A.J. Verhoeven, A. Brand and F.A.C. van Engelenburg, Differential sensitivities of pathogens in red cell concentrates to Tri-P(4)-photoinactivation, *Vox Sang.*, 2006, **91**, 111.
56. T. Maisch, J. Baier, B. Franz, M. Maier, M. Landthaler, R.M. Szeimies and W. Baumler, The role of singlet oxygen and oxygen concentration in photodynamic inactivation of bacteria, *Proc. Natl. Acad. Sci. USA*, 2007, **104**, 7223.
57. C. Bombelli, F. Bordi, S. Ferro, L. Giansanti, G. Jori, G. Mancini, C. Mazzuca, D. Monti, F. Ricchelli, S. Sennato and M. Venanzi, New cationic liposomes as vehicles of m-tetrahydroxyphenylchlorin in photodynamic therapy of infectious diseases, *Mol. Pharmaceutics*, 2008, **5**, 672.
58. M. Grinholc, B. Szramka, K. Olender and A. Graczyk, Bactericidal effect of photodynamic therapy against methicillin-resistant *Staphylococcus aureus* strain with the use of various porphyrin photosensitizers, *Acta Biochim. Pol. (Engl. Transl.)*, 2007, **54**, 665.
59. K. O'Riordan, O.E. Akilov and T. Hasan, The potential for photodynamic therapy in the treatment of localized infections, *Photodiagn. Photodyn. Ther.*, 2005, **2**, 247.
60. A.K. Debnath, S. Jiang, N. Strick, K. Lin, P. Haberfield and A.R. Neurath, 3-Dimensional structure-activity analysis of a series of porphyrin derivatives with anti-HIV-1 activity targeted to the v3 loop of the gp120 envelope glycoprotein of the human-immunodeficiency-virus type 1, *J. Med. Chem.*, 1994, **37**, 1099.
61. A.R. Neurath, P. Haberfield, B. Joshi, I.K. Hewlett, N. Strick and S. Jiang, Rapid prescreening for antiviral agents against HIV-1 based on their inhibitory activity in site-directed immunoassays I. The V3 loop of gp120 as target, *Antiviral Chem. Chemother.*, 1991, **2**, 303.
62. D.L. DeCamp, L.M. Babé, R. Salto, J.L. Lucich, M.-S. Koo, S.B. Kahl and C.S. Craik, Specific-inhibition of HIV-1 protease by boronated porphyrins, *J. Med. Chem.*, 1992, **35**, 3426.
63. A.R. Neurath, N. Strick and S. Jiang, Rapid prescreening for antiviral agents against HIV-1 based on their inhibitory activity in site-directed immunoassays – approaches applicable to epidemic HIV-1 strains, *Antiviral Chem. Chemother.*, 1993, **4**, 207.
64. F.H.E. Schagen, A.C.E. Moor, S.C. Cheong, S. J. Cramer, H. van Ormondt, A.J. van der Eb, T.M.A.R. Dubbelman and R.C. Hoeben, Photodynamic treatment of adenoviral vectors with visible light: an easy and convenient method for viral inactivation, *Gene Ther.*, 1999, **6**, 873.
65. J.L. Matthews, F. Sogandaresbernal, M. Judy, K. Gulliya, J. Newman, T. Chanh and A. Marengorowe, Inactivation of viruses with photoactive compounds, *Blood Cells*, 1992, **18**, 75.

66. J. North, S. Freeman, J. Overbaugh, J. Levy and R. Lansman, Photodynamic inactivation of retrovirus by benzoporphyrin derivative: a feline leukemia virus model, *Transfusion*, 1992, **32**, 121.
67. E.G. Argyris, J.M. Vanderkooi and Y. Paterson, Mutagenesis of key residues identifies the connection subdomain of HIV-1 reverse transcriptase as the site of inhibition by heme, *Eur. J. Biochem.*, 2001, **268**, 925.
68. R. Vanyur, K. Heberger and J. Jakus, Prediction of anti-HIV-1 activity of a series of tetrapyrrole molecules, *J. Chem. Inf. Comput. Sci.*, 2003, **43**, 1829.
69. J. Dairou, C. Vever-Bizet and D. Brault, Interaction of sulfonated anionic porphyrins with HIV glycoprotein gp120: photodamages revealed by inhibition of antibody binding to V3 and C5 domains, *Antiviral Res.*, 2004, **61**, 37.
70. Y. Cheng, L.K. Tsou, J. Cai, T. Aya, G.E. Dutschman, E.A. Gullen, S.P. Grill, A. P.-Ç. Chen, B.D. Lindenbach, A.D. Hamilton and Y.-C. Cheng, Novel class of *meso*-tetrakis-porphyrin derivatives exhibits potent activities against *Hepatitis C* virus genotype 1b replicons in vitro, *Antimicrob. Agents Chemother.*, 2010, **54**, 197.
71. R. Bonnett, M.A. Krysteva, I.G. Lalov and S.V. Artarsky, Water disinfection using photosensitizers immobilized on chitosan, *Water Res.*, 2006, **40**, 1269.
72. M. Magaraggia, F. Faccenda, A. Gandolfi and G. Jori, Treatment of microbiologically polluted aquaculture waters by a novel photochemical technique of potentially low environmental impact, *J. Environ. Monitor.*, 2006, **8**, 923.
73. C.M.B. Carvalho, A.T.P.C. Gomes, S.C.D. Fernandes, A.C.B. Prata, M.A. Almeida, M.A. Cunha, J.P.C. Tomé, M.A.F. Faustino, M.G.P.M.S. Neves, A.C. Tomé, J.A.S. Cavaleiro, Z. Lin, J.P. Rainho and J. Rocha, Photoinactivation of bacteria in wastewater by porphyrins: Bacterial beta-galactosidase activity and leucine-uptake as methods to monitor the process, *J. Photochem. Photobiol., B*, 2007, **88**, 112.
74. C. Arrojado, Evaluation of antimicrobial photodynamic therapy by porphyrins to inactivate fish pathogenic bacteria in aquaculture system, MSc Thesis, University of Aveiro, 2009.
75. *The Porphyrin Handbook*, ed. K.M. Kadish, K.M. Smith and R. Guilard, vol. 6 – Applications: Past, Present and Future, Academic Press, San Diego, 2000.
76. K.M. Smith, Synthesis and preparation of porphyrin compounds, in *Porphyrins and Metalloporphyrins*, ed. K.M. Smith, Elsevier, New York, ch. 2, 1975.
77. R.K. Dinello and C.K. Chang, Isolation and modification of natural porphyrins, in *The Porphyrins*, ed. D. Dolphin, Academic Press, New York, vol. 1, ch. 7, 1978.
78. K.M. Smith, Strategies for the synthesis of ocataalkylporphyrin systems, in *The Porphyrin Handbook*, ed. K.M. Kadish, K.M. Smith and R. Guilard, Academic Press, San Diego, vol. 1, ch. 1, 2000.
79. M. Grandadam, D. Ingrand, J.M. Huraux, B. Aveline, O. Delgado, C. Vever-Bizet and D. Brault, Photodynamic inactivation of cell-free HIV strains by a red-absorbing chlorin-type photosensitizer, *J. Photochem. Photobiol., B*, 1995, **31**, 171.
80. J.S. Lindsey, Synthesis of meso-substituted porphyrins, in *The Porphyrin Handbook*, ed. K. M. Kadish, K.M. Smith and R. Guilard, Academic Press, San Diego, vol. 1, ch. 2, 2000.
81. Z. Luksiene, I. Buchovec and E. Paskeviciute, Inactivation of food pathogen *Bacillus cereus* by photosensitization in vitro and on the surface of packaging material, *J. Appl. Microbiol.*, 2009, **107**, 2037.
82. S. Oriel and Y. Nitzan, Photoinactivation of Candida albicans by its own endogenous porphyrins, *Curr. Microbiol.*, 2010, **60**, 117.

83. K. Szocs, F. Gábor, G. Csik and J. Fidy, delta-Aminolaevulinic acid-induced porphyrin synthesis and photodynamic inactivation of *Escherichia coli B*, *J. Photochem. Photobiol., B*, 1999, **50**, 8.
84. Y. Nitzan and M. Kauffman, Endogenous porphyrin production in bacteria by-aminolaevulinic acid and subsequent bacterial photoeradication, *Lasers Med. Sci.*, 1999, **14**, 269.
85. Y. Nitzan, M. Salmon-Divon, E. Shporen and Z. Malik, ALA induced photodynamic effects on Gram positive and negative bacteria, *Photochem. Photobiol. Sci.*, 2004, **3**, 430.
86. N. Fotinos, M. Convert, J.C. Piffaretti, R. Gurny and N. Lange, Effects on gram-negative and gram-positive bacteria mediated by 5-aminolevulinic acid and 5-aminolevulinic acid derivatives, *Antimicrob. Agents Chemother.*, 2008, **52**, 1366.
87. F.W. Van der Meulen, K. Ibrahim, H.J.C.M. Sterenborg, L.V. Alphen, A. Maikoe and J. Dankert, Photodynamic destruction of Haemophilus parainfluenzae by endogenously produced porphyrins, *J. Photochem. Photobiol., B*, 1997, **40**, 204.
88. Z. Smetana, Z. Malik, A. Orenstein, E. Mendelson and E. BenHur, Treatment of viral infections with 5-aminolevulinic acid and light, *Lasers Surg. Med.*, 1997, **21**, 351.
89. R.F. Donnelly, P.A. McCarron, J.M. Lightowler and A.D. Woolfson, Bioadhesive patch-based delivery of 5-aminolevulinic acid to the nail for photodynamic therapy of onychomycosis, *J. Controlled Release*, 2005, **103**, 381.
90. S.K. Bisland, C. Chien, B.C. Wilson and S. Burch, Pre-clinical in vitro and in vivo studies to examine the potential use of photodynamic therapy in the treatment of osteomyelitis, *Photochem. Photobiol. Sci.*, 2006, **5**, 31.
91. M. Sharma, H. Bansal and P.K. Gupta, Photodynamic inactivation of antibiotic resistant strain of Pseudomonas aeruginosa by porphyrins induced by delta-aminolaevulinic acid, *Indian J. Med. Res.*, 2002, **116**, 99.
92. R. Sailer, W.S.L. Strauss, K. Konig, A. Ruck and R. Steiner, Correlation between porphyrin biosynthesis and photodynamic inactivation of *Pseudomonas aeruginosa* after incubation with 5 aminolaevulinic acid, *J. Photochem. Photobiol., B*, 1997, **39**, 236.
93. H. Ashkenazi, Z. Malik, Y. Harth and Y. Nitzan, Eradication of *Propionibacterium acnes* by its endogenic porphyrins after illumination with high intensity blue light, *FEMS Immunol. Med. Microbiol.*, 2003, **35**, 17.
94. I. Buchovec, Z. Vaitonis and Z. Luksiene, Novel approach to control Salmonella enterica by modern biophotonic technology: photosensitization, *J. Appl. Microbiol.*, 2009, **106**, 748–754.
95. S. Karrer, R.M. Szeimies, S. Ernst, C. Abels, W. Baumler and M. Landthaler, Photodynamic inactivation of staphylococci with 5-aminolaevulinic acid or photofrin, *Lasers Med. Sci.*, 1999, **14**, 54.
96. J. Walther, M.J. Brocker, D. Watzlich, M. Nimtz, M. Rohde, D. Jahn and J. Moser, Protochlorophyllide: a new photosensitizer for the photodynamic inactivation of Gram-positive and Gram-negative bacteria, *FEMS Microbiol. Lett.*, 2009, **290**, 156.
97. T. Yamamoto, T. Iriuchishima, S. Aizawa, T. Okano, B. Goto, Y. Nagai, T. Horaguchi, J. Ryu and A. Saito, Bactericidal effect of photodynamic therapy using Na-Pheophorbide a: evaluation of adequate light source, *Photomed. Laser Surg.*, 2009, **27**, 849.

98. Z. Drulis-Kawa, A. Bednarkiewicz, G. Bugla-Ploskonska, W. Strek and W. Doroszkiewicz, The susceptibility of anaerobic bacteria isolated from periodontal diseases to photodynamic inactivation with Fotolon (chlorin e6), *Pol. J. Microbiol.*, 2005, **54**, 305.
99. A. Fomichev, V.P. Zorin, T.E. Zorina and S.N. Cherenkevich, Photoinduced damage of gram-positive and gram-negative bacterial cells in the presence of e6 chlorine derivatives, *Mikrobiologia*, 1991, **60**, 507.
100. A. Ulatowska-Jarza, J. Zychowicz, I. Holowacz, J. Bauer, J. Razik, A. Wieliczko, H. Podbielska, G. Müller, W. Strek and U. Bindig, Antimicrobial PDT with chlorophyll-derived photosensitizer and semiconductor laser, *Med. Laser Appl.*, 2006, **21**, 177.
101. S. Schastak, B. Gitter, R. Handzel, R. Hermann and P. Wiedemann, Improved photoinactivation of Gram-negative and Gram-positive methicillin-resistant bacterial strains using a new near-infrared absorbing *meso*-tetrahydroporphyrin: A comparative study with a chlorine e6 photosensitizer photolon, *Methods Find. Exp. Clin. Pharmacol.*, 2008, **30**, 129.
102. P.M. Go, R.N. Reed, R.C. Straight and M. Waner, Laser photodynamic therapy for papilloma viral lesions, *Arch. Otolaryngol., Head Neck Surg.*, 1990, **116**, 1177.
103. A. Pfitzner, B.W. Sigusch, V. Albrecht and E. Glockmann, Killing of periodontopathogenic bacteria by photodynamic therapy, *J. Periodontol.*, 2004, **75**, 1343.
104. B.W. Sigusch, A. Pfitzner, V. Albrecht and E. Glockmann, Efficacy of photodynamic therapy on inflammatory signs and two selected periodontopathogenic species in a beagle dog model, *J. Periodontol.*, 2005, **76**, 1100.
105. C.R. Rovaldi, A. Pievsky, N.A. Sole, P.M. Friden, D.M. Rothstein and P. Spacciapoli, Photoactive porphyrin derivative with broad-spectrum activity against oral pathogens in vitro, *Antimicrob. Agents Chemother.*, 2000, **44**, 3364.
106. I. Stojiljkovic, V. Kumar and N. Srinivasan, Non-iron metalloporphyrins: potent antibacterial compounds that exploit haem/Hb uptake systems of pathogenic bacteria, *Mol. Microbiol.*, 1999, **31**, 429.
107. M.L. Embleton, S.P. Nair, B.D. Cookson and M. Wilson, Selective lethal photosensitization of methicillin-resistant *Staphylococcus aureus* using an IgG-tin(IV) chlorin e6 conjugate, *J. Antimicrob. Chemother.*, 2002, **50**, 857.
108. M. L. Embleton, S.P. Nair, W. Heywood, D.C. Menon, B.D. Cookson and M. Wilson, Development of a novel targeting system for lethal photosensitization of antibiotic-resistant strains of *Staphylococcus aureus*, *Antimicrob. Agents Chemother.*, 2005, **49**, 3690.
109. F. Gad, T. Zahra, K.P. Francis, T. Hasan and M.R. Hamblin, Targeted photodynamic therapy of established soft-tissue infections in mice, *Photochem. Photobiol. Sci.*, 2004, **3**, 451.
110. C.K. Hope and M. Wilson, Induction of lethal photosensitization in biofilms using a confocal scanning laser as the excitation source, *J. Antimicrob. Chemother.*, 2006, **57**, 1227.
111. T.H. Dai, G.P. Tegos, Z.S. Lu, L.Y. Huang, T. Zhiyentayev, M. J. Franklin, D.G. Baer and M.R. Hamblin, Photodynamic therapy for Acinetobacter baumannii burn infections in mice, *Antimicrob. Agents Chemother.*, 2009, **53**, 3929.
112. N. S. Soukos, L.A. Ximenez-Fyvie, M.R. Hamblin, S.S. Socransky and T. Hasan, Targeted antimicrobial photochemotherapy, *Antimicrob. Agents Chemother.*, 1998, **42**, 2595.
113. T.N. Demidova and M.R. Hamblin, Photodynamic inactivation of bacillus spores, mediated by phenothiazinium dyes, *Appl. Environ. Microbiol.*, 2005, **71**, 6918.

114. B.B. Fuchs, G.P. Tegos, M.R. Hamblin and E. Mylonakis, Susceptibility of Cryptococcus neoformans to photodynamic inactivation is associated with cell wall integrity, *Antimicrob. Agents Chemother.*, 2007, **51**, 2929.
115. T.N. Demidova, F. Gad, T. Zahra, K.P. Francis and M.R. Hamblin, Monitoring photodynamic therapy of localized infections by bioluminescence imaging of genetically engineered bacteria, *J. Photochem. Photobiol., B*, 2005, **81**, 15.
116. A.S. Garcez, M.S. Ribeiro, G.P. Tegos, S.C. Nunez, A.O. Jorge and M.R. Hamblin, Antimicrobial photodynamic therapy combined with conventional endodontic treatment to eliminate root canal biofilm infection, *Lasers Surg. Med.*, 2007, **39**, 59.
117. T.N. Demidova and M.R. Hamblin, Effect of cell-photo sensitizer binding and cell density on microbial photoinactivation, *Antimicrob. Agents Chemother.*, 2005, **49**, 2329.
118. H.M. Tang, M.R. Hamblin and C.M. Yow, A comparative in vitro photoinactivation study of clinical isolates of multidrug-resistant pathogens, *J. Infect. Chemother.*, 2007, **13**, 87.
119. M.R. Hamblin, D.A. O'Donnell, N. Murthy, K. Rajagopalan, N. Michaud, M.E. Sherwood and T. Hasan, Polycationic photosensitizer conjugates: effects of chain length and Gram classification on the photodynamic inactivation of bacteria, *J. Antimicrob. Chemother.*, 2002, **49**, 941.
120. G.P. Tegos, M. Anbe, C. Yang, T.N. Demidova, M. Satti, P. Mroz, S. Janjua, F. Gad and M.R. Hamblin, Protease-stable polycationic photosensitizer conjugates between polyethyleneimine and chlorin(e6) for broad-spectrum antimicrobial photoinactivation, *Antimicrob. Agents Chemother.*, 2006, **50**, 1402.
121. T. Dai, G.P. Tegos, T. Zhiyentayev, E. Mylonakis and M.R. Hamblin, Photodynamic therapy for methicillin-resistant *Staphylococcus aureus* infection in a mouse skin abrasion model, *Lasers Surg. Med.*, 2010, **42**, 38.
122. A. Nazarova, A. Ignatova, A. Feofanov, T. Karmakova, A. Pljutinskaya, O. Mass, M. Grin, R. Yakubovskaya, A. Mironov and J.C. Maurizot, 13,15-N-cycloimide derivatives of chlorin p6 with isonicotinyl substituent are photosensitizers targeted to lysosomes, *Photochem. Photobiol. Sci.*, 2007, **6**, 1184.
123. B. Bayarmaa, B. Bayarmaa and Y.K. Shim, Photodynamic inactivation of wound-associated bacteria with new troponyl (pyro)pheophobides, *J. Porphyrins Phthalocyanines*, 2009, **13**, 832.
124. G. Fraikin, M.G. Strakhovskaya and A.B. Rubin, The role of membrane-bound porphyrin-type compound as endogenous sensitizer in photodynamic damage to yeast plasma membranes, *J. Photochem. Photobiol., B*, 1996, **34**, 129.
125. H. Podbielska, A. Ulatowska-Jaria, J. Bauer, A. Wieliczko, J. Razik and W. Strck, The comparison of photodynamic activity of photolon and photoporphyrine on pathogenic bacteria in vitro, *Polish J. Environ. Stud.*, 2006, **15**, 147.
126. Y. Nitzan, H.M. Wexler and S.M. Finegold, Inactivation of anaerobic-bacteria by various photosensitized porphyrins or by hemin, *Curr. Microbiol.*, 1994, **29**, 125.
127. C.E. Millson, M. Wilson, A.J. Macrobert, J. Bedwell and S.G. Bown, The killing of Helicobacter pylori by low-power laser light in the presence of a photosensitiser, *J. Med. Microbiol.*, 1996, **44**, 245.
128. M.R. Hamblin, J. Viveiros, C. Yang, A. Ahmadi, R. A. Ganz and M.J. Tolkoff, Helicobacter pylori accumulates photoactive porphyrins and is killed by visible light, *Antimicrob. Agents Chemother.*, 2005, **49**, 2822.
129. M. Asanaka, T. Kurimura, H. Toya, J. Ogaki and Y. Kato, Anti-HIV activity of protoporphyrin, *AIDS*, 1989, **3**, 403.

130. K. Tsutsui and G.C. Mueller, Hemin inhibits virion-associated reverse-transcriptase of murine leukemia-virus, *Biochem. Biophys. Res. Commun.*, 1987, **149**, 628.
131. A.D. Izzo and J.T. Walsh, Light-induced modulation of Porphyromonas gingivalis growth, *J. Photochem. Photobiol., B*, 2004, **77**, 63.
132. K. Konig, M. Teschke, B. Sigusch, E. Glockmann, S. Eick and W. Pfister, Red light kills bacteria via photodynamic action, *Cell Mol. Biol. (Noisy-le-grand)*, 2000, **46**, 1297.
133. H. Ladan, Y. Nitzan and Z. Malik, The antibacterial activity of hemin compared with cobalt, zinc and magnesium protoporphyrin and its effect on potassium-loss and ultrastructure of *Staphylococcus aureus*, *FEMS Microbiol. Lett.*, 1993, **112**, 173.
134. M. Grinholc, B. Szramka, J. Kurlenda, A. Graczyk and K.P. Bielawski, Bactericidal effect of photodynamic inactivation against methicillin-resistant and methicillin-susceptible *Staphylococcus aureus* is strain-dependent, *J. Photochem. Photobiol., B*, 2008, **90**, 57.
135. R. Shawar and B.H. Cooper, Comparative kinetics of hematoporphyrin derivative uptake and susceptibility of Bacillus subtilis and Streptococcus faecalis to photodynamic-action, *Photochem. Photobiol.*, 1990, **52**, 825.
136. C.E. Millson, W. Thurrell, G. Buonaccorsi, M. Wilson, A.J. Macrobert and S.G. Bown, The effect of low-power laser light at different doses on gastric mucosa sensitized with methylene blue, haematoporphyrin derivative or toluidine blue, *Lasers Med. Sci.*, 1997, **12**, 145.
137. M. Perlin, J.C.H. Mao, E.R. Otis, N.L. Shipkowitz and R.G. Duff, Photodynamic inactivation of influenza and herpes viruses by hematoporphyrin, *Antiviral Res.*, 1987, **7**, 43.
138. J.L. Matthews, J.T. Newman, F. Sogandares-Bernal, M.M. Judy, H. Skiles, J.E. Leveson, A.J. Marengo-Rowe and T.C. Chanh, Photodynamic therapy of viral contaminants with potential for blood banking applications, *Transfusion*, 1988, **28**, 81.
139. T. Zoladek, B.N. Nguyen, I. Jagiello, A. Graczyk and J. Rytka, Diamino acid derivatives of porphyrins penetrate into yeast cells, induce photodamage, but have no mutagenic effect, *Photochem. Photobiol.*, 1997, **66**, 253.
140. M. Szpakowska, K. Lasocki, J. Grzybowski and A. Graczyk, Photodynamic activity of the haematoporphyrin derivative with rutin and arginine substituents (HpD-Rut(2)-Arg(2)) against *Staphylococcus aureus* and *Pseudomonas aeruginosa*, *Pharmacol. Res.*, 2001, **44**, 243.
141. M. Szpakowska, J. Reiss, A. Graczyk, S. Szmigielski, K. Lasocki and J. Grzybowski, Susceptibility of *Pseudomonas aeruginosa* to a photodynamic effect of the arginine hematoporphyrin derivative, *Int. J. Antimicrob. Agents*, 1997, **8**, 23.
142. H. Ashkenazi, Y. Nitzan and D. Gal, Photodynamic effects of antioxidant substituted porphyrin photosensitizers on gram-positive and-negative bacteria, *Photochem. Photobiol.*, 2003, **77**, 186.
143. M. Kreitner, K.H. Wagner, G. Alth, R. Ebermann, H. Foissy and I. Elmadfa, Haematoporphyrin- and sodium chlorophyllin-induced phototoxicity towards bacteria and yeasts – a new approach for safe foods, *Food Control*, 2001, **12**, 529.
144. K. Lasocki, M. Szpakowska, J. Grzybowski and A. Graczyk, Examination of antibacterial activity of the photoactivated arginine haematoporphyrin derivative, *Pharmacol. Res.*, 1999, **39**, 181.

145. G. Bertoloni, F.M. Lauro, G. Cortella and M. Merchat, Photosensitizing activity of hematoporphyrin on *Staphylococcus aureus* cells, *Biochim. Biophys. Acta, Gen. Subj.*, 2000, **1475**, 169.
146. S. Ferro, F. Ricchelli, G. Mancini, G. Tognon and G. Jori, Inactivation of methicillin-resistant *Staphylococcus aureus* (MRSA) by liposome-delivered photo sensitising agents, *J. Photochem. Photobiol., B*, 2006, **83**, 98.
147. M.J. Shikowitz, A.L. Abramson, K. Freeman, B.M. Steinberg and M. Nouri, Efficacy of DHE photodynamic therapy for respiratory papillomatosis: immediate and long-term results, *Laryngoscope*, 1998, **108**, 962.
148. T.C. Chanh, J.S. Allan, J.L. Matthews, F. Sogandares-Bernal, M.M. Judy, H. Skiles, J. Leveson, A. Marengo-Rowe and J.T. Newman, Photodynamic inactivation of simian immunodeficiency virus, *J. Virol. Methods*, 1989, **26**, 125.
149. L.N. Dovigo, A.C. Pavarina, D.G. Ribeiro, C.S. Adriano and V. Salvador, Photodynamic inactivation of four Candida species induced by photogem®, *Braz. J. Microbiol.*, 2010, **41**, 42.
150. J.M. Bliss, C.E. Bigelow, T.H. Foster and C.G. Haidaris, Susceptibility of *Candida* species to photodynamic effects of photofrin, *Antimicrob. Agents Chemother.*, 2004, **48**, 2000.
151. C.D. Lytle, P. G. Carney, R.P. Felten, H.F. Bushar and R.C. Straight, Inactivation and mutagenesis of herpes-virus by photodynamic treatment with therapeutic dyes, *Photochem. Photobiol.*, 1989, **50**, 367.
152. J.S.M. Giusti, L. Santos-Pinto, A.C. Pizzolitto, C. Kurachi and V.S. Bagnato, Effectiveness of Photogem (R) activated by LED on the decontamination of artificial carious bovine dentin, *Laser Phys.*, 2006, **16**, 859.
153. S. Wood, D. Metcalf, D. Devine and C. Robinson, Erythrosine is a potential photosensitizer for the photodynamic therapy of oral plaque biofilms, *J. Antimicrob. Chemother.*, 2006, **57**, 680.
154. M. Krouit, R. Granet, P. Branland, B. Verneuil and P. Krausz, New photo-antimicrobial films composed of porphyrinated lipophilic cellulose esters, *Bioorg. Med. Chem. Lett.*, 2006, **16**, 1651.
155. Y. Nitzan, A. Balzam-Sudakevitz and H. Ashkenazi, Eradication of *Acinetobacter baumannii* by photosensitized agents in vitro, *J. Photochem. Photobiol., B*, 1998, **42**, 211.
156. Y. Nitzan and H. Ashkenazi, Photoinactivation of *Deinococcus radiodurans*: An unusual gram-positive microorganism, *Photochem. Photobiol.*, 1999, **69**, 505.
157. Y. Nitzan, M. Gutterman, Z. Malik and B. Ehrenberg, Inactivation of Gram-negative bacteria by photosensitized porphyrins, *Photochem. Photobiol.*, 1992, **55**, 89.
158. B.C.M. Yap, G.L. Simpkins, C.A. Collyer, N. Hunter and M.J. Crossley, Porphyrin-linked nitroimidazole antibiotics targeting Porphyromonas gingivalis, *Org. Biomol. Chem.*, 2009, **7**, 2855.
159. A. Orenstein, D. Klein, J. Kopolovic, E. Winkler, Z. Malik, N. Keller and Y. Nitzan, The use of porphyrins for eradication of *Staphylococcus aureus* in burn wound infections, *FEMS Immunol. Med. Microbiol.*, 1997, **19**, 307.
160. T.G. Smijs, R.N.S. van der Haas, J. Lugtenburg, Y. Liu, R.L.P. De Jong and H. J. Schuitmaker, Photodynamic treatment of the dermatophyte *Trichophyton rubrum* and its microconidia with porphyrin photosensitizers, *Photochem. Photobiol.*, 2004, **80**, 197.
161. A.M. Richter, S. Yip, H. Meadows, A.K. Jain, H. Neyndorff, G. Moreno, C. Salet and J.G. Levy, Photosensitising potencies of the structural analogues of

benzoporphyrin derivative in different biological test systems, *J. Clin. Laser Med. Surg.*, 1996, **14**, 335.
162. J. North, H. Neyndorff, D. King and J.G. Levy, Viral inactivation in blood and red cell concentrates with benzoporphyrin derivative, *Blood Cells*, 1992, **18**, 129.
163. A. Oliveira, A. Almeida, C.M. Carvalho, J.P. Tomé, M.A. Faustino, M.G. Neves, A.C. Tomé, J.A. Cavaleiro and A. Cunha, Porphyrin derivatives as photosensitizers for the inactivation of Bacillus cereus endospores, *J. Appl. Microbiol.*, 2009, **106**, 1986.
164. M.P. Cormick, M.G. Alvarez, M. Rovera and E.N. Durantini, Photodynamic inactivation of *Candida albicans* sensitized by tri- and tetra-cationic porphyrin derivatives, *Eur. J. Med. Chem.*, 2009, **44**, 1592.
165. Y. Nitzan and H. Ashkenazi, Photoinactivation of *Acinetobacter baumannii* and *Escherichia coli* B by a cationic hydrophilic porphyrin at various light wavelengths, *Curr. Microbiol.*, 2001, **42**, 408.
166. S. Banfi, E. Caruso, L. Buccafurni, V. Battini, S. Zazzaron, P. Barbieri and V. Orlandi, Antibacterial activity of tetraaryl-porphyrin photosensitizers: An in vitro study on Gram negative and Gram positive bacteria, *J. Photochem. Photobiol., B*, 2006, **85**, 28.
167. Y. Nitzan, R. Dror, H. Ladan, Z. Malik, S. Kimel and V. Gottfried, Structure-activity relationship of porphines for photoinactivation of bacteria, *Photochem. Photobiol.*, 1995, **62**, 342.
168. L. Benov, I. Batinic-Haberle, I. Spasojevic and I. Fridovich, Isomeric N-alkyl-pyridylporphyrins and their Zn(II) complexes: inactive as SOD mimics but powerful photosensitizers, *Arch. Biochem. Biophys.*, 2002, **402**, 159.
169. N.A. Romanova, L.Y. Brovko, L. Moore, E. Pometun, A.P. Savitsky, N.N. Ugarova and M.W. Griffiths, Assessment of photodynamic destruction of *Escherichia coli* O157:H7 and *Listeria monocytogenes* by using ATP bioluminescence, *Appl. Environ. Microbiol.*, 2003, **69**, 6393.
170. Z. Alouini and M. Jemli, Destruction of helminth eggs by photosensitized porphyrin, *J. Environ. Monitor.*, 2001, **3**, 548.
171. C. Kasturi and M.S. Platz, Inactivation of lambda-phage with 658 nm light using a DNA-binding porphyrin sensitizer, *Photochem. Photobiol.*, 1992, **56**, 427.
172. M.J. Casteel, K. Jayaraj, A. Gold, L.M. Ball and M.D. Sobsey, Photoinactivation of hepatitis A virus by synthetic porphyrins, *Photochem. Photobiol.*, 2004, **80**, 294
173. E. Feese and R.A. Ghiladi, Highly efficient in vitro photodynamic inactivation of *Mycobacterium smegmatis*, *J. Antimicrob. Chemother.*, 2009, **64**, 782.
174. S.A.G. Lambrechts, M.C.G. Aalders, D.H. Langeveld-Klerks, Y. Khayali and J.W.M. Lagerberg, Effect of monovalent and divalent cations on the photoinactivation of bacteria with *meso*-substituted cationic porphyrins, *Photochem. Photobiol.*, 2004, **79**, 297.
175. L.L. Trannoy, J.W.M. Lagerberg, B.M.A.R. Dubbelman, H.J. Schuitmaker and A. Brand, Positively charged porphyrins: a new series of photosensitizers for sterilization of RBCs, *Transfusion*, 2004, **44**, 1186.
176. E. Reddi, M. Ceccon, G. Valduga, G. Jori, J. Bommer, F. Elisei, L. Latterini and U. Mazzucato, Photophysical properties and antibacterial activity of *meso*-substituted cationic porphyrins, *Photochem. Photobiol.*, 2002, **75**, 462.
177. S. Ferro, F. Ricchelli, D. Monti, G. Mancini and G. Jori, Efficient photoinactivation of methicillin-resistant *Staphylococcus aureus* by a novel porphyrin incorporated into a poly-cationic liposome, *Int. J. Biochem. Cell. Biol.*, 2007, **39**, 1026.

178. K.G. Yu, D.H. Li, C.H. Zhou and J.L. Dia, Study on the synthesis and antimicrobial activity of novel cationic porphyrins, *Chinese Chem. Lett.*, 2009, **20**, 411.
179. C.F. Xing, Q.L. Xu, H.W. Tang, L.B. Liu and S. Wang, Conjugated polymer/porphyrin complexes for efficient energy transfer and improving light-activated antibacterial activity, *J. Am. Chem. Soc.*, 2009, **131**, 13117.
180. D.A. Caminos, M.B. Spesia and E.N. Durantini, Photodynamic inactivation of Escherichia coli by novel *meso*-substituted porphyrins by 4-(3-N,N,N-trimethylammoniumpropoxy)phenyl and 4-(trifluoromethyl)phenyl groups, *Photochem. Photobiol. Sci.*, 2006, **5**, 56.
181. M.B. Spesia, D. Lazzeri, L. Pascual, M. Rovera and E.N. Durantini, Photoinactivation of *Escherichia coli* using porphyrin derivatives with different number of cationic charges, *FEMS Immunol. Med. Microbiol.*, 2005, **44**, 289.
182. D. Lazzeri, M. Rovera, L. Pascual and E.N. Durantini, Photodynamic studies and photoinactivation of *Escherichia coli* using *meso*-substituted cationic porphyrin derivatives with asymmetric charge distribution, *Photochem. Photobiol.*, 2004, **80**, 286.
183. S.A.G. Lambrechts, M.C.G. Aalders and J. Van Marle, Mechanistic study of the photodynamic inactivation of *Candida albicans* by a cationic porphyrin, *Antimicrob. Agents Chemother.*, 2005, **49**, 2026.
184. J.P.C. Tomé, E.M.P. Silva, A.M.V.M. Pereira, C.M.A. Alonso, M.A.F. Faustino, M.G.P.M.S. Neves, A.C. Tomé, J.A.S. Cavaleiro, S.A.P. Tavares, R.R. Duarte, M.F. Caeiro and M.L. Valdeira, Synthesis of neutral and cationic tripyridylporphyrin-D-galactose conjugates and the photoinactivation of HSV-1, *Bioorg. Med. Chem.*, 2007, **15**, 4705.
185. K. Steenkeste, F. Tfibel, M. Perreìe-Fauvet, R. Briandet and M.-P. Fontaine-Aupart, Tracking the photosensitizing antibacterial activity of mono(acridyl)-bis(arginyl)porphyrin (MABAP) by time-resolved spectroscopy, *J. Phys. Chem. A*, 2010, **114**, 3334.
186. S.A.G. Lambrechts, M.C.G. Aalders, F.D. Verbraak, J.W.M. Lagerberg, J.B. Dankert and J.J. Schuitmaker, Effect of albumin on the photodynamic inactivation of microorganisms by a cationic porphyrin, *J. Photochem. Photobiol., B*, 2005, **79**, 51.
187. J.P.C. Tomé, M.G.P.M.S. Neves, A.C. Tomé, J.A.S. Cavaleiro, M. Soncin, M. agaraggia, S. Ferro and G. Jori, Synthesis and antibacterial activity of new poly-S-lysine-porphyrin conjugates, *J. Med. Chem.*, 2004, **47**, 6649.
188. L.L. Trannoy, Y. van Hensbergen, J.W.M. Lagerberg and A. Brand, Photodynamic treatment with mono-phenyl-tri-(N-methyl-4-pyridyl)-porphyrin for pathogen inactivation in cord blood stem cell products, *Transfusion*, 2008, **48**, 2629.
189. S.A.G. Lambrechts, T.N. Demidova, M.C.G. Aalders, T. Hasan and M.R. Hamblin, Photodynamic therapy for Staphylococcus aureus infected burn wounds in mice, *Photochem. Photobiol. Sci.*, 2005, **4**, 503.
190. M. Krouit, R. Granet and P. Krausz, Photobactericidal plastic films based on cellulose esterified by chloroacetate and a cationic porphyrin, *Bioorg. Med. Chem.*, 2008, **16**, 10091.
191. M. Krouit, R. Granet and P. Krausz, Photobactericidal films from porphyrins grafted to alkylated cellulose – synthesis and bactericidal properties, *Eur. Polym. J.*, 2009, **45**, 1250.
192. J.P.C. Tomé, M.G.P.M.S. Neves, A.C. Tomé, J.A.S. Cavaleiro, A.F. Mendonça, L.S.N. Pegado, R. Duarte and M.L. Valdeira, Synthesis of glycoporphyrin

derivatives and their antiviral activity against herpes simplex virus types 1 and 2, *Bioorg. Med. Chem.*, 2005, **13**, 3878.
193. F. Gábor, J. Szolnoki, K. Toth, A. Fekete, P. Maillard and G. Csik, Photoinduced inactivation of T7 phage sensitized by symmetrically and asymmetrically substituted tetraphenyl porphyrin: Comparison of efficiency and mechanism of action, *Photochem. Photobiol.*, 2001, **73**, 304.
194. J. Schwiertz, A. Wiehe, S. Grafe, B. Gitter and M. Epple, Calcium phosphate nanoparticles as efficient carriers for photodynamic therapy against cells and bacteria, *Biomaterials*, 2009, **30**, 3324.
195. D.A. Caminos and E.N. Durantini, Interaction and photodynamic activity of cationic porphyrin derivatives bearing different patterns of charge distribution with GMP and DNA, *J. Photochem. Photobiol., A*, 2008, **198**, 274.
196. R.M. Ion and M.A. Calin, Comparative study of some nano- and micro-sensitizers in photodynamic inactivation of microorganisms, *J. Optoelectr. Adv. Mater.*, 2007, **9**, 1933.
197. F. Vargas, C. Rivas, T. Zoltan, L. Padron, C. Izzo, V. Lopez, L. Gomez, F. Pujol, H. Rangel, D. Garzaro and R. Fabbro, Comparative antiviral (HIV) photoactivity of metalized *meso*-tetraphenylsulfonated porphyrins, *Med. Chem.*, 2008, **4**, 138.
198. A. Kubin, F. Wierrani, R.H. Jindra, H.G. Loew, W. Grunberger, R. Ebermann and G. Alth, Antagonistic effects of combination photosensitization by hypericin, meso-tetrahydroxyphenylchlorin (mTHPC) and photofrin II on Staphylococcus aureus, *Drugs Exp. Clin. Res.*, 1999, **25**, 13.
199. R. Ackroyd, C. Kelty, N. Brown and M. Reed, The history of photodetection and photodynamic therapy, *Photochem. Photobiol.*, 2001, **74**, 656.
200. G.H. Cassell and J. Mekalanos, Development of antimicrobial agents in the era of new and reemerging infectious diseases and increasing antibiotic resistance, *J. Am. Med. Assoc.*, 2001, **285**, 601.
201. S. Gross, A. Brandis, L. Chen, V. Rosenbach-Belkin, S. Roehrs, A. Scherz and Y. Salomon, Protein-A-mediated targeting of bacteriochlorophyll-IgG to *Staphylococcus aureus*: a model for enhanced site-specific photocytotoxicity, *Photochem. Photobiol.*, 1997, **66**, 872.
202. F. Van Der Meulen, K. Ibrahim, H. Sterenborg, L. Alphen, A. Maikoe and J. Dankert, Photodynamic destruction of Haemophilus parainfluenzae by endogenously produced porphyrins, *J. Photochem. Photobiol. B: Biol.*, 1997, **40**, 204.
203. V. Sol, P. Branland, R. Granet, C. Kaldapa, B. Verneuil and P. Krausz, Nitroglycosylated *meso*-arylporphyrins as photoinhibitors of Gram positive bacteria, *Bioorg. Med. Chem. Lett.*, 1998, **8**, 3007.
204. D.O. Lapotko, V.P. Zharov, T.R. Romanovskaya and G.S. Kuchinskii, Investigation of the influence of the photodynamic effect on micro-organisms using the laser photothermal cytometry method, *Quant. Electron.*, 1999, **29**, 1060.
205. F. Wierrani, Experimental investigations and clinical use of photodynamic therapy (PDT) in the Rudolf Foundation Hospital, *Gynakol. Geburt. Rundsch.*, 1999, **39**, 217.
206. T.O. Philippova, B.N. Galkin, O. Yu. Zinchenko, M. Yu. Rusakova, V.A. Ivanitsa, Z.I. Zhilina, S.V. Vodzinskii and Y.V. Ishkov, The antimicrobial properties of new synthetic porphyrins, *J. Porphyrins Phthalocyanines*, 2003, **11**, 755.
207. V. Sol, P. Branland, V. Chaleix, R. Granet, M. Guilloton, F. Lamarche, B. Verneuil and P. Krausz, Amino porphyrins as photoinhibitors of Gram-positive and-negative bacteria, *Bioorg. Med. Chem. Lett.*, 2004, **14**, 4207.

208. M.L. Embleton, S.P. Nair, B.D. Cookson and M. Wilson, Antibody-directed photodynamic therapy of methicillin resistant *Staphylococcus aureus*, *Microb. Drug Resist.*, 2004, **10**, 92.
209. D.A. Caminos, M.B. Spesia and E.N. Durantini, Photodynamic inactivation of *Escherichia coli* by novel *meso*-substituted porphyrins by 4-(3-N,N,N-trimethylammoniumpropoxy)phenyl and 4-(trifluoromethyl)phenyl groups, *Photochem. Photobiol. Sci.*, 2006, **5**, 56.
210. K. O'Riordan, D.S. Sharlin, J. Gross, S. Chang, D. Errabelli, O.E. Akilov, S. Kosaka, G.J. Nau and T. Hasan, Photoinactivation of mycobacteria in vitro and in a new murine model of localized *Mycobacterium bovis* BCG-induced granulomatous infection, *Antimicrob. Agents Chemother.*, 2006, **50**, 1828.
211. D. Caminos and E. Durantini, A simple experiment to show photodynamic inactivation of bacteria on surfaces, *Biochem. Mol. Biol. Educ.*, 2007, **35**, 64.
212. M. Drábková, B. Marsálek and W. Admiraal, Photodynamic therapy against cyanobacteria, *Environ. Toxicol.*, 2007, **22**, 112.
213. A.A. El-Adly, Photoactive anionic porphyrin derivative against Gram-positive and Gram-negative bacteria, *J. Appl. Sci. Res.*, 2008, **4**, 1817.
214. R.F. Donnelly, C.M. Cassidy, R.G. Loughlin, A. Brown, M.M. Tunney, M.G. Jenkins and P.A. McCarron, Delivery of methylene blue and *meso*-tetra(N-methyl-4-pyridyl)porphine tetra tosylate from cross-linked poly (vinyl alcohol) hydrogels: A potential means of photodynamic therapy of infected wounds, *J. Photochem. Photobiol. B: Biol.*, 2009, **96**, 223.
215. M.D. Funes, D.A. Caminos, M.G. Alvarez, F. Fungo, L.A. Otero and E.N. Durantini, Photodynamic properties and photoantimicrobial action of electrochemically generated porphyrin polymeric films, *Environ. Sci. Technol.*, 2009, **43**, 902.
216. N. Ooi, K. Miller, C. Randall, W. Rhys-Williams, W. Love and I. Chopra, XF-70 and XF-73, novel antibacterial agents active against slow-growing and non-dividing cultures of *Staphylococcus aureus* including biofilms, *J. Antimicrob. Chemother.*, 2010, **65**, 72.
217. A. Tavares, E. Alves, C. Carvalho, J.P.C. Tomé, M.A.F. Faustino, M.G.P.M.S. Neves, A.C. Tomé, J.A.S. Cavaleiro, A. Cunha and A. Almeida, Mechanisms of photoinactivation of Gram-negative recombinant bioluminescent bacteria by cationic porphyrins, submitted.
218. D.W. Dixon, L.G. Marzilli and R.F. Schinazi, Porphyrins as agents against the human-immunodeficiency-virus, *Ann. NY Acad. Sci.*, 1990, **616**, 511.
219. H. Neyndorff, D. Bartel, F. Tufaro and J. Levy, Development of a model to demonstrate photosensitizer-mediated viral inactivation in blood, *Transfusion*, 1990, **30**, 485.
220. R.D. Levere, Y.F. Gong, A. Kappas, D.J. Bucher, G. Wormser and N.G. Abraham, Heme inhibits human immunodeficiency virus 1 replication in cell cultures and enhances the antiviral effect of zidovudine, *Proc. Natl. Acad. Sci.*, 1991, **88**, 1756.
221. D.W. Dixon, M.S. Kim, V. Kumar, G. Obara, L.G. Marzilli and R.F. Schinasi, Amino-and hydroxytetraphenylporphyrins with activity against the human immunodeficiency virus, *Antiviral Chem.*, 1992, **3**, 279.
222. J. North, H. Neyndorff and J.G. Levy, New trends in photobiology: Photosensitizers as virucidal agents, *J. Photochem. Photobiol.*, 1993, **14**, 99.
223. R. Staudinger, N. Abraham, R. Levere and A. Kappas, Inhibition of human immunodeficiency virus-1 reverse transcriptase by heme and synthetic heme analogs, *Proc. Assoc. Am. Physicians*, 1996, **108**, 47.

224. E. Argyris, J. Vanderkooi, P. Venkateswaran, B. Kay and Y. Paterson, The connection domain is implicated in metalloporphyrin binding and inhibition of HIV reverse transcriptase, *J. Biol. Chem.*, 1999, **274**, 1549.
225. A. Debnath, S. Jiang, N. Strick, K. Lin, S. Kahl and A. Neurath, Anti-HIV-1 activity of carborane derivatives of porphyrins, *Med. Chem. Res.*, 1999, **9**, 267.
226. D. Lim, S. Ko, S. Kim, Y. Park, J. Park and W. Lee, Photoinactivation of vesicular stomatitis virus by a photodynamic agent, chlorophyll derivatives from silkworm excreta, *J. Photochem. Photobiol. B: Biol.*, 2002, **67**, 149.
227. A. Chen-Collins, D. Dixon, A. Vzorov, L. Marzilli and R. Compans, Prevention of poxvirus infection by tetrapyrroles, *BMC Infect. Dis.*, 2003, **3**, 9.
228. K. Zupán, L. Herenyi, K. Toth, Z. Majer and G. Csik, Binding of cationic porphyrin to isolated and encapsidated viral DNA analyzed by comprehensive spectroscopic methods, *Biochemistry (Mosc.)*, 2004, **43**, 9151.
229. J. Liu, S. Shi and L.-N. Ji, Investigation on DNA binding and photo-cleavage properties of water-soluble porphyrin and metalloporphyrins, *Transition Metal Chem.*, 2005, **30**, 684.
230. A. Vzorov, J. Bozja, D. Dixon, L. Marzilli and R. Compans, Parameters of inhibition of HIV-1 infection by small anionic microbicides, *Antiviral Res.*, 2007, **73**, 60.
231. L. Lin and J. Hu, Inhibition of hepadnavirus reverse transcriptase RNA interaction by porphyrin compounds, *J. Virology*, 2008, **82**, 2305.
232. G. Csík, M. Egyeki, L. Herényi, Z. Majer and K. Tóth, Role of structure-proteins in the porphyrin-DNA interaction, *J. Photochem. Photobiol. B: Biol.*, 2009, **96**, 207.
233. M.L. Sagristá, F. Postigo, M.A. Madariaga, R.M. Pintó, S. Caballero, A. Bosch, M.A. Vallés and M. Mora, Photodynamic inactivation of viruses by immobilized chlorin-containing liposomes, *J. Porphyrins Phthalocyanines*, 2009, **13**, 578.
234. W. Wen, M. Lin, C. Su, S. Wang, Y. Cheng, J. Fang and C. Wong, Synergistic effect of Zanamivir-Porphyrin conjugates on inhibition of neuraminidase and inactivation of influenza virus, *J. Med. Chem.*, 2009, **52**, 4903.
235. J.S. Friedberg, C. Skema, E.D. Baum, J. Burdick, S.A. Vinogradov, D.F. Wilson, A.D. Horan and I. Nachamkin, In vitro effects of photodynamic therapy on Aspergillus fumigatus, *J. Antimicrob. Chemother.*, 2001, **48**, 105.
236. T. Philippova, B. Galkin, O. Zinchenko, M. Rusakova, V. Ivanitsa, Z. Zhilina, S. Vodzinskii and Y. Ishkov, The antimicrobial properties of new synthetic porphyrins, *J. Porphyrins Phthalocyanines*, 2004, **7**, 755.
237. J. Bliss, C. Bigelow, T. Foster and C. Haidaris, Susceptibility of Candida species to photodynamic effects of Photofrin, *Antimicrob. Agents Chemother.*, 2004, **48**, 2000.
238. Z. Luksiene, D. Peciulyte and A. Lugauskas, Inactivation of fungi in vitro by photosensitization: preliminary results, *Ann. Agr. Env. Med.: AAEM*, 2004, **11**, 215.
239. Y. Chabrier-Rosello, T.H. Foster, N. Perez-Nazario, S. Mitra and C.G. Haidaris, Sensitivity of *Candida albicans* germ tubes and biofilms to Photofrin-mediated phototoxicity, *Antimicrob. Agents Chemother.*, 2005, **49**, 4288.
240. Z. Luksiene, D. Peciulyte, S. Jurkoniene and R. Puras, Inactivation of possible fungal food contaminants by photosensitization, *Food Technol. Biotechnol.*, 2005, **43**, 335.
241. G. Monfrecola, E.M. Procaccini, A. Bevilacqua, A. Manco, G. Calabro and P. Santoianni, In vitro effect of 5-aminolaevulinic acid plus visible light on *Candida albicans*, *Photochem. Photobiol. Sci.*, 2004, **3**, 419.

242. T. Smijs, J. Bouwstra, H. Schuitmaker, M. Talebi and S. Pavel, A novel ex vivo skin model to study the susceptibility of the dermatophyte *Trichophyton rubrum* to photodynamic treatment in different growth phases, *J. Antimicrob. Chemother.*, 2007, **59**, 433.
243. R. Ghazaryan, L. Sahakyan, A. Tovmasyan, L. Movsisyan and A. Hambardzumyan, Novel antimicrobial agents on the base of natural and synthetic metalloporphyrins, *New Armenian Med. J.*, 2008, **2**, 40.
244. T. Ito, Photodynamic action of hematoporphyrin on yeast cells – a kinetic approach, *Photochem. Photobiol.*, 2008, **34**, 521.
245. A. Vorobey and S. Pinchuk, Photodamage to spores of Fusarium fungi sensitized by protoporphyrin IX, *Biophysics*, 2008, **53**, 386.
246. T.G.M. Smijs, S. Pavel, M. Talebi and J.A. Bouwstra, Preclinical studies with 5,10,15-tris(4-methylpyridinium)-20-phenyl-[21H,23H]-porphine trichloride for the photodynamic treatment of superficial mycoses caused by Trichophyton rubrum, *Photochem. Photobiol.*, 2009, **85**, 733.
247. S.J. Mora, M.P. Cormick, M.E. Milanesio and E.N. Durantini, The photodynamic activity of a novel porphyrin derivative bearing a fluconazole structure in different media and against *Candida albicans*, *Dyes Pigm.*, 2010, **87**, 234.
248. P.G. Calzavara-Pinton, M. Venturini, R. Capezzera, R. Sala and C. Zane, Photodynamic therapy of interdigital mycoses of the feet with topical application of 5-aminolevulinic acid, *Photodermatol. Photoimmunol. Photomed.*, 2004, **20**, 144.
249. B.M. Piraccini, G. Rech and A. Tosti, Photodynamic therapy of onychomycosis caused by Trichophyton rubrum, *J. Am. Acad. Dermatol.*, 2008, **59**, S75.
250. P. Grellier, R. Santus, E. Mouray, V. Agmon, J.-C. Mazière, D. Rigomier, A. Dagan, S. Gatt and J. Schrével, Photosensitized inactivation of Plasmodium falciparum and Babesia divergens infected erythrocytes in whole blood by lipophilic pheophorbide derivatives, *Vox Sang.*, 1997, **72**, 211.
251. K. Kassab, T. Amor, G. Jori and O. Coppellotti, Photosensitization of Colpoda inflata cysts by meso-substituted cationic porphyrins, *Photochem. Photobiol. Sci.*, 2002, **1**, 560.
252. C.-A. Bristow, R. Hudson, T.A. Paget and R.W. Boyle, Potential of cationic porphyrins for photodynamic treatment of cutaneous leishmaniasis, *Photodiagn. Photodyn. Ther.*, 2006, **3**, 162.
253. O.E. Akilov, S. Kosaka, K. O'Riordan and T. Hasan, Photodynamic therapy for cutaneous leishmaniasis: the effectiveness of topical phenothiaziniums in parasite eradication and Th1 immune response stimulation, *Exp. Dermatol.*, 2007, **16**, 651.
254. J.B. Morgenthaler, S.J. Peters, D.L. Cedeño, M.H. Constantino, K.A. Edwards, E.M. Kamowski, J.C. Passini, B.E. Butkus, A.M. Young, T.D. Lash and M.A. Jones, Carbaporphyrin ketals as potential agents for a new photodynamic therapy treatment of leishmaniasis, *Bioorg. Med. Chem.*, 2008, **16**, 7033.
255. K. Gardlo, Z. Horska, C.D. Enk, L. Rauch, M. Megahed, T. Ruzicka and C. Fritsch, Treatment of cutaneous leishmaniasis by photodynamic therapy, *J. Am. Acad. Dermatol.*, 2003, **48**, 893.
256. K. Gardlo, S. Hanneken, T. Ruzicka and N.J. Neumann, Photodynamische therapie bei kutaner leishmaniose, *Der Hautarzt*, 2004, **55**, 381.
257. A. Asilian and M. Davami, Comparison between the efficacy of photodynamic therapy and topical paromomycin in the treatment of Old World cutaneous leishmaniasis: a placebo-controlled, randomized clinical trial, *Indian J. Dermatol.*, 2006, **51**, 182.

258. E.M. van der Snoek, D.J. Robinson, J.J. van Hellemond and H.A.M. Neumann, A review of photodynamic therapy in cutaneous leishmaniasis, *J. Eur. Acad. Dermatol. Venereol.*, 2008, **22**, 918.
259. K. König, H. Meyer, H. Schneckenburger and A. Rück, The study of endogenous porphyrins in human skin and their potential for photodynamic therapy by laser induced fluorescence spectroscopy, *Lasers Med. Sci.*, 1993, **8**, 127.
260. W. Hongcharu, C. Taylor, Y. Chang, D. Aghassi, K. Suthamjariya and R. Anderson, Topical ALA-photodynamic therapy for the treatment of acne vulgaris, *J. Invest. Dermatol.*, 2000, **115**, 183.
261. Y. Itoh, Y. Ninomiya, S. Tajima and A. Ishibashi, Photodynamic therapy for acne vulgaris with topical 5-aminolevulinic acid, *Arch. Dermatol.*, 2000, **136**, 1093.
262. Y. Itoh, Y. Ninomiya, S. Tajima and A. Ishibashi, Photodynamic therapy of acne vulgaris with topical aminolaevulinic acid and incoherent light in Japanese patients, *Br. J. Dermatol.*, 2001, **144**, 575.
263. B. Pollock, D. Turner, M. Stringer, R. Bojar, V. Goulden, G. Stables and W. Cunliffe, Topical aminolaevulinic acid-photodynamic therapy for the treatment of acne vulgaris: a study of clinical efficacy and mechanism of action, *Br. J. Dermatol.*, 2004, **151**, 616.
264. C. Borelli, K. Merk, M. Schaller, K. Jacob, M. Vogeser, G. Weindl, U. Berger and G. Plewig, In vivo porphyrin production by P. acnes in untreated acne patients and its modulation by acne treatment, *Acta Derm. Venereol.*, 2006, **86**, 316.
265. C. Hörfelt, J. Funk, M. Frohm-Nilsson, D. Edström and A. Wennberg, Topical methyl aminolaevulinate photodynamic therapy for treatment of facial acne vulgaris: results of a randomized, controlled study, *Br. J. Dermatol.*, 2006, **155**, 608.
266. S. Wiegell and H. Wulf, Photodynamic therapy of acne vulgaris using methyl aminolaevulinate: a blinded, randomized, controlled trial, *Br. J. Dermatol.*, 2006, **154**, 969.
267. T. Maisch, C. Bosl, R. Szeimies, B. Love and C. Abels, Determination of the antibacterial efficacy of a new porphyrin-based photosensitizer against MRSA ex vivo, *Photochem. Photobiol. Sci.*, 2007, **6**, 545.
268. F. Berthiaume, S. Reiken, M. Toner, R. Tompkins and M. Yarmush, Antibody-targeted photolysis of bacteria in vivo, *Nat. Biotechnol.*, 1994, **12**, 703.
269. M.R. Hamblin, T. Zahra, C.H. Contag, A.T. McManus and T. Hasan, Optical monitoring and treatment of potentially lethal wound infections in vivo, *J. Infectious Diseases*, 2003, **187**, 1717.
270. M.G. Hurtuk, L.K. He, A. Szilagyi, R.L. Gamelli, D.W. Hecht, R.H. Kennedy, W. Rhys-Williams, W.G. Love and R. Shankar, The novel antibacterial drug XF-70 is a potent inhibitor of Staphylococcus aureus infection of the burn wound, *J. Burn Care Res.*, 2010, **31**, 462.
271. C. Henry, M. Judy, B. Dyer, M. Wagner and J. Matthews, Sensitivity of Porphyromonas and Prevotella species in liquid media to argon laser, *Photochem. Photobiol.*, 1995, **61**, 410.
272. J. Dobson and M. Wilson, Sensitization of oral bacteria in biofilms to killing by light from a low-power laser, *Arch. Oral Biol.*, 1992, **37**, 883.
273. K. König, M. Teschke, B. Sigusch, E. Glockmann, S. Eick and W. Pfister, Red light kills bacteria via photodynamic action, *Cell. Mol. Biol. (Noisy-le-Grand, France)*, 2000, **46**, 1297.
274. N. Soukos, S. Mulholland, S. Socransky and A. Doukas, Photodestruction of human dental plaque bacteria: enhancement of the photodynamic effect by photomechanical waves in an oral biofilm model, *Lasers Surg. Med.*, 2003, **33**, 161.

275. C. Lee, C. Lee, C. Chen and C. Huang, [delta]-Aminolaevulinic acid mediated photodynamic antimicrobial chemotherapy on Pseudomonas aeruginosa planktonic and biofilm cultures, *J. Photochem. Photobiol. B: Biol.*, 2004, **75**, 21.
276. C.E. Millson, Photodynamic therapy for the treatment of *Helicobacter pylori* infection, PhD Thesis, University of London, 1995, 107.
277. M.R. Hamblin, A. Ahmadi, M.J. Tolkoff and T. Zahra, Light-mediated killing of *Helicobacter pylori* in vitro and ex vivo, 30th Annual Meeting Am. Soc. Photobiol., Quebec City, Canada, Am. Soc. Photobiol., Washington, DC, 2002, p. 30.
278. H. Wolfsen and K. Wang, Tolerance of normal swine gastric epithelium to photodynamic therapy using topically applied haematoporphyrin derivative, *Lasers. Surg. Med.*, 1993, **5**, 18.
279. C. Wilder-Smith, P. Wilder-Smith, P. Grosjean, H. Van Den Bergh, A. Woodtli, P. Monnier, G. Dorta, F. Meister and G. Wagnières, Photoeradication of Helicobacter pylori using 5-aminolevulinic acid: preliminary human studies, *Lasers Surg. Med.*, 2002, **31**, 18.
280. R. Ganz, J. Viveiros, A. Ahmad, A. Ahmadi, A. Khalil, M. Tolkoff, N. Nishioka and M. Hamblin, *Helicobacter pylori* in patients can be killed by visible light, *Lasers Surg. Med.*, 2005, **36**, 260.
281. A. Lembo, R. Ganz, S. Sheth, D. Cave, C. Kelly, P. Levin, P. Kazlas, P. Baldwin III, W. Lindmark and J. Mcgrath, Treatment of *Helicobacter pylori* infection with intra-gastric violet light phototherapy: A pilot clinical trial, *Lasers Surg. Med.*, 2009, **41**, 337.
282. V.M. Mullooly, A.L. Abramson and M.J. Shikowitz, Dihemato- porphyrin ether-induced photosensitivity in laryngeal papilloma patients, *Lasers Surg. Med.*, 1990, **10**, 349.
283. N.M. Bauman and R.J. Smith, Recurrent respiratory papillomatosis, *Pediatric Clinics North America*, 1996, **43**, 1385.
284. R.G. Frank and J.D. Bos, Photodynamic therapy for condylomata acuminata with local application of 5-aminolevulinic acid, *Genitourin Med.*, 1996, **72**, 70.
285. M.K. Fehr, R. Hornung, A. Degen, V.A. Schwarz, D. Fink, U. Haller and P. Wyss, Photodynamic therapy of vulvar and vaginal condyloma and intraepithelial neoplasia using topically applied 5-aminolevulinic acid, *Lasers Surg. Med.*, 2002, **30**, 273.
286. I.M. Stender, R. Na and H. Fogh, Recalcitrant hand and foot warts successfully treated with photodynamic therapy with topical 5-aminolevulinic acid, *A pilot study. Clin. Exp. Dermatol.*, 1999, **24**, 154.
287. I.M. Stender, R. Na and H. Fogh, Photodynamic therapy with 5-aminolaevulinic or placebo for recalcitrant foot and hand warts: randomized double-blind trial, *Lancet*, 2000, **355**, 963.
288. G. Fabbrocini, M. Costanzo and A. Ricardo, Photodynamic therapy with topical aminolevulinic acid for the treatment of plantar warts, *J. Photochem. Photobiol. B.*, 2001, **61**, 30.
289. I.M. Stefanaki, S. Georgiou, G.C. Themelis, E.M. Vazgiouraki and A.D. Tosca, In vivo fluorescence kinetics and photodynamic therapy in condylomata acuminata, *Br. J. Dermatol.*, 2003, **149**, 972.
290. P. Schleier, P. Hyckel, A. Berndt, H.P. Bode, V. Albrecht, W. Hindermann, H. Kosmehl, W. Zenk and D. Schumann, Photodynamic therapy of virus-associated epithelial tumours of the face in organ transplant recipients, *J. Cancer Res. Clin. Oncol.*, 2004, **130**, 279.

291. X.L. Wang, H.W. Wang, H.S. Wang, S.Z. Xu, K.H. Liao and P. Hillemanns, Topical 5-aminolaevulinic acid-photodynamic therapy for the treatment of urethral condylomata acuminata, *Br. J. Dermatol.*, 2004, **151**, 880.
292. J.M. Shikowitz, A.L. Abramson, B.M. Steinberg, J. DeVoti, V.R. Bonagura, V. Mullooly, M. Nouri, A.M. Ronn, A. Inglis, J. McClay and K. Freeman, Clinical trial of photodynamic therapy with *meso*-tetra (hydroxyphenyl) chlorin for respiratory papillomatosis, *Arch. Otolaryngol. Head Neck Surg.*, 2005, **13**, 99.
293. T. Herzinger, P. Wienecke, P. Weisenseel, C. Borelli, C. Berking and K. Degitz, Photodynamic therapy of genital condylomata in men, *Clin. Exp. Dermatol.*, 2006, **31**, 51.
294. C.A. Schroeter, J. Pleunis, V. Nispen and C. Pannerden, Photodynamic therapy: New treatment for therapy-resistant plantar warts, *Dermatol. Surg.*, 2005, **31**, 71.
295. K. Chen, B. Chang, M. Ju, X. Zhang and H. Gu, Comparative study of photodynamic therapy vs. CO2 laser vaporization in treatment of condylomata acuminata, a randomized clinical trial, *Br. J. Dermatol.*, 2007, **156**, 516.
296. D. Kacerovska, K. Pizinger, M. Kumpova and P. Cetkovska, Genital warts treated by photodynamic therapy, *SKINmed: Dermatology for the Clinician*, 2007, **6**, 295.
297. V. Nucci, D. Torchia and P. Cappugi, Treatment of anogenital condylomata acuminata with topical photodynamic therapy: report of 14 cases and review, *Clin. Infect. Dis.*, 2009, **48**, 1330.
298. J. Liu, S. Shi and L.-N. Ji, Investigation on DNA binding and photo-cleavage properties of water-soluble porphyrin and metalloporphyrins, *Transition Metal Chem.*, 2005, **30**, 684.
299. A.L. Abramson, M. Waner and J. Brandsma, The clinical treatment of laryngeal papillomas with hematoporphyrin therapy, *Arch. Otolaryngol. Head Neck Surg.*, 1988, **114**, 795.
300. A. Debnath, L. Radigan and S. Jiang, Anti-HIV-1 activity of carborane derivatives of porphyrins, *J. Med. Chem.*, 1999, **42**, 3203.
301. P.G. Calzavara-Pinton, M. Venturini and R. Sala, A comprehensive overview of photodynamic therapy in the treatment of superficial fungal infections of the skin, *J. Photochem. Photobiol. B: Biol.*, 2005, **78**, 1.
302. R.F. Donnelly, P.A. Mccarron and M.M. Tunney, Antifungal photodynamic therapy, *Microbiol. Res.*, 2008, **163**, 1.
303. F.P. Gonzales, S.H. Da Silva, D.W. Roberts and G.U. Braga, Photodynamic inactivation of conidia of the fungi *Metarhizium anisopliae* and *Aspergillus nidulans* with methylene blue and toluidine blue, *Photochem. Photobiol.*, 2010, **86**, 653.
304. E. Latorre-Esteves, O.E. Akilov, P. Rai, S.M. Beverley and T. Hasan, Monitoring the efficacy of antimicrobial photodynamic therapy in a murine model of cutaneous leishmaniasis using L. major expressing GFP, *Biophotonics*, 2010, **3**, 328.
305. T. Smith and K. Kain, Inactivation of Plasmodium falciparum by photodynamic excitation of heme cycle intermediates derived from aminolevulinic acid, *J. Infect. Dis.*, 2004, **190**, 184.
306. F. Käsermann and C. Kempf, Buckminsterfullerene and photodynamic inactivation of viruses, *Rev. Med. Virology*, 1998, **8**, 143.
307. M.A. Almeida, J.A.S. Cavaleiro, J. Rocha, C.M.B. Carvalho, L.A.S. Costa, E.S.C.F. Alves, M.A.S.D.A. Cunha, J.P.C. Tomé, M.A.F. Faustino, M.G.P.M.S. Neves, A.C. Tomé, Z. Lin and J.P.J. Rainho, Materiais híbridos nanomagnete-

porfirina, processo para a sua síntese e respectiva aplicação em desinfecção de águas (Nanomagnet-porphyrin hybrid materials: synthesis and water desinfection application), Patent PT103828, 2009.

308. C. Carvalho, A. Gomes, M.A. Faustino, J.P. Tomé, M.G. Neves, A.C. Tomé, Z. Lin, J.P. Raínho, S.C. Fernandes, A. Almeida, A. Cunha, J. Rocha and J.A. Cavaleiro, Functional cationic nanomagnet-porphyrin hybrids for the photoinactivation of microorganisms, *ACS Nano.*, accepted.
309. E. Alves, C. Carvalho, J.P.C. Tomé, M.A.F. Faustino, M.G.P.M.S. Neves, A.C. Tomé, J.A.S. Cavaleiro, A. Cunha and A. Almeida, Antimicrobial photodynamic therapy in aquaculture: effect of pH, temperature, salinity, oxygen and suspended solids on the photoinactivation of Vibrio fischeri, submitted.
310. K.S. Baker and R.C. Smith, in *The Role of Solar Radiation in Marine Ecosystems*, ed. J. Calkins, Plenum Press, New York, 1982, p. 233.
311. E. Alves, Evaluation of antimicrobial photodynamic therapy by porphyrins using recombinant bioluminescent *Escherichia coli*, MSc Thesis, University of Aveiro, 2007.
312. D. Faust, K. Funken, G. Horneck, B. Milow, J. Ortner, M. Sattlegger, M. Schäfer and C. Schmitz, Immobilized photosensitizers for solar photochemical applications, *Sol. Energy*, 1999, **65**, 71.
313. L. Costa, Sewage bacteriophage photoinactivation by free and immobilized porphyrins, MSc Thesis, University of Aveiro, 2007.
314. Z. Malik, J. Hanania and Y. Nitzan, New trends in photobiology bactericidal effects of photoactivated porphyrins – An alternative approach to antimicrobial drugs, *J. Photochem. Photobiol. B: Biol.*, 1990, **5**, 281.
315. H. Nikaido, Prevention of drug access to bacterial targets: permeability barriers and active efflux, *Science*, 1994, **264**, 382.
316. E.S. Nyman and P.H. Hynninen, Research advances in the use of tetrapyrrolic photosensitizers for photodynamic therapy, *J. Photochem. Photobiol.: B*, 2004, **73**, 1.
317. M.E. Milanesio, M.G. Alvarez, J.J. Silber, J.V. Rivarola and E.N. Durantini, Photophysical characterization and photodynamic activity of metallo 5-(4-(trimethylammonium)phenyl)-10,15,20-tris(2,4,6-trimethoxyphenyl)porphyrin in homogeneous and biomimetic media, *Photochem. Photobiol. Sci.*, 2003, **2**, 926.
318. O.E. Akilov, S. Kosaka, K. O'Riordan, X. Song, M. Sherwood, T.J. Flotte, J.W. Foley and T. Hasan, The role of photosensitizer molecular charge and structure on the efficacy of photodynamic therapy against Leishmania parasites, *Chem. Biol.*, 2006, **13**, 839.
319. C.D. Enk, K. Gardlo, M. Hochberg, A. Ingber and T. Ruzicka, Cutaneous leishmaniasis, *Hautarzt*, 2003, **54**, 506.
320. D. Nowis, M. Makowski and T. Stoklosa, Direct tumor damage mechanisms of photodynamic therapy, *Acta Biochim. Pol.*, 2005, **52**, 339.
321. E. Reddi and G. Jori, Steady-state and time-resolved spectroscopic studies of photodynamic sensitizers: porphyrins and phthalocyanines, *Res. Chem. Intermed.*, 1988, **10**, 241.
322. G.I. Stables and D.V. Ash, Photodynamic therapy, *Cancer Treat. Rev.*, 1995, **21**, 311.
323. L. Brancaleon and H. Moseley, Laser and non-laser light sources for photodynamic therapy, *Lasers Med Sci.*, 2002, **17**, 173.
324. G. Whitelam and G. Codd, Damage to microorganisms by light, *Special Publications of the Society for General Microbiology*, 1986, **17**, 129.

325. A. Schindl, B. Rosado-Sholosser and F. Trautinger, Reciprocity regulation in photobiology. An overview, *Hautarzt*, 2001, **52**, 779.
326. S.C. Souza, J.C. Junqueira, I. Balducci, C.Y. Koga-Ito, E. Munin and A.O. Jorge, Photosensitization of different Candida species by low power laser light, *J. Photochem. Photobiol. B*, 2006, **83**, 34.
327. E. Munin, L.M. Giroldo, L.P. Alves and M.S. Costa, Study of germ tube formation by Candida albicans after photodynamic antimicrobial chemotherapy (PACT), *J. Photochem. Photobiol. B*, 2007, **18**, 16.

Chapter 6

Fullerene Derivatives as Antimicrobial Photosensitizing Agents

M. Elisa Milanesio and Edgardo N. Durantini

Departamento de Química, Facultad de Ciencias Exactas Físico-Químicas y Naturales, Universidad Nacional de Río Cuarto, Agencia Postal Nro. 3, X5804BYA Río Cuarto, Córdoba, Argentina

Table of Contents

Photodynamic Inactivation of Microbial Pathogens: Medical and Environmental Applications
Editors Michael R. Hamblin and Giulio Jori

Published by the Royal Society of Chemistry, www.rsc.org

6.1. Introduction

Since their first discovery[1] and isolation,[2] the fullerenes (originally buckminsterfullerenes) have attracted the attention of an increasing number of scientists. The most abundant form of fullerenes is C_{60} with 60 carbon atoms arranged in a spherical structure. The shape of the molecule, known as truncated icosahedron containing 12 pentagons and 20 hexagons, in which every carbon atom forms bonds to three other adjacent atoms through sp^2 hybridization.[3] However, the popularity of fullerenes in science is not merely due to aesthetics. The chemical and physical features of fullerene C_{60}, together with its spherical shape, have aroused the hope of successful use in many different fields, in both biological and material chemistry.[4–6] The condensed aromatic rings present in the compound lead to an extended π conjugation of molecular orbitals causing significant absorption of visible light. Also, the facile electron acceptability of up to six electrons makes them good candidates as electron acceptors.[7]

Potential biological activities of fullerenes have been investigated with the aim of using them in the field of medicine.[8–10] An important inconvenience for this application is the low solubility of fullerenes in polar solvents and the consequent formation of aggregates in aqueous solutions.[4] However, the development of the covalent chemistry of C_{60} has opened the possibility to attach this sphere structure with several groups, which allows the biological activity to be incremented.[3,4,7] Therefore, the most versatile methodology is based on the chemical modification of the sphere that leads to a wide variety of C_{60} derivatives, possessing different physical and chemical properties.[11]

Interesting biological properties were reported for various functionalized fullerenes.[9,11] Amide derivatized fullerenes acted as antiretroviral drugs in cells,[12] while cationic functionalized fullerenes showed improved action against HIV-1 and HIV-2 strains.[13,14] An amino acid-type fullerene derivative was the most efficient in human immunodeficiency virus-reverse transcriptase inhibition.[15] It was reported that water-soluble trimalonic acid derivative of fullerene prevented death from a group A *Streptococcus* infection in mice.[16] Polyhydroxylated fullerenes were proposed as antioxidants and free radical scavengers.[17–19] Antibacterial and antiproliferative activities of cationic fullerenes were investigated and the results indicated that the mechanism of antimicrobial activity is a respiratory chain inhibition.[20,21]

The combination of visible light absorption and a long lifetime of triplet excited state allows fullerenes to act as photosensitizers. Irradiation of solubilized fullerenes in the presence of oxygen leads to the generation of reactive oxygen species (ROS).[22] Therefore, they are interesting candidates for use in photosensitization processes, especially in situations where absorption in the red part of the spectra is not completely required. Because of their photochemical properties, fullerenes were proposed for photodynamic therapy (PDT) of malignant cancer cells.[8,23–25] A recent review of Hamblin and coworkers contains an interesting compilation of PDT with fullerenes.[26]

The antimicrobial chemotherapy field is undergoing constant change, because of the great variety of pathogenic species found and their rapid evolutionary changes.[27] The great successes in the war against microorganisms are probably coming to an end. For this reason, the development of new drugs and therapies is imperative. The new antibiotics methods include bacteriophages, bacterial phenotype modification, a genomic library to find unknown target sites for new drugs and a non-oncologic application of PDT, named photodynamic inactivation (PDI) of microorganisms.[28,29] Essentially, PDI is based on the administration of a photosensitizer, which is preferentially accumulated in the microbial cells. The subsequent irradiation with visible light, in the presence of oxygen, specifically generates a cascade of biochemical events that produce cell damage, which inactivates the microorganisms.[30,31] Because of their photochemical properties, fullerene derivatives have been used as efficient agents to mediate PDI of various classes of microbial cells.[25,32] Therefore, the following sections cover the existing literature on fullerenes for PDI applications, as shown in Scheme 6.1, that have been reported for fullerenes either pristine or functionalized with various solubilizing groups.

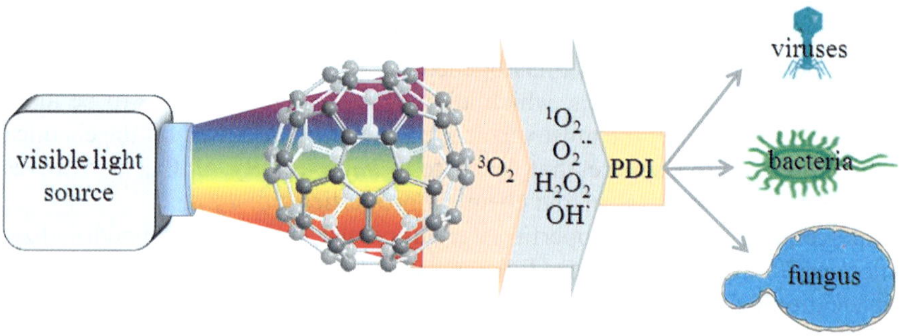

Scheme 6.1 Possible applications of fullerenes as antimicrobial photosensitizing agents.

6.2. Structures of Fullerene C_{60} Derivatives

Fullerenes are hydrophobic molecules and their potential biomedical applications have been hampered by their extremely poor solubility in polar solvents. Therefore, different approaches have been developed for the transfer of fullerenes to water, such as: a) simply adding dry C_{60} to water and vigorously mixing for an extended period.[33–35] While this method enables the investigation of pristine C_{60} in water, its disadvantages include formation of large aggregates and low fullerene concentration. b) Solvent exchange method that uses volatile water-miscible organic solvents to dissolve fullerene, while in the next step water is added and the solvent evaporated, leaving fullerene aggregates in water suspension.[36–38] Alternatively, transfer of fullerene can be performed by dissolving C_{60} in a water-immiscible organic solvent and applying ultrasound to gradually transfer C_{60} into water as the organic solvent is evaporated.[34,39] The structure of fullerene is not changed and the fullerene core can freely collide with water molecules. c) Incorporation of fullerenes into water-soluble supramolecular structures using surfactants, liposomes or other modifying agents, such as calixarenes or cyclodextrins.[40–43] In most cases the fullerene core is completely covered by the modifying agent and can hardly come in contact with water. d) Chemical modification of the fullerene carbon cage by attachment of various functional hydrophilic groups, which allow fullerene molecules to establish bonds with water *via* functional adducts.[44–46]

Original and well-established synthetic methodologies applied to fullerenes have produced a wide variety of novel derivatives, whose properties are continuously being investigated. An important subject when dealing with unmodified fullerenes is their insolubility in biologically compatible solvents, limiting their use in biological applications. Therefore, fullerenes have been chemically modified or functionalized by the introduction of addends in order to acquire aqueous solubility.[47–49] The newly reported fullerene derivatives were designed to be amphiphilic in nature, according to the general structure shown in Scheme 6.2. Modifications in the structures of fullerenes were performed to obtain an increase in the efficiency and selectivity of the agents. The combination of hydrophobic and hydrophilic substituents in the sensitizer structure results in an intramolecular polarity axis. This effect can facilitate membrane penetration and produces a better accumulation in sub-cellular compartments, enhancing the effective photosensitization.[50]

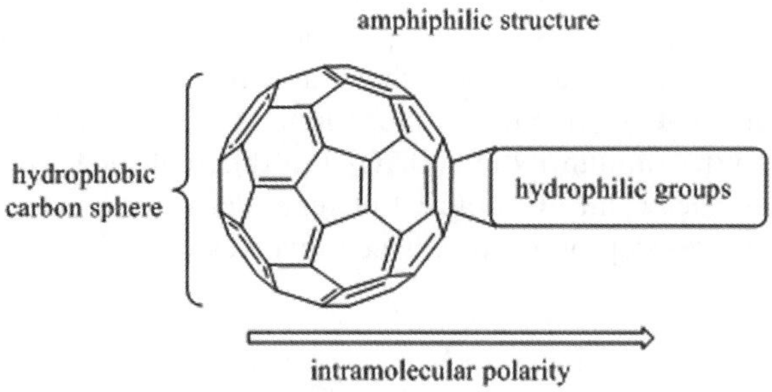

Scheme 6.2 Amphiphilic structure of fullerenes.

In the last years, several fullerene derivatives were evaluated as potential photosensitizers to inactivate microorganisms.[51–53] Tegos *et al.* have reported on the relative ability of two series of functionalized fullerenes to mediate photoinactivation of Gram-positive and Gram-negative bacteria, and fungi (Scheme 6.3).[51] One series of three compounds had 1, 2 or 3 polar diserinol groups (**BF1-3**), and a second series of three compounds had 1, 2 or 3 quaternary pyrrolidinium groups (**BF4-6**). The synthesis of **BF1-3** was based on Bingel additions to C_{60} and the malonodiserinolamide fullerenes were designed to provide water solubility and biological compatibility.[54] The synthesis of

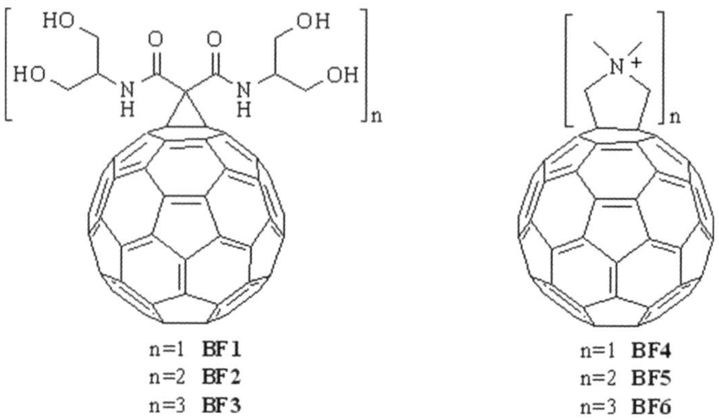

Scheme 6.3 Structures of **BFI-6** fullerenes.

BF4-6 was established on the fulleropyrrolidine construct, first developed by Prato *et al.*[55] There are 8 possible regioisomers of the bis-substituted fullerenes and 46 possible regioisomers of the tris-substituted fullerenes, and it was not practical to separate these mixtures of regioisomers into individual pure compounds; therefore, **BF2**, **BF3**, **BF5** and **BF6** were studied as mixtures of regioisomers. These compounds were designed to increase either the lipophilic or the cationic character of already synthesized compounds for comparison.

Spesia *et al.* have synthesized a novel *N*,*N*-dimethyl-2-(4′-*N*,*N*,*N*-trimethylaminophenyl) fulleropyrrolidinium iodide (**DTC$_{60}^{2+}$**) bearing two cationic groups (Scheme 6.4).[52] The spectroscopic and photodynamic properties of the **DTC$_{60}^{2+}$** were compared with a non-charged *N*-methyl-2-(4′-acetamidophenyl) fulleropyrrolidine (**MAC$_{60}$**) and a monocationic *N*,*N*-dimethyl-2-(4′-acetamidophenyl)fulleropyrrolidinium iodide (**DAC$_{60}^{+}$**) in different media and in a typical Gram-negative bacterium. Cationic fullerene derivatives were conveniently synthesized in a two-step procedure, which involves a 1,3-dipotar cycloaddition

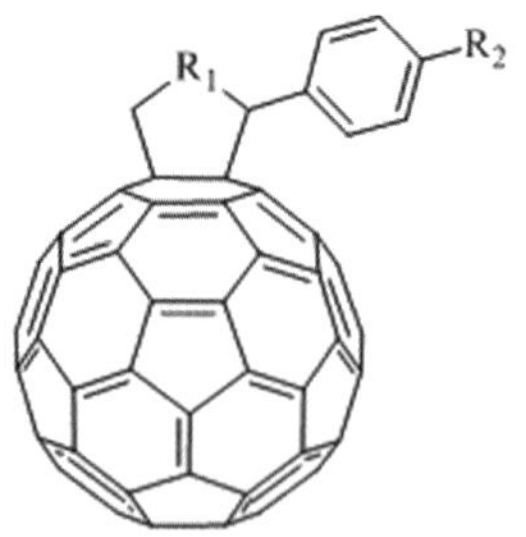

Scheme 6.4 Structures of **MAC$_{60}$**, **DAC$_{60}^{+}$** and **DTC$_{60}^{2+}$** fullerenes.

of azomethine ylides to C_{60} and an exhaustive methylation. The compound $\mathbf{DTC_{60}^{2+}}$ bears a hydrophobic carbon sphere substituted by two cationic groups forming one amphiphilic monoadduct. To evaluate the effect produced by the last groups upon the intramolecular polarity, the dipole moments of fullerene derivatives were calculated giving values of 3.4 D, 17.3 D and 46.0 D for $\mathbf{MAC_{60}}$, $\mathbf{DAC_{60}^{+}}$ and $\mathbf{DTC_{60}^{2+}}$, respectively. As expected, the presence of a dicationic moiety in the periphery of fullerene derivative considerably enhances the dipole moment with respect to the non-charged structures. This effect could help fullerene derivatives to pass through or accumulate in biomembranes, enhancing the effective photosensitization.[56]

Huang *et al.* have described the synthesis and characterization of several new cationic fullerenes (Scheme 6.5).[53] Compounds **BF1′** and **BF2′** were designed to increase cationic charge per addend, whereas compounds **BF3′–BF5′** were designed to increase overall lipophilicity of the photosensitizer. The fullerene

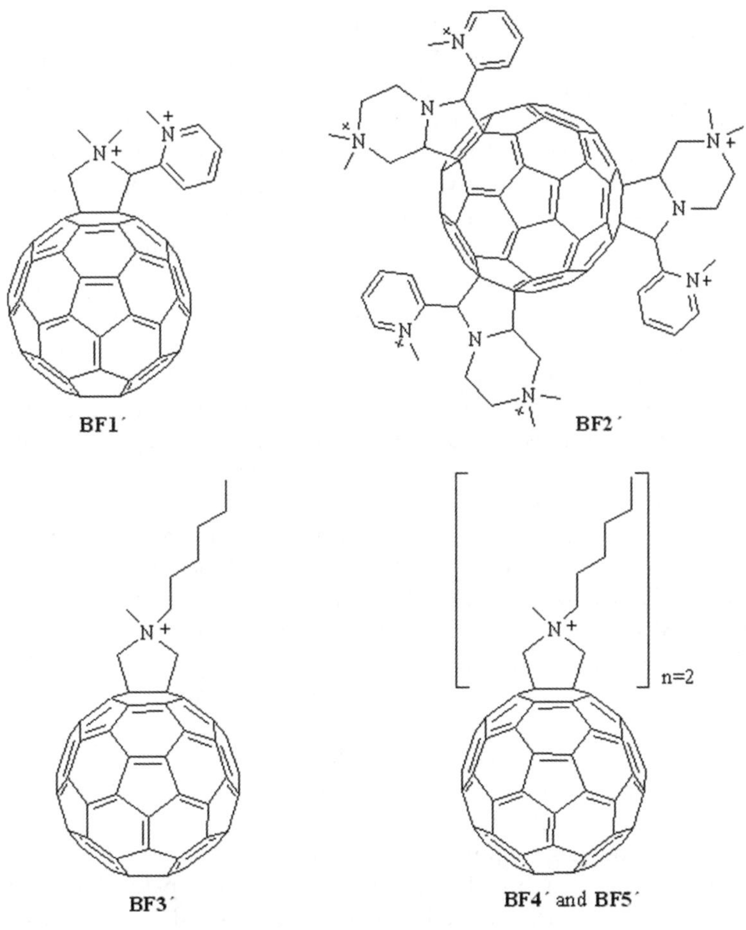

Scheme 6.5 Structures of **BF1′-5′** fullerenes.

BF2′ was used as a mixture of bis- and tris-substitution ratios, while **BF4′** and **BF5′** were identified as regioisomeric bis-products in order of increasing polarity. The authors found a good linear correlation between log *P* and the number of cationic charges, as might be expected considering that the remainder of the fullerene molecules (and the hydrophobicity attributed to the side chain functionality) remains relatively similar.

6.3. Spectroscopic Properties of Fullerenes

Strong absorptions in the UV region and a weaker but significant band in the visible region characterize C_{60}, the most abundant representative of the fullerene family.[57] The symmetry-forbidden transitions between 410 and 620 nm are responsible for its purple colour.[32] When functionalized, C_{60} retains the same basic characteristics but, in addition, the absorption of the derivatives extends further into the near-IR region. This implies that C_{60} and its derivatives are very easily excited by low-energy light, which combined with the high electron affinities of fullerenes makes these materials attractive partners in photoinduced electron transfer and for applications in medicinal chemistry.[57]

Perturbation of the fullerene π-system, which stems from removal of a C═C double bond upon functionalization, brings about noticeable blue shifts of the excited-state absorptions relative to those of C_{60}.[58] In the visible region, functionalized C_{60}, such as fulleropyrrolidines, show a broader range of absorption up to almost 800 nm, with a sharp peak at 430 nm ($\varepsilon \sim 4 \times 10^4$ M^{-1} cm^{-1}) and a broader band at around 700 nm.[57] In polar solvents, the strong hydrophobicity of the fullerenes causes these spheres to stick together, leaving the hydrophilic headgroups on the outside. The results are clusters of undefined sizes, which can be evidenced by UV-visible spectrophotometry. The typical fullerene bands at 215, 260 and 330 nm become broader than in organic solvents, and their intensities decrease. In addition, the fine structures are lost due to significant light scattering by the particles formed, which reach sizes as large as hundreds of nanometres in diameter.

The absorption spectra of fullerenes **BF4-6** were compared in DMSO : water (1 : 9).[51] The overall extinction coefficients of the fullerenes were in the following order: **BF4** > **BF5** > **BF6**. The shoulder in the UVA range moved from 340 nm for **BF4** to 310 nm for **BF5** and disappeared altogether for **BF6**. The blue shift observed for the excited states on going from a monoadduct to various bis adducts is similar to that observed on going from C_{60} to the monoadduct.[57]

The UV-visible absorption spectra of $\mathbf{MAC_{60}}$, $\mathbf{DAC_{60}^{+}}$ and $\mathbf{DTC_{60}^{2+}}$ fullerene derivatives are typical of most C_{60} monoadducts (Figure 6.1, Table 6.1).[52] Similarly to C_{60}, the electronic spectra of the pyrrolidine derivatives are dominated by strong absorptions mainly in the UV region. In the visible region, these fullerenes show a broader range of absorption up to almost 710 nm (Table 6.1). A low intensity and a broadening of the bands are observed in the spectrum of $\mathbf{DTC_{60}^{2+}}$ in *N*,*N*-dimethylformamide (DMF), probably due to

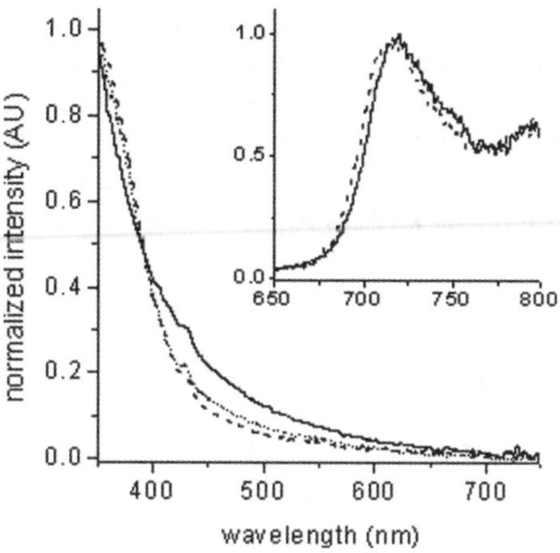

Figure 6.1 Absorption spectra of **MAC_{60}** (solid line), **DAC_{60}^{+}** (dotted line) and **DTC_{60}^{2+}** (dashed line) in DMF. Inset: fluorescence emission spectra of **MAC_{60}** (solid line) and **DTC_{60}^{2+}** (dashed line) in DMF, $\lambda_{exc} = 450$ nm.

Table 6.1 Spectroscopic properties and quantum yield of $O_2(^1\Delta_g)$ production (Φ_Δ) of fullerene derivatives.

Fullerenes	λ_{max}^{Abs} (nm)^a		λ_{max}^{em} (nm)^a	Φ_F^{b}	Φ_Δ^{c}
C_{60}	–	700	712	2.3×10^{-4}	1.00
MAC_{60}	430	704	715	3.0×10^{-4}	0.72
DAC_{60}^{+}	429	704	715	2.0×10^{-4}	0.69
DTC_{60}^{2+}	429	708	716	5.4×10^{-4}	0.07

^{a}DMF
^{b}DMF, from ref. 52
^{c}DMF/water (10%), from ref. 52

aggregation of dicationic fullerene in this medium. Similar behaviour was also found for **DTC_{60}^{2+}** in toluene and phosphate-buffered saline solution, although it is partially dissolved as monomer in benzene/benzyl-*n*-hexadecyldimethyl ammonium chloride (BHDC)/water reverse micelles.[52]

Fullerene C_{60} derivatives show a little fluorescence with moderate quantum yield (1.0×10^{-4} and $\sim 6.0 \times 10^{-4}$ for pristine C_{60} and for monofunctionalized fullerene derivatives, respectively).[59,60] The fluorescence spectrum of *N*-methylfulleropyrrolidine is in excellent agreement with the mirror-imaged absorption features.[58,61] This reflects the fact that the force constants of vibrational levels in the first singlet excited state resemble those in the singlet ground state. The good match of the longest-wavelength absorption and the shortest-wavelength emission and the fact that they exhibit the highest oscillator strengths are convincing evidence for an assignment to $0 \rightarrow 0$ transition bands. The fluorescence emission spectra of fullerenes **MAC_{60}** and **DTC_{60}^{2+}** in DMF show a band centred at $\sim$712 nm, which is characteristic for similar fulleropyrrolidines (Figure 6.1, inset).[52] In accordance with the photophysical properties of fulleropyrrolidines, low values of Φ_F are expected for these fullerenes (Table 6.1).[7,22] In these cases, a small Stokes shift ($\sim$6 nm) was observed, indicating that the spectroscopic energy is nearly identical to the relaxed energy of the singlet state. The minor Stokes shift is well compatible with the highly symmetric and rigid structure of the fullerenes and their derivatives.

6.4. Photodynamic Activity of Fullerenes

One of the most exciting developments of fullerene chemistry is related to their excited-state properties. The excitation of fullerene C_{60} ground state ($^1C_{60}$) with UV or visible light leads to the population of the singlet excited state ($^1C_{60}{}^*$). The lifetime of this intermediate state is relatively short (below the nanosecond level). In addition to losing energy by fluorescence emission or internal conversion, the $^1C_{60}{}^*$ rapidly undergoes intersystem crossing (ISC) to the long-lived (tens to hundreds of microseconds) excited triplet state ($^3C_{60}{}^*$) with nearly unit yield. The process is a spin-forbidden ISC with a high rate of 5.0×10^8 s^{-1} driven by an efficient spin-orbit coupling.[22,57] Fullerenes with their triply degenerate low lying LUMO are excellent electron acceptors capable of accepting as many as six electrons.[62] There is some evidence that fullerene excited states, in particular the $^3C_{60}{}^*$, are even better electron acceptors than the ground state.[57] This leads to the formation of charge transfer states with a variety of strong electron donating substances. Therefore, fullerenes can undergo a different process, being easily reduced to C_{60} radical anion ($C_{60}^{\cdot-}$) by electron transfer.[58]

Under aerobic conditions, the $^3C_{60}{}^*$ can interact with ground state molecular oxygen to form ROS. This process can occur by energy transfer from the $^3C_{60}{}^*$ to produce singlet molecular oxygen, $O_2(^1\Delta_g)$, (or simply $^1O_2$) or by electron transfer to form superoxide anion radical ($O_2^{\cdot-}$).[63] Fullerenes are extremely efficient $^1O_2$ generators with a quantum yield that is near unity.[64–66] It is thought that the $^3C_{60}{}^*$ or $C_{60}^{\cdot-}$ can transfer an electron to molecular oxygen forming $O_2^{\cdot-}$. In contrast to $^1O_2$ generation, the electron transfer type of reaction preferentially occurs in polar solvents, particularly in the presence of reducing agents such as NADH.[67] These two pathways, yielding $^1O_2$ and $O_2^{\cdot-}$, are analogous to the two main photochemical reaction types known as type II and type I photochemical mechanisms, respectively (Scheme 6.6).

The $^1O_2$ production of several laser-excited fullerene derivatives was compared by measuring the 1268 nm emissions.[66] The efficiency was independent of the kind of addends, but it decreased with an increase in the number of addends. Moreover, addends at adjacent positions caused a greater decrease than remote addends. 1,2-Epoxy[60]fullerene and bis-(ethoxycarbonyl)[61]fullerene were compared with respect to fullerene C_{60} and a significant difference was observed between these functionalized fullerenes in $^1O_2$-producing ability. It appears that functionalization of the fullerene core reduced the photodynamic activity, and

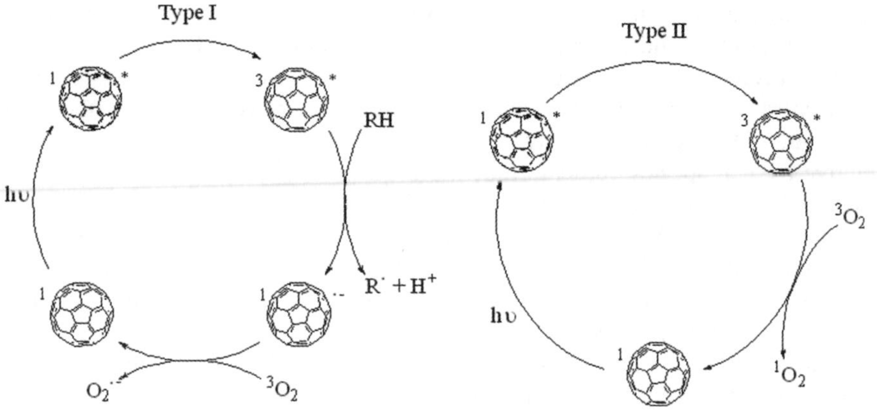

Scheme 6.6 Schematic representation of type I (charge transfer) and type II (energy transfer) photochemical mechanisms.

the effect was not significantly dependent on the kind of addend, oxygen or diethyl malonate.

The photodynamic activities of C_{60}, $\mathbf{MAC_{60}}$, $\mathbf{DAC_{60}^{+}}$ and $\mathbf{DTC_{60}^{2+}}$ were studied in different media.[52] The $^1O_2$ production quantum yields (Φ_Δ) of these fullerene derivatives were calculated in DMF/water (10% v/v) (Table 6.1). Non-charged $\mathbf{MAC_{60}}$ and $\mathbf{DAC_{60}^{+}}$ efficiently produced $^1O_2$ with a Φ_Δ of $\sim$0.7. On the other hand, $\mathbf{DTC_{60}^{2+}}$ was one order of magnitude lower than that of non-charged fullerene derivative. This low $^1O_2$ production of $\mathbf{DTC_{60}^{2+}}$ was probably due to an incomplete monomerization of the cationic fullerene in the DMF/water medium. However, an increase in the $^1O_2$ production of $\mathbf{DTC_{60}^{2+}}$ was found in a biomimetic microenvironment formed by BHDC micelles.[68] This enhancement in the microheterogeneous medium was facilitated by a better solubilization as monomer of $\mathbf{DTC_{60}^{2+}}$.

An interesting behaviour was observed in fullerene C_{60} linked to an electron donor molecular structure, such as porphyrins.[69,70] In these dyads, the values of Φ_Δ were strongly dependent on the solvent polarity. These dyads present higher capacity to form photoinduced charge-separated state and the electron transfer process competes with the $^1O_2$ production depending on the solvent polarity. In a more polar solvent, such as DMF, the $^1O_2$ generated by dyads was significantly diminished due to stabilization of charge-transfer state. Thus, depending on the microenviroment where the sensitizer is localized, this compound could produce biological photodamage through either a $^1O_2$-mediated photoreaction process or a free radical mechanism under low oxygen concentration. Thus, these dyads have offered a promising molecular architecture for photosensitizer agents that can still produce cellular inactivation under anoxic conditions.[69,70]

To characterize fullerenes as photosensitizers in biological systems, the generation of ROS, through energy transfer and electron transfer, was studied by a combination of methods.[67] Whereas $^1O_2$ was generated effectively by photoexcited C_{60} in non-polar solvents such as benzene and benzonitrile, it was found that $O_2^{\cdot -}$ and hydroxyl radical ($HO^{\cdot}$) were produced instead of $^1O_2$ in polar solvents such as water, especially in the presence of a physiological concentration of reductants including NADH. The above results, together with those of a DNA cleavage assay in the presence of various scavengers of specific active oxygen species, indicate that the ROS primarily responsible for photo-induced DNA cleavage by C_{60} under physiological conditions are reduced species, such as $O_2^{\cdot -}$ and $HO^{\cdot}$. Also, visible-light irradiation of solubilized fullerenes into water, with polyvinylpyrrolidone as a detergent, resulted in the formation of $O_2^{\cdot -}$, when experiments were performed in the presence of NADH as a reductant and molecular oxygen.[71]

Therefore, it can be noted that ROS production is very dependent of the medium and considerably diminishes when the photosensitizer is aggregated. Also, the biological microenvironment of the sensitizer can induce important modifications in the photophysics of the fullerene derivatives established in solution.[72] In consequence, there are limitations to predict photodynamic efficiencies of sensitizers in biological systems on the basis of photophysical investigations in homogeneous solution.

6.5. Fullerenes for Photoinactivation of Viruses

Because of their photochemical properties, fullerene derivatives were used to inactivate viruses in biological fluids.[32,73,74] The potential of pristine C_{60} to inactivate viruses by a $^1O_2$-mediated mechanism was investigated in enveloped model viruses belonging to two different families, Semliki Forest virus (SFV, family: *Togaviridae*) and vesicular stomatitis virus (VSV, family: *Rhabdo-viridae*). Illumination with visible light for up to 5 h resulted in up to 7 logs of loss of infectivity. Viral inactivation was oxygen dependent and equally efficient in solutions containing protein.[32] Also, VSV was inactivated upon illumination with a water-soluble methoxy-polyethylene glycol amine conjugated fullerene, in a concentration- or dose-dependent manner.[75] The amount of VSV was reduced by more than 5 logs of plaque forming units in the presence of 400 μM fullerene derivative with 120 J cm^{-2} white light irradiation. The concentration needed for virus inactivation was higher than that of other sensitizers such as methylene blue.[75]

Light-dependent and light-independent inactivation of dengue-2 and other enveloped viruses were compared using two regioisomers of carboxyfullerene.[74] The asymmetric isomer with a bipolar structure showed a greater dark activity, indicating the possible involvement of a light-independent mechanism due to its interaction with the lipid envelope of the virus.

Virus inactivating properties of fullerene C_{60} suspension was investigated regarding influenza virus in allantoic fluid of chicken embryos.[76] The virus activity was reduced in saline and allantoic fluid in the presence of fullerene and oxygen after 2 h irradiation with 400–850 nm light. Morphological study revealed numerous defects of surface membranes leading to the loss of infectious properties of the virus.

Inactivation of MS-2 bacteriophage was investigated in the presence of various fullerene derivatives, when irradiated with a blacklight blue lamp.[77] Under identical conditions, $0.15\,g\,L^{-1}$ of a cationic aminofullerene hexakis derivative led to 2 logs inactivation of MS-2 phage within only 4 min. irradiation, whereas it took 3 h to achieve a similar inactivation level with $0.2\,g\,L^{-1}$ of nano-TiO_2.

6.6. Fullerenes for Antimicrobial Photoinactivation

Derivatized fullerenes with functional groups become aqueous soluble and can act as efficient photosensitizers.[25] Tegos *et al.* investigated the broad-spectrum antimicrobial photodynamic activity of functionalized C_{60}, **BF1-3** and **BF4-6** (Scheme 6.3). Gram-positive bacteria, Gram-negative bacteria and fungi were tested, and the photodynamic killing was compared with that mediated by a widely used antimicrobial photosensitizer, toluidine blue O (TBO). Initial screening experiments carried out against the Gram-positive *Staphylococcus aureus* after 10 min. incubations with 100 μM concentrations showed that the C_{60} substituted with pyrrolidinium groups behaved very differently from the series substituted with di-serinol groups. The cationic fullerenes **BF5** and **BF6** gave high levels of dark toxicity (>99.9999%), while the compound **BF4** gave a dark toxicity of 99%. Fullerenes **BF1-3** showed a minor dark toxicity (60–80%) and when irradiated with white light a light dose-dependent loss of colony-forming ability was observed. However, it needed concentrations as high as 100 μM and white-light fluences as high as $120\,J\,cm^{-2}$ to achieve significant killing (2–3 logs or up to 99.9%) of the Gram-positive *S. aureus*. Even with these relatively high doses of both photosensitizer and light, the Gram-negative *Escherichia coli* was only slightly killed (less than 1 log or 90%). In contrast the cationic fullerenes were highly effective photosensitizers at lower concentrations and lower light doses. Thus, **BF4–6** were all surprisingly effective in causing light-mediated killing of *S. aureus*. Sensitizers **BF5** and **BF6** needed only 1 μM concentration and 1 or $2\,J\,cm^{-2}$ of white light to kill 4–5 logs. Also, **BF4–6** were tested against *E. coli* at 10 μM, and **BF5** and **BF6** were highly effective with $2\,J\,cm^{-2}$ giving 4 and 6 logs of killing, respectively. Fullerene **BF4** was much less potent, needing $8\,J\,cm^{-2}$ to give 3 logs of killing. Similar results were obtained with the yeast *Candida albicans*, in which **BF6** was slightly more effective than **BF5** and both were much better than **BF4**. The Gram-negative bacterium *Pseudomonas aeruginosa* was more resistant than the other organisms tested. Authors doubled the maximum light dose delivered to $16\,J\,cm^{-2}$ in order to obtain more killing, but this had only a minimal effect. Fullerenes **BF5** and **BF6** were able to kill 3–5 logs, while **BF4** only gave 2 logs of killing of *P. aeruginosa*.

These findings agree with numerous reports in the literature that demonstrate that photosensitizers with cationic groups are efficient antimicrobial agents even in the absence of additives.[78–85] Microbial cells possess overall negative charges and it is thought that cationic agents bind to these groups on the outer layers of the cell surface. Gram-positive and fungal cells have relatively permeable outer layers of peptidoglycan and lipoteichoic acid or β-glucan, respectively, although the mannan layer of *Candida* species can present a permeability barrier. This allows cationic and to a lesser extent non-cationic photosensitizers to diffuse inwards to the plasma membrane where the generation of ROS under illumination can damage the membrane structure producing leakage of essential components and causing cell death. Cationic compounds are able to displace divalent cations (C_a^{2+} and M_g^{2+}) that play a role in the attachment of lipopolysaccharide to the outer membrane.[86] This displacement weakens the structure of the outer permeability allowing the photosensitizer to penetrate further in a process that has been termed "*self promoted uptake*".[87]

The fact that this mechanism requires cationic compounds explains why **BF1–3** were relatively non-effective against the Gram-negative *E. coli* and such findings have been reported with numerous other non-cationic photosensitizers.[30] Although *C. albicans* does not have the permeability barriers associated with Gram-negative bacteria, the eukaryotic cells are very much larger than bacteria.[85] This means that there is correspondingly more membrane and proteins per cell to react with the ROS produced by illuminated fullerenes and, consequently, a higher concentration is necessary to achieve killing compared to the smaller *S. aureus*. *P. aeruginosa* is known to have even more efficient permeability barriers than the majority of other Gram-negative species.[88,89] Although 10 μM **BF5** and **BF6** with a wash did give 4–5 logs of killing, it was necessary to use significantly more light (up to 16 J cm^{-2}) compared to what was used with *E. coli*. Even though **BF6** was generally more effective than **BF5** in killing all species, the difference between these two compounds was much smaller than the difference between **BF4** and **BF5**. Presumably, the addition of a third cationic group to the fullerene molecule increased the binding and/or penetration into the cells sufficiently to more than compensate for the reduced light absorption. In order to make comparisons between the effectiveness and selectivity of the cationic fullerenes **BF4–5** with an established antimicrobial photosensitizer, authors studied the phenothiazinium dye TBO under the same conditions. However, TBO did not kill more than 90% of any of the microbial species. Therefore, **BF5** and **BF6** are many orders of magnitude more effective than TBO. Also, in order to assess the selectivity of light-mediated killing for microbes over mammalian cells, mouse L929 fibroblasts were incubated with **BF4–6** and with TBO under the same conditions. The cationic fullerenes showed a greater level of selectivity for microbes over mammalian cells than observed for TBO.

Spesia *et al.* studied the PDI of fullerene derivatives, $\mathbf{MAC_{60}}$, $\mathbf{DAC_{60}^{+}}$ and $\mathbf{DTC_{60}^{2+}}$ (Scheme 6.4), as photodynamic agents *in vitro* on *E. coli* cells. The cell toxicity induced by these fullerenes was first analyzed in the absence of light at different photosensitizer concentrations (1–5 μM). No toxicity was detected for $\mathbf{MAC_{60}}$, whereas $\mathbf{DTC_{60}^{2+}}$ was completely toxic to *E. coli* incubated with 5 μM sensitizer. Dark toxicity of $\mathbf{DTC_{60}^{2+}}$ was reduced to 23% of cellular inactivation using 2.5 μM sensitizer and no toxicity was found when cell cultures were treated with 1 μM of this dicationic fullerene. The viability of *E. coli* cells after irradiation was dependent upon both the fullerene derivative used in the treatment and the light exposure level (Figure 6.2). As is expected for non-charged photosensitizers, no inactivation effect was found for cultures treated with 1 μM of $\mathbf{MAC_{60}}$ after 30 min. of irradiation. Similar behaviour was also found using 5 μM of $\mathbf{MAC_{60}}$. This result is in agreement with that reported before for non-charged fullerene derivatives, indicating that these non-cationic sensitizers are unsuccessful sensitizers for Gram-negative bacteria under these conditions.[51] In contrast, *E. coli* cells are rapidly photoinactivated when the unwashed cultures treated with dicationic $\mathbf{DTC_{60}^{2+}}$ are exposed to visible light.

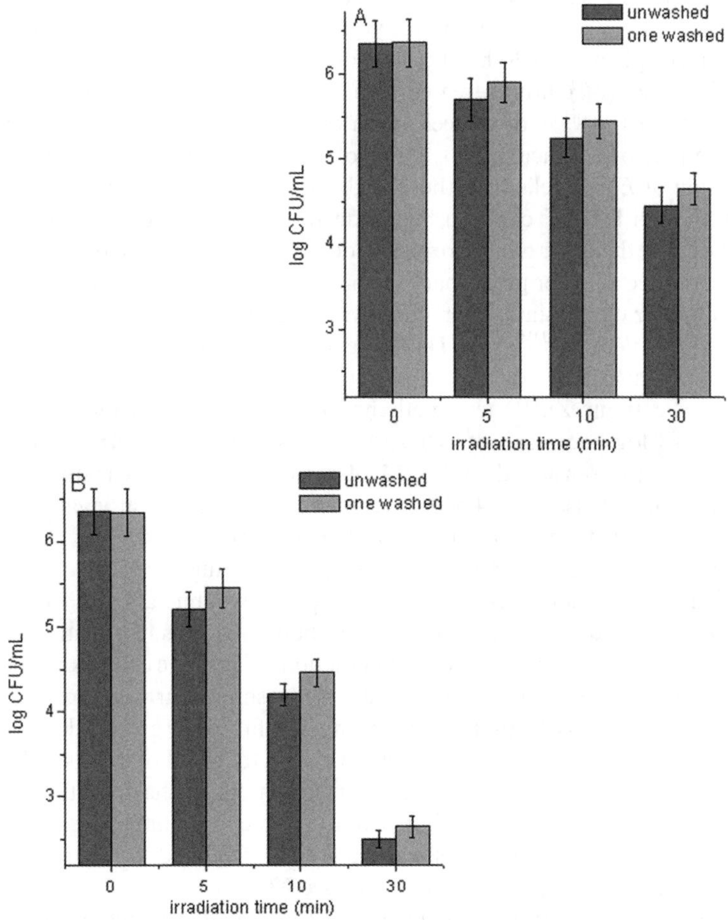

Figure 6.2 Survival of *E. coli* cells (~10^6 CFU ml^{-1}) incubated with 1 μM of A) $\mathbf{MAC_{60}}$ and B) $\mathbf{DTC_{60}^{2+}}$ for 30 min at 37 °C in dark (unwashed or washing once) and exposed to visible light for different irradiation times.

In particular, the dicationic fullerene exhibits a photosensitizing activity causing a ~3.5 logs decrease of cell survival, when the cultures are irradiated for 30 min. These results represent a value greater than 99.97% of cellular inactivation. The photodynamic activity sensitized by $\mathbf{DTC_{60}^{2+}}$ can be directly compared with those of other different families of cationic sensitizers under similar irradiation conditions and using the same *E. coli* strain. Thus, when *E. coli* cultures were treated with 1 μM of a *cis* dicationic porphyrin, 5,10-bis (4-methylphenyl)-15,20-bis(*N*,*N*,*N*-trimethylammoniumphenyl)porphyrin iodide, 70% of cellular inactivation was found after 5 min. of irradiation with visible light, whereas under comparable conditions $\mathbf{DTC_{60}^{2+}}$ inactivates 86% of bacterial cells.[80] When the *E. coli* cultures were incubated with 5 μM of *cis* dicationic porphyrin and irradiated for 30 min., the PDI effect produces a ~3.2 logs decrease of cell survival.[90] This value is comparable with that obtained using 1 μM of $\mathbf{DTC_{60}^{2+}}$. Also, the behaviour of dicationic fullerene derivative can be compared with that of a *trans* dicationic porphyrin, 5,15-bis[(4-3-*N*,*N*,*N*-trimethylammoniumpropoxy)phenyl]-10,20-bis(4-trifluoromethyl phenyl)porphyrin iodide.[91] Irradiation of *E. coli* cultures treated with 1 μM of this *trans* dicationic porphyrin photoinactivates ~87% of the cells even after 20 min. of irradiation. Therefore these comparisons indicate that 1 μM of $\mathbf{DTC_{60}^{2+}}$ produces a similar photoinactivation of *E. coli* cells than those of dicationic porphyrin after 5 min. of irradiation; though $\mathbf{DTC_{60}^{2+}}$ causes a higher decrease of cell survival after 30 min. of irradiation. On the other hand, porphyrin derivatives bearing three of four cationic charges produce a higher photoinactivation than that found for $\mathbf{DTC_{60}^{2+}}$, probably due to a tighter electrostatic interaction with negatively charged sites at the outer surface of *E. coli* cells.[80,91] Also, $\mathbf{DTC_{60}^{2+}}$ can be compared with cationic phthalocyanine derivatives. For example, 1 μM of dicationic fullerene is more effective than 10 μM of a cationic Zn(II) tetramethyltetrapyridinoporphyrazinium salt, which produces ~2 logs decrease of cell survival after 30 min. of irradiation.[83] The photoinactivation sensitized by 10 μM of a cationic Zn(II) *N*-methylpyridyloxyphthalocyanine causes a ~4.5 logs decrease of *E. coli* cell survival; even so, it required ten times higher concentration that that used with $\mathbf{DTC_{60}^{2+}}$.[84] The photosensitized inactivation produced by $\mathbf{DTC_{60}^{2+}}$ remains high even after one washing step (Figure 6.2). Under this condition, the photodynamic effect is mainly associated with fullerenes that are more tightly bound to cells. Thus, the enhanced photodynamic activity imputed to the interactions of positive charges on the *E. coli* cellular surface is stable enough to survive the washing. Furthermore, the photodynamic activity of this dicationic fullerene was confirmed by growth delay of *E. coli* cultures. This experiment was performed to ensure that PDI of cells is still possible when the cultures were not under starvation conditions or the potential damaging effects of phosphate buffer washing. Even in these conditions, the growth was suppressed when *E. coli* was exposed to 2 μM of $\mathbf{DTC_{60}^{2+}}$ and irradiated, whereas a negligible effect was found for the non-charged $\mathbf{MAC_{60}}$.

Huang *et al.* have reported the testing of functionalized fullerene derivatives **BF1′-5′** (Scheme 6.5) with cationic charges as broad-spectrum antimicrobial photosensitizers against a panel of human pathogens.[53] The results showed that **BF2′** was overall the most powerful of the five fullerenes tested. More than

3 logs of *S. aureus* was killed at the low concentration of 100 nM with no dark toxicity, and cells were eliminated at 1 μM. Also, *E. coli* was eliminated at 10 μM with 3 logs of dark toxicity. In the case of **BF2′**, *P. aeruginosa* was more susceptible to PDI than *E. coli*, with over 4 logs of killing at 1 μM with no dark toxicity and elimination at 10 μM. On the other hand, *C. albicans* had over 5 logs of killing at 10 μM and was eliminated at 100 μM, with considerable (>5 logs) dark toxicity. Photosensitizer **BF3′** was the least effective photosensitizer of those fullerenes tested. Furthermore, *C. albicans* was not killed with either **BF4′** or **BF5′** at concentrations up to 100 μM. The authors plotted the concentration of fullerenes necessary to kill 2 logs (LD99) of microbial cells (after illumination with 10 J cm^{-2} of white light), against the calculated values of log *P*. For these cationic fullerenes their effectiveness as antimicrobial photosensitizers against both Gram-positive and Gram-negative bacteria negatively correlated with their calculated log *P* values.

6.7. Mechanism of Action of Fullerenes

At present, the mechanism of microbial inactivation with photoactivated fullerenes is not completely established. Reaction of $^1O_2$ with various biological targets, including unsaturated lipids, proteins and nucleic acids, resulting in cell damage or functional alterations is thought to be responsible for cell inactivation. Also, in a complex environment such as intracellular space, which contains reducing agents (*e.g.* NADH), the electron transfer from the $C_{60}^{\bullet-}$ to molecular oxygen (type I photochemical mechanism) or directly from the reducing agent to $^1O_2$ could also yield $O_2^{\bullet-}$ and, subsequently, hydrogen peroxide (H_2O_2) and $HO^{\bullet}$.[92] It is therefore an interesting strategy to employ ROS-generating/quenching capacity of C_{60} in designing biologically active agents with diverse functions. Type I mechanisms can be similarly effective or even more effective than type II. This is because hydroxyl radicals are the most reactive and potentially the most cytotoxic of all ROS. It is assumed that hydroxyl radicals are formed from H_2O_2 by Fenton chemistry reactions catalyzed by Fe^{2+} or Cu^+ ions, and that the H_2O_2 is produced by dismutation of $O_2^{\bullet-}$ either by enzyme catalysis or naturally.[67] Another possible mechanism of cytotoxicity is the diffusion-controlled reaction between superoxide and nitric oxide to form peroxynitrite, a highly toxic species.[93]

Photoinactivation of VSV with fullerene conjugated with methoxy polyethylene glycol amine was investigated under different conditions. VSV inactivation was inhibited by oxygen removal or by the addition of sodium azide, a known $^1O_2$ quencher. The substitution of H_2O by D_2O, which is known to prolong the lifetime of singlet oxygen, promoted the virucidal activity. These results indicate that $^1O_2$ may play a major role in VSV photoinactivation by the water-soluble fullerene derivative.[75]

Several cationic fullerenes demonstrated high levels of dark toxicity against microorganisms, such as *E. coli* and *S. aureus*.[51,52] It was reported that cationic fullerenes can inhibit the growth of *E. coli* and *S. aureus* by interfering with the respiratory chain.[20,21] The inactivation of microorganisms growth was attributed to the inhibition of energy metabolism by two opposite dose-dependent mechanisms. At low fullerene concentration the oxygen uptake is decreased, on the contrary at high concentration the oxygen uptake increases and oxygen is converted to H_2O_2. Besides, these fullerenes concentrations inhibit the respiratory chain activity without photoirradiation.

The photochemical mechanism studies with cationic fullerenes **BF4–6** confirmed that depending on the precise conditions of the experiment, illuminated fullerenes can produce both $O_2^{\cdot-}$ and $^1O_2$.[25] The $^1O_2$ production of the more hydrophobic **BF4** dropped to almost zero when the solvent was changed from organic to aqueous. This is consistent with the aggregation of the compound in aqueous media, while the more polar **BF6** remained completely in solution. The production of $O_2^{\cdot-}$ was as expected much higher in the presence of a reducing agent (NADH) than in the presence of a $^1O_2$ trap (histidine). In both cases **BF6** gave more $O_2^{\cdot-}$ than **BF4**, and the difference was significant in the presence of NADH. The fact that the reduction of oxygen consumption by both **BF4** and **BF6** was almost completely inhibited by azide in the presence of histidine confirmed that in the absence of an electron donor and in the presence of a $^1O_2$ substrate the mechanism is almost all type II for both fullerenes. However, in the presence of NADH the relative reduction of oxygen consumption by azide was much less, and proportionately lower for **BF6** than for **BF4** suggesting that about 40% of the oxygen was transformed into $O_2^{\cdot-}$, and the overall production was higher for **BF6** than for **BF4**.

6.8. Conclusions and Perspectives

Despite great enthusiasm about the potential use of fullerenes in biomedicine, the progress in their application as possible antimicrobial photosensitizing agents is still under development. Fullerenes have certain unique properties that make them favourable candidates and simultaneously some disadvantages that go against them as photosensitizers for PDI.

The lack of solubility in biological environments is the major obstacle in the development of this field. However, the cationic modifications have demonstrated that with the correct functionalities present on the fullerene cage these difficulties can be overcome. The results have shown that the presence of one or more cationic charges greatly increases the antimicrobial photoinactivation efficacy of fullerene-based photosensitizers. The charge increases the association of the photosensitizer with negatively charged pathogen membranes, whereas the hydrophobic character increases association with or penetration into the lipid components of the membrane.

The main disadvantage of fullerenes is possibly their optical absorption properties. The absorption spectrum of fullerenes is higher in the blue regions of the spectrum. Even so, the molar absorption coefficients of fullerenes are relatively high and the tail of absorption is extended into the red regions of the visible spectrum. Therefore, fullerenes are interesting photosensitizers mainly in those situations in which red light activation is not important for the light to penetrate deep into tissue.

The most important favourable property is their rather unusual photochemical mechanism. In aqueous solutions and particularly in the presence of reducing agents, these compounds produce a substantial amount of $O_2^{\cdot -}$ in a type I photochemical process. The ROS producing ability of covalently functionalized water soluble fullerenes mainly depends on the number, structure and substitution pattern of the addends. It was demonstrated that not only the number of substituents on the fullerene, but also the relative positions of the substituents, can affect the photochemical production of ROS. The full exploration of the ROS-dependent biological effects of different fullerene derivatives represents one of the principal aims for their potential development in PDI applications.

Functionalized fullerenes have showed several attractive features as antimicrobial photosensitizers. They have a high degree of photostability compared with tetrapyrrole-based compounds, they can produce more ROS through type I photoprocess and they show selectivity for microbial cells over mammalian cells. Also, cationic fullerenes appear to be as efficient as some cationic porphyrins and phthalocyanines with the same number of charges. The studies on series of cationic fullerenes demonstrated that increasing number of cationic charges correlated with increasing efficiency in light-mediated killing of bacteria and fungal yeast. Therefore, cationic fullerene derivatives are interesting photosensitizers with potential applications in PDI of microorganisms growing in cellular suspensions, such as disinfection of water and *in vivo* as localized foci of infection on skin or on accessible mucous membrane. Also, fullerenes represent attractive molecular structures to form permanent antimicrobial surfaces activated by visible light.

References

1. H.W. Kroto, J.R. Heath, S.C. O'Brien, R.F. Curl and R.E. Smalley, C_{60}: Buckminsterfullerene, *Nature*, 1985, **318**, 162.
2. W. Krätschmer, L.D. Lamb, K. Fostiropoulos and D.R. Huffman, Solid C_{60}: a new form of carbon, *Nature*, 1990, **347**, 354.
3. F. Diederich, L. Isaacs and D. Philp, Syntheses, structures, and properties of methanofullerenes, *Chem. Soc. Rev.*, 1994, **23**, 243.
4. M. Prato, [60] Fullerene chemistry for materials science applications, *J. Mater. Chem.*, 1997, **7**, 1097.
5. T. Da Ros and M. Prato, Medicinal chemistry with fullerenes and fullerene derivatives, *Chem. Commun.*, 1999, **663**.
6. M.E. El-Khouly, O. Ito, P.M. Smith and F. D'Souza, Intermolecular and supramolecular photoinduced electron transfer processes of fullerene–porphyrin/phthalocyanine systems, *J. Photochem. Photobiol. C: Photochem. Rev.*, 2004, **5**, 79.
7. M. Prato and M. Maggini, Fulleropyrrolidines: a family of full-fledged fullerene derivatives, *Acc. Chem. Res.*, 1998, **31**, 519.
8. Y. Tabata and Y. Ikada, Biological functions of fullerene, *Pure Appl. Chem.*, 1999, **71**, 2047.

9. N. Tagmatarchis and H. Shinohara, Fullerenes in medicinal chemistry and their biological applications, *Mini Rev. Med. Chem.*, 2001, **1**, 339.
10. Y. Tabata, Y. Murakami and Y. Ikada, Photodynamic effect of polyethylene glycol-modified fullerene on tumor, *Jpn J. Cancer Res.*, 1997, **88**, 1108.
11. S. Bosi, T. Da Ros, G. Spalluto and M. Prato, Fullerene derivatives: an attractive tool for biological applications, *Eur. J. Med. Chem.*, 2003, **38**, 913.
12. R.F. Schinazi, R. Sijbesma, G. Srdanov, C.L. Hill and F. Wudl, Synthesis and virucidal activity of a water-soluble, configurationally stable, derivatized C_{60} fullerene, *Antimicrob. Agents Chemother.*, 1993, **37**, 1707.
13. S. Bosi, T. Da Ros, G. Spalluto, J. Balzarini and M. Prato, Synthesis and anti-HIV properties of new water-soluble bis-functionalized[60]fullerene derivatives, *Bioorg. Med. Chem. Lett.*, 2003, **13**, 4437.
14. S. Marchesan, T. Da Ros, G. Spalluto, J. Balzarini and M. Prato, Anti-HIV properties of cationic fullerene derivatives, *Bioorg. Med. Chem. Lett.*, 2005, **15**, 3615.
15. T. Mashino, K. Shimotohno, N. Ikegami, D. Nishikawa, K. Okuda, K. Takahashi, S. Nakamura and M. Mochizuki, Human immunodeficiency virus-reverse transcriptase inhibition and hepatitis C virus RNA-dependent RNA polymerase inhibition activities of fullerene derivatives, *Bioorg. Med. Chem. Lett.*, 2005, **15**, 1107.
16. N. Tsao, T.Y. Luh, C.K. Chou, J.J. Wu, Y.S. Lin and H.Y. Lei, Inhibition of group A Streptococcus infection by carboxyfullerene, *Antimicrob. Agents Chemother.*, 2001, **45**, 1788.
17. L.L. Dugan, J.K. Gabrielsen, S.P. Yu, T.S. Lin and D.W. Choi, Buckminsterfullerenol free radical scavengers reduce excitotoxic and apoptotic death of cultured cortical neurons, *Neurobiol. Dis.*, 1996, **3**, 129.
18. H. Jin, W.Q. Chen, X.W. Tang, L.Y. Chiang, C.Y. Yang, J.V. Schloss and J.Y. Wu, Polyhydroxylated C_{60}, fullerenols, as glutamate receptor antagonists and neuroprotective agents, *J. Neurosci. Res.*, 2000, **62**, 600.
19. M.C. Tsai, Y.H. Chen and L.Y. Chiang, Polyhydroxylated C_{60}, fullerenol, a novel free-radical trapper, prevented hydrogen peroxide-and cumene hydroperoxide-elicited changes in rat hippocampus *in-vitro*, *J. Pharm. Pharmacol.*, 1997, **49**, 438.
20. T. Mashino, D. Nishikawa, K. Takahashi, N. Usui, T. Yamori, M. Seki, T. Endo and M. Mochizuki, Antibacterial and antiproliferative activity of cationic fullerene derivatives, *Bioorg. Med. Chem. Lett.*, 2003, **13**, 4395.
21. T. Mashino, N. Usui, K. Okuda, T. Hirota and M. Mochizuki, Respiratory chain inhibition by fullerene derivatives: hydrogen peroxide production caused by fullerene derivatives and a respiratory chain system, *Bioorg. Med. Chem.*, 2003, **11**, 1433.
22. K.G. Thomas, V. Biju, M.V. George, D.M. Guldi and P.V. Kamat, Excited-state interactions in pyrrolidinofullerenes, *J. Phys. Chem. A*, 1998, **102**, 5341.
23. F. Rancan, S. Rosan, F. Boehm, A. Cantrell, M. Brellreich, H. Schoenberger, A. Hirsch and F. Moussa, Preparation of PEG-conjugated fullerene containing Gd(3+) ions for photodynamic therapy, *J. Photochem. Photobiol. B: Biol.*, 2002, **67**, 157.
24. J. Liu, S. Ohta, A. Sonoda, M. Yamada, M. Yamamoto, N. Nitta, K. Murata and Y. Tabata, Preparation of PEG-conjugated fullerene containing Gd3+ ions for photodynamic therapy, *J. Controlled Release*, 2007, **117**, 104.
25. P. Mroz, A. Pawlak, M. Satti, H. Lee, T. Wharton, H. Gali, T. Sarna and M.R. Hamblin, Functionalized fullerenes mediate photodynamic killing of cancer cells: Type I versus Type II photochemicalmechanism, *Free Radical Biol. Med.*, 2007, **43**, 711.

26. P. Mroz, G.P. Tegos, H. Gali, T. Wharton, T. Sarna and M.R. Hamblin, Photodynamic therapy with fullerenes, *Photochem. Photobiol. Sci.*, 2007, **6**, 1139.
27. C.A. Diaz Granados, D.M. Cardo and J.E. McGowan Jr, Antimicrobial resistance: international control strategies, with a focus on limited-resource settings, *Int. J. Antimicrob. Agents*, 2008, **32**, 1.
28. P.W. Taylor, P.D. Stapleton and J.P. Luzio, New ways to treat bacterial infections, *Drug Discov. Today*, 2002, **7**, 1086.
29. G. Jori and S.B. Brown, Photosensitized inactivation of microorganisms, *Photochem. Photobiol. Sci.*, 2004, **5**, 403.
30. M.R. Hamblin and T. Hasan, Photodynamic therapy: a new antimicrobial approach to infectious disease?, *Photochem. Photobiol. Sci.*, 2004, **3**, 436.
31. E.N. Durantini, Photodynamic inactivation of bacteria, *Curr. Bioact. Compd*, 2006, **2**, 127.
32. F. Käsermann and C. Kempf, Buckminsterfullerene and photodynamic inactivation of viruses, *Rev. Med. Virol.*, 1998, **8**, 143.
33. R.G. Alargova, S. Deguchi and K. Tsujii, Stable colloidal dispersions of fullerenes in polar organic solvents, *J. Am. Chem. Soc.*, 2001, **123**, 10460.
34. J.A. Brant, J. Labille, J. Bottero and M.R. Wiesner, Characterizing the impact of preparation method on fullerene cluster structure and chemistry, *Langmuir*, 2006, **22**, 3878.
35. E. Oberdorster, S. Zhu, T.M. Blickley, P. McClellan-Green and M.L. Haasch, Ecotoxicology of carbon-based engineered nanoparticles: effects of fullerene (C_{60}) on aquatic organisms, *Carbon*, 2006, **44**, 1112.
36. S. Deguchi, R.G. Alargova and K. Tsujii, Stable dispersions of fullerenes, C_{60} and C_{70}, in water. Preparation and characterization, *Langmuir*, 2001, **17**, 6013.
37. D.Y. Lyon, L.K. Adams, J.C. Falkner and P.J. Alvarez, Antibacterial activity of fullerene water suspensions: effects of preparation method and particle size, *Environ. Sci. Technol.*, 2006, **40**, 4360.
38. J.D. Fortner, D.Y. Lyon, C.M. Sayes, A.M. Boyd, J.C. Falkner, E.M. Hotze, L.B. Alemany, Y.J. Tao, W. Guo, K.D. Ausman, V.L. Colvin and J. B. Hughes, C_{60} in water: nanocrystal formation and microbial response, *Environ. Sci. Technol.*, 2005, **39**, 4307.
39. G.V. Andrievsky, M.V. Kosevich, O.M. Vovk, V.S. Shelkovsky and L.A. Vashchenko, On the production of an aqueous colloidal solution of fullerenes, *J. Chem. Soc. Chem. Commun.*, 1995, **12**, 1281.
40. T. Braun, A. Buvari-Barcza, L. Barcza, I. Konkoly-Thege, M. Fodor and B. Migali, Water soluble buckminsterfullerene-γ-cyclodextrin inclusion complexes via a solid-solid reaction, *Solid State Ionics*, 1994, **74**, 47.
41. J.-M. Janot, E. Bienvenüe, P. Seta, R.V. Bensasson, A.C. Tomé, R.F. Enes, J.A.S. Cavaleiro, S. Leach, X. Camps and A. Hirsch, [60]Fullerene and three [60]fullerene derivatives in membrane model environments, *J. Chem. Soc., Perkin Trans.*, 2000, **2**, 301.
42. M. Makha, A. Purich, C.L. Raston and A.N. Sobolev, Structural diversity of host-guest and intercalation complexes of fullerene C_{60}, *Eur. J. Inorg. Chem.*, 2006, **3**, 507.
43. S. Deguchi, S.A. Mukai, M. Tsudome and K. Horikoshi, Facile generation of fullerene nanoparticles by hand-grinding, *Adv. Mater.*, 2006, **18**, 729.
44. L.Y. Chiang, L.Y. Wang, J.W. Swirczewski, S. Soled and S. Cameron, Efficient synthesis of polyhydroxylated fullerene derivatives via hydrolysis of polycyclosulfated precursors, *J. Org. Chem.*, 1994, **59**, 3960.

45. S. Bosi, L. Feruglio, T. Da Ros, G. Spalluto, B. Gregoretti, M. Terdoslavich, G. Decorti, S. Passamonti, S. Moro and M. Prato, Hemolytic effects of water-soluble fullerene derivatives, *J. Med. Chem.*, 2004, **47**, 6711.
46. L.O. Husebo, B. Sitharaman, K. Furukawa, T. Kato and L.J. Wilson, Fullerenols revisited as stable radical anions, *J. Am. Chem. Soc.*, 2004, **126**, 12055.
47. M. Brettreich and A. Hirsch, A highly water-soluble dendro[60]fullerene, *Tetrahedron Lett.*, 1998, **39**, 2731.
48. T. Da Ros, M. Prato, F. Novello, M. Maggini and E. Banfi, Easy access to water-soluble fullerene derivatives via 1,3-dipolar cycloadditions of azomethine ylides to C_{60}, *J. Org. Chem.*, 1996, **61**, 9070.
49. S. Foley, C. Crowley, M. Smaihi, C. Bonfils, B.F. Erlanger, P. Seta and C. Larroque, Cellular localisation of a water-soluble fullerene derivative, *Biochem. Biophys. Res. Commun.*, 2002, **294**, 116.
50. M.E. Milanesio, M.G. Alvarez, J.J. Silber, V. Rivarola and E.N. Durantini, Photodynamic activity of monocationic and non-charged methoxyphenylporphyrin derivatives in homogeneous and biological media, *Photochem. Photobiol. Sci.*, 2003, **2**, 926.
51. G.P. Tegos, T.N. Demidova, D. Arcila-Lopez, H. Lee, T. Wharton, H. Gali and M.R. Hamblin, Cationic fullerenes are effective and selective antimicrobial photosensitizers, *Chem. Biol.*, 2005, **12**, 1127.
52. M.B. Spesia, M.E. Milanesio and E.N. Durantini, Synthesis, properties and photodynamic inactivation of Escherichia coli by novel cationic fullerene C_{60} derivatives, *Eur. J. Med. Chem.*, 2008, **43**, 853.
53. L. Huang, M. Terakawa, T. Zhiyentayev, Y.-Y. Huang, Y. Sawayama, A. Jahnke, G.P. Tegos, T. Wharton and M.R. Hamblin, Innovative cationic fullerenes as broad-spectrum light-activated antimicrobials, *Nanomedicine: Nanotech. Biol. Med.*, 2010, **6**, 442.
54. T. Wharton, V.U. Kini, R.A. Mortis and L.J. Wilson, New non-ionic, highly water-soluble derivatives of C_{60} designed for biological compatibility, *Tetrahedron Lett.*, 2001, **42**, 5159.
55. M. Maggini, G. Scorrano and M. Prato, Addition of azomethine ylides to C_{60}: synthesis, characterization, and functionalization of fullerene pyrrolidines, *J. Am. Chem. Soc.*, 1993, **115**, 9798.
56. R.W. Boyle and D. Dolphin, Structure and biodistribution relationships of photodynamic sensitizers, *Photochem. Photobiol.*, 1996, **64**, 469.
57. D.M. Guldi and M. Prato, Excited-state properties of C_{60} fullerene derivatives, *Acc. Chem. Res.*, 2000, **33**, 695.
58. D.M. Guldi and K.-D. Asmus, Photophysical properties of mono- and multiplyfunctionalized fullerene derivatives, *J. Phys. Chem. A*, 1997, **101**, 1472.
59. Y.-P. Sun, P. Wang and N.B. Hamilton, Fluorescence spectra and quantum yields of buckminsterfullerene (C_{60}) in room-temperature solutions. No excitation wavelength dependence, *J. Am. Chem. Soc.*, 1993, **115**, 6378.
60. D.M. Guldi, M. Maggini, G. Scorrano and M. Prato, Intramolecular electron transfer in fullerene/ferrocene based donor-bridge-acceptor dyads, *J. Am. Chem. Soc.*, 1997, **119**, 974.
61. J.-F. Eckert, J.-F. Nicoud, J.-F. Nierengarten, S.-G. Liu, L. Echegoyen, F. Barigelletti, N. Armaroli, L. Ouali, V. Krasnikov and G. Hadziioannou, Fullerene-oligophenylenevinylene hybrids: synthesis, electronic properties, and incorporation in photovoltaic devices, *J. Am. Chem. Soc.*, 2000, **122**, 7467.

62. R. Koeppe and N.S. Sariciftci, Photoinduced charge and energy transfer involving fullerene derivatives, *Photochem. Photobiol. Sci.*, 2006, **5**, 1122.
63. Y. Yamakoshi, S. Sueyoshi, K. Fukuhara, N. Miyata, T. Masumizu and M. Kohno, $^{\bullet}OH$ and $O_2^{\bullet-}$ generation in aqueous C_{60} and C_{70} solutions by photoirradiation: an EPR study, *J. Am. Chem. Soc.*, 1998, **120**, 12363.
64. J.A. Arbogast, A.P. Darmanyan, C.S. Foote, F.N. Diederich, R.L. Whetten, Y. Rubin, M.M. Alvarez and S.J. Anz, Photophysical properties of C_{60}, *J. Phys. Chem.*, 1991, **95**, 11.
65. R. Hung and J. Grabowski, A precise determination of the triplet energy of C_{60} by photoacoustic calorimetry, *J. Phys. Chem.*, 1991, **95**, 6073.
66. T. Hamano, K. Okuda, T. Mashino, M. Hirobe, K. Arakane, A. Ryu, S. Mashiko and T. Nagano, Singlet oxygen production from fullerene derivatives: effect of sequential functionalization of the fullerene core, *J. Chem. Soc., Chem. Commun.*, 1997, **1**, 21.
67. Y. Yamakoshi, N. Umezawa, A. Ryu, K. Arakane, N. Miyata, Y. Goda, T. Masumizu and T. Nagano, Active oxygen species generated from photoexcited fullerene (C_{60}) as potential medicines: $O2^{-\bullet}$ versus $^1O_2$, *J. Am. Chem. Soc.*, 2003, **125**, 12803.
68. I. Scalise and E.N. Durantini, Photodynamic effect of metallo 5-(4-carboxyphenyl)-10,15,20-tris(4-methylphenyl) porphyrins in biomimetic media, *J. Photochem. Photobiol. A: Chem.*, 2004, **162**, 105.
69. M.E. Milanesio, M.G. Alvarez, V. Rivarola, J.J. Silber and E.N. Durantini, Porphyrin-fullerene C_{60} dyads with high ability to form photoinduced charge-separated state as novel sensitizers for photodynamic therapy, *Photochem. Photobiol.*, 2005, **81**, 891.
70. M.G. Alvarez, C. Prucca, M.E. Milanesio, E.N. Durantini and V. Rivarola, Photodynamic activity of a new sensitizer derived from porphyrin-C_{60} dyad and its biological consequences in a human carcinoma cell line, *Int. J. Biochem. Cell Biol.*, 2006, **38**, 2092.
71. N. Miyata, Y. Yamakoshi and I. Nakanishi, Reactive species responsible for biological actions of photoexcited fullerenes, *J. Pharm. Soc. Jpn*, 2000, **120**, 1007.
72. B.M. Aveline and R.W. Redmond, Can cellular phototoxicity be accurately predicted on the basic of sensitizer photophysics?, *Photochem. Photobiol.*, 1999, **69**, 306.
73. F. Kasermann and C. Kempf, Photodynamic inactivation of enveloped viruses by buckminsterfullerene, *Antiviral Res.*, 1997, **34**, 65.
74. Y.L. Lin, H.Y. Lei, Y.Y. Wen, T.Y. Luh, C.K. Chou and H.S. Liu, Light-independent inactivation of dengue-2 virus by carboxyfullerene C3 isomer, *Virology*, 2000, **275**, 258.
75. J. Hirayama, H. Abe, N. Kamo, T. Shinbo, Y. Ohnishi-Yamada, S. Kurosawa, K. Ikebuchi and S. Sekiguchi, Photoinactivation of vesicular stomatitis virus with fullerene conjugated with methoxy polyethylene glycol amine, *Biol. Pharm. Bull.*, 1999, **22**, 1106.
76. V.V. Zarubaev, I.M. Belousova, O.I. Kiselev, L.B. Piotrovsky, P.M. Anfimov, T.C. Krisko, T.D. Muraviova, V.V. Rylkov, A.M. Starodubzev and A.C. Sirotkin, Photodynamic inactivation of influenza virus with fullerene C_{60} suspension in allantoic fluid, *Photodiag. Photodyn. Ther.*, 2007, **4**, 31.
77. J. Lee, Y. Mackeyev, M. Cho, D. Li, J.-H. Kim, L.J. Wilson and P.J.J. Alvarez, Photochemical and antimicrobial properties of novel C_{60} derivatives in aqueous systems, *Environ. Sci. Technol.*, 2009, **43**, 6604.

78. M. Merchat, G. Bertoloni, P. Giacomini, A. Villanueva and G. Jori, *meso*-Substituted cationic porphyrins as efficient photosensitizers of Gram-positive and Gram-negative bacteria, *J. Photochem. Photobiol. B: Biol.*, 1996, **32**, 153.
79. E. Reddi, M. Ceccon, G. Valduga, G. Jori, J.C. Bommer, F. Elisei, L. Latterini and U. Mazzucato, Photophysical properties and antibacterial activity of meso-substituted cationic porphyrin, *Photochem. Photobiol.*, 2002, **75**, 462.
80. M.B. Spesia, D. Lazzeri, L. Pascual, M. Rovera and E.N. Durantini, Photoinactivation of *Escherichia coli* using porphyrin derivatives with different number of cationic charges, *FEMS Immunol. Med. Microbiol.*, 2005, **44**, 289.
81. A. Minnock, D.I. Vernon, J. Schofield, J. Griffiths, J.H. Parish and S.B. Brown, Photoinactivation of bacteria. Use of a cationic water-soluble zinc phthalocyanines to photoinactivate both Gram-negative and Gram-positive bacteria, *J. Photochem. Photobiol. B: Biol.*, 1996, **32**, 159.
82. A. Segalla, C.D. Borsarelli, S.E. Braslavsky, J.D. Spikes, G. Roncucci, D. Dei, G. Chiti, G. Jori and E. Reddi, Photophysical, photochemical and antibacterial photosensitizing propertirs of a novel octacationic Zn(II)-phthalocyanine, *Photochem. Photobiol. Sci.*, 2002, **1**, 641.
83. E.A. Dupouy, D. Lazzeri and E.N. Durantini, Photodynamic activity of cationic and non-charged Zn(II) tetrapyridinoporphyrazine derivatives: biological consequences in human erythrocytes and *Escherichia coli*, *Photochem. Photobiol. Sci.*, 2004, **3**, 992.
84. I. Scalise and E.N. Durantini, Synthesis, properties, and photodynamic inactivation of *Escherichia coli* using a cationic and noncharged Zn(II) pyridyloxyphthalocyanine derivatives, *Bioorg. Med. Chem.*, 2005, **13**, 3037.
85. T.N. Demidova and M.R. Hamblin, Effect of cell-photosensitizer binding and cell density on microbial photoinactivation, *Antimicrob. Agents Chemother.*, 2005, **49**, 2329.
86. S.A. Lambrechts, M.C. Aalders, D.H. Langeveld-Klerks, Y. Khayali and J.W. Lagerberg, Effect of monovalent and divalent cations on the photoinactivation of bacteria with *meso*-substituted cationic porphyrins, *Photochem. Photobiol.*, 2004, **79**, 297.
87. R.E. Hancock and A. Bell, Antibiotic uptake into Gram-negative bacteria, *Eur. J. Clin. Microbiol. Infect. Dis.*, 1988, **7**, 713.
88. R.E. Hancock and P.G. Wong, Compounds which increase the permeability of the *Pseudomonas aeruginosa* outer membrane, *Antimicrob. Agents Chemother.*, 1984, **26**, 48.
89. R.E. Hancock, Intrinsic antibiotic resistance of *Pseudomonas aeruginosa*, *J. Antimicrob. Chemother.*, 1986, **18**, 653.
90. D. Lazzeri, M. Rovera, L. Pascual and E.N. Durantini, Photodynamic studies and photoinactivation of *Escherichia coli* using *meso*-substituted cationic porphyrin derivatives with asymmetric charge distribution, *Photochem. Photobiol.*, 2004, **80**, 286.
91. D.A. Caminos, M.B. Spesia and E.N. Durantini, Photodynamic inactivation of *Escherichia coli* by novel meso-substituted porphyrins by 4-(3-*N,N,N*-trimethylammoniumpropoxy) phenyl and 4-(trifluoromethyl)phenyl groups, *Photochem. Photobiol. Sci.*, 2006, **5**, 56.
92. Z. Markovic and V. Trajkovic, Biomedical potential of the reactive oxygen species generation and quenching by fullerenes (C_{60}), *Biomaterials*, 2008, **29**, 3561.
93. J.S. Beckman and W.H. Koppenol, Nitric oxide, superoxide, and peroxynitrite: the good, the bad, and ugly, *Am. J. Physiol. Cell Physiol.*, 1996, **271**, 1424.

Chapter 7

Photosensitizer Delivery for Photodynamic Antimicrobial Chemotherapy

Ryan F. Donnelly, Corona M. Cassidy and Michael M. Tunney

School of Pharmacy, Queen's University Belfast, Medical Biology Centre, 97 Lisburn Road, Belfast BT9 7BL, UK

Table of Contents

Photodynamic Inactivation of Microbial Pathogens: Medical and Environmental Applications
Editors Michael R. Hamblin and Giulio Jori

Published by the Royal Society of Chemistry, www.rsc.org

7.1. Introduction

For photodynamic antimicrobial chemotherapy (PACT) to be of use clinically, effective methods for delivery of both light and photosensitizer to the site of action are necessary. Due to limited light penetration through tissue, clinical PACT will be limited to areas of the body where light can be delivered relatively easily, such as the skin and body cavities. Light delivery to the skin is simple and has been frequently employed in photodynamic therapy (PDT) of dysplastic and neoplastic diseases. In PDT, however, the photosensitizer or its precursor is typically administered orally or intravenously. Due to disordered metabolism and blood flow peculiar to dysplastic or neoplastic tissue, photosensitizing concentrations of drug accumulate in the target lesions relatively quickly. Targeting of photosensitizers to wound infections in this way is not possible and so the drug must be applied topically. Factors such as physicochemical properties of the photosensitizer, photosensitizer dose to be delivered, barrier properties of the target site and patient acceptability all impact upon the method of drug delivery.

7.2. Drug Delivery and Diffusion

The clinical success of topical PACT is in part influenced by pharmaceutical and physico-chemical considerations, such as tissue penetration characteristics of photosensitizers, their ease of application and stability in the applied dosage form. These, in turn, are critically dependent on formulation factors, notably the design of the delivery system. Before any formulation design can be considered, it is important to appreciate the permeation process within a dosage form matrix, as this is the fundamental mechanism by which most formulations control the drug delivery process.

Molecular diffusion is a passive transport mechanism that always directs a chemical system toward a state of thermodynamic equilibrium. It is a random process and the driving force behind the mass transfer that occurs from applied topical dosage systems. If applied to skin, for example, the depth of drug penetration can be such that the drug is localized in cutaneous sites or it may extend deeper to the underlying capillary network and bring about transdermal delivery into the systemic circulation. This would be undesirable in PACT. Diffusion of drug molecules occurs from areas of high concentration to those where the concentration is lower and is described by Fick's first law of mass flow. The movement of drug through a cross section of area, S, in unit time, t, is known as the flux, J. This is a vector that gives the direction and magnitude of the transport. Considering the one-dimensional form along the x component, the flux is shown to be proportional to the concentration gradient, as shown in eqn (1):

$$J = \frac{dM}{dt} \cdot \frac{1}{S} = -D\frac{\partial C}{\partial x} \tag{1}$$

where M is the amount of drug diffusing through a distance, x, and D is the diffusion coefficient

Most drug release experiments use a receptor and donor phase separated by a membrane, as shown in Figure 7.1. The membrane can be a biological barrier, such as excised skin, or a model membrane, such as silicone sheeting. During the initial part of the experiment, a non-steady state exists where Fick's second law describes how the concentration of diffusant varies at a distance, x, through the membrane with respect to time. After a period of time, a quasi-stationary state (pseudosteady-state) is established where the concentration gradient (*äC*/δx) across the membrane remains constant and does not vary with respect to time. During this state, it is possible to express eqn (1) as eqn (2), namely:

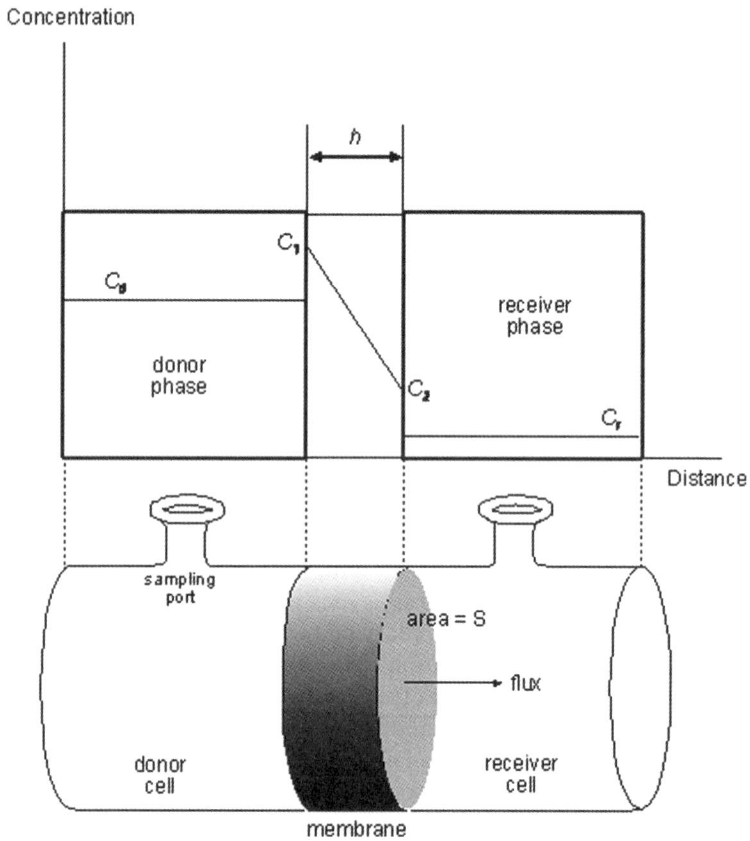

Figure 7.1 Schematic representation of a typical side-by-side diffusion cell apparatus. The concentrations across the cells are represented in the top part of the figure. The concentration gradient across the membrane of thickness h is shown at steady state. C_d is not expected to equal C_l unless the partition coefficient of the drug into the membrane from the donor phase is unity.

$$\frac{dM}{dt} = \frac{DSK(C_d - C_r)}{h} \tag{2}$$

where C_d and C_r are the concentrations in the donor and receptor phases, respectively, h is the thickness of the membrane and K is the partition coefficient into and out of the membrane from the phases.

Drug release from a cream or ointment can be evaluated using a diffusion cell, such as that shown in Figure 7.1, with some modification. Diffusion of the drug through the semi-solid matrix becomes an important consideration. If the donor phase is replaced by a sample of the formulation, the membrane can either be rate limiting in terms of drug permeating to the receiver phase or simply act as a means to separate the phases. In the latter case, the membrane poses little resistance to drug diffusion. The membrane may be dispensed with altogether in cases where a hydrophobic formulation is poorly miscible in an aqueous receiver phase.

Incorporating a drug with limited solubility in a semi-solid matrix results in a suspension type formulation. Even moderate loadings will inevitably possess a substantial fraction of drug in the solid phase. In this case, the Higuchi equation is applicable, as shown in eqn (3):

$$Q = [D(2A - C_S)C_S t]^{1/2} \tag{3}$$

where Q is the amount of drug depleted per unit area of matrix, C_S is the solubility of drug in the matrix and A is the total drug per unit volume. When $A \gg C_S$, which is the case here, eqn (3) can be differentiated with respect to t to yield eqn (4):

$$\frac{dQ}{dt} = \sqrt{\frac{ADC_S}{2t}} \tag{4}$$

It is clear from eqn (4) that the rate of release can be increased by increasing the total amount of drug in the formulation, its solubility and its diffusion coefficient. Ordinarily, drug release that is under matrix (bulk formulation) control will display linear $\sqrt{t}$ kinetics.

Under conditions where the drug release greatly exceeds diffusional resistance at any interface between the donor and receiver phases and the drug has significant solubility in the vehicle, then eqn (3) is expected to hold. When the solubility of a photosensitizer in the formulations is high, $A \gg C_S$ is no longer considered appropriate. However, if a biological barrier exerts a significant resistance to diffusion, then barrier resistance must be considered similarly. In these cases, when the biological barrier (*e.g.* the skin's *stratum corneum*) is believed to be the rate-controlling factor, eqn (2) can be used, providing the thickness of the barrier, h, is known and remains constant.

Many antimicrobial photosensitizers are particularly water-soluble molecules. Even when loading as high as 20% w/w, no solid drug exists and all drug is present in the dissolved state. In most topical systems used in PACT, it is clear that $A < C_S$ and the following solution to Fick's law can be used when a pseudosteady-state is assumed:

$$M_t \approx 2A\sqrt{\frac{Dt}{\pi}} \tag{5}$$

and differentiating with respect to t gives the rate of release:

$$\frac{dM_t}{dt} \approx A\sqrt{\frac{D}{\pi t}} \tag{6}$$

where M_t is the amount of solute released in time t per unit area and providing the skin acts as a perfect sink and is not a barrier to permeation. This is unlikely to be true when hydrophilic photosensitizers are applied to intact skin. However, the disordered structure associated with many superficial infected lesions and the lack of a keratinized barrier in certain mucous membranes, such as the oral cavity and vulval epithelium, may present situations where eqn (5) is valid.

It is clear in eqn (5), as with eqn (3), that M_t *vs.* $\sqrt{t}$ is linear. This demonstrates that the mass of a photosensitizer released from a semi-solid matrix, such as a cream, does not remain constant with respect to time. Thus, the mass of photosensitizer released during the initial application phase is more than that after a period of time has elapsed. It is also clear that drug release can be enhanced by increasing A and D. Indeed, these strategies have been manifest in the high loadings characteristic of cream-based preparations and the use of alkyl esters of 5-aminolevulinic acid, the latter prodrug approach increasing the diffusivity of the parent compound through lipid-rich regions of the *stratum corneum*.

Equation (6) defines the relationship between the diffusivity of the drug and the rate of release. The diffusivity of a hydrophilic photosensitizer through a matrix with less hydrophilic character, such as an emulsion system with a high proportion of lipid, is expected to be low. This will not be the case when using a more hydrophilic vehicle, such as a structured aqueous hydrogel, where the water-solubility of the photosensitizer will permit a more rapid flux through the matrix. Permeation through a biological barrier arising from a crystalline photosensitizer deposition on the barrier surface is most unlikely, unless dissolution occurs beforehand.

7.3. Photosensitizer Delivery Strategies

As with PDT, the field of drug delivery for PACT is at a relatively early stage of development. However, various delivery strategies have been employed to date, with the aim of enhancing PACT of relevant bacterial, fungal and viral infections. In this chapter, we discuss in detail the approaches taken by researchers in academia and industry, including topical delivery systems, targeted delivery and site-specific administration, as well as fabrication of devices with in-built photodynamic antimicrobial activity for use in ophthalmology, urology and disinfection of water and blood products.

7.3.1. Topical Delivery Systems

Topical application of a photosensitizer followed by irradiation is a viable application of PACT in treatment of infection, due to the accessibility of the target site. Several studies have investigated the use of PACT in the treatment of acne,[1–4] warts,[5–8] infected wounds[9–12] and the oncychomycosis-infected nail.[13] In these studies, the method of drug delivery depended on a number of different factors, but particularly the photosensitizer used.

In the treatment of acne vulgaris, PACT is usually mediated by 5-aminolevulinic acid (ALA) formulated in an oil-in-water cream at a loading of 20% w/w or in the Levulan Kerastick® (Dusa Pharma, Tarrytown, NJ, USA).[1,3,4,13,15] The latter is a two-component system, where the aqueous/alcoholic carrier (ethanol, isopropyl alcohol, water, a surfactant and propylene glycol) is mixed with a solid form of ALA prior to use. Both formulations are applied for varying time periods under occlusive dressings to aid absorption and retention at the application site prior to red light illumination. ALA creams are made up immediately before use, as the stability of ALA in aqueous formulations where the pH is greater than 6 is poor.[16] However, it is not practical for the clinician administering PACT to manufacture creams before therapy. Use of the Levulan Kerastick® is preferable to the manufacture of cream by clinicians, as ingredients are pre-measured, so there is less scope for error, and it is less time-consuming to prepare. However, when the solution is prepared, it must be used within two hours, and the method of delivery to the skin is *via* a swab, which should be "uniformly wet" prior to application.[15] Thus, it is unknown what the exact dose of ALA applied to the skin is. In addition, it is virtually impossible to keep a liquid vehicle in place on the skin for any reasonable time period, which is not ideal in the clinical situation.

It is notable that while clinical improvements were seen in acne patients treated with PACT,[1,3,4,14,15] the *Propionibacterium acnes* load on the skin before and after treatment was not reported. As a result, it is difficult to determine whether treatment success was obtained as a result of a bactericidal effect or a photodynamic effect on the sebum producing machinery in the follicles.

A small number of patents exist on delivery systems for ALA that are formulated with the intention of enhancing ALA stability and ease of application. Lee *et al.* described the formulation of an adhesive patch, loaded with a defined amount of ALA.[17] Crystalline ALA was dispersed in a non-aqueous pressure sensitive adhesive (PSA) matrix to enhance stability. In another formulation, described by Zhang *et al.*,[18] ALA was mixed with a hydrogel (5% w/w carboxymethylcellulose, 20% w/w polyvinyl alcohol and 75% w/w water) in a ratio of 5 : 95. Hydrochloric acid was added to the mixture to reduce the pH to approximately 3 for storage. For both of these patents, no clinical data were presented and so their potential usefulness in the treatment of acne vulgaris is unclear.

Several studies have demonstrated the efficacy of PACT for the treatment of recalcitrant warts[6–8] and genital warts.[19,20] Photosensitizers used include ALA,[6,7] methylene blue,[8] proflavine and neutral red.[21] Again, topical creams and solutions were the preferred drug delivery methods.[6,20,22] Similar to the studies on acne vulgaris, good clinical results were obtained.[6,20,22] However, human papilloma virus load in infected lesions and treated skin was not assessed. Therefore, it is difficult to comment on whether there was a true antiviral effect leading to lesion regression, or whether the accumulated tissue of the lesions was simply destroyed by mechanisms similar to destruction of neoplastic tissue in PDT.

As a result of the inherent difficulties in treating wound infections caused, in particular, by antibiotic-resistant pathogens, much recent effort has focused on PACT of wound infections.[23–26,9–11] Artificially created wounds in mice exposed to *Vibrio vulnficus* were treated *in vivo* using toluidine blue O-mediated PACT.[24] Toluidine blue O (TBO) was dissolved in 0.9% w/w sodium chloride to achieve concentrations ranging from 1 to 100 $\mu g\, ml^{-1}$.[24] For each concentration tested, 50 μl of the TBO solution was added into the wound 2 minutes prior to irradiation and 30 minutes post *V. vulnificus* exposure. The light source used was a Waldmann PDT-1200 lamp, delivering a total light dose of 80 J cm^{-2}. The results showed that TBO-PACT protected mice from *V. vulnificus* lethal wound infection, where 53% of mice exposed to 100 $\mu g\, ml^{-1}$ TBO and irradiation survived, in comparison to the dark control, where only 4% survived. Obvious problems with this method of drug delivery include a lack of homogeneity in the amount of drug delivered across the surface area of the wound, and total dose delivered when factors such as loss of solution from the wound are considered. A similar method was used by Lambrechts *et al.* to treat infected burn wounds in mice.[9] A 500 μM photosensitizer solution composed of 5-phenyl-10,15,20-tris(*N*-methyl-4-pyridyl)porphyrin chloride (PTMPP), a cationic porphyrin, dissolved in either phosphate buffered saline pH 7.4 (PBS) or PBS containing 25% w/w dimethyl sulfoxide (DMSO) was prepared prior to use. PTMPP solution (100 μl) was applied to or injected into the wound 24 hours after exposure to bioluminescent *Staphylococcus aureus*, by which time infection had been established. The wound was then illuminated using a non-coherent light source with interchangeable filtered fibre bundles, emitting at 635 nm and delivering a total light dose of either 211 or 423 J cm^{-2}. The results showed that DMSO aided penetration of the photosensitizer through the damaged skin, though in no case was complete sterilization of wounds achieved. Skin healing was slower in PACT treated mice than in controls. The authors suggested that enhancing the specificity of PACT could limit tissue damage and accelerate wound healing.

Jung *et al.*[27] investigated the possibility of PACT of otitis media. *Streptococcus pneumoniae* and *Haemophilus influenza* were separately injected into the bullae of gerbils. Upon development of otitis media 2 days later, a solution of a proprietary haematoporphyrin derivative (Photogem®) was injected and transcranial irradiation was performed with a 632 nm laser (90 J). PACT was successful in eradicating 84% of *S. pneumoniae* and 50% of *H. Influenza* from bullae, leading the authors to suggest that PACT may be useful in recalcitrant otitis media.

Engelhardt *et al.*[28] suggested that water-soluble formulations of hypericin (hypericin-poylvinylpyrrolidone) and *m*-tetrahydroxyphenylchlorin (Fospeg®) could be useful in PACT of wound and burn infections, since they could be applied using a spray applicator and then washed away after a suitable incubation period. They postulated that this would prevent self-shielding by photosensitizer not in close proximity to infecting bacteria, leading to an enhanced PACT effect. Both formulations required short (5 minutes) incubation times *in vitro* to give a 4–5 log reduction in *S. aureus* upon irradiation (25 mW cm^{-2}, 30 minutes).

Hirao *et al.*[29] showed that dual wavelength photoacoustic monitoring could be successfully employed to determine photosensitizer distribution in deep burns in rats *in vivo*. They postulated that this technique would allow the optimal application time prior to irradiation to be accurately determined.

Photopharmica, a university spin-out company based in Leeds (UK), completed a placebo-controlled phase II clinical trial in July 2007 on the use of PACT in treatment of infected leg ulcers mediated by a novel phenothiazinium-based photosensitizer, 3,7-bis(di-*n*-butylamino)phenothiazin-5-ium bromide (PPA904).[23] In this study, the first of its kind worldwide in human patients, PACT treatment of infected chronic leg and diabetic foot ulcers was compared to placebo. PPA904 (500 μM) was dissolved in a cream composed of Unguentum Merck® and water in a ratio of 1 : 2, and the placebo consisted of the same cream without photosensitizer. The cream was applied to the ulcer surface and surrounding ulcer margin, to a thickness of 1.0 mm, and covered with an occlusive dressing for 15 minutes. Excess cream was then removed, the ulcer exposed to 51 J cm^{-2} of light from a light source, emitting broadband red light and then dressed. In all cases, there was a significant reduction in bacterial load in patients when comparing pre-treatment and post-treatment colony forming units per ml, with single treatments reducing bacterial load by 90% compared to placebo. Problems with this delivery method include the incomplete removal of the carrier cream from the wound before irradiation. Furthermore, the cream could absorb light during irradiation, leading to incomplete kill and the requirement for multiple treatments. Moreover, the amount of cream applied would be difficult to reproduce in other studies, as the dose was only given in thickness of cream applied, and not total amount applied per unit area.

Poly(vinyl alcohol)-borate complexes were evaluated as a potentially novel drug delivery platform suitable for *in vivo* use in PACT of wound infections.[30] An optimized formulation (8.0% w/w PVA, 2.0% w/w borax) was loaded with 1.0 mg ml^{-1} of methylene blue (MB) or meso-tetra (*N*-methyl-4-pyridyl) porphine tetra tosylate (TMP). Both drugs were released, using the apparatus shown in Figure 7.2, to yield receiver compartment concentrations (>5.0 mg ml^{-1}) found to be phototoxic to both planktonic and biofilm-grown meticillin-resistant *Staphylococcus aureus* (MRSA), a common cause of wound infections in hospitals. Newborn calf serum, used to simulate the conditions prevalent in an exuding wound, did not adversely affect the properties of the hydrogels and had no significant effect on the rate of TMP-mediated photodynamic kill of MRSA, despite

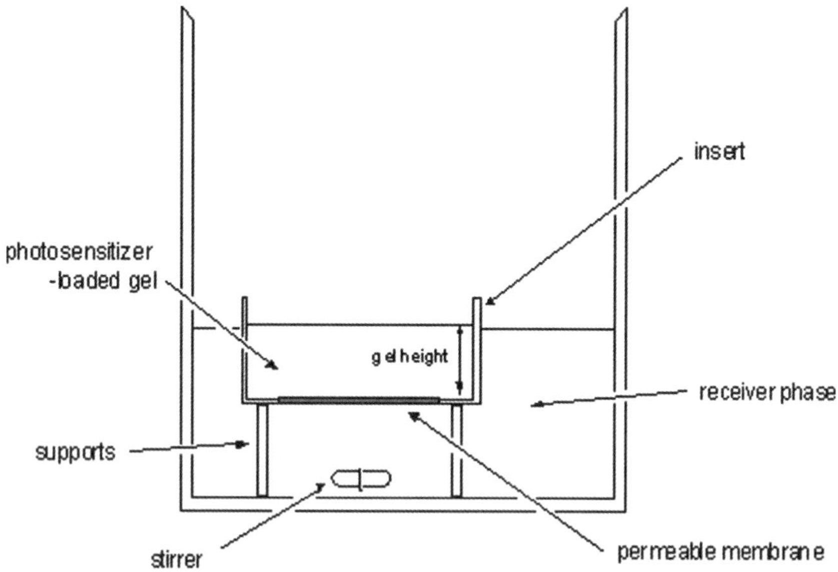

Figure 7.2 Diagrammatic representation of the drug release compartment used by Donnelly *et al.*[30] to quantify release of PS from a novel PVA-borate hydrogel.

appreciably reducing the fluence rate of incident light. However, MB-mediated photodynamic kill of MRSA was significantly reduced in the presence of calf serum and when the clinical isolate was grown in a biofilm. Results supported the contention that delivery of MB or TMP using gel-type vehicles as part of PACT could make a contribution to the photodynamic eradication of MRSA from infected wounds. Importantly, the dosage forms described in this study had unique flow properties, enabling them to be poured into a wound, but then removed intact.

Photodynamic therapy of fungal nail infections would be an attractive option, given the extremely long duration of conventional treatment with oral and/or topical antifungal drugs. Donnelly *et al.* developed a bioadhesive patch containing 50 mg ALA cm^{-2}.[13] Patch application for 24 hours allowed an ALA concentration of 2.8 mM to be achieved on the ventral side of excised human nail. Application for 48 hours resulted in a concentration of 6.9 mM at the site of infection. Application time had no significant effect on the ALA concentration at mean depths of 2.4 mm in neonate porcine, with application times of 24, 48 and 72 hours all producing concentrations of 0.1 mM. Incubation of *Candida albicans* and *Trichophyton interdigitale* with ALA concentrations of 10.0 mM for 30 minutes and 6 hours, respectively, caused reductions in viability of 87% and 42%, respectively, following irradiation with red light. Incubation with 0.1 mM ALA for 30 minutes and 6 hours, respectively, caused reductions in viability of 32% for *Candida albicans* and 6% for *Trichophyton interdigitale*, following irradiation. The authors suggested that drug penetration across nail may be improved using penetration enhancers, or by filing of the impenetrable dorsal surface of the nail. Iron chelators could be used to increase PpIX production for a given ALA dose and ALA stability upon storage could be improved by preparing the patch using cubic phase gel formulations consisting of thermodynamically stable, three-dimensional, liquid crystalline phases.[31,32]

7.3.2. *Targeted Photosensitizer Delivery*

Methods of targeting photosensitizers specifically to a certain type of microorganism include antibody conjugation,[33] attachment of polycationic peptides[11] and employing bacteriophages.[12] Berthiaume *et al.* investigated the efficacy of antibody-conjugated tin (IV) chlorin e_6 in photodynamic killing of *Pseudomonas aeruginosa* in a mouse model of wound infection.[33] Untreated wounds were compared to those treated with specific and non-specific antibody photosensitizer-conjugates plus visible light irradiation. Treatment using antibody-photosensitizer conjugates was more effective than the untreated control and treatment with the non-specific conjugate. Wound infection is defined as the presence of replicating organisms within a wound resulting in host injury.[34] The wound infection model used by Berthiaume *et al.* did not meet this definition, as the bacteria were injected into the mouse immediately prior to treatment and, therefore, had caused no injury to the host. In addition, the antibody photosensitizer-conjugate was added at the same time as the bacterial load, so transport of conjugates through host tissues and targeting of bacterial cells *in vivo* was not investigated. In order to prepare the photosensitizer-conjugate and bacterial suspension, approximately 2×10^6 cells ml^{-1} of *Pseudomonas aeruginosa* suspended in 50 µl PBS were added to an equal volume of tin (IV) chlorin e_6-specific monoclonal antibody conjugate in PBS, to make a final conjugate concentration of 20 or 200 nM.[33] The suspension was injected intradermally, and the resulting bubble irradiated under a 630 nm dye laser, administering a total light dose of 160 J cm^{-2}. The selective conjugates were able to kill over 75% of injected bacteria.

Wu *et al.*[35] showed that "dynamic nanoplatforms", composed of poly (acrylamide) hydrogels and loaded with methylene blue, were able to eradicate both planktonic and biofilm-grown Gram-positive and Gram-negative bacteria *in vitro*, with reduced dark toxicity a notable feature of the system. Nisnevitch *et al.*[36] encapsulated the water soluble photosensitizers methylene blue and neutral red in liposomes and found that photodynamic activity against a range of prokaryotes was enhanced relative to the free photosensitizers. Cahan *et al.*[37] conjugated Rose Bengal to penicillanic acid and kanamycin, with and without a linker, and found that the photosensitizer-antibiotic conjugates had enhanced activity against *S. aureus* and *Escherichia coli* upon irradiation (100 J cm^{-2}) than both unconjugated photosensitizer and antibiotics. Gil-Thomas and colleagues[38] showed that gold nanoparticles (GNPs) functionalized with tiopronin and TBO yielded a four-fold increase in photodynamic activity against *S. aureus in vitro*. The complex conjugate was stable in water, air and phosphate-buffered saline, with the authors suggesting that the enhanced effect was due to the tiopronin GNPs acting as a support for TBO.

Hamblin's group have completed extensive work using poly-L-lysine chlorin e_6 conjugates for PACT treatment of wounds in BALB/c mice.[11,25,26] The conjugate was prepared in all studies by dissolving 220 mg of poly-L-lysine hydrobromide in 3 ml DMSO, to which was added 67 mg c_{e6} and 61 mg 1-ethyl-3-(3-dimethylaminopropyl)carbodiimide hydrochloride, followed by 20 µl triethylamine.[26] This mixture was stirred for 24 hours and, on completion of the reaction, as detected by thin layer chromatography, 1.0 ml of methanol and 1.0 ml of water were added and the mixture evaporated to dryness under vacuum. The residue was dissolved in 3 ml of 10 mM sodium acetate buffer, and applied to a column of Sephadex G25 and eluted using the same buffer at a flow rate of 3 ml h^{-1}, from which 3 ml fractions were taken, and fractions 14 to 21 combined to form poly-L-lysine chlorin$_{e6}$ conjugate.

Initial studies tested the susceptibility of non-invasive *E. coli* in one-hour-old wounds,[26] whilst a further study used pathogenic *P. aeruginosa* in one-hour-old wounds.[25] Whilst neither of these conditions meet the definition of an infected wound, they did show that the conjugate can be used to prevent infection in wounds exposed to bacteria.

In the first experiment, a 50 µl suspension of *E. coli* containing approximately 5×10^6 cells in PBS was inoculated into a fresh wound along the dorsum of the mice.[26] After 30 minutes' incubation, 50 µl of the conjugate in PBS, equivalent to 100 µM c_{e6}, was added to the wound. After a further 30 minutes' incubation, wounds were irradiated using a 1.0 W diode laser coupled with a 200 µm fibre to give a spot of 3.0 cm diameter. The mice were exposed to a total light dose of 160 J cm^{-2} in four aliquots of 40 J cm^{-2}, delivered over a period of 27 minutes. The study used bioluminescent *E. coli* to enumerate bacteria in the wound, as the luminescence signal was linearly proportional to bacterial colony forming units. A 99% reduction in luminescence following irradiation demonstrated that conjugate-mediated PACT caused a substantial reduction in bacterial cell count. However, as the species introduced into the wound was non-invasive, the luminescence from non-treated controls reduced markedly over 24 hours, indicating that the bacteria in the wound had died even in absence of PACT.

In the second study, a strain of *P. aeruginosa* (ATCC 19660) that is known to cause wound infection was used, after transformation into a bioluminescent variant.[25] Wounds were inoculated with a 50 µl suspension of *P. aeruginosa* in PBS containing approximately 5×10^6 cells and, after 30 minutes, 50 µl of the conjugate in PBS, equivalent to 200 µ*M* c_{e6}, was added to the wound. After a further 30 minutes' incubation, wounds were irradiated using a 1.0 W diode laser coupled with a 200 µm fibre to give a spot of 2.0 cm diameter. This covered the wound and 5.0 mm of surrounding unbroken skin. Mice were exposed to a total light dose of 240 J cm^{-2} in four aliquots, delivered over 30 minutes. All the mice in the control groups ($n = 10$) died of lethal wound infection within 5 days of inoculation, whilst 90% of mice in the conjugate-PACT treatment group ($n = 10$) survived. In the treatment group, 99% reduction in luminescence was noted after light delivery.

In a third study, the group employed the conjugate in PACT of established wound infection.[11] In this set of experiments, a solution of the conjugate was injected, using a Hamilton syringe and 28-gauge needle, at five points over the surface area of the wound, in comparison to the earlier studies, where a topical solution was applied.[11,25,39] The use of an injection means that the total dose delivered can be accurately determined, although the dose will not be evenly distributed over the total surface area of the wound. In the experiments, the mice were rendered neutropenic *via* cyclophosphamide therapy before injection to an area 2.0 mm beneath the surface of the thigh, with 50 μl of a suspension of *S. aureus* in PBS, containing approximately 1×10^6 cells. Mice were monitored over a 24-hour period for signs of infection, and were then injected with 50 μl of conjugate, corresponding to 1.0 mM c_{e6} concentration, in 10 μl aliquots, at the centre and at the edges of the four quadrants of the wound.[11] After 30 minutes, the wounds were irradiated using a 1.0 W 665 nm diode laser and 200 μm fibre, either with a plain polish distal end to deliver a spot of 1.5 cm diameter or a diffuser tip, to deliver light interstitially. Luminescence of *S. aureus* within the wound was measured after each 40 J cm^{-2} light dose, with a total dose of 160 J cm^{-2} delivered. Delivering the light dose in three fractionated steps composed of interstitial followed by two surface illuminations was associated with the maximum reduction in bacterial count, with a 99% reduction in bioluminescence observed. However, re-growth of bacteria after PACT was a noted problem in this study. Interstitial light delivery reduces the amount of light scattered by the overlying skin, though this may be unacceptable in a clinical situation. Increasing the duration of light exposure may achieve the same total light dose in a less invasive manner.

Embleton *et al.* employed the use of *Staphylococcus aureus* bacteriophage 75 attached to tin (IV) chlorine e_6 to target meticillin-resistant strains.[12] The conjugate was prepared by dissolving 4 mg tin(IV) chlorin$_{e6}$ (Snc$_{e6}$) in 2.4 ml of conjugation buffer, consisting of 0.1 M 2-(*N*-morpholino)ethanesulfonic acid pH 6 and 0.5 M NaCl, and mixing with 2 mM 1-ethyl-3-(3-dimethylamino-propyl)carbodiimide hydrochloride (EDAC) and 5 mM sulfo-*N*-hydroxy-succinimide. After a 4-hour mixing period, β-mercaptoethanol was added to a final concentration of 20 mM to quench EDAC, and the pH adjusted to 7.6. Phage 75 (3.0 ml) in 0.85% NaCl was added and incubated for 16 hours with mixing, after which time the reaction was quenched with 20 mM ethanolamine. The phage-Snc$_{e6}$ conjugate was precipitated using 20% polyethylene glycol 8000 containing 1.0 M NaCl, and dialysed in at least four changes of 0.85% NaCl to remove any remaining free Snc$_{e6}$. The ability of the conjugate to photosensitize several drug-resistant *S. aureus* strains, including EMRSA-16, was tested *in vitro*. Aliquots (50 μl) of phage-Snc$_{e6}$ in 0.85% NaCl, corresponding to 5 μg ml^{-1} Snc$_{e6}$, were added to 50 μl aliquots of bacterial suspension in PBS in a 96-well plate, each aliquot containing approximately 10^7 bacterial cells. After 30 minutes' incubation with stirring at 37 °C, the samples were irradiated using Stabilite helium-neon laser, emitting at a wavelength of 632.8 nm and delivering a total light dose of 21 J cm^{-2}. After serial diluting and plating, any survivors were enumerated. Percentage kill attained with all strains tested was greater than 99% when compared to a negative control. Furthermore, exposure to phage-mediated PACT required to cause a 4 $\log_{10}$ reduction in bacterial cell count had no detrimental effect on human epithelial (Hep-2) cells. Potential barriers to this therapy may be the high specificity of such phages, which would render them unsuitable for treating plymicrobial wound infection. Similarly, they would not be effective against bacteria which are intracellular pathogens.[40] However, Hope *et al.*[41] have recently shown that the inability of a bacteriophage to infect *S. aureus* does not prevent it from effectively delivering photosensitizer *in vitro*.

The rationale behind the use of targeting devices in PACT is that they increase specificity for bacterial over human cells. As mentioned previously, Zeina *et al.* demonstrated that PACT conditions yielding a reduction in bacterial load of seven log cycles had no effect on mammalian keratinocytes.[42] Consequently, to add a targeting moiety to a photosensitizer may be viewed as an expensive and unnecessarily complicated exercise. In addition, the increase in specificity and the resulting decrease in the number of target sites in the bacterial cell could potentially promote development of resistance to PACT.

7.3.3. Intranasal Delivery

Genina *et al.* carried out a clinical study in 140 patients with either acute or chronic maxillary sinusitis, comparing methylene blue-mediated PACT of the infected sinus with traditional treatment as the control group.[43] For delivery of photosensitizer, a 1.0 mm catheter was inserted into the sinus *via* a standard aspirating needle puncture, and 1.5 ml of a 0.1% w/v aqueous methylene blue solution delivered to the maxillary sinus. The solution was left in place for 30 minutes, before being flushed out with physiological saline. The sinus was then irradiated using a helium-neon laser, *via* an optical fibre and diffuser, working at a power of 40 mW, and emitting at a wavelength of 632.8 nm. The method of drug delivery appears invasive, but puncture and lavage of the maxillary sinus is a common procedure in ear, nose and throat (ENT) practice.[44] The diffuser tip ensured homogeneous irradiation of the sinus.[43] In patients treated with PACT, full recovery was noted in 54.3%, whereas in patients undergoing traditional therapy, consisting of intramuscular injections of lincomycin (0.6–0.9 g day^{-1}) or cephazolin (1–2 g day^{-1}) for 7 days, full recovery was noted in 50.0%.[43]

Destiny Pharma has, in phase II clinical trials, a cationic porphyrin-based photosensitizer code-named XF-73, formulated in an undisclosed gel, which is applied to the nasal passages for decolonization of MRSA on irradiation with an undisclosed light source.[45]

7.3.4. Oral/Buccal Drug Delivery

If the delivery of photosensitizer were optimized, PACT could prove a viable alternative treatment regimen for oral candidiasis. A study carried out by Teichert *et al.* investigated the potential of methylene blue-mediated PACT in treatment of oral candidiasis in an immunosuppressed murine model, mimicking the conditions in an immunodeficient human.[46] Complete eradication of *C. albicans* infection was achieved after methylene blue-PACT. In the study, the mice were inoculated with *C. albicans*-coated swabs on a thrice weekly basis for four weeks prior to the experiment, with drinking water also inoculated. At the beginning of the PACT experiment, mice were swabbed to confirm infection status, before instillation of 0.05 ml of aqueous methylene blue solution in concentrations ranging between 250 and 500 $\mu g\,ml^{-1}$. After 10 minutes' incubation, the oral cavity of the mice was irradiated with a diode laser light coupled with a 1.0 cm cylindrical diffuser, emitting at 664 nm and delivering a total light dose of 275 $J\,cm^{-2}$, with culturing taking place immediately after irradiation. Complete eradication was noted after 450 and 500 $\mu g\,ml^{-1}$ MB-mediated PACT, with lower MB concentrations associated with a reduction in colony forming units isolated, when compared to negative control and pre-treatment samples. Application of a photosensitizer solution would cause staining of the buccal cavity and teeth, which would be cosmetically unacceptable in human therapy. Therefore, development of a more aesthetically pleasing method of photosensitizer delivery is important before the therapy could be used clinically.

Donnelly *et al.* have formulated a patch containing TBO, a phenothiazinium compound structurally similar to methylene blue, for treatment of oral candidiasis.[47] The patch was identical to that containing ALA used in the onychomycosis study described above with TBO, at a drug loading between 10 and 100 $mg\,cm^{-2}$ replaced ALA. The patches were capable of resisting dissolution when immersed in artificial saliva. When releasing directly into an aqueous sink, patches containing 50 and 100 mg TBO cm^{-2} both generated receiver compartment concentrations exceeding the concentration (2.0–5.0 $mg\,ml^{-1}$) required to produce high levels of kill (>90%) of both planktonic and biofilm-grown *C. albicans* upon illumination. However, the concentrations of TBO in the receiver compartments separated from patches by membranes intended to mimic biofilm structures were an order of magnitude below those inducing high levels of kill, even after 6 hours' release. Therefore, the authors suggested that short application times of TBO-containing mucoadhesive patches should allow treatment of recently acquired oropharyngeal candidiasis, caused solely by planktonic cells. Longer patch application times may be required for persistent disease where biofilms are implicated.

Lin *et al.*[48] showed that TBO-mediated PACT (red light, 48 $J\,cm^{-2}$) eradicated 97% of microorganisms from oral wound infections in rats. Wound size post-PACT was also significantly reduced compared to controls, further emphasizing the potential of pheothiazinium-based PACT in the treatment of oral infections.

Many studies have also reported the suitability of PACT for treatment of periodontitis, which is an inflammatory disease caused by bacteria and affects the gums, bone and other supporting tissues of the teeth.[49] Pathogens isolated from patients with periodontal disease have been successfully killed by PACT. For example, TBO-mediated PACT has been determined, *in vitro*, to be lethal to planktonic and biofilm cultures of *Streptococcus sanguis*, *Porphyromonas gingivalis*, *Actinobacillus actinomycetemcomitans* and *Fusobacterium nucleatum* under specific conditions.[50] Photosensitizer conjugates have also been used to induce photodynamic killing in several pathogens involved in periodontal disease.[51] The resulting increased specificity of photosensitizer and potential narrowing spectrum of activity could be an issue in clinical therapy where, as mentioned previously, it would be of benefit for a photosensitizer to be active against a multitude of oral pathogens. Bhatti *et al.* demonstrated that the presence of serum adversely affected the killing ability of toluidine blue O-PACT over *P. gingivalis in vitro.*[52] *In vitro* PACT treatment of plaque biofilms formed *in vivo* determined that the thickness of biofilms could be halved after pyridinium Zn (II) phthalocyanine-mediated PACT.[53]

Kömerik *et al.* inoculated the oral cavities of rats with a 25 µl suspension of *Porphyromonas gingivalis*, corresponding to 10^{10} colony forming units per ml of bacteria.[54] The oral cavity was either treated immediately after inoculation or after disease development, by exposure to 25 µl TBO in 0.85% w/v NaCl, at a concentration between 0.01 and 1.0 mg ml^{-1}, followed by exposure to 630 nm laser light from a fibre-optic cable at a total dose between 6 and 48 J. In all cases, almost 100% kill was achieved. Where the disease was allowed to progress before treatment, treated rats had markedly reduced bone loss in comparison with controls.

A similar study completed by Sigusch *et al.* used a beagle dog model to determine *in vivo* susceptibility of oral periodontal pathogens to PACT mediated by two photosensitizers chlorin$_{e6}$ and BLC 1010.[55] Animals were inoculated with *P. gingivalis* and *F. nucleatum* and, 4 weeks later, when infection was established, a non-disclosed quantity of 10 µM photosensitizer, dissolved in PBS, was applied to the sites of infection. An optical fibre was used to deliver light from a diode laser, emitting at 662 nm and delivering a total light dose of 12.7 J cm^{-2} to each infection site. Treatment resulted in a significant reduction in both periodontal inflammation and bacterial load.

The method of drug delivery in both studies is simplistic and would not be aesthetically acceptable in human patients. Photosensitizers are highly coloured compounds, and so their delivery to the site of infection should be targeted. One such drug delivery device is the Periowave® system. Currently a treatment option in Canada for patients with periodontitis, the system allows targeted delivery of methylene blue. The treatment kit contains 0.005 % w/v methylene blue in phosphate buffered saline, and hydroxymethylcellulose as a mucoadhesive.[56] The solution is delivered *via* an irrigating needle. This system is targeted in that the needle allows mechanical targeting of the solution to the desired site. After irrigation of the periodontal pocket, a fibre-optic tube is used to deliver a total light dose (670 nm) of between 10 and 20 J cm^{-2} over a 60-second period. Use of the Periowave® system in combination with scaling and root planing was found by Andersen *et al.* to increase the clinical attachment level (CAL) of the gum by three-fold in comparison to scaling and root planing alone. Another system, the SaveDent® light delivery system, is used in conjunction with a solution of pharmaceutical grade TBO, both provided by Denfotex Ltd, a UK-based company.[57] Again the treatment area, in this case a root canal, is irrigated with an aqueous solution of the photosensitizer at a concentration of 12.7 μg ml^{-1} using an endodontic micro-needle and, after a 60-second incubation period, is irradiated with the SaveDent® device for 60 seconds at 100 mW. An *in vivo* study by Bonsor *et al.*, using the SaveDent® system described above for disinfection of root canals; this system led to culture-free canals in 93% of cases, compared to 76% of patients whose root canals were disinfected by conventional methods, namely irrigation with 2.25% aqueous sodium hypochlorite solutions.

Thermoresponsive hydrogels composed of *N*-isopropylacrylamide and hydroxyethylmethacrylate loaded with zinc tetraphenylporphyrin have been found, by Jones *et al.*, to demonstrate drug release properties making them suitable for implant devices.[58] To prepare the gel, required masses of monomer (2 g in total), namely *N*-isopropylacrylamide (NIPAA) or mixtures of NIPAA and 2-hydroxyethyl methacrylate (HEMA) were dissolved in 8 g of deionized water with stirring, after which, 12.5 mg of potassium persulfate and 0.125 ml *N*,*N*′,*N*″,*N*-tetramethylethylenediamine (TEMED) were added and stirred until dissolution had occurred. Polymerization was then allowed to occur for 16 h at 20 °C and, when required, zinc tetraphenylporphyrin (Zn-TPP, 0.04% w/w) was dissolved in the monomer solution prior to polymerization. Studies on the gel consisted of characterization of mechanical properties such as hardness and compressibility of the hydrogel and drug release. To determine drug release, sections of the various hydrogels, contained within a 10 cm^3 circular mould, were immersed into and anchored to the surface of beakers containing deionized water at either 20 or 30 °C and the beakers incubated at 20 or 37 °C in a water bath, shaking at 100 oscillations min^{-1}. The release of the photosensitizer was dependent on formulation of gel and temperature, for example, the time required for the release of 15% w/w of the original loading of drug from *p*(HEMA) at 20 °C was 35.75 minutes, whereas for the *p*(NIPAA-*co*-HEMA) hydrogels this value increased to approximately 47 minutes. The release of the photosensitizer may be controlled by heating or cooling of the gel, which swells in response to reduction of temperature and releases drug.

Possible uses suggested by the author for this system include the treatment of periodontal disease or infected wounds. The hydrogel may be syringed into the periodontal pocket, where, at the temperature within the buccal cavity, no drug would be released. Reduction of temperature by 3 °C by irrigation with cold water would cause photosensitizer release. The released photosensitizer could then be irradiated using a fibre-optic tube. One of the benefits of using this hydrogel is that, once the photosensitizer is released from the matrix, the hydrogel could be reloaded by immersion in photosensitizer solution at a temperature below the lower critical solution temperature. This novel method of controlled drug delivery could be useful in this arena, although extensive work is required before it could be used routinely in the clinical situation.

Lulic *et al.*[59] have recently reported the outcome of a one-year follow-up study where ten patients with periodontitis were treated with PACT five times over a two-week period. They found greater reductions in the depth of periodontal pockets in PACT-treated patients than in control patients, which is indicative of healing. The use of PACT in treatment of periodontitis is now gaining significant momentum, with treatment shown to reduce bacterial load in periodontal pockets, inactivate bacterial virulence factors and host cytokines that impair periodontal restoration. Furthermore, there is minimal chance of resistance development and the adjacent host tissue and microflora are not adversely affected.[59–62]

7.3.5. Pulmonary Drug Delivery

Donnelly *et al.* investigated the potential use of PACT in treatment of *P. aeruginosa* cystic fibrosis (CF) lung infection.[63] The *in vitro* study focused on susceptibility of clinical *P. aeruginosa* isolates to TBO- and meso-tetra (*N*-methyl-4-pyridyl) porphine tetra tosylate (TMP)-mediated PACT, as well as light and drug penetration through mucus. For photosensitizer (PS) diffusion studies, 0.1 ml solutions of PS dissolved in PBS at a concentration between 5 and 15 mg ml^{-1} were placed on top of 0.34 ml of synthetic CF mucus in a specially designed glass washer (thickness 3.0 mm, inside circular diameter 12.0 mm) in the donor compartment of a modified Franz cell, as shown in Figure 7.3. Permeation of PS through mucus was determined over a 6-hour period to be 47% of total drug loading for TMP and between 20 and 25% for TBO. For light diffusion through mucus, a Paterson lamp (Phototherapeutics, Manchester, UK) was used as the light source, emitting at 635 nm and 100 W cm^{-2}. Light penetration through increasing concentrations of dissolved PS in a defined thickness of mucus was also investigated using the apparatus shown in Figure 7.4. No significant reduction in light penetration was observed in increasing depths of blank mucus or mucus containing 0.05 mg ml^{-1} TMP, even when the depth was increased to 3.0 mm. In contrast, the addition of TBO to mucus caused a significant reduction in fluence at all depths studied. Percentage kill of *P. aeruginosa* isolates ranged from 0%, in biofilm cultures of the

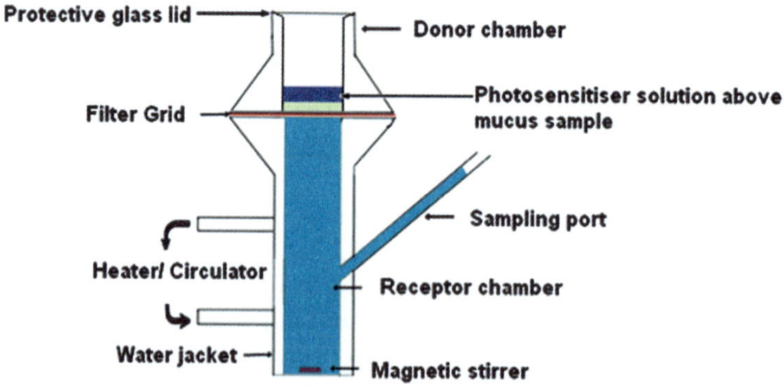

Figure 7.3 Apparatus used by Donnelly *et al.*[63] in PS diffusion studies across artificial CF mucus.

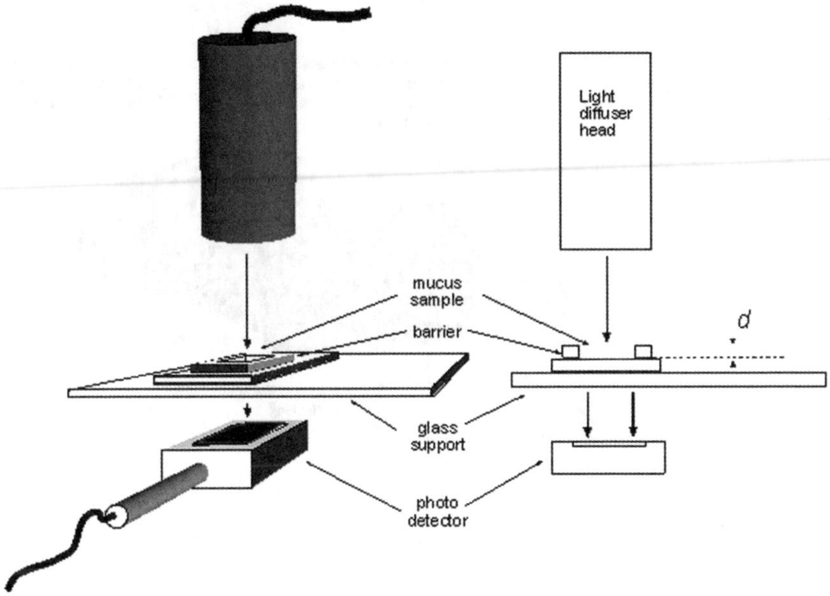

Figure 7.4 Apparatus used in light diffusion studies across artificial CF mucus.[63]

laboratory strain PAO1 exposed to $0.5\,\mathrm{mg\,ml^{-1}}$ TMP and $100\,\mathrm{J\,cm^{-2}}$ of 635 nm light, to 100%, in planktonic cultures of a clinical strain after exposure to $2.5\,\mathrm{mg\,ml^{-1}}$ TMP and the same light dose. This study showed that, although drug and light diffusion through mucus was impaired, sufficient light and concentrations of both drugs could be delivered in photodynamically effective concentrations to a target site. The authors suggested that future work would need to overcome the difficulties inherent in delivering light to the lower airways before clinical treatment would be possible. A further study has assessed the delivery of light and photosensitizer to the target site within the lung by determining penetration of 635 nm light through porcine lung tissue and fine particle characterization of nebulized photosensitizer solution. Figure 7.5 demonstrates the experimental set-up for light delivery through inflated porcine lung. The Mass Median Aerodynamic diameter (MMAD) of methylene blue particles generated using a nebulizer was 4.40 μm, which is a suitable diameter for targeting the site of infection within the CF lung. Using a fibre-optic light delivery device coupled to a helium neon laser, it was found that up to 11% of the total light dose penetrated through inflated porcine lung. The results of this study demonstrate the ability of light and methylene blue to be delivered to the site of infection in the CF lung (unpublished data).

7.3.6. Other Delivery Systems

A patent has been filed for the range of photosensitizers manufactured by Photopharmica, and their incorporation into topical delivery systems based on

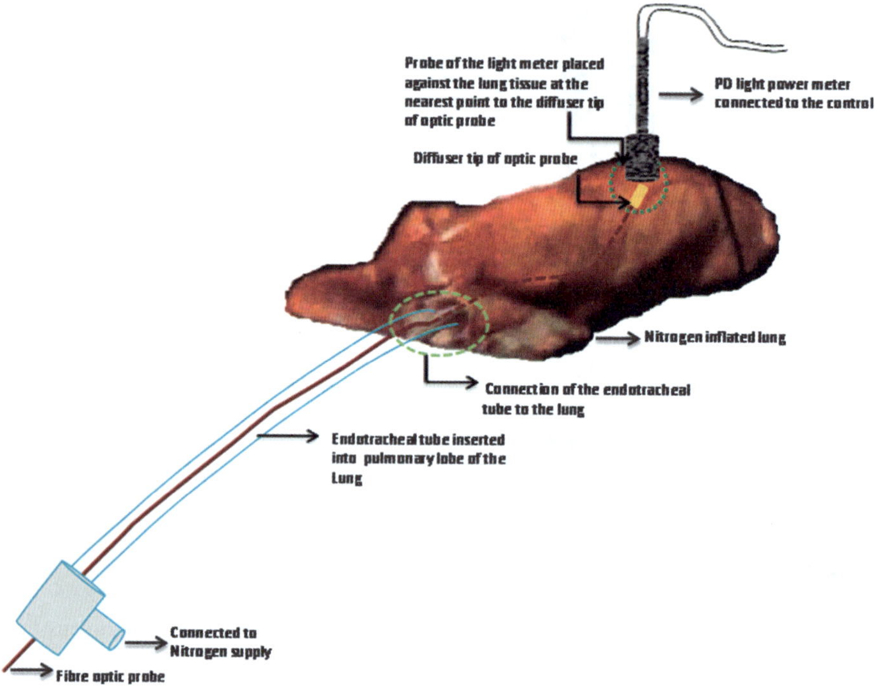

Figure 7.5 Experimental set-up for light delivery through porcine lung tissue.

simple gels or creams or, alternatively, systems based on particulates, such as liposomes or nanoparticles.[64] While the antimicrobial effectiveness of the photosensitizers was comprehensively demonstrated, the use of the described drug delivery systems was not. For example, the patent states that the percentage cell survival of *P. aeruginosa* decreased by over 99.9% on exposure to 10 μM tetra-butyl phenothiazine and light (0.1 W) at a wavelength of 664 nm and 10 minutes' irradiation.

Destiny Pharma also has a patent in place that described the formulation of various porphyrin-based delivery systems for use in photodynamic therapy of bacterial infections.[65] Formulations including solution-, gel-, powder- and lozenge-based systems are described, although the patent covers a wide range of potential formulations, stating that "the compounds of the invention will generally be administered in admixture with a suitable pharmaceutical excipient, diluent or carrier selected with regard to the intended route of administration and standard pharmaceutical practice".[65] The patent states that one of the compounds in formulation by Destiny Pharma, 5,10,15-tris-[4-(3-trimethylammonio-propyloxy)-phenyl]-20-(4-undecyloxy-phenyl0-porphyrin trichloride, in a concentration of 10 μM in combination with irradiation from a Waldmann light source, emitting in the range of 417–420 nm and 13.68 J cm^{-2}, causes a 6 log reduction in *S. aureus* (strain ATCC BAA-44) *in vitro*.

7.4. Other Applications of PACT

7.4.1. Blood Disinfection

Though not technically a clinical treatment, PACT of blood products is a use which has been widely accepted. Methylene blue-mediated PACT has been an established means of blood product disinfection in some parts of Europe for over five years.[66] Blood products required for transfusion include platelets, erythrocytes and plasma. In the Maco Pharma Plasma Membrane filtration Methylene Blue Illumination and MB Depletion Set®, plasma is leuco-reduced *via* filtration using a Plasmaflex® filter, before being passed through an MB dry pill. This results in an MB-plasma concentration of 1 μM. The plasma then passes into an illumination bag, where it is irradiated at a wavelength of 590 nm and receives a total light dose of 180 J cm^{-2}. After this, it is filtered through a Blueflex® MB depletion filter and into a storage bag.[67,68] As these components are anucleate, photosensitizers such as methylene blue, which bind preferentially to nucleic acids, would be of use in the elimination of nucleate cells, which are contaminants and could potentially cause disease if transfused. The benefit of using PACT over conventional methods is the non-specific nature of the antimicrobial target of PACT with fungi, viruses and bacteria targeted simultaneously.[69] In addition, PACT could potentially target nucleate white blood cells, which have the potential to cause an immune response in the recipient. A potential problem with PACT of blood products could be host toxicity, if, for example, the photosensitizing agent was not completely removed prior to transfusion.[69] Stringent monitoring of samples disinfected *via* PACT should limit the likelihood of such an event. Drug delivery in this circumstance is not an issue and the topic is covered in great detail by Wainwright and so will not be discussed further in this chapter.[70–73]

7.4.2. Anti-infective Medical Devices

Brady *et al.* have developed a novel hydrogel drug delivery system, composed of 2-hydroxyethyl methacrylate and methacrylic acid. The system was prepared by mixing the required amounts of HEMA, 2-methacrylic acid (MAA), 1% w/w ethylene glycol dimethacrylate (EGDMA) and 0.4% w/w benzoyl peroxide in a flat-bottomed beaker.[74] The mixture was mechanically stirred and injected into a mould and maintained at 90 °C for 2 hours, then the films were washed and immersed in deionized water for 14 days, after which time tetrakis(4-sulfonatophenyl)porphyrin and tetrakis(4-*N*-methylpyridyl)porphyrin loaded into the polymer films by immersing for 60 seconds in 100 μg ml^{-1} of the porphyrin solution. The porphyrin binds electrostatically to the gel, and so is locked within the gel matrix. Potential use for such materials, suggested by the author, is as an intraocular lens. Such hydrogels are already employed as artificial lenses, used in lens replacement after cataract removal.[75] The authors hypothesize that the porphyrin will remain at the hydrogel surface, where it is activated by ambient room light. It will then produce reactive oxygen species, which will induce cell death in any adherent microorganisms present, but also prevent colonization of the device. As post-operative infection is a complication of cataract replacement, this novel hydrogel could potentially be used to reduce its incidence.[74] One potential problem could be the patient acceptability of a lens impregnated with a highly coloured photosensitizer. Results show that the cumulative tetrakis(4-*N*-methylpyridyl)porphyrin release from a 10:90 MAA: HEMA copolymer over a 10-week period is 7.4%, which indicates that release from the material is slow and coloration will be present long-term. No clinical data, microbiological studies or conditions for PACT exposure were presented in this paper.

Wainwrigtht *et al.*[76] have suggested that catheter-mediated PACT would be a potentially useful "local" approach to management of urinary tract infections in older patients. Love *et al.* have a patent in place for the use of porphyrin derivatives in antimicrobial coatings for stents, catheters and other currently employed medical devices.[77] In the patent, no microbiological or clinical data is given, although potential mechanisms of attaching photosensitizers to insoluble supports are provided.

For implantable biomaterials, the selectivity of photosensitizers for microbial cells over mammalian cells is removed, as the photosensitizer is bound to the surface of the material. Upon illumination a "sheath" of reactive oxygen species will exist at the material surface. As a result, any cells or tissue coming into contact with the material will be damaged. This could be a particular problem if an indwelling device is illuminated when in place.

7.4.3. *Surface Disinfection*

Bonnett *et al.* considered, as early as 1993, the development of materials with photodynamic activity, with a view to creating a surface that, in ambient light, would maintain sterility with regards to microbial growth.[78] The group prepared cellulose-porphyrin and phthalocyanine-based polymers. The group made two copolymer films, the first prepared by dissolving 10 mg tetra(*tert*-butyl)phthalocyanine and 71.8 g cellulose diacetate in 360 g acetone containing 7 g water and 7.1 g diethyl phthalate. The second was composed of 1 mg tetrakis-pyridinium quaternary salt and 15.26 g poly(methylmethacrylate) in 10 g acetone also containing 0.5 g water and 0.5 g diethyl phthalate. The λ_{max} associated with the first film was 615 nm, and the second was 429 nm. They found a cationic porphyrin embedded in cellulose to be a suitable photobactericidal but not bactericidal surface against *S. aureus*, but gave no experimental data.

Wilson also investigated the potential of cellulose acetate loaded with toluidine blue O in the lethal photosensitization of MRSA and *P. aeruginosa*.[79] Experimental data of film manufacture is not disclosed, but surfaces were coated in a cellulose acetate layer containing 0.005 mg ml^{-1} TBO, to which suspensions of EMRSA-16 and *P. aeruginosa* containing approximately 10^6 bacterial cells were added, and exposed to light from a 60 W domestic lamp bulb, emitting over the visible region of 400 to 700 nm, for 8, 16 or 24 hours. Greater than 90% kill was achieved with both organisms when exposed to light over a 24-hour period.

Importantly, Wilson's group have also reported the efficacy of cellulose acetate films containing toluidine blue O and Rose Bengal in the photodynamic killing of *Clostridium difficile*.[80] This is the causative organism in *C. difficile*-associated diarrhoea (CDAD), an infection which has reached epidemic levels in recent years.[81] Irradiation of photosensitizer-containing films previously inoculated with *C. difficile*, *S. aureus*, *E. coli* and *C. albicans* resulted in a significant decrease in viable count of each organism. The films were prepared by dissolving cellulose acetate in acetone to achieve a concentration of 50 mg ml^{-1} cellulose, and stock solutions of 100 μg ml^{-1} photosensitizers in acetone were added to give a final concentration of 25 μM of photosensitizer. Aliquots (450 μl) of the resulting solution were poured into flat-bottomed containers and the acetone allowed to evaporate overnight. Bacterial and fungal suspensions (250 μl) corresponding to approximately 10^6 cells were added to the films, which were then irradiated for between 2 and 16 hours using a General Electric 28 W Biax 2D compact fluorescent lamp, which emits across the visible spectrum. Percentage kill of *C. difficile* on exposure to the TBO-loaded films and 4 hours' irradiation was 100%.[80] This group have also shown[82] that silicone elastomers impregnated with 2 nm gold nanoparticles (GNPs) and methylene blue were highly active against both MRSA and *E. coli* when irradiated at 660 nm. However, the mechanism of enhancement relative to similar systems without GNPs was unclear.

Wainwright *et al.* have incorporated new methylene blue (NMB) into aqueous-based urethane-acrylate and styrene-butadiene copolymers to form films that have proven photobactericidal activity against *Staphylococcus epidermidis* and *E. coli*.[83] Films were prepared by stirring 240 ml of the copolymer (aqueous-based 40% w/w urethane-acrylate and styrene-butadiene) at room temperature and slowly adding 10 ml of either a 25 mM or a 0.25 mM aqueous solution of NMB. Approximately 50 ml of the resulting solution was used to make a patch of 20×80 cm dimensions, which was left to dry at 25 °C before curing overnight at 60 °C. Bacterial suspensions (10 μl) containing 10^4 cells were inoculated onto squares of the film and spread, before irradiation with a Paterson lamp, emitting at 615–645 nm and delivering a total light dose of between 1.5 and 11.5 J cm^{-2} of light.[83] Complete kill of *S. epidermidis* and *E. coli* was achieved with a light dose of 11.5 J cm^{-2} at all concentrations of NMB tested.

Bozja *et al.* have synthesized a nylon-based material containing protoporphyrin IX and its zinc derivative, attached *via* a poly(acrylic acid) (PAA) scaffold.[84] The formed material was found to be active against *S. aureus* and, to a lesser extent, *E. coli* at normal room illumination levels. The authors suggested potential uses including privacy curtains or laboratory coats in hospitals. To prepare the fibres, 300 mg of nylon-6,6 fibres were wound onto a glass ladder. PAA (25 mg) was dissolved in 30 ml of deionized water, and the glass ladder submerged in the PAA/water solution. *N*-hydroxysuccinimide (NHS) (40 mg) and 200 mg of 1-ethyl-3-(3-dimethylaminopropyl) carbodiimide hydrochloride (EDC) were added and allowed to run overnight with air bubbling through for stirring. The grafted fibre was then rinsed with deionized water for 8.0 hours, before addition of 30 ml of deionized water, in which 25 mg of the ethylene diamine derivative of protoporphyrin IX (PPIX) was dissolved. NHS (5 mg) and EDC (20 mg) were then added, and the reaction run overnight, with a similar method employed for zinc PPIX addition. Pieces of the fibre (4 cm) were then immersed in 3.0 ml cultures of *S. aureus* and *E. coli*, corresponding to an optical density of 0.6 at 600 nm, and exposed to 30 minutes of light from an incandescent source of undisclosed wavelength at an intensity of between 10 000 and 60 000 lux.[84] The fibres were more active against *S. aureus* than *E. coli*, with exposure to the zinc-PPIX fibre in combination with a light intensity of 40 000 lux causing 94% kill of *S. aureus*, but only 30% kill of *E. coli*.

Recently, Zerdin *et al.*[85] covalently bound an anthraquinone derivative (AQ) to the acid groups of a commercially available ethylene-acrylic acid copolymer. Films prepared from this material (35% w/w AQ) exhibited photodynamic action against *Bacillus cereus* spores when exposed to low-power UVA light due to singlet oxygen production at the film surface.

Loading of surfaces with photosensitizers with a view to attaining surface sterility is obviously an attractive option in this field, with a decrease in susceptibility of organisms to biocides routinely used in surface disinfection.[86] However, the possibility of leaching of photosensitizers out of the polymers over prolonged periods, or photobleaching of the photosensitizers leading to their inactivity, are two important aspects that must be considered. Moreover, some of the materials required 24 hours' illumination to achieve acceptable levels of kill. This could mean that surfaces may not be able to be used for days at a time, if sterility is to be achieved. This is not practical in a busy hospital environment. Furthermore, some of the materials tested demonstrated increased activity for some organisms over others, which could potentially lead to selection and overgrowth of less susceptible organisms. Extensive further investigation is, therefore, required in this area before photobactericidal surfaces may be employed on a widespread basis.

7.4.4. Water Disinfection

The use of PACT for the disinfection of water is a concept over 30 years old. Rose Bengal, immobilized by covalent bonding to poly(styrene) beads, was found, by Bezman *et al.* in 1977, to be effective in killing 99.99 % of *E. coli* in a contaminated water sample.[87] The loaded beads were formed by adding 1.55 mmol of Rose Bengal to 2.0 g of chloromethylated styrene-divinylbenzene copolymer beads in 60 ml of dimethylformamide. The mixture was stirred magnetically, heated at 60 °C for 20 hours, and then cooled to 25 °C, filtered and the residue washed thoroughly with 200 ml portions of solvents in the following order: ethyl acetate, ethanol, ethanol-water (1 : 1), water, methanol-water (l:l) and methanol. The polymer beads were then extracted with methylene chloride and methanol in a Soxhlet® extractor for 20 hours until no visible colour appeared in the solvent. The polymer-sensitizer was then dried in a vacuum oven at 60 °C for 12 hours. Approximately 16.5% of thc Rose Bengal was polymerized in the reaction, as analysed by combustion for iodine content, in order to determine the percent reaction of the chloromethyl groups with Rose Bengal, and the beads were 15–20% Rose Bengal by weight.[87,88] For the microbiological study, 200 mg of Rose Bengal beads were added to 75 ml of PBS, to which a suspension of *E. coli* was added to achieve a final cell concentration of $3–6\times10^6$ cells per ml.[87] The culture was then irradiated for 30 seconds using four 15 W fluorescent bulbs, positioned 8 cm from the sample. The kill achieved was attributed to immobilized Rose Bengal, as a photobactericidal study of the supernatant when the beads were stored for 2 hours in PBS demonstrated no activity.[87]

In 1984, Savino *et al.* found that water samples contaminated with *E. coli* could be disinfected to acceptable levels for drinking, by methylene blue-PACT, where the photosensitizer again was covalently immobilized, this time on 2% poly(styrene) copolymer by nitration, reduction and diazotization.[89]

Bonnett *et al.* further developed this concept by formulating polymeric membranes with covalently bound phthalocyanine and porphyrin compounds. The chitosan-based membrane was used as the photosensitizing surface for a model of a large-scale water-flow system, and significantly reduced *E. coli* viability from 2×10^5 to 8.3×10^2 cfu ml^{-1} after 160 minutes in a circulating water photoreactor system irradiated using a 500 W 50 Hz halogen lamp.[90] Manjon *et al.*[91] showed that aggregated tris(4,4′,-diphenyl-2,2′,-bipyridine) ruthenium(II) and tris(4,7-diphenyl-1,10-phenanthroline)ruthenium (II) photosensitizers in porous silicone were effective in photodynamic killing of water-borne *Enterococcus faecilis* using a solar simulator and a compound parabolic collector-based photoreactor.

These types of systems could each potentially be of use in developing countries where quality of drinking water is an issue and high-intensity illumination could be provided by the Sun. However, difficulties, such as longevity of the membrane and photobleaching of photosensitizer, would need to be investigated before routine use, as would the spectrum of activity of the materials.

7.5. Discussion

Photodynamic antimicrobial chemotherapy is a potentially viable treatment option for localized infection. However, factors such as acceptable delivery of photosensitizers to their site of action may impede the development of PACT and reduce its clinical applicability. Aspects of drug delivery that are important in PACT include the site of action, the physico-chemical characteristics of the photosensitizer and the characteristics of the formulation.

From the oral mucosa to the skin surface, the structure and, thus, characteristics of the target site vary markedly. A drug delivery device should be able to withstand high moisture without degradation in the case of delivery to the oral cavity. A device should provide the moist environment essential for diffusion of photosensitizer to the site of action in the case of delivery to the skin or onychomycosis-infected nails. In some cases, the temperature and pH of the site may also affect release of drugs from dosage forms. Factors such as the size of the molecule, its pKa and lipophilicity will affect transport of the molecule to the site of action. With respect to drug delivery devices, the solubility and diffusivity of a drug in the dosage form and its chemical stability are vitally important.

Factors such as drug concentration within the formulation, rate of drug delivery and required contact time with site of action for adequate drug release are governed by the formulation. Issues such as stability and ease of application and removal after use are important with respect to usability in a clinical situation.

To date, current research in PACT has focused on achieving significant kill *in vitro*, with little emphasis on drug delivery. Therefore, simplistic methods of drug delivery have been employed as research progresses from experimental lab-based studies, through animal experiments to the clinical environment. Preparation of photosensitizer solutions immediately prior to use is not possible in the clinical scenario, where most patients would require treatment in one day. In this case, the scope for error in preparation of solution and dosing, in addition to the time constraint placed on clinicians, would make PACT a non-viable option. Thus, more complex drug delivery systems that facilitate administration of photosensitizer to the site of action are required.

Significant achievements have been made to date in the field of drug delivery for PACT. The Levulan Kerastick® is a device that can reduce preparation time for clinicians, by removing the need for measuring photosensitizer and solvent prior to use. However, solution-based delivery systems will be almost impossible to keep in place for sufficient times to allow drug absorption. For more complex and controlled drug delivery, patch-based systems in current development provide an exciting alternative in this field. Doses may be controlled by duration of contact with these patches. Formulations which adhere to the site of action and deliver highly coloured photosensitizer only to this site would be attractive for the patient.

Despite the vast number of studies published in the area of PACT, a rational approach to formulation design has not taken place. This may be because this field is dominated by clinicians and basic scientists, rather than those involved in pharmaceutical formulation development. Studies published to date in PACT have used aqueous solutions, oil in water creams, hydrogels, organogels, cubic phase formulations and aqueous and solvent-based patches. These dosage forms, which in many cases seem to have been selected at random with little regard to their nature, possess a multitude of different physico-chemical properties. When formulating a drug delivery system, the aim should be to maximize the thermodynamic activity of the drug substance in the vehicle, so as to maximize the concentration drive for diffusion and the partition coefficient between the target site in the body and vehicle. The situation is, of course, different when photosensitizers are immobilized on insoluble supports for use in water disinfection or as implantable biomaterials. In these cases, there is a real need to demonstrate broad spectrum antimicrobial activity, while ensuring damage to human tissue is minimized. From a practical viewpoint, systems that can eradicate high levels of both adhering microorganisms and those in close proximity in relatively short periods of time (hours rather than days) without being subject to significant photobleaching or detachment of photosensitizer from the insoluble support should be the ultimate goal. It is difficult to see the real benefit of using targeting devices for PACT photosensitizers, given their already-demonstrated affinity for microbial cells over their mammalian counterparts. This is especially so as one oft-cited reason for lack of development of resistance to PACT is its multi-factorial mechanism of action, which could be compromised by targeting attachment to a specific microbial structure.

In recent times, PACT has been the subject of intense investigation, in the laboratory especially, but now also to some extent in the clinic. Although still considered as an experimental technique, the status of PACT and its value within modern clinical practice is likely to continue to grow, especially considering the increasing incidence of infection caused by antibiotic-resistant microorganisms. It is hoped that small-scale clinical trials carried out by small- to medium-sized enterprises, such as Destiny Pharma and Photopharmica, will lead to more widespread use of PACT, to the ultimate benefit of patients worldwide.

References

1. C. Horfelt, B. Stenquist, O. Larko, J. Faergemann and A.M. Wennberg, Photodynamic therapy for acne vulgaris: a pilot study of the dose-response and mechanism of action, *Acta Derm. Venereol.*, 2007, **87**, 325.
2. S.R. Wiegell and H.C. Wulf, Photodynamic therapy of acne vulgaris using methyl aminolaevulinate: a blinded, *randomized, controlled trial, Br. J. Dermatol.*, 2006, **154**, 969.
3. W. Hongcharu, C.R. Taylor, Y. Chang, D. Aghassi, K. Suthamjariya and R.R. Anderson, Topical ALA-photodynamic therapy for the treatment of acne vulgaris, *J. Invest. Dermatol.*, 2000, **115**, 183.
4. Y. Itoh, Y. Ninomiya, S. Tajima and A. Ishibashi, Photodynamic therapy for acne vulgaris with topical 5-aminolevulinic acid, *Arch. Dermatol.*, 2000, **136**, 1093.
5. I. M. Stender, R. Na, H. Fogh, C. Gluud and H.C. Wulf, Photodynamic therapy with 5-aminolaevulinic acid or placebo for recalcitrant foot and hand warts: randomised double-blind trial, *Lancet*, 2000, **355**, 963.
6. I.M. Stender, J. Lock-Andersen and H.C. Wulf, Recalcitrant hand and foot warts successfully treated with photodynamic therapy with topical 5-aminolaevulinic acid: a pilot study, *Clin. Exp. Dermatol.*, 1999, **24**, 154.
7. Y.S. Wang, Y.K. Tay, C. Kwok and E. Tan, Photodynamic therapy with 20% aminolevulinic acid for the treatment of recalcitrant viral warts in an Asian population, *Int. J. Dermatol.*, 2007, **46**, 1180.

8. D. Stahl, N.K. Veien and H.C. Wulf, Photodynamic inactivation of virus warts: a controlled clinical trial, *Clin. Exp. Dermatol.*, 1979, **4**, 81.
9. S.A. Lambrechts, T.N. Demidova, M.C. Aalders, T. Hasan and M.R. Hamblin, Photodynamic therapy for Staphylococcus aureus infected burn wounds in mice, *Photochem. Photobiol. Sci.*, 2005, **4**, 503.
10. A. Orenstein, D. Klein, J. Kopolovic, E. Winkler, Z. Malik and N. Keller, The use of porphyrins for eradication of Staphylococcus aureus in burn wound infections, *FEMS Immunol. Med. Microbiol.*, 1997, **19**, 307.
11. F. Gad, T. Zahra, K.P. Francis, T. Hasan and M.R. Hamblin, Targeted photodynamic therapy of established soft-tissue infections in mice, *Photochem. Photobiol. Sci.*, 2004, **3**, 451.
12. M.L. Embleton, S.P. Nair, W. Heywood, D.C. Menon, B.D. Cookson and M. Wilson, Development of a novel targeting system for lethal photosensitization of antibiotic-resistant strains of Staphylococcus aureus, *Antimicrob. Agents Chemother.*, 2005, **49**, 3690.
13. R.F. Donnelly, P.A. McCarron, J.M. Lightowler and A.D. Woolfson, Bioadhesive patch-based delivery of 5-aminolevulinic acid to the nail for photodynamic therapy of onychomycosis, *J. Control. Rel.*, 2005, **103**, 381.
14. A.F. Taub, Photodynamic therapy for the treatment of acne: a pilot study, *J. Drugs Dermatol.*, 2004, **3**, S10–4.
15. DUSA pharmaceuticals, Product information for Levulan Kerastick topical solution, http://www.dusapharma.com/levulan-prescribing-information.html. Accessed 15/06/2010.
16. P.A. McCarron, R.F. Donnelly, G.P. Andrews and A.D. Woolfson, Stability of 5-aminolevulinic acid in novel non-aqueous gel and patch-type systems intended for topical application, *J. Pharm. Sci.*, 2005, **94**, 1756–1771.
17. G. Lee, R. Szeimies and U. Kosciessa, Dermal application system for aminolaevulinic acid-derivatives, Patent Number WO/2003/061621, 2005.
18. J. Zhang, L. Rigby and W. Hull, Methods and formulations for photodynamic therapy, Patent Number 10208662, 2002.
19. D. Kacerovska, K. Pizinger, M. Kumpova and P. Cetkovska, Genital warts treated by photodynamic therapy, *Skinmed.*, 2007, **6**, 295.
20. X.L. Wang, H.W. Wang, H.S. Wang, S.Z. Xu, K.H. Liao and P. Hillemanns, Topical 5-aminolaevulinic acid-photodynamic therapy for the treatment of urethral condylomata acuminata, *Br. J. Dermatol.*, 2004, **151**, 880.
21. N.K. Veien, J. Genner, H. Brodthagen and G. Wettermark, Photodynamic inactivation of verrucae vulgares, *Acta Derm. Venereol.*, 1977, **57**, 445.
22. G. Fabbrocini, M.P. Di Costanzo, A.M. Riccardo, M. Quarto, A. Colasanti and G. Roberti, Photodynamic therapy with topical delta-aminolaevulinic acid for the treatment of plantar warts, *J. Photochem. Photobiol. B: Biol.*, 2001, **61**, 30–34.
23. Photopharmica website, www.photopharmica.com. Photopharmica targets infected wounds with its photodynamic therapy. Accessed 28/06/2010.
24. T.W. Wong, Y.Y. Wang, H.M. Sheu and Y.C. Chuang, *Antimicrob. Agents Chemother.*, 2005, **49**, 895–902.
25. T.W. Wong, Y.Y. Wang, H.M. Sheu and Y.C. Chuang, Bactericidal effects of toluidine blue-mediated photodynamic action on *Vibrio vulnificus, Antimicrob. Agents Chemother.*, 2005, **49**, 895–902.
26. M.R. Hamblin, T. Zahra, C.H. Contag, A.T. McManus and T. Hasan, Optical monitoring and treatment of potentially lethal wound infections *in vivo, J. Infect. Dis.*, 2003, **187**, 1717.

27. J.Y. Jung, P.S. Kwon, J.C. Ahn, R.F. Ge, M.W. Suh and C.K. Rhee, *In vitro* and *in vivo* photodynamic therapy of otitis media in gerbils, *Laryngoscope*, 2009, **119**, 1781.
28. V. Engelhardt, B. Krammer and K. Plaetzer, Antibacterial photodynamic therapy using water-soluble formulations of hypericin or mTHPC is effective in inactivation of Staphylococcus aureus, *Photochem Photobiol. Sci.*, 2010, **9**, 365.
29. A. Hirao, S. Sato, D. Saitoh, N. Shinomiya, H. Ashida and M. Obara, *In vivo* photoacoustic monitoring of photosensitizer distribution in burned skin for antibacterial photodynamic therapy, *Photochem. Photbiol.*, 2010, **86**, 426.
30. R.F. Donnelly, C.M. Cassidy, R.G. Loughlin, A. Brown, M.M. Tunney, M.G. Jenkins and P.A. McCarron, Delivery of Methylene Blue and meso-tetra (N-methyl-4-pyridyl) porphine tetra tosylate from crosslinked poly(vinyl alcohol) hydrogels: A potential means of photodynamic therapy of infected wounds, *J. Photochem. Photobiol. B: Biol.*, 2009, **96**, 223.
31. C. Valenta, B.G. Auner and I. Loibl, Skin permeation and stability studies of 5-aminolevulinic acid in a new gel and patch preparation, *J. Control. Rel.*, 2005, **107**, 495–501.
32. J. Bender, M.B. Ericson, N. Merclin, V. Iani, A. Rosen, S. Engstrom and J. Moan, Lipid cubic phases for improved topical drug delivery in photodynamic therapy, *J. Control. Rel.*, 2005, **106**, 350–360.
33. F. Berthiaume, S.R. Reiken, M. Toner, R.G. Tompkins and M.L. Yarmush, Antibody-targeted photolysis of bacteria in vivo, *Biotechnology (NY)*, 1994, **12**, 703–706.
34. S.G. Jones, R. Edwards and D.W. Thomas, Inflammation and wound healing: the role of bacteria in the immuno-regulation of wound healing, *Int. J. Low Extrem. Wounds*, 2004, **3**, 201–208.
35. J.F. Wu, X. Hao, T. Wei, R. Kopelman, M.A. Philbert and C.W. Xi, Eradication of bacteria in suspension and biofilms using Methylene Blue-loaded dynamic nanoplatforms, *Antimicrob. Agents Chemother.*, 2009, **53**, 3042.
36. M. Nisnevitch, F. Nakonechy and Y. Nitzan, Photodynamic antimicrobial chemotherapy by liposome-encapsulated water-soluble photosensitizers, *Rus. J. Bioorg. Chem.*, 2010, **36**, 363.
37. R. Cahan, N. Swissa, G. Gellerman and Y. Nitzan, Photosensitizer–antibiotic conjugates: a novel class of antibacterial molecules, *Photochem. Photobiol.*, 2010, **86**, 418.
38. J. Gil-Tomás, S. Tubby, I.P. Parkin, N. Narband, L. Dekker, S.P. Nair, M. Wilson and C. Street, Lethal photosensitisation of Staphylococcus aureus using a toluidine blue O–tiopronin–gold nanoparticle conjugate, *J. Mat. Chem.*, 2007, **17**, 3739.
39. M.R. Hamblin and T. Hasan, Photodynamic therapy: a new antimicrobial approach to infectious disease? *Photochem. Photobiol. Sci.*, 2004, **3**, 436.
40. P.A. Barrow and J.S. Soothill, Bacteriophage therapy and prophylaxis: rediscovery and renewed assessment of potential, *Trends Microbiol.*, 1997, **5**, 268–271.
41. C.K. Hope, S. Packer, M. Wilson and S.P. Nair, The inability of a bacteriophage to infect *Staphylococcus aureus* does not prevent it from specifically delivering a photosensitizer to the bacterium enabling its lethal photosensitization, *J. Antimicrob. Chemother.*, 2009, **64**, 59.
42. B. Zeina, J. Greenman, D. Corry and W.M. Purcell, Antimicrobial photodynamic therapy: assessment of genotoxic effects on keratinocytes in vitro, *Br. J. Dermatol.*, 2003, **148**, 229–232.

43. E.A. Genina, A.N. Bashkatov, E.E. Chikina, A.B. Knyazev, O.V. Mareev and V.V. Tuchin, Methylene blue mediated laser therapy of maxillary sinusitis, *Laser Phys.*, 2006, **16**, 1128.
44. Y. Zohar, Y.P. Talmi, Y. Finkelstein and Y. Shem-Tov, Use of a Braunula for maxillary sinus irrigation, *J. Laryngol. Otol.*, 1989, **103**, 279–280.
45. B. Love, Destiny Pharma website, www.destinypharma.com. Accessed 20/06/2010.
46. M.C. Teichert, J.W. Jones, M.N. Usacheva and M.A. Biel, Treatment of oral candidiasis with methylene blue-mediated photodynamic therapy in an immunodeficient murine model, *Oral Surg. Oral Med. Oral Pathol. Oral Radiol. Endod.*, 2002, **93**, 155.
47. R.F. Donnelly, P.A. McCarron, M.M. Tunney and A.D. Woolfson, Potential of photodynamic therapy in treatment of fungal infections of the mouth. Design and characterisation of a mucoadhesive patch containing toluidine blue O, *J. Photochem. Photobiol. B: Biol.*, 2007, **86**, 59–69.
48. J. Lin, L.J. Bi, Z.G. Zhang, Y.M. Fu and T.T. Dong, Toluidine blue-mediated photodynamic therapy of oral wound infections in rats, *Las. Med. Sci.*, 2010, **25**, 233.
49. British Society of Periodontology website, www.bsperio.org.uk/. Periodontal disease and treatment. Accessed 26/6/2010.
50. J. Dobson and M. Wilson, Sensitization of oral bacteria in biofilms to killing by light from a low-power laser, *Arch. Oral Biol.*, 1992, **37**, 883.
51. F.M. Lauro, P. Pretto, L. Covolo, G. Jori and G. Bertoloni, Photoinactivation of bacterial strains involved in periodontal diseases sensitized by porphycene-polylysine conjugates, *Photochem. Photobiol. Sci.*, 2002, **1**, 468–470.
52. M. Bhatti, A. MacRobert, S. Meghji, B. Henderson and M. Wilson, Effect of dosimetric and physiological factors on the lethal photosensitization of Porphyromonas gingivalis in vitro, *Photochem. Photobiol.*, 1997, **65**, 1026.
53. S. Wood, B. Nattress, J. Kirkham, R. Shore, S. Brookes and J. Griffiths, An *in vitro* study of the use of photodynamic therapy for the treatment of natural oral plaque biofilms formed *in vivo*, *J. Photochem. Photobiol. B: Biol.*, 1999, **50**, 1.
54. N. Komerik, H. Nakanishi, A.J. MacRobert, B. Henderson, P. Speight and M. Wilson, *In vivo* killing of Porphyromonas gingivalis by toluidine blue-mediated photosensitization in an animal model, *Antimicrob. Agents Chemother.*, 2003, **47**, 932–940.
55. B.W. Sigusch, A. Pfitzner, V. Albrecht and E. Glockmann, Efficacy of photodynamic therapy on inflammatory signs and two selected periodontopathogenic species in a beagle dog model, *J. Periodontol.*, 2005, **76**, 1100.
56. R. Andersen, N. Loebel, D. Hammond and M. Wilson, Treatment of periodontal disease by photodisinfection compared to scaling and root planing, *J. Clin. Dent.*, 2007, **18**, 34–38.
57. S.J. Bonsor, R. Nichol, T.M. Reid and G.J. Pearson, Microbiological evaluation of photo-activated disinfection in endodontics (an *in vivo* study), *Br. Dent. J.*, discussion 329, 2006, **200**, 337–341.
58. D.S. Jones, C.J. Lorimer, G.P. Andrews, C.P. McCoy and S.P. Gorman, An examination of the thermorheological and drug release properties of zinc tetraphenylporphyrin-containing thermoresponsive hydrogels, designed as light activated antimicrobial implants, *Chem. Eng. Sci.*, 2007, **62**, 990–999.
59. M. Lulic, I.L. Gorog, G.E. Salvi, C.A. Ramseier, N. Mattheos and N.P. Lang, One-year outcomes of repeated adjunctive photodynamic therapy during periodontal

maintenance: a proof-of-principle randomized-controlled clinical trial, *J. Clin. Periodont.*, 2009, **36**, 661.

60. M. Raghavendra, A. Koregol and S. Bhola, Photodynamic therapy: a targeted therapy in periodontics, *Austral. Dent. J.*, 2009, **54**, S102.
61. S.L. Pinheiro, J.M. Donega, L.M.S. Seabra, M.D. Abado, T. Lopes, T.H.D. do Carmo, M.C. Ribeiro and P.F.R. Bertolini, Capacity of photodynamic therapy for microbial reduction in periodontal pockets, *Las. Med. Sci.*, 2010, **25**, 87.
62. P. Braham, C. Herron, C. Street and R. Darveau, Antimicrobial photodynamic therapy may promote periodontal healing through multiple mechanisms, *J. Periodont.*, 2009, **80**, 1790.
63. R.F. Donnelly, P.A. McCarron, C.M. Cassidy, J.S. Elborn and M.M. Tunney, Delivery of photosensitisers and light through mucus: investigations into the potential use of photodynamic therapy for treatment of Pseudomonas aeruginosa cystic fibrosis pulmonary infection, *J. Control. Rel.*, 2007, **117**, 217–226.
64. S.B. Brown, C.C.O'Grady, J. Griffiths, K.J. Mellish and D.I. Vernon, Biologically active methylene blue derivatives, 10/723,420 (13/5/2008) 1–37.
65. W. Love, D. Brundish, W. Rhys-Williams, X.D. Feng and B. Pugin, Porphyrin derivatives and their use in photodynamic therapy, Patent Number 10744863, 2003.
66. C. Politis, L. Kavallierou, S. Hantziara, P. Katsea, V. Triantaphylou and C. Richardson, Quality and safety of fresh-frozen plasma inactivated and leucoreduced with the Theraflex methylene blue system including the Blueflex filter: 5 years' experience, *Vox Sang.*, 2007, **92**, 319.
67. L.M. Williamson, R. Cardigan and C.V. Prowse, Methylene blue-treated fresh-frozen plasma: what is its contribution to blood safety? *Transfusion*, 2003, **43**, 1322.
68. F. Depasse, L. Sensebé, J. Seghatchian, G. Andreu and M. Samama, The influence of methylene blue light treatment and methylene blue removal filter on fibrinogen activity states and fibrin polymerisation indices, *Transfus. Apheresis Sci.*, 2005, **33**, 63.
69. M. Wainwright, The emerging chemistry of blood product disinfection, *Chem. Soc. Rev.*, 2002, **31**, 128.
70. M. Wainwright, Photodynamic antimicrobial chemotherapy (PACT), *J. Antimicrob. Chemother.*, 1998, **42**, 13.
71. M. Wainwright, H. Mohr and W.H. Walker, Phenothiazinium derivatives for pathogen inactivation in blood products, *J. Photochem. Photobiol. B: Biol.*, 2007, **86**, 45.
72. M. Wainwright, Pathogen inactivation in blood products, *Curr. Med. Chem.*, 2002, **9**, 127.
73. M. Wainwright, Methylene blue derivatives – suitable photoantimicrobials for blood product disinfection? *Int. J. Antimicrob. Agents*, 2000, **16**, 381.
74. C. Brady, S.E. Bell, C. Parsons, S.P. Gorman, D.S. Jones and C.P. McCoy, Novel porphyrin-incorporated hydrogels for photoactive intraocular lens biomaterials, *J. Phys. Chem. B.*, 2007, **111**, 527–534.
75. B. Baroli, Hydrogels for tissue engineering and delivery of tissue-inducing substances, *J. Pharm. Sci.*, 2007, **96**, 2197.
76. M. Wainwright, A. Stanforth, R. Jones, C. Loughran and K. Meegan, Photoantimicrobials as a potential local approach to geriatric UTIs, *Lett. Appl. Micro.*, 2010, **50**, 486.
77. W.G. Love, M.J. Cook and D.A. Russell, Porphyrin derivatives: Their use in photodynamic therapy and the medical devices containing them, Patent number 09/786, 139 (7/10/2003) 1–32.

78. R. Bonnett, D.G. Buckley, T. Burrow, A.B. Galia, B. Saville and S.P. Songca, Photobactericidal materials based on porphyrins and phthalocyanines, *J. Mater. Chem.*, 1993, **3**, 323.
79. M. Wilson, Light-activated antimicrobial coating for the continuous disinfection of surfaces, *Infect. Control Hosp. Epidemiol.*, 2003, **24**, 782.
80. V. Decraene, J. Pratten and M. Wilson, Cellulose acetate containing toluidine blue and rose bengal is an effective antimicrobial coating when exposed to white light, *Appl. Environ. Microbiol.*, 2006, **72**, 4436.
81. R. Cunningham and S. Dial, Is over-use of proton pump inhibitors fuelling the current epidemic of *Clostridium difficile*-associated diarrhoea? *J. Hosp. Infect.*, 2008, **70**, 1.
82. S. Perni, C. Piccirillo, J. Pratten, P. Prokopovich, W. Chrzanowski, I.P. Parkin and M. Wilson, The antimicrobial properties of light-activated polymers containing methylene blue and gold nanoparticles, *Biomaterials*, 2009, **30**, 89.
83. M. Wainwright, M.N. Byrne and M.A. Gattrell, Phenothiazinium-based photobactericidal materials, *J. Photochem. Photobiol. B: Biol.*, 2006, **84**, 227.
84. J. Bozja, J. Sherrill, S. Michielsen and I. Stojiljkovic, Porphyrin-based, light-activated antimicrobial materials, *J. Polym. Sci. Polym. Chem.*, 2003, **41**, 2297.
85. K. Zerdin, M.A. Horsham, R. Durham, P. Wormell and A.D. Scully, Photodynamic inactivation of bacterial spores on the surface of a photoactive polymer, *React. Funct. Polym.*, 2009, **69**, 821.
86. C.E. DeMarco, L.A. Cushing, E. Frempong-Manso, S.M. Seo, T.A. Jaravaza and G.W. Kaatz, Efflux-related resistance to norfloxacin, dyes, and biocides in bloodstream isolates of *Staphylococcus aureus*, *Antimicrob. Agents Chemother.*, 2007, **51**, 3235.
87. S.A. Bezman, P.A. Burtis, T.P. Izod and M.A. Thayer, Photodynamic inactivation of E. coli by rose bengal immobilized on polystyrene beads, *Photochem. Photobiol.*, 1978, **28**, 325–329.
88. A.P. Schaap, A.L. Thayer, E.C. Bloosey and D.C. Neckers, Polymer-based sensitizers for photooxidations II, *J. Am. Chem. Soc.*, 1975, **97**, 3741–3745.
89. A. Savino and G. Angeli, Photodynamic inactivation of *E. coli* by immobilized or coated dyes on insoluble supports, *Water Res.*, 1985, **19**, 1465.
90. R. Bonnett, M.A. Krysteva, I.G. Lalov and S.V. Artarsky, Water disinfection using photosensitizers immobilized on chitosan, *Water Res.*, 2006, **40**, 1269.
91. F. Manjon, M. Santana-Magana, D. Garcia-Fresnadillo and G. Orellana, Singlet oxygen sensitizing materials based on porous silicone: photochemical characterization, effect of dye reloading and application to water disinfection with solar reactors, *Photochem. Photobiol. Sci.*, 2010, **6**, 838.

Chapter 8

Photodynamic Therapy of Localized Infections in Animal Models

Tianhong Dai[a,b] and Michael R. Hamblin[a,b,c]

[a] Wellman Center for Photomedicine, Massachusetts General Hospital, Boston, MA 02114, USA; [b] Department of Dermatology, Harvard Medical School, Boston, MA 02114, USA; [c] Harvard-MIT Division of Health Sciences and Technology, Cambridge, MA 02139, USA

Table of Contents

Photodynamic Inactivation of Microbial Pathogens: Medical and Environmental Applications
Editors Michael R. Hamblin and Giulio Jori

Published by the Royal Society of Chemistry, www.rsc.org

8.1. Introduction

Despite advances in wound care and management, infections remain a leading cause of mortality, morbidity and economic upheaval in millions of patients with surgical wounds, traumatic lesions and other conditions around the world.[1–2] Challenges remain in countering microorganisms even where antibiotics are available.

Photodynamic therapy (PDT) was discovered over 100 years ago by observing the killing of microorganisms when harmless photosensitizers (PS) and visible light were combined *in vitro*. Since then it has primarily been developed as a treatment for cancer, ophthalmologic disorders and in dermatology. However, in recent years interest in the antimicrobial effects of PDT has been revived and it has been proposed as a therapy for a large variety of localized infections. This revival of interest has largely been driven by the inexorable increase in drug resistance amongst many classes of pathogen. Advantages of PDT include equal killing effectiveness regardless of antibiotic resistance, and a lack of induction of PDT resistance.

In order to develop new therapies such as PDT, we need to have a detailed understanding of basic aspects of the host-pathogen relationship and the complexity of the host response to infection. Animal models have become standard tools for the study of a wide array of localized infections. In general, animal models of wound infections attempt to reflect human wound infection problems. Although the final judgment of the efficacy of antimicrobial PDT must depend on the results of controlled clinical trials, the findings made in experimental models can permit reasonably accurate predictions of clinical efficacy, provided that the limitations of these models are clearly realized. In this chapter, we will discuss the use of animal models for investigating PDT of various types of localized infections.

8.2. PDT of Localized Infections in Animal Models

8.2.1. PDT of Soft-tissue Infections

Complicated soft-tissue infections remain a common reason for hospitalization. The incidence of soft-tissue infections have been increasing in recent years due to increased numbers of critically ill, elderly and immune-compromised patients including those with AIDS, cancer patients and organ transplant recipients taking immune-suppressive therapy.

8.2.1.1. PDT of Staphylococcus aureus Soft Tissue Infections in Mice

Gad *et al*[3] reported on the establishment of a soft-tissue infection model using engineered bioluminescent *Staphylococcus aureus* in mice who have been rendered temporarily neutropenic by cyclophosphamide administration. The use of luminescent bacteria together with a bioluminescence imaging camera allowed the real-time monitoring of the extent of bacterial infection in living animals. These infections were treated with PDT by direct injection of poly-L-lysine chlorine e6 (pL-ce6) conjugate into the infected area, followed by illumination with red laser light.

Male BALB/c mice were shaved on one or both back legs and depilated. Mice were pre-treated with two separate doses of cyclophosphamide in order to create a temporary state of neutropenia. One million (10^6) mid-log phase bioluminescent *S. aureus* cells were injected 2 mm beneath the surface of the thigh muscle in mice. Twenty-four hours later, mice were injected with pL-ce6 at a dose of 50 μL of a 1 mM ce6 equivalent solution into the area of subcutaneous and/or intramuscular infection. Thirty minutes after pL-ce6 injection, red light at 665 nm from a diode laser was delivered as a surface spot or by interstitial fibre into the infection.

Typical time course sets of bioluminescence images from a representative infected mouse are shown in Figure 8.1a. The mouse shows a greater than 100-fold increase in signal 24 h after infection and this remains at a very high level for a further 3 days. Bioluminescence was still present in the mouse thighs for up to 16 days after infection.

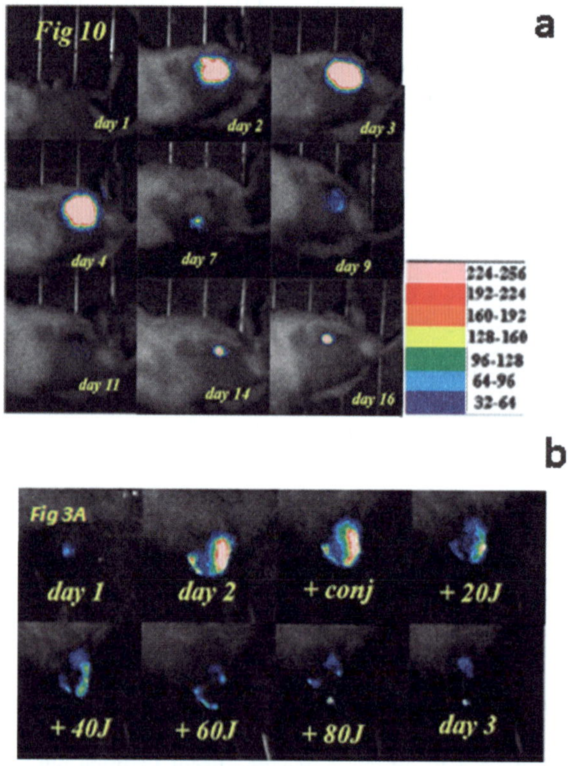

Figure 8.1 a) A series of bioluminescence images resulting from the injection of 10^6 log phase *S. aureus* 8352-4 into the thigh muscle of a neutropenic mouse. b) A series of bioluminescence images from a mouse infected on day 1 and treated with PDT on day 2.

When pL-ce6 was injected into the infected area, there was a slight reduction in bacterial bioluminescence observed immediately after the conjugate was injected into the infection. Luminescence was further reduced after the 30 min. incubation period in the dark. When illumination was commenced, there was a light dose-dependent decrease in luminescence after each 20 J cm^{-2} increment of red light (Figure 8.1b). In two of the mice, at the completion of delivery of 160 J cm^{-2}, the bioluminescence had declined to the limit of detection and there was no re-growth of the bacteria in these animals, as evidenced by imaging for the following days.

8.2.1.2. PDT of *Pseudomonas aeruginosa* Soft-tissue Infections in Mice

Berthiaume *et al.*[4] evaluated the efficacy of antibody targeted photolysis to kill bacteria *in vivo* using immunconjugates against *Pseudomonas aeruginosa.* Initially, they mixed the bacteria with the tin(IV) chlorin e6-monoclonal antibody conjugate *in vitro* and injected the mixture into the subcutaneous dorsal area in mice. After infection, both specific and non-specific conjugates were injected at the infection site. After a 15 min. incubation period, the site was exposed to 630 nm light with a power density of 100 mW cm^{-2} for 1600 s (total light 160 J cm^{-2}). Illumination resulted in a greater than 75% decrease in the number of viable bacteria at sites treated with a specific conjugate, whereas normal bacterial growth was observed in animals that were untreated or treated with a non-specific conjugate.

8.2.2. PDT of Burn Infections

The effect of burns in destroying the cutaneous barrier, rendering the affected tissue non-perfused, and depressing immune defences, means that they very commonly become infected. In past years the majority of patients with serious burns died from infections.

8.2.2.1. PDT of *S. aureus* Burn Infections in Mice[5]

Male BALB/c mice were depilated on their backs. Burn wounds were induced in anesthetized mice by pressing the ends of two preheated brass rods ($\approx$95 °C) against opposite sides of a raised dorsal skin fold for 10 s, producing third-degree burns of 1–1.5 cm^2 area (1.5–2.2% of body surface area). Ten minutes after burning, bacterial suspension containing 10^8 CFU was applied to the eschar of the burns. PDT was performed 24 h after the bacterial inoculation when the bacteria had been allowed to multiply in the wound and establish an infection. The PS used was 5-phenyl-10,15, 20-tris(*N*-methyl-4-pyridyl)-porphyrin chloride (PTMPP), a cationic porphyrin that has proved to be an effective and versatile PS. The PTMPP concentration was 500 μM and the solvent was phosphate buffered saline (PBS) or a 25% DMSO/PBS mixture; 100 μL of the PS was applied both topically and injected under the burn. The burns were illuminated directly after the application of the PTMPP. The total illumination time was 42 min. (short PDT), which gives a delivered light dose of 211 J cm^{-2}, or 84 min. (long PDT), which results in a delivered light dose of 423 J cm^{-2}. Control groups consisted of mice with infected burns that were subsequently illuminated in the absence of PS, mice with infected burns treated with PTMPP but kept in the dark or mice with infected burns that received no treatment at all.

More than 98% of the bacteria were eradicated after a light dose of 210 J cm^{-2} in the presence of PTMPP. However, bacterial re-growth was observed. Light alone or PDT both delayed the wound healing. These data suggest that PDT has the potential to rapidly reduce the bacterial load in infected burns.

8.2.2.2. PDT of Acinetobacter baumannii Burn Infections in Mice

Using the combination of polyethylenimine chlorin(e6) (PEI-ce6) conjugate and non-coherent red light, Dai *et al.*[6] investigated PDT for *Acinetobacter baumannii* burn infections in mice. Adult female BALB/c mice were used in the study. The burns were created using the same method as used in ref. (5). Five minutes after the creation of the burns, a suspension of luminescent *A. baumannii* containing 10^8 cells was inoculated onto the eschar of each burn.

PEI-ce6 was added at 30 min., 24 h or 48 h after infection. The conjugate was added as 50 µL of a solution in PBS-dimethyl sulfoxide (DMSO) cocktail (10% or 20% DMSO, 800 to 900 µM ce6 equivalent), which was added to the burns that would be treated by the use of PDT or with PS only (dark controls). After a further 15 min. to allow the conjugate to bind to and penetrate the bacteria, the mice were illuminated with 660 ± 15 nm light delivered by a non-coherent light source. The mice were given a total light exposure of up to 240 J cm^{-2} in aliquots.

Application of 10^8 *A. baumannii* cells to the surface of 10-s burns made on the mouse backs led to chronic infections that lasted, on average, 22 days and that were characterized by a remarkably stable bacterial bioluminescence (Figure 8.2a). PDT carried out on day 0 soon after application of the bacteria gave over 3 $\log_{10}$ units of loss of bacterial luminescence in a light exposure-dependent manner, while PDT carried out on day 1 and day 2 gave an approximately 1.7-$\log_{10}$ reduction (Figure 8.2b). The application of PS dissolved in 10% or 20% DMSO without light gave only a modest reduction in the bacterial luminescence from mouse burns. Some bacterial re-growth in the treated burn was observed but was generally modest. It was also found that PDT did not lead to the inhibition of wound healing. The data suggest that PDT may be an effective new treatment for multi-drug resistant localized *A. baumannii* infections.

8.2.2.3. PDT of S. aureus Burn Infections in Guinea Pigs

Using a guinea pig model, Orenstein *et al.*[7] studied the effect of porphyrins on the eradication of *S. aureus* in burns. Guinea pigs were shaved on their backs. A pre-heated copper plate (150 °C) of 10 mm×10 mm×30 mm was placed on the shaved back of each guinea pig for 10 s, resulting in a third-degree burn. Wounds were infected with 10^8 CFU of *S. aureus* 15 min. after the burning. It

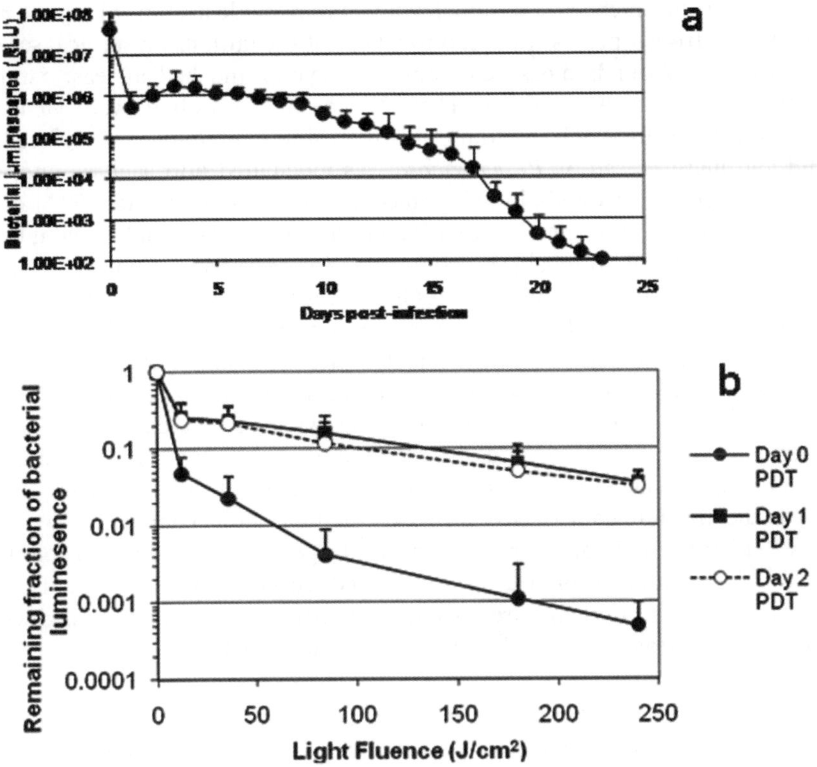

Figure 8.2 a) Time course of mean bacterial luminescence value of the infected mouse burn $(n=19)$; b) Mean dose response of bacterial luminescence of the mouse burns infected with *A. baumannii* and treated with PDT at 30 minutes, 24 hours and 48 hours after infection, respectively.

was found that deuteroporphyrin alone was strongly bactericidal only after photosensitization. On the other hand, hemin alone was moderately bactericidal but light independent. A combination of both deuterophyrin and hemin was extremely potent even in the dark and did not require illumination to eradicate the bacteria. It was further noted that a reduction of 99% of the viable bacteria was achieved after the porphyrin was dropped on the eschar or injected into the eschar.

8.2.3. PDT of Full-thickness Excisional Wound Infections

Full thickness excisional wounds involve the removal of a significant volume of the target tissue.

8.2.3.1. PDT of E. coli/P. aeruginosa Excisional Wound Infections in Mice

Hamblin *et al.*[8] developed a mouse model of excisional wound infections. Male BALB/c mice were anesthetized, shaved on the back and depilated. Four rectangular full-thickness excisional wounds were made in a line along the back (numbered from 1 nearest the head to 4 nearest the tail), using surgical scissors and forceps. Wounds measured 8 mm × 12.5 mm and had at least 5 mm of unbroken skin between them. The bottom of the wound was *panniculus carnosus*, with no visible bleeding. A suspension containing 5×10^6 cells of mid-log phase *E. coli* or *P. aeruginosa* was inoculated into each wound. The next day, infected wounds in living mice had lost, on an average, 90% of the original luminescence signal but with considerable inter-animal variability.

Polycationic photosensitizer conjugate (pL-ce6) was applied topically followed by red-light illumination at 665 nm for up to 240 J cm^{-2}. A rapid light dose-dependent loss of luminescence was observed as measured by image analysis. Figure 8.3 shows the effect of topical application of conjugate and successive applications on the bacterial luminescence (*E. coli*). These data were obtained from a mouse in which bacteria were inoculated in all wounds, conjugate was added to wounds 1 and 4 and wounds 3 and 4 were illuminated with red light. Therefore, wound 1 was the dark control with conjugate, wound 2 was the absolute control, wound 3 was the light-alone control and wound 4 was PDT treated. Topical application of a targeted polycationic PS conjugate followed by illumination led to a 99% reduction in luminescence. For the *P. aeruginosa* infection, all three groups of non-treated control mice died within 5 days; in contrast, 90% of the PDT-treated mice survived.

8.2.3.2. PDT of V. vulnificus Excisional Wound Infections in Mice

Using the same mouse model of excisional wound, Wong *et al.*[9] studied the anti-microbial effects of toluidine blue O (TBO)-mediated PDT for *V. vulnificus* wound infections in mice. Mice with single wounds on the middle of the back with an area of 8 mm×12.5 mm received inocula of mid-log-phase *V. vulnificus* suspended to appropriate concentrations in 50 μL of PBS. After a 30-min interval to allow *V. vulnificus* to penetrate the wounds, 50 μL of TBO at various concentrations was added to the wounds and was retained at the edges of the wounds for 2 min.

Fifty-three percent (10 of 19) of mice treated with 100 g of TBO per ml and exposed to broad-spectrum red light (150 J cm^{-2} at 80 mW cm^{-2}) survived, even

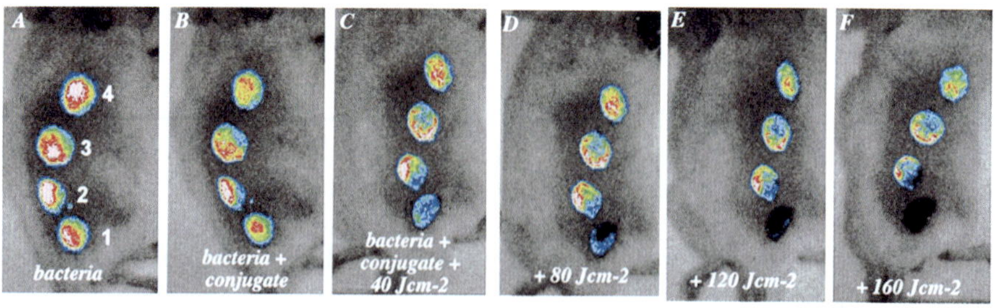

Figure 8.3 Successive overlaid luminescence images of a mouse with four excisional wounds infected with equal numbers of *E.coli* (A). Wounds 1 (nearest tail) and 4 (nearest head) receive topical application of conjugate (B). Wounds 1 and 2 (two nearest tail) were then illuminated with successive fluences (40–160 J cm^{-2}) of 665 nm light (C–F)

though systemic septicaemia had been established with a bacterial inoculum 100 times the 50% lethal dose. The study has demonstrated that PDT can cure mice with otherwise fatal *V. vulnificus* wound infections.

8.2.3.3. PDT for MRSA Excisional Wound Infections in Mice

Using methylene blue (MB) as a PS, Zolfaghari *et al.*[10] evaluated the effect of PDT on the survival of an epidemic strain of MRSA in excisional wounds in mice. Excision wounds were created by pinching and lifting the skin of the back using sterile forceps and cutting a 6 mm circular area using sharp scissors to cut down to the subcutaneous areolar tissue. Bacterial suspension containing 10^8 CFU was then added to the wound, and incubated for one hour prior to treatment. MRSA was found to be the predominant bacterium colonizing the wound at day 5.

PDT was performed 1 hour after inoculating the wounds with the bacterial suspension. The excision wounds received 25 μL of MB (100 μg ml^{-1}) solely at the start of irradiation. The wounds were irradiated immediately after the application of MB and continued for 30 minutes, equating to a total delivered light dose of 360 J cm^{-2}. Irradiation of the wounds in the presence of MB resulted in a significant reduction in the number of viable bacteria recovered from the wounds. This reduction was 25 fold (1.40 $\log_{10}$ CFU/wound).

8.2.4. PDT of Skin Abrasion Infections

Skin abrasion is a type of partial skin wound involving superficial damage of the stratum corneum and epidermis allowing microbial cells to bypass the main skin barrier and form an infection.

8.2.4.1. PDT of MRSA Skin Abrasion Infections in Mice

A mouse model of skin abrasion infected with genetically engineered bioluminescent MRSA was recently reported by Dai *et al.*[11]. Female BALB/c mice were used. At days 4 and 1 before the infection, mice were administered two doses of cyclophosphamide, fostering a more vulnerable environment in the mice to infection. Skin abrasion wounds were made on the backs of anesthetized mice using 28-gauge needles by creating 6×6 crossed scratch lines within a defined 1×1 cm^2 area. The scratches were made in such a manner that they only damaged the stratum corneum and upper layer of the epidermis but not the dermis. Five minutes after wounding, an aliquot of bacterial suspension containing 10^8 CFU of bioluminescent MRSA in PBS was inoculated over each defined area containing the crossed scratches.

A stable infection was developed in the mouse wounds as characterized by bioluminescence imaging. Figure 8.4a shows the successive bacterial luminescence images of a representative mouse skin abrasion infected with MRSA. After infection, the bacterial luminescence remained strong and stable until day 6 and detectable until day 9. Gram-staining of the histological section of a representative skin abrasion specimen (harvested at day 3 post-infection) (Figure 8.4b) showed that MRSA was localized superficially in the epidermis.

PDT was carried out at 30 min. post-infection. Figure 8.4C shows the PDT dose response of bacterial luminescence of a representative mouse skin abrasion wound infected with MRSA and treated with PDT. PDT gave a light-dose-dependent reduction of bacterial luminescence from the mouse wound until over 99% was gone at 360 J cm^{-2}.

8.2.4.2. *PDT of MRSA Skin Abrasion Infections in Mice*

Zolfaghari *et al.*[10] studied MB-mediated PDT for skin abrasion wounds infected with MRSA. Twenty-five mm^2 square shaped wounds were created in

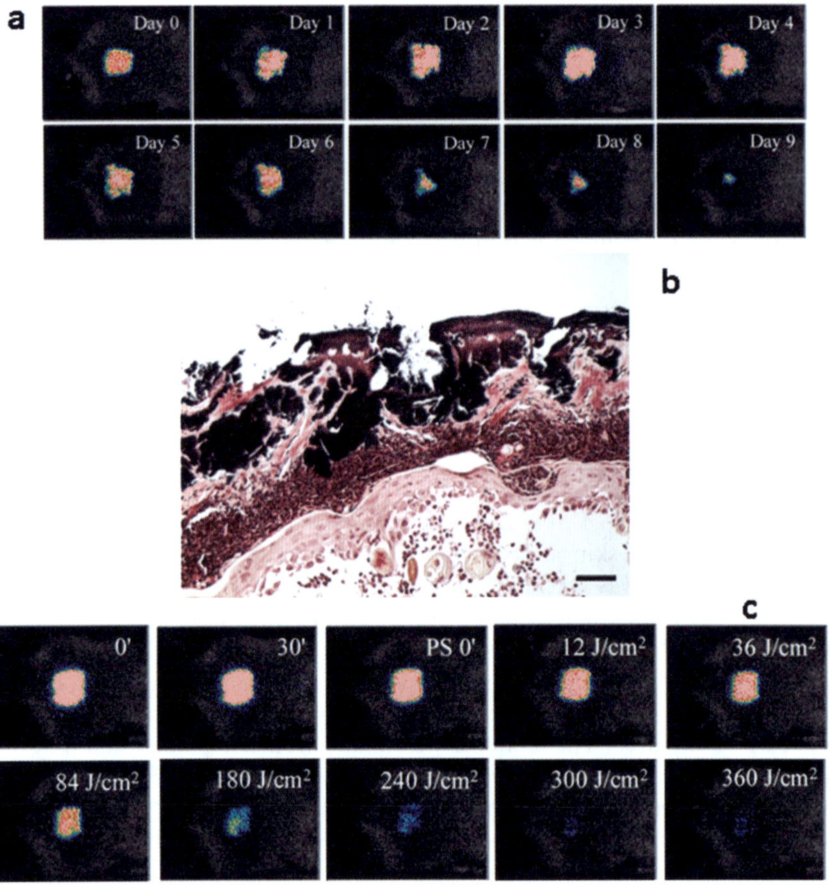

Figure 8.4 a) Successive bioluminescence images of a representative mouse skin abrasion wound infected with luminescent MRSA. b) A Gram-stained section of a mouse skin abrasion specimen showing the biofilms formed by Gram-positive MRSA near the skin surface. Dark blue area: biofilms of MRSA. The mouse skin abrasion specimen was harvested at day 3 post-infection. c) Dose response of bacterial luminescence of a mouse skin abrasion infected with MRSA and treated with PDT. PDT was carried out 30 minutes after infection.

the skin of the back by scarification using a 27G needle, run ten times parallel in one direction and another ten times perpendicular to the original tracks. Bacterial suspension containing 4×10^7 CFU was placed on the wound, and incubated for one hour prior to treatment. This method also resulted in a reproducible MRSA wound colonization model, which persisted for up to 5 days post inoculation.

PDT was performed 1 hour after inoculating the wounds with the bacterial suspension. The wounds received 10 μL of MB just before the start of irradiation and a further 10 μL after 15 minutes of irradiation. The wounds were irradiated immediately after the application of MB and continued for 30 minutes. This equated to a total delivered light dose of $360\,J\,cm^{-2}$. The results showed that irradiation of the wounds in the presence of MB resulted in a significant reduction in the number of viable bacteria recovered from the wounds. This reduction was 14-fold (1.15 $\log_{10}$ CFU/wound) in the scarified wounds.

8.2.5. PDT of Osteomyelitis

Osteomyelitis refers to an acute or chronic inflammation of bone and bone marrow secondary to contamination of pathogenic microorganisms. Bisland *et al.*[12] investigated PDT of osteomyelitis in rats. Anaesthetized female Sprague-Dawley CD rats were shaved on their flanks. The knee joint was flexed to outline the proximal anterior margin of the tibial epicondyle. Bacterial inoculation was performed for both the left and right tibia on the same animal. A scalpel was used to open the skin (1 cm incision) and expose the underlying bone. The margins of adhered fascia were cleared using the scalpel and a small bore hole was created into the intramedullary space of the tibia using a high-speed drill. A sterile needle (25G) was used to create a small channel through the medullary space, into which was placed a *S. aureus* biofilm-coated kirschner-wire (0.5 cm long). Five μL of arachidonic acid (50 $\mu g\,mL^{-1}$ in 0.9% NaCl solution) was subsequently injected 30 s after the bacterial inoculum. Arachidonic acid acts as a sclerosing agent to facilitate the bacterial infection in bone. The burn hole was closed with bone wax and the skin sutured closed and disinfected.

Test animals received an intraperitoneal (i.p.) injection of ALA (300 $mg\,kg^{-1}$ in saline). Animals were, thereafter, kept in dim light conditions for 4 h. Laser light (635 nm) was delivered transcutaneously *via* an optical fibre directed towards the infected area within the tibia. The output of the laser was adjusted to deliver fluence of $75\,J\,cm^{-2}$ at $250\,mW\,cm^{-2}$. Colony forming studies had confirmed that 48 h was sufficient time post treatment for cell death to occur. Results showed that ALA-mediated PDT inhibited biofilm implants in bone.

8.2.6. PDT of MRSA Arthritis

Tanaka *et al.*[13] studied the efficacy as well as side effects of PDT on MRSA arthritis in mice. The bilateral knees of anesthetized male C57BL / 6Cr mice were shaved.

MRSA suspension containing 10^6 CFU was injected into the left knee through the midline of the patellar ligament. The right knee was not inoculated with MRSA.

Photofrin solution was intravenously administered into the left knee at a concentration of 5, 20 or 80 mg kg^{-1} 2 days after MRSA administration. At 24 h after PS administration, the left knee was irradiated using a 635-nm CW laser for 700 s (70 J cm^{-2}).

The number of MRSA in synovial fluid was then examined at 24 h after PDT. All of the mice that had been inoculated with 80 mg kg^{-1} Photofrin died after PDT. There was no death in other PDT treated mice inoculated with 20 or 5 mg kg^{-1} Photofrin. However, an inhibitory effect of PDT on MRSA proliferation was not seen; rather, a significant increase in MRSA was seen compared to that in control groups. The results suggested that the cytotoxic effect of PDT on intra-articular neutrophils may impair the antimicrobial efficacy of PDT.

8.2.7. PDT of Oral Infections

8.2.7.1. PDT of Oral Candidiasis in Rats

Teichert *et al.*[14] evaluated the efficacy of using methylene blue (MB)-mediated PDT to treat oral candidiasis in an immunosuppressed murine model, mimicking what is found in human patients.

Experimental beige nude mice with severe combined immunodeficiency disease (SCID) were inoculated 3 times a week by swabbing the oral cavity with a *C. albicans*-coated swab for a period of 4 weeks. On treatment day, mice were cultured for baseline fungal growth and received a topical oral cavity administration of 0.05 mL MB solution at varying concentrations from 250 to 500 μg mL^{-1}. After 10 minutes the mice underwent light activation with a 664-nm diode laser light with a cylindrical diffuser.

CFU counting of post-PDT culture indicated that MB concentrations from 250 to 400 μg mL^{-1} reduced fungal growth but did not eliminate *C. albicans*. MB concentrations of 450 and 500 μg mL^{-1} totally eradicated *C. albicans* from the oral cavity, resulting in reductions from 2.5 $\log_{10}$ and 2.74 $\log_{10}$ to 0, respectively. These results suggest that MB-mediated PDT can potentially be used to treat oral candidiasis in immunodeficient patients.

8.2.7.2. PDT of Oral Implantitis in Mongrel Dogs

In a pilot study, Shibli *et al.*[15] reported the results on lethal photosensitization on ligature-induced peri-implantitis in male mongrel dogs with different implant surfaces. A total of 36 dental implants with 4 different surface coatings (9 commercially pure titanium surface; 9 titanium plasma-sprayed; 9 hydroxyapatite; and 9 acid-etched) were inserted in 6 male mongrel dogs 3 months after extraction of mandibular premolars. After 14 months, dogs underwent surgical debridement of the dental implant sites and lethal photosensitization by combination of toluidine blue O (100 μg ml^{-1}) and illumination from a

685-nm diode laser at the energy of 200 J cm^{-2}. Five months later, biopsies of the implant sites were dissected and prepared for ground sectioning and analysis. The percentage of bone fill was 26.70 to 48.28, and the percentage of reosseointegration was 15.83 to 25.25, depending on the different surface coating used.

8.2.7.3. PDT of Oral Wound Infections in Rats

A rat model was presented by Lin *et al.*[16] to investigate toluidine blue-mediated PDT for oral wounds infected with *Streptococcus* spp. and *Actinomyces viscosus*. To induce an oral wound in rat, a 1×2 mm^2 excisional wound was made on the palate of the first maxillary molar with the gingiva and connective tissue beneath cut off. After the blood had been blocked, bacterial suspension containing 2×10^8 CFU was injected into the wound. Twenty-four hours after infection, toluidine blue solution was topically applied to the wound and retained in the infected wounds in the dark for 10 min. The stained areas were subsequently irradiated with a 635-nm diode laser at 12 J cm^{-2}, 24 J cm^{-2} and 48 J cm^{-2}, respectively. In the group that was given the highest light dose of 48 J cm^{-2}, an average killing of approximately 97% was achieved. A lesser killing effect was achieved in the group that was subjected to the lowest light dose of 12 J cm^{-2}, where an average of approximately 25% of the bacteria survived. After PDT, the lesions were allowed to develop, and the peak sizes of the lesions were recorded. It was observed that in the 24 J cm^{-2} and 48 J cm^{-2} groups the lesions were of significantly smaller size than in the 12 J cm^{-2} and non-treated control groups.

8.2.8. PDT of Otitis Media

Otitis media with effusion (OME) is the most common disease of childhood with the exception of viral upper respiratory infections. One of the most important factors that contribute to the development of OME is bacterial infection. Despite the popular use of antibiotics, complications of otitis media with significant morbidity still occur.

Using a gerbil model, Jung *et al.*[17] studied the antibacterial effects of PDT *in vivo* on *Haemophilus influenzae* and *Streptococcus pneumoniae*, the common bacterial species causing OME. Bacteria solution containing 2×10^5 CFU was injected through the bullae under sterile conditions. PDT was performed two days after the infection by injecting 20 μL of Photogem (1 mg mL^{-1}, a haematoporphyrin derivative) solution into the bullae followed by 632-nm laser illumination from a sterilized fibre tip for 90 J. PDT was effective in killing *S. pneumoniae* in 87% of the infected bullae with OME, whereas it was effective in eradicating *H. influenzae* in 50% of the infected bullae with OME.

In summary, Table 8.1 depicts the animal infection models discussed in this chapter.

Table 8.1 Photodynamic therapy for localized infections in animal models: literature review.

Models	Animal species	Microorganism	Methods used to produce infections	*PDT*		*Refs*
				PS	Light	
Soft tissue infection	Mouse	*Staphylococcus aureus*	Intraperitoneal inoculation; Neutropenic animals	pL-ce6	665 nm diode laser	3
Soft tissue infection	Mouse	*Pseudomonas qaeruginosa*	Intraperitoneal inoculation	IV-ce6	630 nm dye laser	4
Burn infection	Mouse	*Staphylococcus aureus*	Topical inoculation; Intradermal inoculation	PTMPP	Non-coherent 635 nm	5
Burn infection	Mouse	*Acinetobacter qbaumannii*	Topical inoculation	PEI-ce6	Non-coherent 660 nm	6
Burn infection	Guinea pig	*Staphylococcus aureus*	Topical inoculation	Porphyrins	Incande-scent lamps	7
Excisional wound infection	Mouse	*Escherichia coli*; *Pseudomonas aeruginosa*	Topical inoculation	pL-ce6	665 nm diode laser	8
Excisional wound infection	Mouse	*V. vulnificus*	Topical inoculation	Toluidine blue O	Broadband red light	9
Excisional wound infection	Mouse	MRSA	Topical inoculation	Methylene blue	665 nm diode laser	10

Skin abrasion infection	Mouse	MRSA	Topical inoculation; Neutropenic animals	PEI-ce6	Non-coherent 660 nm	11
Skin abrasion infection	Mouse	MRSA	Topical inoculation	Methylene blue	665 nm diode laser	10
Osteomyelitis	Rat	*Staphylococcus aureus*	*S. aureus*-coated K-wires	ALA	635 nm laser	12
Arthritis	Mouse	MRSA	Intraperitoneal inoculation	Photofrin	635 nm laser	13
Oral candidiasis	Rat	*C. albicans*	Topical inoculation; immune-suppressed animals	Methylene blue	664 nm diode laser	14
Oral implantitis	Mongrel dog		Ligature	Toluidine blue O	685 nm diode laser	15
Oral wound infection	Rat	*Streptococcus* spp. *Actinomyces viscosus*	Intraperitoneal inoculation	Toluidine blue	635 nm diode laser	16
Otitis media	Gerbil	*Haemophilus influenza; Streptococcus pneumoniae*	Intraperitoneal inoculation	Photogem	632 nm laser	17

References

1. W.G. Cheadle and M. Turina, Infection and organ failure in the surgical patient: a tribute to seminal contributions by Hiram C. Polk, Jr, M.D, *Am. J. Surg*, 2005, **190**(2), 173–177.
2. C.K. Murray, Infections in burns, *J. Trauma.*, 2007, **62**(6), S73.
3. F. Gad, T. Zahra, K.P. Francis, T. Hasan and M.R. Hamblin, Targeted photodynamic therapy of established soft-tissue infections in mice, *Photochem. Photobiol. Sci.*, 2004, **3**(5), 451–458.
4. F. Berthiaume, S.R. Reiken, M. Toner, R.G. Tompkins and M.L. Yarmush, Antibody-targeted photolysis of bacteria *in vivo*, *Biotechnology (NY)*, 1994, **12**(7), 703–706.
5. S.A. Lambrechts, T.N. Demidova, M.C. Aalders, T. Hasan and M.R. Hamblin, Photodynamic therapy for Staphylococcus aureus infected burn wounds in mice, *Photochem. Photobiol. Sci.*, 2005, **4**(7), 503–509.
6. T. Dai, G.P. Tegos, Z. Lu, L. Huang, T. Zhiyentayev, M.J. Franklin, D.G. Baer and M.R. Hamblin, Photodynamic therapy for Acinetobacter baumannii burn infections in mice, *Antimicrob. Agents Chemother*, 2009, **53**(9), 3929–3934, PMCID: 2737832.
7. A. Orenstein, D. Klein, J. Kopolovic, E. Winkler, Z. Malik, N. Keller and Y. Nitzan, The use of porphyrins for eradication of Staphylococcus aureus in burn wound infections, *FEMS Immunol. Med. Microbiol*, 1997, **19**(4), 307–314.
8. M.R. Hamblin, D.A. O'Donnell, N. Murthy, C.H. Contag and T. Hasan, Rapid control of wound infections by targeted photodynamic therapy monitored by *in vivo* bioluminescence imaging, *Photochem. Photobiol.*, 2002, **75**(1), 51–57.
9. T.W. Wong, Y.Y. Wang, H.M. Sheu and Y.C. Chuang, Bactericidal effects of toluidine blue-mediated photodynamic action on Vibrio vulnificus, *Antimicrob. Agents Chemother.*, 2005, **49**(3), 895–902, PMCID: 549273.
10. P.S. Zolfaghari, S. Packer, M. Singer, S.P. Nair, J. Bennett, C. Street and M. Wilson, *In vivo* killing of Staphylococcus aureus using a light-activated antimicrobial agent, *BMC Microbiol.*, 2009, **9**, 27, PMCID: 2642833.
11. T. Dai, G.P. Tegos, T. Zhiyentayev, E. Mylonakis and M.R. Hamblin, Photodynamic therapy for methicillin-resistant Staphylococcus aureus infection in a mouse skin abrasion model, *Lasers Surg. Med.*, 2010, **42**(1), 38–44, PMCID: 2820267.
12. S.K. Bisland, C. Chien, B.C. Wilson and S. Burch, Pre-clinical *in vitro* and *in vivo* studies to examine the potential use of photodynamic therapy in the treatment of osteomyelitis, *Photochem. Photobiol. Sci.*, 2006, **5**(1), 31–38.
13. M. Tanaka, M. Kinoshita, Y. Yoshihara, N. Shinomiya, S. Seki, K. Nemoto and Y. Morimoto, Influence of intra-articular neutrophils on the effects of photodynamic therapy for murine MRSA arthritis, *Photochem. Photobiol.*, 2010, **86**(2), 403–409.
14. M.C. Teichert, J.W. Jones, M.N. Usacheva and M.A. Biel, Treatment of oral candidiasis with methylene blue-mediated photodynamic therapy in an immunodeficient murine model, *Oral Surg. Oral Med. Oral Pathol. Oral Radiol. Endod.*, 2002, **93**(2), 155–160.
15. J. A. Shibli, M.C. Martins, F.H. Nociti Jr., V.G. Garcia and E. Marcantonio J.r, Treatment of ligature-induced peri-implantitis by lethal photosensitization and guided bone regeneration: a preliminary histologic study in dogs, *J. Periodontol.*, 2003, **74**(3), 338–345.
16. J. Lin, L.J. Bi, Z.G. Zhang, Y.M. Fu and T.T. Dong, Toluidine blue-mediated photodynamic therapy of oral wound infections in rats, *Lasers Med. Sci.*, 2010, **25**(2), 233–238.
17. J.Y. Jung, P.S. Kwon, J.C. Ahn, R. Ge, M.W. Suh and C.K. Rhee, *In vitro* and *in vivo* photodynamic therapy of otitis media in gerbils, *Laryngoscope*, 2009, **119**(9), 1781–1787.

Chapter 9

Pathogen Reduction in Blood for Transfusion Using Photodynamic Treatments

Ehud Ben-Hur[a] and Raymond P. Goodrich[b]

[a] Consultant in blood safety and photomedicine, 3209 Huckleberry Way, Loveland, CO 80538, USA; [b] Chief Science Officer, CaridianBCT Biotechnologies LLC, Lakewood, CO 80215, USA

Table of Contents

Photodynamic Inactivation of Microbial Pathogens: Medical and Environmental Applications
Editors Michael R. Hamblin and Giulio Jori

Published by the Royal Society of Chemistry, www.rsc.org

9.1. Introduction

9.1.1. Overview

Blood transfusion is a very safe medical procedure. Adverse events are rare and transmission of infectious disease due to blood transfusion is even rarer in developed countries. This was not always the case. The emergence of human immunodeficiency virus (HIV) as a transfusion-transmitted virus in the early 1980s was the first time a serious alarm was raised with regard to the safety of the blood supply. The previous risk of transmission of hepatitis viruses caused less alarm presumably because of the perceived difference in morbidity and mortality between these pathogens. Since that time a number of safety layers were introduced to reduce the risk of transmitting pathogens during blood transfusion. Donor screening and deferral based on a questionnaire are the first line of defence. Donor testing for the presence of a number of pathogens, mainly HIV and the hepatitis viruses HBV and HCV, is the second line. Initially testing was serological in nature. Nucleic acid technology (NAT) is a recent introduction to reduce the window period between the time of infection and the time at which the pathogen is detectable. NAT reduced the risk of transmission by an order of magnitude, from about 1 in 10^5 to about 1 in a million. Finally, a pathogen reduction step was introduced for greatly enhanced safety against pathogen transmission by blood transfusion. For a comprehensive discussion of blood safety see ref 1.

It should be noted that whole blood is rarely transfused nowadays. Blood is usually collected in blood centres where it is separated into three components: plasma, platelets concentrate and red cells concentrate (RCC). Each component is transfused for different indications and is stored differently. Plasma is stored frozen at −80 °C hence its name: fresh-frozen plasma (FFP). Platelets are stored at room temperature with continuous agitation for 5 days and RCC is stored at 1–6 °C for up to 6 weeks. While testing is done on the blood prior to its separation into components, each component undergoes a different pathogen reduction process. A pathogen reduction process applicable for all components and to whole blood is the ideal but is unavailable as yet.

The driving force behind the introduction of a pathogen reduction step is not the very low current risk of transmitting pathogens but rather the real risk of the emergence of new pathogens. Thus, in the early years of the new century the emergence of West Nile virus (WNV) left the blood supply exposed for risk of its transmission for over 2 years until a test was introduced. New viruses are continuously emerging and the blood supply is left exposed until they are recognized and a test is developed. The most recent emerging pathogen is Chikungunya virus.[2] The threat to the blood supply by this virus remains to be determined. A pathogen reduction step is the only way to protect against emerging viruses. It has the added benefit of potentially avoiding the need to introduce new costly tests.

9.1.2. Historical Background

During the Second World War there was an unprecedented need for large quantities of plasma derived from pooled sources, resulting in an increase of hepatitis infections among the troops. Sterilization efforts to prevent the transmission of the pathogen(s) were focused on the use of UVC (254 nm), with initially promising results.[3] This approach was abandoned when it became apparent that adequate inactivation of viruses in plasma could be achieved only by UVC doses that caused an unacceptable level of plasma clotting factors inactivation.[4] Serum albumin, a relatively heat-stable protein, was the only blood product available in a sterile form for a long time. It was heated at 60 °C for 10 hours, a process that is still in use today.

The concerns about the spread of hepatitis viruses, and later HIV, as a result of blood transfusion gave impetus to the successful development of other virus-inactivation processes.[5] These processes, however, are applicable only to plasma and its fractions (blood products) and not to the cellular components of blood (platelets and red cells). In addition, these methods can inactivate lipid-enveloped viruses such as HIV, HBV and HCV, but some non-enveloped viruses, such as HAV and parvovirus B19 are not completely inactivated. The last two viruses are of minor concern.

Virus inactivation procedures offer other potential advantages, in addition to eliminating the risk of virus transmission. The window period may no longer be of concern and efforts to reduce it by the use of NAT may no longer be needed. Errors in testing or the release of blood that tests positive may no longer result in viral transmission. Pathogens that are not tested for, including new viruses, may be eliminated, obviating the need to introduce new tests. The requirements for implementation of new pathogen reduction processes for blood components include that it should not affect it in a way that reduces its therapeutic activity. In addition, the approach should not pose health risks, *e.g.* from new immunogenic structures or toxic residues. The process should also be efficient with respect to yield and ease of implementation and must be reasonably priced.

9.2. Plasma

9.2.1. Methylene Blue and Light

Methylene blue (MB) is a photodynamic sensitizer that can target nucleic acids *via* intercalative binding and produce singlet oxygen-mediated damage, primarily in guanine residues, upon light exposure.[6] MB is well known for its virucidal activity.[7] Photodynamic treatment (PDT) with MB for pathogen reduction in FFP has been in use in Germany since 1992, following the pioneering work of Dr H. Mohr and his associates.[8] The procedure involved addition of 1 μM MB to thawed FFP and irradiation with white fluorescent light for 1 hr. The plasma is then refrozen and stored for later use.[9] Efficacy of virus inactivation depends on light irradiance, duration of irradiation, bag volume and MB concentration. HIV is inactivated by 3 log after 15 min. of irradiation; after 1 hr inactivation is 5–7 log. Unfortunately, no results have been published for human pathogenic hepatitis viruses (HAV, HBV and HCV). In addition, intracellular viruses are not inactivated[10] even though *ca.* 20% of contaminating white cells in FFP are still viable despite freezing and thawing.[11] These cells can be eliminated by introducing a filtration step.

No adverse reactions were observed in a controlled clinical study of MB-treated FFP.[12] Also, neoantigens do not appear to be produced in FFP as a result of the treatment.[13] However, the *in vitro* coagulation capacity of MB-treated FFP was reduced.[14] This was mainly because of reduced fibrinogen and factor VIII activity. Thus, in patients with critically reduced haemostasis, in whom the use of FFP is indicated, MB-treated FFP may be less effective. Consequently, its use has been discontinued for patients with severely reduced ability for synthesizing clotting factors.[14] More recently it has been reported that MB is mutagenic in a cultured mammalian cell system.[15] Considerable investigation will be required, therefore, to assess the genotoxic potential of MB-treated FFP prior to its clinical use in the USA. This may be the reason that MB-treated FFP has been withdrawn from the market in Germany.

9.3. Platelet Concentrates

9.3.1. Riboflavin and UV Light

Riboflavin (vitamin B_2) has absorption maxima at 220, 265, 375 and 446 nm in water and is yellow-orange in colour. When aqueous solutions containing riboflavin are exposed to light the vitamin is converted to lumichrome under neutral conditions. Lumichrome is a well-known metabolic product of riboflavin in the human body.[16] Photosensitization by riboflavin is oxygen-mediated *via* both Type I and Type II mechanisms.[17–19] Because of its selective binding to nucleic acids, well-defined photochemistry and lack of toxicity, riboflavin is an attractive photosensitizer for pathogen reduction in blood, especially for FFP and platelet concentrates where penetration of UV light is not a problem. It is not surprising, therefore, that CaridianBCT pursued the development of a system for pathogen reduction based on riboflavin. The system developed, termed Mirasol PRT,[20] involves transfer of the platelets from the collection bag to an illumination/storage bag using a sterile docking device, addition of riboflavin to a final concentration of 50 μM and illumination of the bag with broadband UV light (285–365 nm) for about 8 min. The platelets are mixed during illumination using a back-and-forth motion at a rate of 120 cycles per min. The system has been recently approved (CE Mark) for clinical use in Europe.

Preclinical studies of Mirasol PRT in platelet concentrates demonstrated 4.5 log reduction of intracellular HIV, 5.9 log reduction of cell-associated HIV, 5.2 log of WNV and >5 log of porcine parvovirus (PPV).[21] *S. epidermidis* and *E. coli* bacteria were reduced by >4 log.[21] Intracellular bacteria, *Orientia tsutsugamushi* (the scrub typhus agent), were reduced by 5 log in an animal model.[22] The blood-borne parasite *Leishmania* was also reduced by 5 log.[23] Thus, the Mirasol system is capable of reducing enveloped and non-enveloped viruses, Gram-positive and Gram-negative bacteria as well as parasites.

Treatment of platelets with the Mirasol system preserves platelet quality during storage.[21] Functional studies indicated that the treated platelets preserved adhesive and cohesive functions to levels compatible with those observed in untreated platelets.[24] In a clinical study the Mirasol-treated apheresis platelets had a somewhat reduced recovery (50% *vs.* 66% of control) and survival was similarly shortened (104 hr *vs.* 142 hr in control).[25] These values were similar to those of pathogen-reduced and licensed platelets that have been shown to have useful clinical applicability (the Intercept system developed by Cerus, using a psoralen as a photosensitizer).

Finally, the presence of white blood cells (WBC) in blood components has been shown to contribute to the development of a variety of immune responses. These include both donor anti-recipient and recipient anti-donor responses. Transfusion-associated graft-*versus* host disease (TA-GVHD) is an example of the former. The introduction of a leukoreduction step during the preparation of blood components reduces but does not eliminate these risks. The Mirasol system is able to functionally inactivate WBC and thus should prevent immunological consequences resulting from the presence of WBC in platelet concentrates.[26]

9.4. Red Cells Concentrates

9.4.1. *Methylene Blue and Red Light*

A major problem in using a photosensitizer and light for pathogen reduction in RCC is the strong light absorption by haemoglobin. The efforts were focused therefore on dyes absorbing in the far red where haemoglobin absorption is negligible. MB has absorption maximum at 665 nm, favourable for use in RCC. There is also some clinical experience for its use for sterilizing FFP (see above). As a result, the American Red Cross has initiated studies in the late 1980s to find out the usefulness of MB for pathogen reduction in RCC.

Three major problems were found in these studies. One was the lack of inactivation of cell-associated HIV.[27] This is not a problem for sterilizing FFP, in which a filtration step removes the leukocytes. Moreover, MB can photosensitize induction of HIV in latently infected cells.[28] The most serious problem was the production of red cell membrane damage at a dose range in which sufficient virus inactivation was achieved.[29]

In an effort to obviate the problems associated with MB, a more hydrophobic derivative, dimethylmethylene blue (DMMB), was tested. DMMB binds strongly to nucleic acids.[30] Binding by intercalation has been demonstrated by reduction of DMMB fluorescence, a red shift in its absorption maximum and induced circular dichroism when DMMB is incubated with

DNA.[31] In the presence of oxygen, DMMB monomers absorb light and inactivate virus by damaging viral nucleic acid.[30,32]

Treatment of 30% haematocrit red cell suspensions with 4 μM DMMB and 13.5 J cm^{-2} white light resulted in >5 log inactivation of VSV, pseudorabies virus (PRV), bovine diarrhea virus (BDV), bacteriophage ϕ6 and bacteriophage R17. Unlike MB, DMMB can photoinactivate intracellular viruses. Thus, >3 log of cell-associated VSV and >4.8 log cell-associated PRV were inactivated under the conditions described above.[33] The ability to inactivate intracellular virus has been attributed to improved plasma membrane permeability due to the higher hydrophobicity of DMMB compared to MB.[33] In additional studies, using RCC of higher haematocrit, duck hepatitis B virus (DHBV) was shown to be considerably more sensitive to DMMB and light than VSV.[34] Another useful target for photoinactivation of RCC are T lymphocytes, which play a crucial role in the development of TA-GVHD. Virucidal treatment of RCC with DMMB and light has been shown to inactivate T lymphocytes by >5 log.[35]

Treatment of RCC with DMMB under virucidal conditions resulted in no detectable immunoglobulin G (IgG) binding, unchanged red cell morphology, normal banding patterns of red cell membrane proteins on sodium dodecyl sulfate-polyacrylamide gel electrophoresis (SDS-PAGE) and unaltered characteristics of red cell antigens during storage as measured by antibody titration.[33] In addition, minimal changes were observed in red cell osmotic fragility, haemolysis, potassium efflux, ATP and 2,3-DPG levels during storage of phototreated samples compared to controls.[33]

In an effort to bring the process to a more practical configuration, compatible with the practice in a blood centre, RCC were treated with DMMB and light under virucidal conditions in a flow cell of 1 mm thickness. The red cells were suspended in two additive solutions, Adsol and Erythrosol, and were then stored for 6 weeks. Erythrosol was able to reduce haemolysis during storage by ten-fold due to the presence of citrate. However, even under these conditions RBC properties were inadequate at the end of storage.[36] Adding quinacrine, an antimalaria drug similar in structure to DMMB but with no absorption of light in the red, reduced DMMB binding to red cells. As a result, virus inactivation was enhanced and haemolysis was reduced after photoinactivation with DMMB in the presence of 0.5 mM quinacrine.[37] However, this approach was not deemed practical for blood banking purposes and the ARC abandoned its DMMB effort in the late 1990s.

9.4.2. *Flexible Dyes*

The ARC next evaluated some chalcogenoxanthylium dyes as photosensitizers for pathogen reduction in RCC.[38,39] None of them proved sufficient for this purpose. Although quite effective, red cell damage under virucidal conditions was unacceptable.[39] The focus was then turned to thiopyrilium (TP), a flexible dye that binds to nucleic acids by intercalation.[40] Most photosensitizers structures are characterized by fused and rigid aromatic ring systems that generate reactive oxygen species (ROS) upon light absorption whether they are bound to substrate or free in solution. Some unique sensitizers are flexible in such a way that their separated aromatic ring systems can rotate about adjoining covalent linkages of partial single bond character. These flexible dyes can act as photosensitizers when rigidly bound to substrate but cannot generate ROS when free in solution. The reason is that the energy from absorbed light is dissipated as heat through bond rotation (rotation does not occur when the dye is bound). Because the lifetime of the excited singlet state is prolonged to the nanosecond time scale in rigidly bound dye, it allows for greater efficiency of fluorescence and increases the probability for inter-system crossing to the triplet state. The triplet state is necessary for singlet oxygen-mediated photochemical reactions. TP is an example of such a flexible dye and was therefore selected for evaluation for use in RCC for pathogen reduction.[41] Because dipyridamol (DP) has a structure analogous to that of TP, it was added as a competitive inhibitor to reduce the binding of TP to red cell membrane. TP and red light inactivated >6 log of extracellular lipid enveloped viruses (VSV, DHBV, PRV, BDV and HIV) as well as intracellular HIV. There was a 25-fold difference in sensitivity of the various viruses to photoinactivation by TP, similar to that observed using DMMB. Unlike DMMB, phototreatment with TP also inactivated 5–7 log of Gram-positive and Gram-negative bateria.[41] Parasites such as *T. cruzi* could also be inactivated in RCC by TP and their sensitivity is much higher than that of viruses.[42] Presumably the higher sensitivity of parasites reflects their much larger genome.

RCC storage properties after TP and red light were evaluated under conditions harsher than those required for pathogen reduction. Haemolysis was limited to 0.3% in RCC suspended in Erythrosol and the presence of 0.2 mM DP. In the absence of DP haemolysis was 0.5%. Considerably more haemolysis occurred when Adsol was used as an additive solution. Since DP reduces TP binding to red cells, the strong protection against haemolysis in the presence of DP suggests that bound rather than free TP contributes to haemolysis. Red cells treated with TP and red light become ion permeable, as evidenced by enhanced K^+ leakage during storage. They are protected against colloidal-osmotic haemolysis in Erythrosol because at ionic equilibrium, the external osmotic pressure that is contributed by the impermeable citrate ion balances the internal osmotic pressure that is contributed by haemoglobin.

TP-phototreated RBC exhibit significantly decreased ATP levels after 6 weeks' storage compared with control. Since ATP production was not impacted, the treated red cells must consume more ATP, perhaps because ion-leakage drives increased Na/K ATPase activity.

Adequate preservation of RBC properties using TP-phototreatment could only be obtained by adding DP as a competitive membrane-binding inhibitor. The use of another flexible dye, thiazole orange (TO), was therefore evaluated. TO is an unsymmetrical cyanine dye that binds to nucleic acids by intercalation and is widely used for fluorescent enumeration of reticulocytes.[43] At 80 μM and 7.9 J cm^{-2} of white light a range of lipid-enveloped viruses were inactivated by

>5 logs when spiked in RCC and treated with TO. Bacteria were also inactivated to a large degree (2–8 logs) but their sensitivity varied in an unpredictable manner.[44]

Compared with the other photosensitizers tested, TO had significantly less binding to red cells. As a result, quenchers of ROS or competitive inhibitors were not required to achieve adequate red cell properties during storage. Thus, following TO phototreatment potassium leakage during storage of RCC was only modestly increased and ATP levels were not affected.[44] Haemolysis in Erythrosol was only 0.4% after 42 day storage, within the guidelines of the FDA. Interestingly, parasites were inactivated by TO even in the absence of light.[45]

While TO appeared to be a suitable photosensitizer for pathogen reduction in RCC based on the *in vitro* results, much work remains to be done before a clinical study using TO can be contemplated. This includes demonstration of adequate circulation of treated red cells in an animal model, lack of toxicity of the treated RCC and no production of new antigens on the treated red cells.

9.4.3. Phthalocyanines and Red Light

Phthalocyanines (Pc) attracted attention as potential photosensitizers for pathogen reduction in RCC in the early 1990s due to their strong absorption in the far red (660–690 nm) where haemoglobin has no absorption. At the time, Pc were actively studied for use in PDT of cancer and their photobiology and lack of toxicity were well characterized.[46] Initial studies focused on aluminium phthalocyanine tetrasulfonate ($AlPcS_4$).[47,48] These studies demonstrated inactivation of a range of lipid-enveloped viruses and minimal damage to RBC. Nevertheless, screening of a number of Pc demonstrated that $AlPcS_4$ was unable to inactivate intracellular HIV.[49] The screening studies demonstrated that the silicon Pc termed Pc 4 (Figure 9.1) at 2 μM and a few J cm^{-2} was the most effective for inactivation of >6 logs of a large number of lipid enveloped

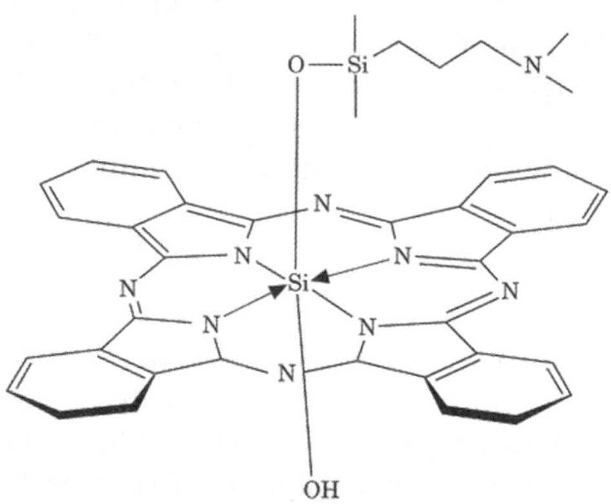

Figure 9.1 The molecular structure of Pc4.

viruses (VSV, HIV, BDV, PRV) including intracellular HIV and the blood parasites *T. cruzi* and *P. falciparum*.[49–52] Cell-associated HIV in the latent form was very sensitive and its inactivation was primarily due to apoptosis.[53]

Because Pc 4 does not specifically target nucleic acids, red cells sustain an unacceptable amount of damage following virucidal treatment of RCC. A number of approaches were designed to enhance the specificity of the treatment. While none of these approaches by itself could result in red cells that can be stored without excessive haemolysis, their combination produced a product that was almost ready for clinical studies.

9.4.3.1. Quenchers of ROS

It appears that virus inactivation using Pc occurs primarily *via* singlet oxygen pathway while RBC damage is mediated by $^1O_2$ as well as other ROS species.[54] As a result, addition of quenchers that specifically remove ROS other than $^1O_2$ selectively protected red cells during Pc-induced virus inactivation in RCC.[54–56] The protection was reflected in improved storability of RBC[55] as well as their increased circulatory survival *in vivo*.[54] Specificity of quenchers can be enhanced if they bind to RBC. Thus, DP, which is a ligand to band 3 protein in the red cell membrane, can selectively protect against RBC damage induced by $AlPcS_4$-PDT.[57] Another specific effect is exerted by SH-containing compounds (cystein and glutathione) which prevent binding of IgG in the plasma of PDT-treated RCC to the red cell membrane.[56] Such binding interferes with blood matching.

9.4.3.2. Light Fluence Rate

Modulation of irradiance proved to be beneficial in enhancing selectivity of pathogen reduction using Pc-PDT. RBC damage was reduced when the same light dose was delivered at higher fluence rates using $AlPcS_4$ or Pc5.[58,59] At the same time VSV inactivation by Pc5 was not affected by changes in irradiance. The effect is not due to oxygen depletion at higher irradiance. A mechanism to explain this effect postulates an intermediate product (TOO) as a result of the reaction between ROS and the biological target (T). TOO can either react to form the oxidized target (T_{ox}) or react with another TOO to regenerate T, ground state oxygen and T_{ox}. This last reaction which reduces the overall efficiency is favoured at high fluence rates because the concentration of TOO increases with irradiance.[59]

9.4.3.3. Magnetic Field Effect

Magnetic fields can exert a dramatic influence on free radical reactions.[60] In the case of pathogen reduction in RCC using Pc4, the application of a static magnetic field of 0.1 T reduced red cell damage without affecting VSV inactivation.[61] These results can be explained if only free radicals but not $^1O_2$ reactions are affected by a magnetic field. Magnetic fields are expected to enhance the yield of photochemical reactions initiated from triplet excited state,[62] but to decrease the yield if the photodamaging free radical reactions are initiated from a singlet excited state. In the case of Pc4-PDT, the reduction of red cell damage in the presence of a magnetic field is consistent with a molecular singlet excited state precursor. A decrease in the coupling between the singlet and triplet radical pair manifold by a magnetic field is expected to favour the recombination of singlet radical pair states to the singlet ground state, thus reducing the overall photodamage.

9.4.3.4. Action Spectra

The absorption peak of Pc4 in the red is at 668 nm in an organic solvent. When bound to a biological target the absorption spectrum of Pc is red shifted, resulting in a shift of their action spectra.[63–65] The extent of the shift may depend on the biological target. In the case of Pc4-PDT, when bound to red cell membranes the absorption peak of Pc4 is shifted to 573 nm. The action spectrum for production of red cell damage parallels the shifted absorption spectrum. The action spectrum for virus kill, on the other hand, was shifted by 10 nm to 678 nm.[66] Thus, by using a light source emitting at wavelengths >680 nm the selectivity of pathogen reduction by Pc4 could be enhanced.[67]

9.4.3.5. Delivery Vehicle

Because Pc4 is water-insoluble it has to be delivered in an organic solvent or another delivery vehicle. The proper selection of a delivery vehicle has a profound effect on the specificity of pathogen reduction with Pc4-PDT. Thus, the use of Cremophor rather than dimethyl sulfoxide as a delivery vehicle decreases the binding of Pc4 to red cells and the resultant RBC damage upon light exposure. At the same time VSV inactivation is not affected.[68] An even more dramatic effect is observed when Pc4 is formulated in negatively charged liposomes, which reduced Pc4 binding to RBC even further. Again, virus kill is the same irrespective of the delivery vehicle.[69] The composition of the liposomes is critical for optimal effect. Some compositions, *e.g.* those containing cholesterol, dramatically enhance red cell damage.[67]

9.4.3.6. Oxygen Partial Pressure

Previous studies have shown that the photodynamic effect of $AlPcS_4$ in killing cultured mammalian cells is saturated at pO_2 of 20 mm Hg and above.[69] It appears, however, that pathogen reduction in RCC by Pc4-PDT has a different dependence on oxygen. Figure 9.2 shows that for maximal effectiveness, pO_2 in the RCC has to exceed 40 mm Hg. Since in freshly prepared RCC pO_2 is around 20 mm Hg and it increases with time of storage, partial oxygen pressure has to be controlled prior to treatment by oxygenation. RBC damage has a similar dependence on oxygen as virus inactivation (not shown).

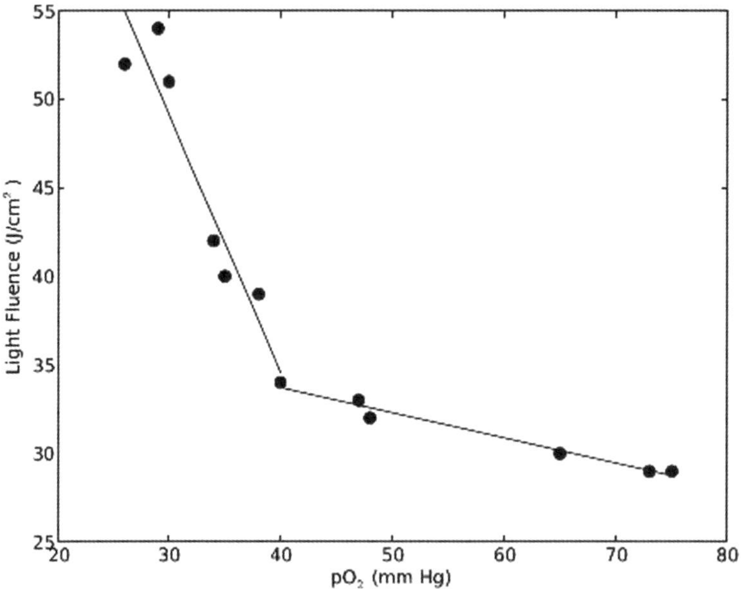

Figure 9.2 Oxygen dependence of VSV inactivation by Pc4. VSV was spiked into 50 ml of RCC in a small blood bag containing 2 μM Pc4, pO_2 level was determined and the blood was exposed to 670 nm light from LED array on a rotary shaker at 125 rpm. The light fluence required to inactivate 5 log VSV was determined from the inactivation kinetics.

9.4.3.7. Mixing

The limiting factor in the penetration of red light into RCC is light scatter, which allows a penetration of only a few mm. As a result, mixing of the red cells during treatment is of critical importance in obtaining a fast and uniform virus inactivation. This is demonstrated in Figure 9.3, where small blood bags containing 50 ml RCC were spiked with VSV and treated with Pc4-PDT on a rotary table. Increasing the speed of mixing enhanced the rate of virus kill and straightened the inactivation curve.

It turned out that for a standard blood bag containing 450 ml of RCC the treatment conditions required a rather large and cumbersome device to achieve the same results as those obtained with a 50 ml bag. Using such a device is not practical in a blood centre setting. Another option is the use of a transparent plastic tube into which a static mixer is inserted. This is a commonly used device in industry to obtain mixing during flow. The use of this approach in which the RCC were exposed to red light at high irradiance during flow resulted in excellent throughput and adequate virus kill of >6 logs.

Some of the approaches described above were combined in a system containing 5 μM Pc4 formulated in liposomes, 4 mM cysteine, 0.8 mM vitamin E and 0.5 mM L-carnitine. L-carnitine is present in red cells and participates in their membrane turnover. Its addition to red cells improves their storability.[70]

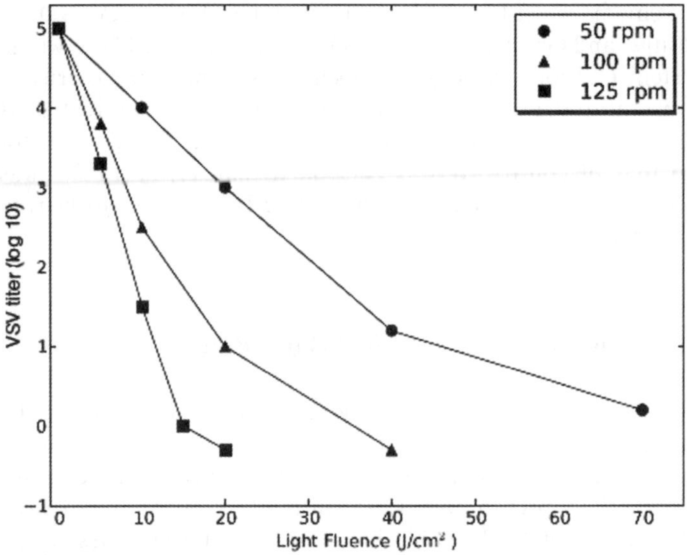

Figure 9.3 Effect of mixing rate on VSV inactivation. VSV was spiked into 50 ml of oxygenated RCC in a small blood bag containing 5 μM Pc4. The bag was exposed to 670 nm light from an LED array on a rotary shaker at different speeds, as indicated.

Exposure of RCC to 40 J cm^{-2} from a light emitting diode array operating at 660–680 nm and 80 mW cm^{-2} resulted in extensive pathogen reduction (>6 log) of a number of lipid enveloped viruses. At the same time RBC were minimally affected as determined by a number of parameters *in vitro* and *in vivo*.[50]

9.4.4. Porphyrins and Red Light

Historically, dihaematoporphyrin ether (DHE) was the first photosensitizer approved for a medical application in the USA (for PDT of cancer). As a result, even though the absorption spectra of porphyrins parallels that of haemoglobin, DHE was the first photosensitizer tested for pathogen reduction in blood using PDT.[71,72] While the preliminary results reported using a flow system were encouraging, there was no follow up to this work. A decade later a series of cationic porphyrins were evaluated with respect to their virucidal activity and ability to cause RBC damage when exposed to red light.[73] The cationic porphyrin mono-phenyl-tri-(*N*-methyl-4-pyridyl-porphyrine), or Tri-P(4) for short, was the most effective in killing VSV while causing the least red cell damage. Haemolysis during storage was 2.3% after 5 weeks. There was IgG binding after treatment, which was reduced by addition of glutathione prior to treatment. Treated Rhesus monkeys RBC had a satisfactory 24-hour recovery and half-life in the circulation. The ability of 25 μM Tri-P(4) to inactivate a series of viruses and bacteria was studied in 60% RCC.[74] Exposure to red light (up to 36 J cm^{-2}) resulted in >5 log inactivation of HIV, PRV, BDV as well as Gram-positive and Gram-negative bacteria. Cell-associated HIV was killed to a limited extent (1.7 log). A non-enveloped virus, canine parvovirus, was completely resistant to the treatment. Subsequent storage of the treated RCC for 5 weeks in AS-3 resulted in 4% haemolysis. ATP reduction during storage was similar to that of control. It was concluded that Tri-P(4)-PDT was neither robust enough nor adequately preserves red cell properties to warrant its use for RCC sterilization.

9.5. Riboflavin and UV Light in Whole Blood

Whole blood was the first blood product to be transfused for therapeutic purposes. In military combat situations, it continues to be a valued therapeutic tool. Haemorrhage resulting from combat-related injuries is the most common cause of potentially preventable deaths in military operations. Evidence indicates that the use of fresh whole blood (FWB) during massive transfusion (>10 units) is independently associated with improved survival.[75] The use of warm FWB currently persists only in emergency life-threatening scenarios when tested, stored blood components are not available, such as combat casualties.[76] Risks of non-leukoreduced FWB transfusions include the transmission of infectious agents due to exposures to these agents either before or during deployment in the field, transfusion-associated graft-*versus*-host-disease (TA-GvHD), alloimmunization to donor antigens and the potential for microchimerism. These risks may be mitigated in large part by a process capable of pathogen reduction and leukocyte inactivation. Such a technique could increase the safety of FWB and thus provide both an adequate and safe blood supply in the combat environment. This section will review work in progress on the development of such a system to treat FWB, which is capable of reducing infectious pathogen loads of bacteria, viruses and parasites and inactivating white blood cells (WBC). Treatment with the system is based on the use of riboflavin (vitamin B2) and UV light and relies on a photochemical process, which is capable of disrupting nucleic acid replication in pathogens and white cells.

9.5.1. Device Description

The prototype Mirasol® System for FWB is adapted from the CE-marked Mirasol pathogen reduction treatment system for platelets and plasma. The device consists of an illuminator and a disposable kit, illustrated in Figure 9.4. For treatment with the system, a unit of whole blood (approximately 460 to 560 mL, with anticoagulant) is transferred to the illumination bag and mixed with the riboflavin solution (35 mL of 500 μM riboflavin in 0.9% NaCl). The mixture is then placed in the illuminator and exposed to broadband (265–370 nm) UV light. The dose of UV light is normalized to account for the

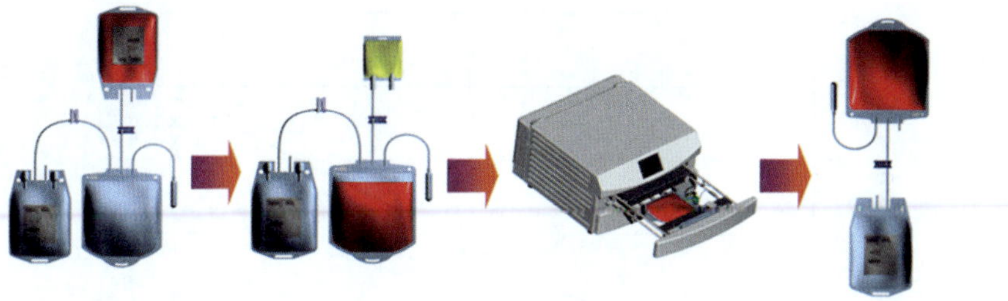

Figure 9.4 Treatment with the Mirasol system for whole blood.

presence of the RBCs and sample volume (weight) and is reported in units of $\mathrm{J\,mL^{-1}}_{\mathrm{RBC}}$. Energy dose, mixing and product temperature are measured and recorded throughout the process on an on-board microprocessor/controller. Treatment times range from 40 to 60 minutes based on product size.

9.5.2. Pre-clinical Results

Proof-of-concept testing for the Mirasol system for whole blood involved evaluations of WBC inactivation, virus reduction, bacteria reduction, component quality and whole blood quality after treatment of the whole blood with varying UV-light energy doses. This information was utilized to select a light energy dose which afforded adequate levels of pathogen reduction and white cell inactivation, while also maintaining cellular and protein component functionality.

9.5.2.1. Virus Inactivation

Several model viruses were used to assess the ability of this system to inactivate viruses in blood products. Efficacy was determined by measuring the residual levels of infectivity present in the samples after treatment and by comparing this to levels of infectivity in the untreated starting material. These values were used to calculate a reduction factor. Units of whole blood were spiked with enveloped (VSV, IBR) and non-enveloped (CPV, BTV, HAV) viruses, as these are representative of the two major virus types. Virus inactivation as a function of energy delivered was evaluated in the range of 40–110 $\mathrm{J\,mL^{-1}}_{\mathrm{RBC}}$. Increasing virus inactivation with increasing energy was observed in the range of 40–110 $\mathrm{J\,mL^{-1}}_{\mathrm{RBC}}$ (Figures 9.5 and 9.6).

9.5.2.2. Bacteria Inactivation

The ability of the system to reduce levels of infectious bacteria load was evaluated by measuring the ability of the system to sterilize products spiked with several strains of pathogenic bacteria. Sterility was assessed by culturing

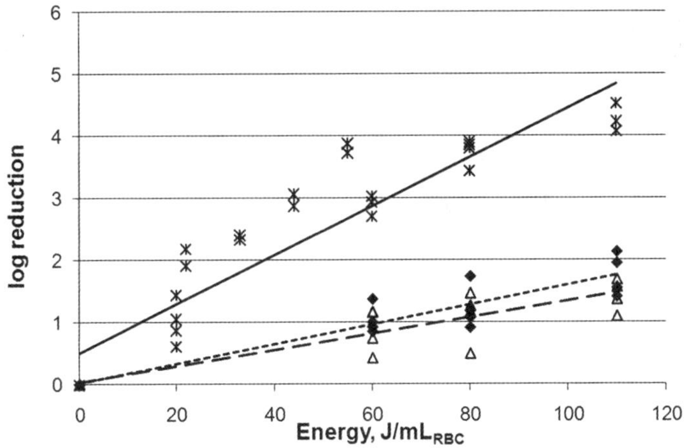

Figure 9.5 Reduction of non-enveloped viruses as a function of light energy. Symbols correspond to data, the lines to linear regressions of the data. Solid line, CPV; dotted line with filled diamonds, HAV; dashed line with open triangles, BTV.

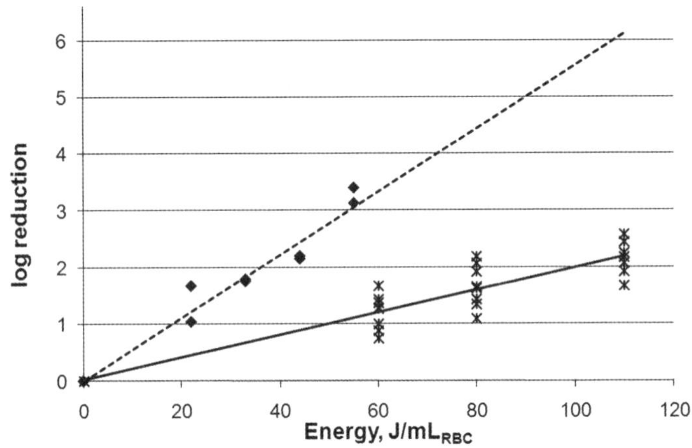

Figure 9.6 Reduction of enveloped viruses as a function of light energy. Solid line, IBR; dotted line, VSV.

samples removed from treated products after storage at room temperature for up to 7 days. Bacterial growth curves, in cases of breakthrough, were also assessed to establish the extent of a bacteriostatic effect (Figure 9.7). Initially, evaluation of bacteria reduction was performed as a function of dose response over the range of 40–110 J mL^{-1}$_{RBC}$. Follow-up experiments focused on evaluating performance after treatment with 80 J mL^{-1}$_{RBC}$. Results are summarized in Table 9.1.

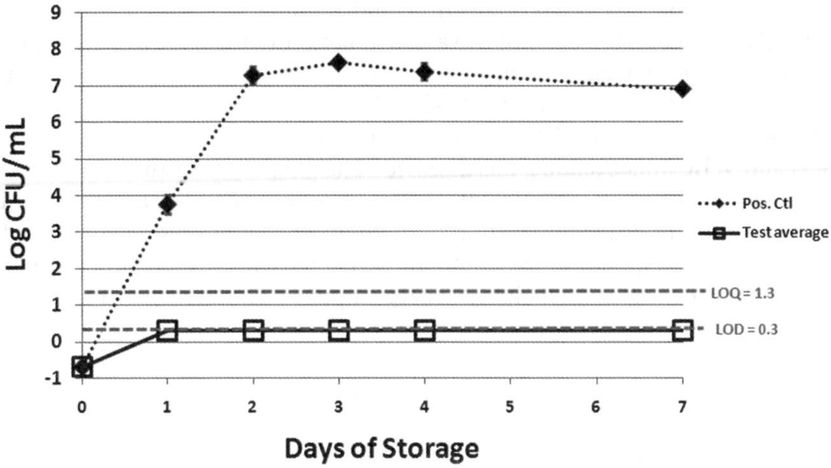

Figure 9.7 Growth rate of *B. cereus* in treated and positive control whole blood. LOD = limit of detection of the assay used to measure bacterial titre.

Table 9.1 Low-titre bacteria reduction after treatment with $80\,\mathrm{J\,mL^{-1}}_{\mathrm{RBC}}$ in the Mirasol System.

Strain of bacteria tested	# of units positive/ # of units tested
Serratia marcescens	0/3
Yersinia enterocolitica	0/15
Escherichia coli	0/2
Klebsiella pneumoniae	0/3
Acinetobacter baumannii	1/9
Serratia liquefaciens	1/10

9.5.2.3. Parasite Inactivation

Several experiments to test the effectiveness of the process for inactivating parasites are ongoing or are planned. These studies are of particular importance because tests for parasitic agents are usually not available due to the lack of suitable diagnostic tests coupled with the unique characteristics of these agents, which can evade detection in blood. Exposure of military donors in the field to these agents can be of particular concern.[77,78] Tests of the reduction of a transfusion-transmissible parasite, *Trypanosmoa cruzi*, with the Mirasol System for Whole Blood are ongoing. Tests of *Babesia microti*, *Leishmania donovani*, *Plasmodium falciparum* and *Anaplasma phagocytophilum* reduction will occur over the next three years.

9.5.2.4. White Blood Cell Inactivation

WBC in blood products can induce multiple adverse effects in the transfusion recipient including life-threatening TA-GvHD, cytokine production and alloimmunization.[79–81] Leukoreduction and gamma irradiation are used to lessen these effects but are not always available or practical for field implementation. Those measures also are only partially effective in alleviating adverse effects due to transfused WBC.[82,83]

To assess inactivation, WBCs were isolated from treated whole blood products and WBC functionality was assessed. The phenotype of the cells does not change immediately after treatment. WBC activation by PMA, as measured by CD69 expression, was completely prevented at all energies tested (Figure 9.8). Proliferation in response to PHA and anti-CD3 plus anti-CD28 antibodies was completely blocked at the highest energy applied (Figure 9.9). In a mixed

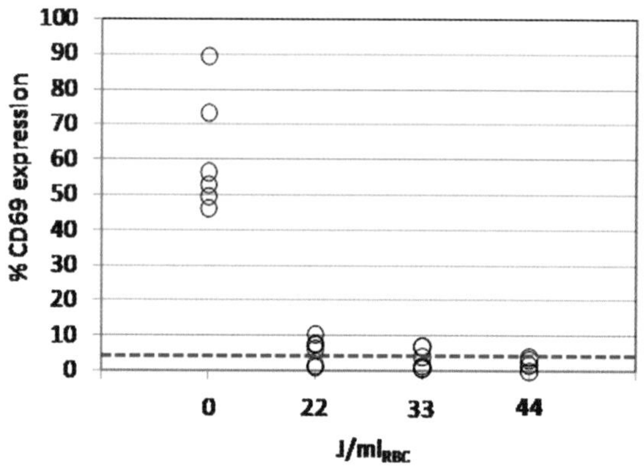

Figure 9.8 Inhibition of WBC activation in response to PHA as a function of light energy. CD69 expression was measured by flow cytometry. The dashed line represents LOD.

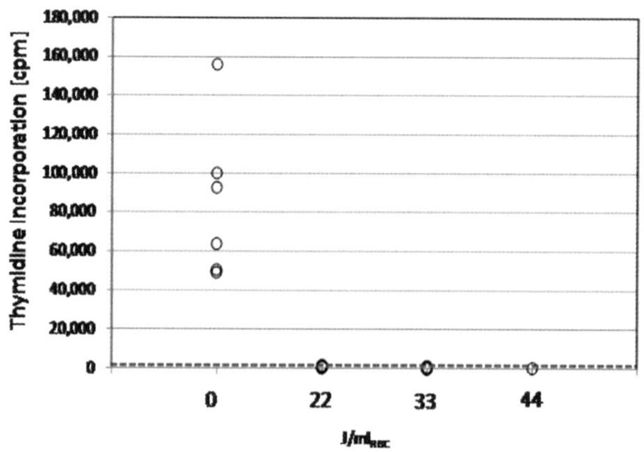

Figure 9.9 Inhibition of WBC proliferation in response to PHA as a function of light energy. Proliferation was measured by ^{3}H-thymidine incorporation. The dashed line represents LOD.

lymphocyte culture proliferation of treated WBCs in response to allogeneic stimulator cells (allorecognition) or proliferation of untreated cells in response to treated stimulator cells (allostimulation) was also inhibited, indicating that Mirasol treatment impairs antigen presentation in addition to cell division (Figures 9.10 and 9.11). Production of inflammatory and TH1/TH2 cytokines in response to LPS and anti-CD3 plus anti-CD28 antibodies was completely inhibited at the higher energies used (data not shown).

The functionality of treated WBCs was also assessed in a mouse model that utilizes xenogeneic GvHD responses of human cells when injected into immunodeficient murine recipients (Rag$^{-/-}\gamma c^{-/-}$ double knockout mice).

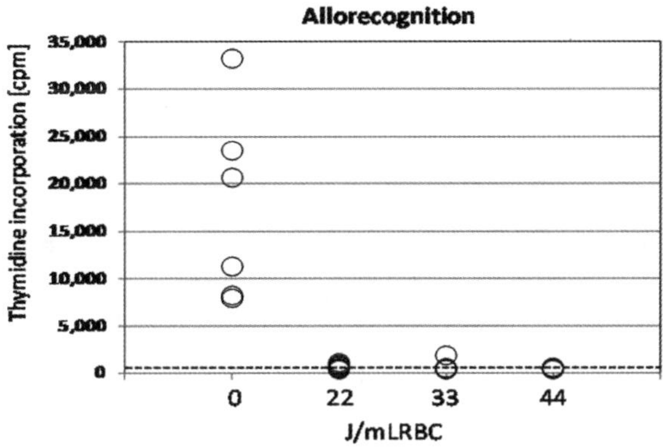

Figure 9.10 Results for allorecognition by WBC as a function of light energy. Proliferation in a mixed lymphocyte culture was measured by ^{3}H-thymidine incorporation. Dashed line represents LOD.

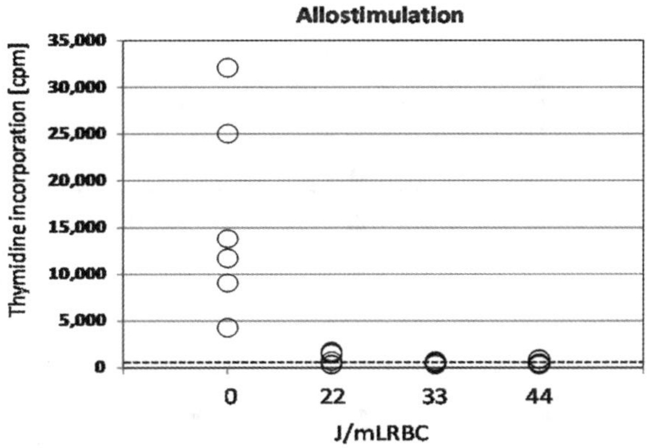

Figure 9.11 Results for allostimulation by WBC as a function of light energy. Proliferation in a mixed lymphocyte culture was measured by ^{3}H-thymidine incorporation. Dashed line represents LOD.

The results of this work showed prevention of GvHD in animals transfused with treated products, as evidenced by lack of GvHD development, human cell engraftment (Figure 9.12), antibody production and increase in human cytokine levels (Figure 9.13).

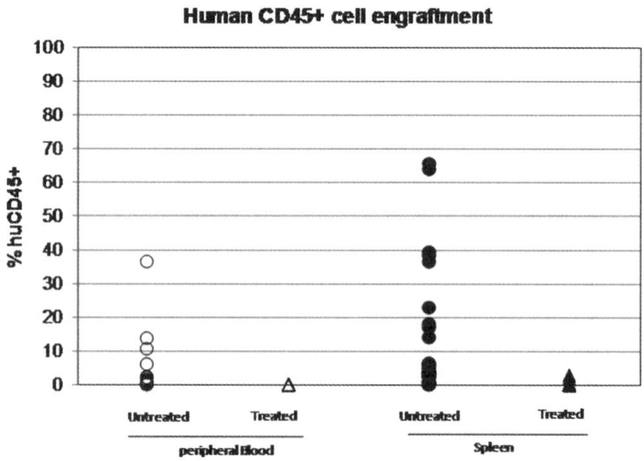

Figure 9.12 Human CD45 + cell engraftment in the mice used in the GvHD study.

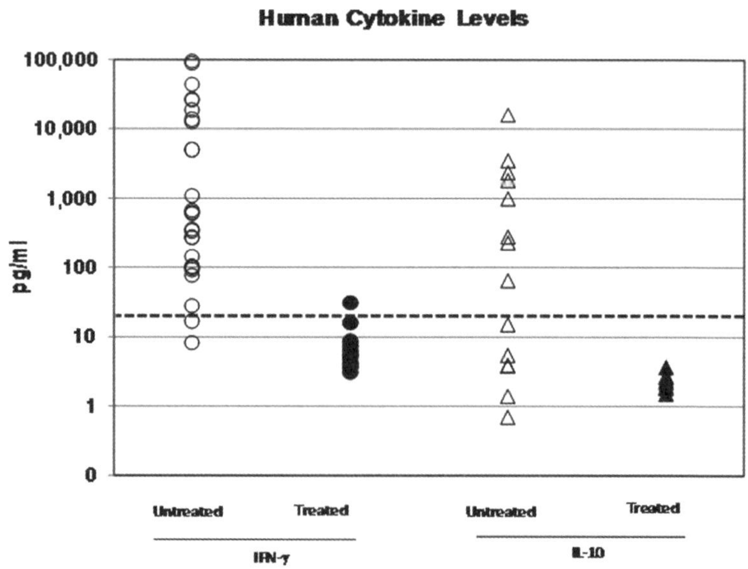

Figure 9.13 Human cytokine levels in the mice used in the GvHD study. The dashed line represents LOD.

9.5.2.5. Whole Blood Quality and Function

In order to assess the quality and functionality of whole blood, units were collected in CPDA-1, and treated with varying energy doses in the Mirasol System and then stored at room temperature, without agitation, for at least 24 hours (range of 1 to 7 days). The results of the studies indicated that $80\,J\,mL^{-1}{}_{RBC}$ would provide a balance of acceptable blood quality and pathogen reduction. Additional studies were conducted to verify the performance of the system with $80\,J\,mL^{-1}{}_{RBC}$ light delivery. Samples were removed to assess haemolysis, osmotic fragility, CBC, clinical chemistry, plasma function (PT, APTT, the activity of representative factors) and platelet function (TEG and ImpactR).

Free haemoglobin remained below the limits of detection ($30\,mg\,dl^{-1}$) in treated and control units throughout 7 days of room-temperature storage. Methaemoglobin values were elevated post-treatment (1.7% + 0.8%), and were at background levels (0.7%) after one day of storage. Table 9.2 displays values for mean osmotic fragility; on Day 0 only, treated units were more fragile than the unpaired controls. On all other storage days, no significant differences in osmotic fragility were observed between treated and control units. Table 9.3 displays values for sodium and potassium concentrations. The potassium concentration increases and the sodium decreases more rapidly in the treated whole blood than in the controls.

Table 9.2 Mean osmotic fragility for treated and control RBCs during storage.

Storage day	MOF for control units (n = 4)	MOF for treated units (n = 8)
Day 0	0.468 ± 0.008	0.490 ± 0.018^a
Day 1	0.503 ± 0.011^b	0.504 ± 0.024^b
Day 2	0.510 ± 0.006^b	0.520 ± 0.019^b
Day 5	0.551 ± 0.015^b	0.549 ± 0.023^b
Day 7	0.569 ± 0.015^b	0.542 ± 0.046^b

[a]Significantly different from control values.
[b]Significantly different from corresponding Day 0.

Table 9.3 Clinical chemistry measurements during storage.

	Storage day	Na+ (mmol L^{-1})	K+ (mmol L^{-1})
Control (n = 4)	Day 0	154 ± 1.4	3.2 ± 0.1
	Day 1	156 ± 1.5	3.8 ± 0.1
	Day 2	157 ± 1.5	5.1 ± 0.2
	Day 5	155 ± 1.5	9.0 ± 0.5
	Day 7	153 ± 1.7	11.3 ± 0.7
Treated (n = 8)	Day 0	156 ± 1.9	3.3 ± 0.4
	Day 1	156 ± 1.6	5.1 ± 0.6^a
	Day 2	155 ± 1.7	7.5 ± 0.8^a
	Day 5	151 ± 2.0^a	14.3 ± 1.2^a
	Day 7	147 ± 2.4^a	18.4 ± 1.5^a

[a]Significantly different from controls.

Post-treatment plasma from eight treated units of whole blood, and plasma from three untreated units, was assayed for Fibrinogen, Factor V, Factor VIII, Factor XI and total protein. In a separate experiment, after illumination the whole blood was transferred to the Mirasol storage bag for room temperature storage. Samples from eight units of treated whole blood and four units of untreated whole blood were assayed post-illumination and after one and two days' storage for prothrombin time (PT) and activated partial thromboplastin time (aPTT).

Table 9.4 displays the data for Fibrinogen concentration, Factor V, VII and XI activity, and total protein concentration; these values are corrected for dilution with riboflavin solution. Total protein concentrations were equivalent in test and control plasma. Post-illumination values for Fibrinogen, Factor V, Factor VIII and Factor XI were higher in the controls than in the treated plasma, but all values were within normal ranges. Table 9.5 displays the data for prothrombin time (PT) and activated partial thromboplastin time (aPTT) on Days 0, 1 and 2 of room-temperature storage. PT and aPTT values were greater in treated units than in controls, post-illumination and after one day of storage. Values for aPTT are in the range of normal values, and for PT are slightly elevated (for treated units).

Platelet function was evaluated in test and control units of whole blood. The platelet function assays used were the ImpactR and thromboelastography (TEG). The Impact-R tests platelet adhesion and aggregation in anti-coagulated whole blood under arterial flow conditions. The results are expressed as the percentage of the well surface covered by platelet aggregates (% SC), which

Table 9.4 Plasma quality: Fibrinogen, Factor V, Factor VIII, Factor XI.

		Fibrinogen (mg dL^{-1})	**Factor V (%)**	**Factor VIII (%)**	**Factor XI (%)**	**Total protein (g dL^{-1})**
Control (n = 3)	Day 0	255 ± 29	86 ± 9	79 ± 7	126 ± 8	5.5 ± 0.7
80 J mL$^{-1}{}_{RBC}$ (n = 8)	Pre-illum	292 ± 25	89 ± 16	98 ± 18	108 ± 19	5.5 ± 0.3
	Day 0	214 ± 15	68 ± 13	70 ± 12[a]	85 ± 15	5.7 ± 0.3

[a]Significant difference (p < 0.05) between treated and control, t-test for unpaired samples, unequal variance.

Table 9.5 Plasma function: PT and APTT.

	Control (n = 4)				**80 J mL$^{-1}{}_{RBC}$ (n = 8)**			
	Pre-illum	**Day 0**	**Day 1**	**Day 2**	**Pre-illum**	**Day 0**	**Day 1**	**Day 2**
PT (sec)	N/A	13.4 ± 0.6	13.6 ± 0.6	13.9 ± 0.5	13.9 ± 0.6[a]	15.6 ± 0.7[a]	16.0 ± 0.9[a]	16.4 ± 0.8[a]
APTT (sec)	N/A	26.4 ± 2.2	27.0 ± 1.6	28.2 ± 2.3	25.2 ± 0.9[a]	31.3 ± 1.3[a]	32.7 ± 1.4[a]	33.1 ± 1.3[a]

[a]Significant difference (p < 0.05) between treated and control, t-test for unpaired samples, unequal variance.

represents platelet adhesion, and the average size of the aggregates (AS in μm^2), which represents platelet aggregation.

TEG analysis evaluates clot formation by rotating a cup, containing the blood sample, to imitate venous flow and activate coagulation. The speed and strength of clot formation are represented with four parameters: R, k, MA and alpha. R (reaction time) is the elapsed time until clot formation in the analyzer – shorter R times are preferred. The alpha angle represents the rate of clot growth – smaller á values are preferred. The k-value is a measure of the speed required to reach a certain level of clot strength, and is closely related to alpha and to MA. MA (Maximal amplitude) is a measure of the strength of the clot formed, and is a measurement of viscosity – larger MA values are preferred.

For these unpaired treated and control articles, no statistically significant differences were observed for the Impact-R measurements throughout storage (Table 9.6). For the Day 1 and Day 7 TEG measurements (Table 9.7), no statistically significant differences were observed in values for MA, k and alpha. R values on Day 7 were significantly greater ($p < 0.05$) for treated units, but were not different on Day 1. Overall, the platelet function measurements indicate that platelets in treated whole blood display function that is equivalent to untreated platelets after 24 hours of room temperature storage. The PT and aPTT measurements in Table 9.5 were obtained from the same units as those

Table 9.6 Platelet function measurements – ImpactR.

		Day 0	Day 1	Day 2	Day 5	Day 7
Surface Coverage, %	Control (n = 4)	8.4 ± 2.2	7.1 ± 1.9	7.2 ± 1.4	3.4 ± 2.3	3.4 ± 2.3
	Treated (n = 8)	9.1 ± 1.8	7.2 ± 1.7	7.8 ± 2.1	3.8 ± 2.3	2.6 ± 2.4
Average Aggregate size, μm^2	Control (n = 4)	29.3 ± 4.3	24.1 ± 1.1	25.6 ± 1.7	26.1 ± 2.7	25.9 ± 2.4
	Treated (n = 8)	34.8 ± 9.6	23.5 ± 1.6	25.0 ± 2.3	23.6 ± 4.0	24.3 ± 2.3

Table 9.7 Platelet function measurements – TEG.

		Day 1	Day 7
R, min	Control (n = 4)	7.8 ± 2.0	13.3 ± 1.7
	Treated (n = 8)	6.6 ± 0.9	16.0 ± 2.0[a]
K, min	Control (n = 4)	1.9 ± 0.2	2.2 ± 0.6
	Treated (n = 8)	1.9 ± 0.3	2.9 ± 0.6
Alpha, degrees	Control (n = 4)	63.0 ± 3.1	57.7 ± 7.1
	Treated (n = 8)	63.3 ± 3.5	51.2 ± 4.9
MA, mm	Control (n = 4)	63.0 ± 5.2	37.4 ± 8.7
	Treated (n = 8)	62.1 ± 2.2	43.2 ± 6.3

[a]Significant difference ($p < 0.05$) between treated and control, t-test for unpaired samples, unequal variance.

evaluated with TEG. The differences in PT and aPTT did not impact clot formation, as evaluated by TEG (Table 9.7).

9.5.3. Clinical Studies: Results from First Human Use Studies

The Mirasol System for Whole Blood has been tested in a feasibility clinical study (the IMPROVE Study) with healthy volunteers, under an FDA-approved Investigational Device Exemption (IDE) and with funds from the US Department of Defense. The overall objective of the IMPROVE Study was to establish the correlation between *in vitro* measures of cell quality and *in vivo* recovery and survival of radiolabelled cells in circulation. These correlations were derived by comparing three different illumination energies and subsequent extended storage for 42 days. In the study, each healthy volunteer donated one unit of whole blood. The unit of whole blood was not leukoreduced, and was treated with the Mirasol System and then separated into components (RCC in AS-3, platelet concentrates (PC) and FFP). Each component was stored at the appropriate storage condition: 4 °C for RCC, 22 °C for PC and –20 °C for FFP. All components were sampled on the day of preparation, and again at the end of storage (Day 42 for RCCs, Day 5 for PCs and Day 28 for FFP).

The *in vitro* testing performed on the units included assessments of WBC inactivation and evaluations of RCC, PC and FFP quality. For the *in vivo* portion of the study, each volunteer received an aliquot (~12 mL) of ^{51}Cr labelled RBCs that had been stored for 42 days after treatment with the Mirasol System. Blood samples from the volunteers were withdrawn during the first day post-infusion to determine the 24-hour recovery values, and for 4 weeks after infusion to determine survival of the labelled cells in circulation.

Spearman correlations of >0.7 were identified between 24-hour recovery values and two of the Day 42 *in vitro* variables: % haemolysis and ATP (Figure 9.14). The highest correlations ($R_s > 0.7$) with T50 survival results were found with ATP and pCO_2 (Day 42 values) (Figure 9.15). These observations were consistent with those reported previously for untreated red cells.[84]

Using the correlations observed in this study, combined with *in vitro* measures of cell quality following treatment with a UV light dose of 80 J $mL^{-1}{}_{RBC}$ and storage at RT, it is possible to predict the performance of the treated red cells and evaluate this relative to FDA standards for red cell performance (Figure 9.16). This standard requires a recovery in circulation of 75% of the transfused cells at 24 hours post-infusion. The accuracy of this predicted outcome will be verified through additional clinical evaluation of the product in both healthy subjects and patients.

9.6. Conclusions

The inactivation of viruses in blood proteins has made the transfusion of these products absolutely safe with respect to transmission of HBV, HCV and HIV. The addition of nanofiltration, when applicable, or UVC irradiation to the

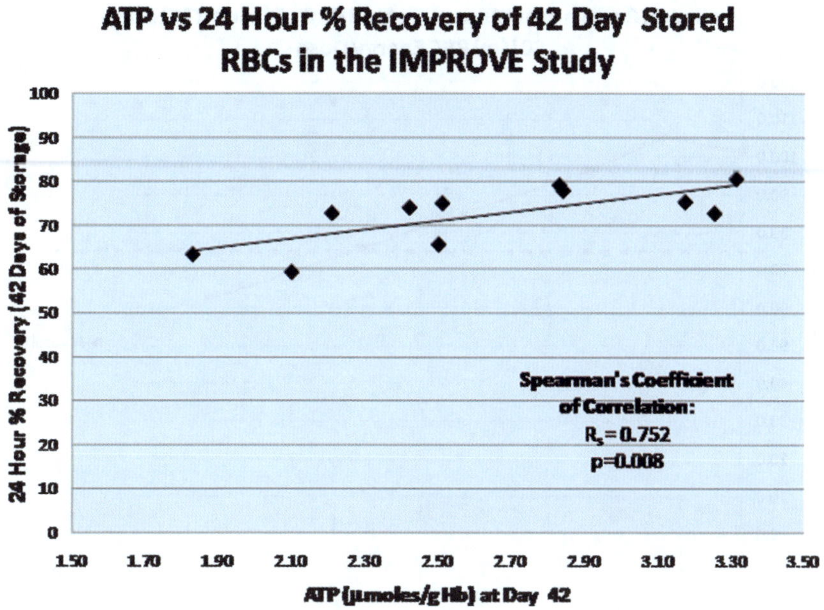

Figure 9.14 Percent recovery of ^{51}Cr-labelled RBCs as a function of ATP content.

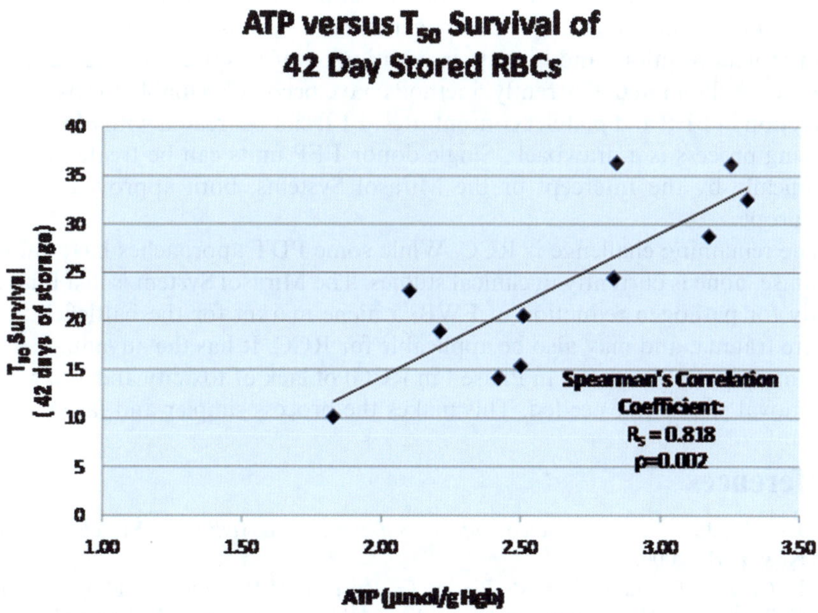

Figure 9.15 Survival half-life of ^{51}Cr-labelled RBCs as a function of ATP content.

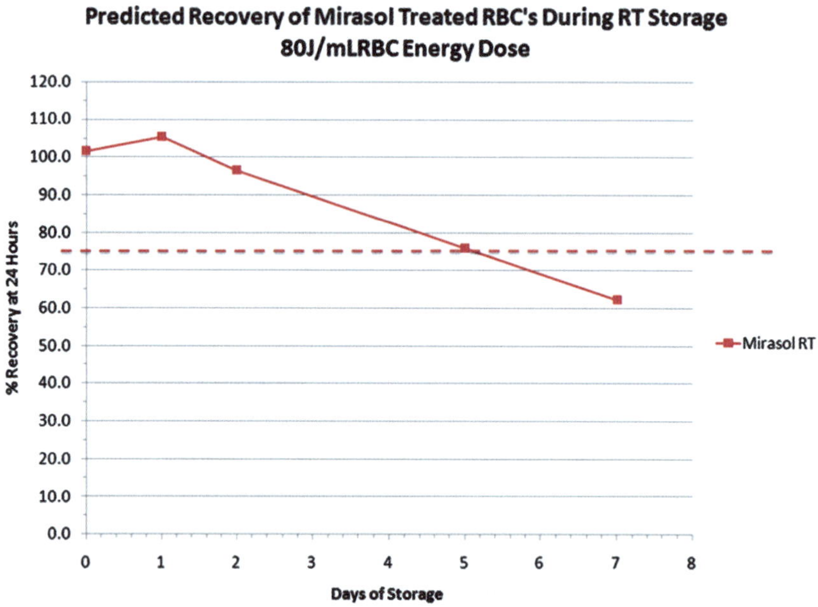

Figure 9.16 Predicted recovery of Mirosal-treated RBCs during room temperature storage after treatment with the Mirasol system for whole blood at 80 J/MI$_{RBC}$ UV light.

established methods (heat and SD treatment) should make these products safe also with respect to HAV and parvovirus B-19, the two non-enveloped viruses transmitted by plasma derivatives. Blood components present a tougher challenge, in particular the cellular components, both because they are more labile than protein solutions and because in addition to viruses bacteria and parasites need to be eliminated. Currently, methods have become available for pathogen reduction in FFP and platelet concentrates. FFP can be treated with SD but the pooling process is a drawback. Single donor FFP units can be treated photochemically by the Intercept or the Mirasol Systems, both approved for use in Europe.

The remaining challenge is RCC. While some PDT approaches have shown promise, none is currently in clinical studies. The Mirasol System is in a Phase I study for pathogen reduction in FWB, a niche market for the battlefield and severe trauma, and may also be applicable for RCC. It has the advantage over the Intercept System (now in Phase I in RCC) of lack of toxicity and as a result a removal step is not needed. This makes the process simpler and faster.

References

1. J.V. Linden and C. Bianco, *Blood Safety and Surveillance*, Marcel Dekker, New York, 2000.
2. F. Cavrini, P. Gaibani, A.M. Pierro, G. Rossini, M.P. Landini and V. Sambro, Chikungunya: an emerging and spreading arthropod-borne viral disease, *J. Infect. Dev. Countries*, 2009, **3**, 744.

3. J.W. Oliphant and A. Hollander, Homologous serum jaundice: experimental inactivation of etiologic agent in serum by ultraviolet irradiation, *Public Health Rep.*, 1946, **61**, 598.
4. R. Murray, J.W. Oliphant, J.T. Tripp, B. Hampi, F. Ratner and W.C.L. Diefenbach, Effect of ultraviolet radiation on the infectivity of icterogenic plasma, *J. Am. Med. Assoc.*, 1955, **157**, 8.
5. J. Hilfenhaus, A. Herrmann, R. Mauler and A.M. Prince, Inactivation of the AIDS-causing retrovirus and other human viruses in antihemophilic plasma protein preparations by pasteurization, *Vox Sang.*, 1986, **50**, 208.
6. M.I. Simon and H. Van Vunakis, The photodynamic reaction of methylene blue with deoxyribonucleic acid, *J. Mol. Biol.*, 1962, **4**, 488.
7. C. Wallis and J.L. Melnick, Photodynamic inactivation of animal viruses: a review, *Photochem. Photobiol.*, 1966, **4**, 159.
8. B. Lambrecht, H. Mohr, J. Knuver-Hopf and H. Shmitt, Photoinactivation of viruses in human fresh plasma by phenothiazine dyes in combination with visible light, *Vox Sang.*, 1991, **60**, 207.
9. J.U. Wieding, P. Hellstern and M. Kohler, Inactivation of viruses in fresh-frozen plasma, *Ann. Hematol.*, 1993, **67**, 259.
10. S.J. Wagner, D. Robinette, J. Storry, X.Y. Chen, J. Shumaker and L. Benade, Differential sensitivities of viruses in red cell suspensions to methylene blue photosensitization, *Transfusion*, 1994, **34**, 521.
11. J.U. Wieding, K. Vehmeyer, J. Dittmann, W. Hiddemann, M. Kohler and G. Lanzer, Contamination of fresh-frozen plasma with white cells and proliferable stem cells, *Transfusion*, 1994, **34**, 185.
12. J.U. Wieding and H. Neumeyer, Initial experience with methylene blue virus inactivated fresh-frozen plasma: results of a clinical and in vitro study, *Infusionther. Transfusionmed.*, 1992, **19**, 84.
13. H. Mohr, J. Knuver-Hopf, B. Lambrecht, H. Scheidecker and H. Shmitt, No evidence for neoantigens in human plasma after photochemical virus inactivation, *Ann. Hematol.*, 1992, **65**, 224.
14. T. Zeiler, H. Riess, G. Wittmann, G. Hintz, R. Zimmermann, C. Muller, H.G. Heuft and D. Huhn, The effect of methylene blue phototreatment on plasma proteins and in vitro coagulation capability of single-donor fresh-frozen plasma, *Transfusion*, 1994, **34**, 685.
15. S.J. Wagner, M.A. Cifone, H. Murli, R.Y. Dodd and B. Myhr, Mammalian genotoxicity assessment of methylene blue in plasma: implications for virus inactivation, *Transfusion*, 1995, **35**, 407.
16. D. Metzler and W. Cairns, Photochemical degradation of flavins. VI. A new photoproduct and its use in studying the photolytic mechanism, *J. Am. Chem. Soc.*, 1971, **93**, 2772.
17. P.C. Joshi, Ultraviolet-radiation-induced photodegradation and $^1O_2$, O_2- production by riboflavin, lumichrome and lumiflavin, *Ind. J. Biochem. Biophys.*, 1989, **26**, 186.
18. W. Speck, S. Rosenkranz and H. Rosenkranz, Further observations on the photooxidation of DNA in the presence of riboflavin, *Biochim. Biophys. Acta*, 1976, **435**, 39.
19. M. Korycka-Dahl and T. Richardson, Photodegradation of DNA with fluorescent light in the presence of riboflavin, and photodegradation by flavin triplet-state quenchers, *Biochim. Biophys. Acta*, 1980, **610**, 229.
20. R.P. Goodrich, R.A. Edrich, J. Li and L. Seghatchia, The Mirasol PRT system for pathogen reduction of platelets and plasma: an overview of current status and future trends, *Transfus. Apher. Sci.*, 2006, **35**, 5.

21. P.H. Ruane, R. Edrich, D. Gampp, S.D. Keil, R.L. Leonard and R.P. Goodrich, Photochemical inactivation of selected viruses and bacteria in platelet concentrates using riboflavin and light, *Transfusion*, 2004, **44**, 877.
22. F. Rentas, R. Harman, C. Gomez, J. Salata, J. Childs, T. Silva, L. Lippert, J. Montgomery, A. Richards, C. Chan, J. Jiang, H. Reddy, J. Li and R. Goodrich, Inactivation of Orientia tsutsugamushi in red blood cells, plasma and platelets with riboflavin and light, as demonstrated in an animal model, *Transfusion*, 2007, **47**, 240.
23. L.J. Cardo, F.J. Rentas, J. Salata, R. Harman, W. Melvin, P.J. Weina, J. Mendez, H. Reddy and R. Goodrich, Pathogen inactivation of Leshmania donovani infantum in plasma and platelet concentrates using riboflavin and light, *Vox Sang.*, 2006, **90**, 85.
24. S. Perez-Pujol, R. Tonda, M. Lozano, B. Fuste, I. Lopez-Vilchez, A.M. Galan, J. Li, R. Goodrich and G. Escolar, Effects of new pathogen-reduction technology (Mirasol PRT) on functional aspects of platelet concentrates, *Transfusion*, 2005, **45**, 911.
25. J.P. AuBuchon, L. Herschel, J. Roger, H. Taylor, P. Whitley, J. Li, R. Edrich and R.P. Goodrich, Efficacy of apheresis platelets treated with riboflavin and ultraviolet light for pathogen reduction, *Transfusion*, 2005, **45**, 1335.
26. L.D. Fast, G. DiLeone, J. Li and R. Goodrich, Functional inactivation of white blood cells by Mirasol treatment, *Transfusion*, 2006, **46**, 642.
27. S.J. Wagner, D. Robinette, J. Storry, X.Y. Chen, J. Shumaker and L. Benade, Differential sensitivities of viruses in red cell suspensions to methylene blue photosensitization, *Transfusion*, 1994, **34**, 521.
28. B. Piet, S. Legrand-Poels, C. Sappety and J. Piette, NF-kappaB transcription factor and human immunodeficiency virus type 1 (HIV-1) activation by methylene blue photosensitization, *Eur. J. Biochem.*, 1995, **228**, 447.
29. S.J. Wagner, J.R. Storry, D.A. Mallory, R.R. Stromber, L.E. Benade and L.I. Friedman, Red cell alterations associated with virucidal methylene blue photosensitization, *Transfusion*, 1993, **33**, 30.
30. S.J. Wagner, A. Skripchenko, D. Robinette, J.W. Foley and L. Cincotta, Factors affecting virus photoinactivation by a series of phenothiazine dyes. *Photochem. Photobiol.*, 1998, **67**, 343.
31. T. Mohammad and H. Morrison, Photonuclease activity of Taylor's blue, *Bioorg. Med. Chem. Lett.*, 1999, **9**, 2249.
32. S.J. Wagner, A. Skripchenko, D. Robinette, D.A. Mallory, J. Hirayama, L. Cincota and J. Foley, The use of dimethylmethylene bluefor virus photoinactivation of red cell suspensions, *Dev. Biol. (Basel)*, 2000, **102**, 125.
33. S.J. Wagner, A. Skripchenko, D. Robinette, D.A. Mallory and L. Cincotta, Preservation of red cell properties after virucidal phototreatment with dimethylmethylene blue, *Transfusion*, 1998, **38**, 729.
34. S.J. Wagner, A. Skripchenko, J.C. Pugh, D. B. Suchmann and M.K. Ijaz, Duck hepatitis B photoinactivation by dimethylmethylene blue photoinactivation, *Transfusion*, 2001, **41**, 1154.
35. A. Skripchenko and S.J. Wagner, Inactivation of WBCs in RBC suspensions by photoactive phenothiazine dyes: comparison of dimethylmethylene blue and MB, *Transfusion*, 2000, **40**, 968.
36. S.J. Wagner, A. Skripchenko and D. Thompson-Montgomery, Use of a flow cell system to investigate virucidal dimethylmethylene blue phototreatment in two RBC additive solutions, *Transfusion*, 2002, **42**, 1200.

37. S.J. Wagner, A. Skripchenko and D. Thompson-Montgomery, Quinacrine enhances vesicular stomatitis inactivation and diminishes hemolysis of dimethylmethylene blue-phototreated red cells, *Photochem. Photobiol.*, 2002, **76**, 514.
38. S.J. Wagner, A. Skripchenko, D.J. Donnelly, K. Ramaswami and A.M. Detty, Chalcogenoxanthylium photosesnsitizer for the photodynamic purging of blood-borne viral and bacterial pathogens, *Bioorg. Medicinal Chem.*, 2005, **13**, 5927.
39. S.J. Wagner, A. Skripchenko, D. Thompson-Montgomery, H. Awatefe, D.J. Donnelly and M.R. Detty, Use of a red cell band 3-ligand/antioxidant to improve red cell storage properties following virucidal phototreatment with chalcogenoxanthylium photosensitizer, *Photochem. Photobiol.*, 2006, **82**, 1595.
40. N. Yamamoto, T. Okamoto and M. Kawaguchi, Novel intercalators of pyrydium dye into double-stranded DNA, *Nucleic Acids Symp. Ser.*, 1993, **29**, 83.
41. S.J. Wagner, A. Skripchenko, L. Cincotta, D. Thompson-Montgomery and H. Awatefe, Use of a flexible thiopyrylium photosensitizer and competitive inhibitor for pathogen reduction of viruses and bacteria with retention of red cells storage properties, *Transfusion*, 2005, **45**, 752.
42. S.J. Wagner, A. Skripchenko, J. Salata and L.J. Cardo, Photoinactivation of Trypanosoma cruzi in red cell suspensions with thiopyrylium, *Transfusion Apher. Sci.*, 2007, **37**, 23.
43. J.X. Corberand, Reticulocyte analysis using flow cytometry, *Hematol. Cell Ther.*, 1996, **38**, 487.
44. A. Skripchenko, S.J. Wagner, D. Thompson-Montgomery and H. Awatefe, Thiozole orange, a DNA binding photosensitizer with flexible structure, can inactivate pathogens in red blood cell suspensions while maintaining red cell storage properties, *Transfusion*, 2006, **46**, 213.
45. S.J. Wagner, A. Skripchenko, J. Sallata, A.M. O'Sullivan and L.J. Card, Inactivation of Leishmania donovani infantum and Trypanosoma cuzi in red cell suspensions with thiazole orange, *Transfusion*, 2008, **48**, 1363.
46. E. Ben-Hur and W.S. Chan, Phthalocyanines in photobiology and their medical applications, in *The Porphyrin Handbook*, ed. K.M. Kadish, K.M. Smith and R. Guilard, Academic Press, New York, 2003, **19**, 1.
47. B. Horowitz, B. Williams, S. Rywkin, A.M. Prince, D. Pascual, N. Geacintov and J. Valinsky, Inactivation of viruses in blood with aluminum phthalocyanine derivatives, *Transfusion*, 1991, **31**, 102.
48. E. Ben-Hur, R.C. Hoeben, H. Van Ormondt, T.M.A.R. Dubbelman and J. Van Steveninck, Photodynamic inactivation of retroviruses by phthalocyanines: the effect of sulphonation, metal ligand and fluoride, *J. Photochem. Photobiol. B*, 1992, **13**, 145.
49. H. Margolis-Nanno, E. Ben-Hur, P. Gottlieb, R. Robinson, J. Oetjen and B. Horowitz, Inactivation by phthalocyanine photosensitization of multiple forms of human immunodeficiency virus in red cell concentrates, *Transfusion*, 1996, **36**, 743.
50. E. Ben-Hur, W.S. Chan, Z. Yim, M.M. Zuk, V. Dayal, N. Roth, E. Heldman, A. Lazo, C.R. Valeri and B. Horowitz, Photochemical decontamination of red blood cell concentrates with the silicon phthalocyanine Pc4 and red light, *Dev. Biol. (Basel)*, 2000, **102**, 149.
51. P. Gottlieb, L.G. Shen, E. Chimezie, S. Bahng, M.E. Kenney, B. Horowitz and E. Ben-Hur, Inactivation of Trypanosoma cruzi trypomastigotes forms in blood components by photodynamic treatment with phthalocyanines, *Photochem. Photobiol.*, 1995, **62**, 869.

52. X.J. Zhao, S. Lustigman, M.E. Kenney and E. Ben-Hur, Structure-activity and mechanism studies on silicon phthalocyanines with Plasmodium falciparum in the dark and under red light, *Photochem. Photobiol.*, 1997, **66**, 282.
53. E. Ben-Hur, J. Oetjen and B. Horowitz, Silicon phthalocyanine Pc4 and red light causes apoptosis in HIV-infected cells, *Photochem. Photobiol.*, 1997, **65**, 456.
54. S. Rywkin, L. Lenny, J. Goldstein, N.E. Geacintov, H. Margolis-Nunno and B. Horowitz, Importance of type I and type II mechanisms in the photodynamic inactivation of viruses in blood with chloroaluminum phthalocyanine derivatives, *Photochem. Photobiol.*, 1992, **56**, 463.
55. E. Ben-Hur, S. Rywkin, I. Rosenthal, N.E. Geacintov and B. Horowitz, Virus inactivation in red cell concentrates by photosensitization with phthalocyanines: protection of red cells but not of vesicular stomatitis virus with water soluble analog of vitamin E, *Transfusion*, 1995, **35**, 401.
56. S. Rywkin, E. Ben-Hur, M.E. Reid, R. Oyen, H. Ralph and B. Horowitz, Selective protection against IgG binding to red cells treated with phthalocyanines and red light for virus inactivation, *Transfusion*, 1995, **35**, 414.
57. J. Van Steveninck, L.L. Trannoy, G.A.J. Besselink, T.M.A.R. Dubbelman, A. Brand, D. de Korte, A.J. Verhoeven and J.W.M. Lagerberg, Selective protection of RBCs against photodynamic damage by the band 3 ligand dipyridamole, *Transfusion*, 2000, **40**, 1330.
58. E. Ben-Hur, N.E. Geacintov, B. Studamire, M.E. Kenney and B. Horowitz, Effect of irradiance on virus sterilization and photodynamic damage in red blood cells sensitized by phthalocyanines, *Photochem. Photobiol.*, 1995, **61**, 190.
59. A.C.E. Moor, J.W.M. Lagerberg, K. Tijssen, S. Foley, G. Truscot, I.E. Kochevar, A. Brand, T.M.A.R. Dubbelman and J. Van Steveninck, In vitro fluence rate effects in photodynamic reactions with AlPcS4 as photosensitizer, *Photochem. Photobiol.*, 1997, **66**, 860.
60. J.C. Scaiano, E.B. Abuin and L.C. Stewart, Photochemistry of benzophenone in micelles. Formation and decay of radical pairs, *J. Am. Chem. Soc.*, 1982, **104**, 5673.
61. E. Ben-Hur, N.E. Geacintov, M.M. Zuk, J. Oetjen and B. Horowitz, photodynamic decontamination of red blood cell concentrates: magnetic field effects, *Curr. Topics Biophys.*, 1996, **20**(Suppl), 5.
62. J.C. Scaiano, F.L. Cozens and J. McLean, Model for the rationalization of magnetic field effects in vivo. Application of the radical-pair mechanism to biological systems, *Photochem. Photobiol.*, 1994, **59**, 585.
63. E. Ben-Hur and I. Rosenthal, Action spectrum (600–700 nm) for chloroaluminum phthalocyanine-induced phototoxicity in Chinese hamster cells, *Lasers Life Sci.*, 1986, **1**, 79.
64. G. Canti, D. Lattuada, E. Leroy, R. Cubbedu, P. Taroni and G. Velentini, Action spectrum of photoactivated phthalocyanine AlS_2Pc in tumor bearing mice, *Anti-Cancer Drugs*, 1992, **3**, 139.
65. J. Griffiths, J. Cruse-Sawyer, S.R. Wood, J. Schofield, S.B. Brown and B. Dixon, On the photodynamic action spectrum of zinc phthalocyanine tetrasulfonic acid in vivo, *J. Photochem. Photobiol. B*, 1994, **24**, 195.
66. E. Ben-Hur, M.M. Zuk, M.E. Kenney, N.L. Oleinick, J. Mulvihill and B. Horowitz, Action spectrum (660–700 nm) for virus inactivation and red cell damage photosensitized by the silicon phthalocyanine Pc4, *Lasers Med. Sci.*, 1996, **11**, 221.

67. E. Ben-Hur, M.M. Zuk, J. Oetjen, W.S. Chan, L. Lenny and B. Horowitz, Photochemical decontamination of red cell concentrates with the silicon phthalocyanine Pc4 and red light, *J. Biomed. Opt.*, 1999, **4**, 292.
68. E. Ben-Hur, M.M. Zuk, S. Chin, D. Banerjee, M.E. Kenney and B. Horowitz, Biodistribution and virus inactivation efficacy of a silicon phthalocyanine in red blood cell concentrates as a function of delivery vehicle, *Photochem. Photobiol.*, 1995, **62**, 575.
69. I. Rosenthal, C. Murali Krishna, P. Riesz and E. Ben-Hur, The role of molecular oxygen in the photodynamic effect of phthalocyanies, *Radiat. Res.*, 1986, **107**, 136.
70. J.D. Sweeney and A. Arduini, L-carnitine and its possible role in red cell and platelet storage, *Transfusion Med. Rev.*, 2004, **18**, 58.
71. T.C. Chanh, J.S. Allan, J.L. Matthews, F. Sogandares-Bernal, M.M. Judy, H. Skiles, J. Leveson, J. Marengo-Rowe and J.T. Newman, Photodynamic inactivation of simian immunodeficiency virus, *J. Virol. Methods*, 1989, **26**, 125.
72. J.L. Matthews, F. Sogandares-Bernal, M.M. Judy, A.J. Marengo-Rowe, J.E. Leveson, H. Skiles, J.T. Newman and T.C. Chanh, Preliminary studies of photoinactivation of human immunodeficiency virus in blood, *Transfusion*, 1991, **31**, 636.
73. L.L. Trannoy, J.W. Lagerberg, T.M. Dubbelman, J.H. Schuitmaker and A. Brand, Positively charged porphyrins: a new series of photosensitizers for sterilization of RBCs, *Transfusion*, 2004, **44**, 1186.
74. L.L. Trannoy, F.G. Terpstra, D. de Korte, J.W. Lagerberg, A.J. Verhoeven, A. Brand and F.A. Engelenburg, Differential sensitivities of pathogens in red cell concentrates to Tri-P(4) photoinactivation, *Vox Sang.*, 2006, **91**, 111.
75. P.C. Spinella, J.G. Perkins, K.W. Grathwohl, A.C. Beekley and J.B. Holcomb, Warm fresh whole blood is independently associated with improved survival for patients with combat-related traumatic injuries, *J. Trauma*, 2009, **66**(4), S69.
76. P.C. Spinella, Warm fresh whole blood transfusion for severe hemorrhage: US military and potential civilian application, *Crit. Care Med.*, 2008, **36**(7), S340.
77. M. Grogl, J.L. Daugirda, K.L. Hoover, A.J. Magill and J. Berman, Survivability and infectivity of viscerotropic Leishmania tropica from Operation Desert Storm participants in human blood products maintained under blood bank conditions, *Am. J. Trop. Med. Hyg.*, 1993, **49**, 308.
78. O. Myles, G. Wortmann, C. Ockenhause, R. Gasser, P. Weina, S. Patel, N. Crum, H. Groff and B.L. Herwaldt, Two cases of visceral leishmaniasis in US military personnel – Afghanistan 2002–2004, *MMWR*, 2004, **53**, 265.
79. M.L. Schroeder, Transfusion-associated graft-versus-host disease, *Br. J. Haematol.*, 2002, **117**, 275.
80. D.S. Gould and H. Auchincloss, Direct and indirect recognition: the role of MHC antigens in graft rejection, *Immunol. Today*, 1999, **20**, 77.
81. M.A. Blajchman, Transfusion-associated immunomodulation and universal white cell reduction: are we putting the cart before the horse?, *Transfusion*, 1999, **39**, 665.
82. The Trial to Reduce Alloimmunization to Platelets Study Group, Leukocyte reduction and ultraviolet B irradiation of platelets to prevent alloimmunization and refractoriness to platelet transfusions, *N. Engl. J. Med.*, 1997, **337**, 1861.
83. N.M. Heddle, Universal leukoreduction and acute transfusion reactions: putting the puzzle together, *Transfusion*, 2004, **44**, 1.
84. W.A.L. Heaton, Evaluation of posttransfusion recovery and survival of red cells, *Transfusion Med. Rev.*, 1992, **6**, 153.

Chapter 10

PDT for Acne

Michael H. Gold

Medical Director, Gold Skin Care Center, Tennessee Clinical Research Center, Clinical Assistant Professor, Vanderbilt University School of Medicine, Department of Dermatology, Vanderbilt University School of Nursing, Nashville, TN, USA; Visiting Professor of Dermatology, Huashan Hospital, Fudan University, Shanghai, China; No. 1 Hospital of China Medical University, Shenyang, China

Table of Contents

Photodynamic Inactivation of Microbial Pathogens: Medical and Environmental Applications
Editors Michael R. Hamblin and Giulio Jori

Published by the Royal Society of Chemistry, www.rsc.org

10.1. Introduction

Acne vulgaris is the most common dermatologic disorder dermatologists see on a regular basis in our day-to-day clinics. Up to 33% of all dermatology visits are for the treatment of acne vulgaris.[1] Dermatologists need to know all the tools of the trade available to them to continue to be the leaders in this field. Photodynamic therapy (PDT) is a treatment option which dermatologists can use to our advantage to deal with many individuals suffering from this most common of skin disorders.

Acne vulgaris is very prevalent in our society. Clinical studies looking at epidemiologic data suggest that almost 94% of all females will have acne at some point in their lifetime.[2] It has been curious to watch the nature of our patients over the past 20 years or so. Acne therapy has changed during this time and clinicians treating those with acne must be aware of the new medications and the new trends in acne care. This includes the use of lasers and light sources, including the use of PDT, a novel concept which has shown some promise in the treatment of acne vulgaris, although it remains to be seen whether PDT for acne will ever become mainstream.

10.2. Acne Vulgaris – Etiology, Pathogenesis and Medical Therapy

Acne vulgaris is defined as a multifactorial disease. There is usually a strong genetic history of acne vulgaris within families of those suffering from acne. Hormonal factors are also involved in the formation of the acne lesion, as evidenced by the fact that most acne occurs at the time of puberty where, in susceptible individuals, one sees an increase in sebum production and the formation of the follicular "plug", one of the first inciting events in the formation of the acne lesion. As well, environmental concerns can play a role in the acne process.[3]

Acne vulgaris is defined as a disorder of the sebaceous glands, where hormonal activity causes dilatation followed by obstruction of the sebaceous glands. Non-inflammatory acne vulgaris lesions, called comedones, are produced at this early phase in the development of the acne lesion. Two kinds of comedonal lesions may form – the open or closed comedone. The open comedone is commonly referred to as white heads while the closed comedone is commonly known as black heads. The obstruction of the sebaceous glands leads to the production and proliferation of the bacteria within those glands. The predominant bacterium is known as *Propionobacterium acnes* (*P. acnes*). Proliferation of the *P. acnes* bacteria results in an inflammatory response and the production of inflammatory acne lesions, known as papules, pustules and cystic acne lesions.

Medical therapy for acne vulgaris remains the gold standard for acne therapy. A variety of topical and systemic therapies exist which successfully treat acne lesions. Topical medical therapy has been a mainstay of acne vulgaris therapy for years and numerous medicines now exist which can benefit our patients. Topical antibiotics, topical benzoyl peroxides and topical retinoids are the most common of the topical therapies being utilized regularly in our clinics. As dermatologists, we have recently seen many new compounds become available for us. Most of these are combination products, including clindamycin-tretinoin, clindamycin-benzoyl peroxides and benzoyl peroxide-tretinoin preparations. A topical anti-inflammatory medicine, known as dapsone, has been approved for the topical treatment of inflammatory acne vulgaris. It is interesting to note as well that, as of the time of this writing, there is the real possibility that topical benzoyl peroxides will all become over-the-counter in the year 2011.

Systemic therapy remains popular as well in treating inflammatory acne vulgaris. Older medicines, including erythromycin and tetracycline, have been, for the most part, replaced by newer derivatives, especially in the tetracycline family, known as minocycline and doxycycline. Newer dose regimens for minocycline and doxycycline as well as sustained released drug delivery are new modifications for these agents. We also have isotretinoin available to us for those patients suffering from recalcitrant cystic acne vulgaris. Isotretinoin has been called a "wonder" drug for many, although the potential for teratogenicity and psychiatric concerns continue to be a barrier to the use of this product.[3] And in the US, several initiatives are now in place for the clinician, patient and pharmacy to ensure patient compliance and adherence to proper birth control methods for those taking isotretinoin.

There are other important side bars of note for those using systemic therapy in patients with inflammatory acne vulgaris. This includes resistance to common systemic antibiotics, with reports noting resistance as high as 60%,[4,5] a report linking breast cancer to long-term antibiotic use[6] and a report leading to an increase in the risk of pulmonary infections as a result of long-term antibiotic use.[7]

Although medications remain the gold standard for the treatment of acne vulgaris, lasers and light sources may also be useful for the treatment of inflammatory acne vulgaris, and the remainder of this chapter will summarize what is the current state of the art in this field.

10.3. Lasers and Light Sources for Inflammatory Acne Vulgaris – A Summary

Lasers and light sources can treat inflammatory acne vulgaris lesions *via* two principle mechanisms. One is through the destruction of the *P. acnes* bacteria within the sebaceous gland and the second is through the destruction of the sebaceous gland itself. A third mechanism of action, acting through the partial destruction of the sebaceous gland, is the presumed mechanism for PDT.

10.4. Lasers and Light Sources that Destroy the *P. acnes* Bacteria

The premise by which lasers and light sources destroy the bacteria responsible for the inflammatory phase in the formation of the acne vulgaris lesion is through the endogenous PDT response already known for *P. acnes*.

P. acnes have a photodynamic response naturally occurring in the skin. During its growth cycle, *P. acnes* naturally produces porphyrins, predominately Protoporphyrin IX (PpIX) and Coproporphyrin III. These porphyrins have an absorption spectrum which is shown in Figure 10.1 for PpIX. From this graph, one notes the major peak for PpIX (Coproporphyrin III is similar) is in the blue light range at 415 nm (called the Soret Band). A second smaller peak is seen at 630 nm, corresponding to red light. Both blue and red light are routinely used for the treatment of acne. There are also smaller peaks seen in the absorption spectrum of PpIX, corresponding to peaks of energy where a variety of lasers and light sources can also have a positive impact on acne's treatment. These include the green light lasers at 532 nm, the yellow light pulsed dye lasers at 585–595 nm and the intense pulsed light (IPL) device, which emits polychromatic light *via* flashlamps and "cutoff" filters in the range of 400–1200 nm.

Let's now review the various lasers and light sources commercially available in the US which work through the mechanism of destruction of *P. acnes*. The first of these are the blue light sources, shown in Figures 10.2–10.4.

The first blue light device approved by the US Food and Drug Administration (FDA) for the treatment of inflammatory acne vulgaris was a high-intensity, narrow-band blue light source, known commercially as the ClearLight Acne Photoclearing System (CureLight, Yokneam, Israel). It received FDA approval for the treatment of mild to moderate inflammatory acne vulgaris. Its use and success in the treatment of inflammatory acne vulgaris is well documented in the medical literature. The majority of patients received blue

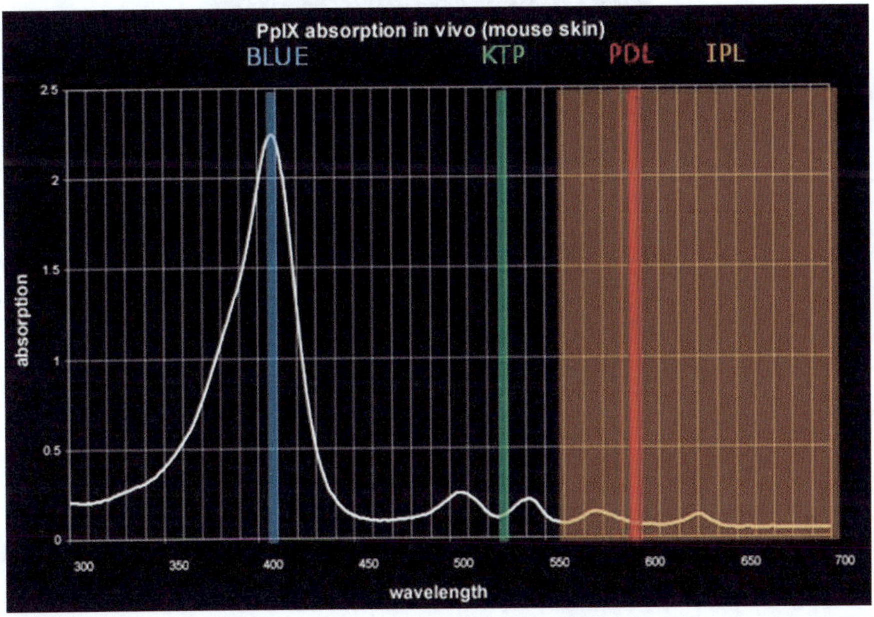

Figure 10.1 PpIX absorption spectrum.

Figure 10.2 ClearLight Acne Photoclearing System (CureLight, Yokneam, Israel).

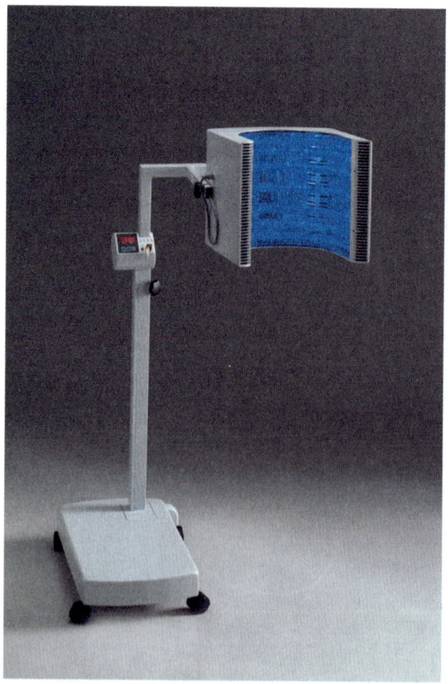

Figure 10.3 Blu-U (DUSA Pharmaceuticals, Wilmington, MA).

Figure 10.4 Omnilux Blue Phototherapeutics (OmniLux, Manchester, UK; PhotoMedix, Philadelphia, PA).

light therapy twice a week for upwards of 18 minutes for a total of 8 treatments, or for four weeks in total. Kawada *et al.*[8] reported a 64% reduction in mild to moderate inflammatory acne vulgaris in 30 patients receiving twice weekly blue light therapy for five weeks. Shalita *et al.*[9] reported on 35 patients with the high-intensity blue light system twice a week for four weeks; they found that 80% noted significant improvement at the end of the therapy. Continued improvement was noted after the therapy was discontinued. Adverse events were not seen, and all skin types were amenable to the therapy. Gold *et al.*[10] treated 40 patients and found a 43% reduction in inflammatory acne vulgaris lesion counts after twice weekly therapy for four weeks and followed for upwards of three months following the last treatment. Continued improvement, in many, was seen. Adverse events were not seen. One of the differentiators between this clinical trial and the previous reported blue light trial was that this study included both responders and non-responders thus lowering the overall response rates. There are some patients, reported to be in the 20% range, that do not respond to the blue light source. Including them in the percent calculations will have an effect on the percent improvements. At the time of this writing, the ClearLight is currently not commercially available in the US; many of us still have it in our clinics and use it daily. An updated version, known as the I Clear, replaced the ClearLight, but it is not readily available in 2010.

Other blue light trials also appeared in the literature. Papageorgiou *et al.*[11] used a mixed blue and red light system (415 and 660 nm), blue light therapy alone and white light therapy. They showed that the combination of blue and red light decreased inflammatory acne vulgaris lesions by 76% *versus* 58% in the blue light alone group, both of which were better than white light (25%). Meffert *et al.*[12] used a high-energy broad spectrum blue light source combining blue light and UVA and noted marked improvement in patients with pustular acne vulgaris after 10 treatments.

The second blue light source approved for use in inflammatory acne vulgaris in the US is known as the Blu-U (DUSA Pharmaceuticals, Wilmington, MA). This is a low-intensity blue light source. Goldman *et al.*[13] first reported its use in inflammatory acne vulgaris in 12 patients. Inflammatory acne lesion showed a 40% reduction in papular acne lesions, 65% reduction in pustular acne lesions and 62% reduction in comedonal acne lesions two weeks after a series of treatments. Gold *et al.*[14] compared topical 1% clindamycin to this blue light source; they showed that the blue light therapy was more effective than the topical clindamycin in reducing inflammatory acne vulgaris lesions. A clinical example of response to this blue light source is shown in Figure 10.5.

A third blue light system, the OmniLux Blue Phototherapeutics (OmniLux, Manchester, UK; PhotoMedix, Philadelphia, PA) also is useful in the treatment of inflammatory acne vulgaris lesions. It uses LED blue lights and has been shown to be effective in reducing inflammatory acne vulgaris lesions by an average of 74%.[15]

A variety of red light sources is also available. Clinical studies with red light alone for inflammatory acne lesion reduction are not common in the medical literature.[16,17] Red lights are very popular devices in association with the methyl form of aminolevulinic acid, which will be discussed later in this chapter. Red light, by virtue of its longer wavelength as compared to blue light, penetrates deeper and thus may have some additional benefits in the treatment of acne vulgaris. The most common of the red light sources currently available on the market is known as the Aktilite (Galderma, Ft. Worth, TX) and is shown in Figure 10.6.

Several laser systems have shown to be effective in treating inflammatory acne vulgaris. Green light KTP lasers (532 nm) are effective in treating inflammatory acne vulgaris lesions. Bowes *et al.*[18] treated 11 patients with mild to moderate inflammatory acne in a split-face prospective, randomized clinical trial with a KTP laser. With four treatments and a one-month follow up evaluation, lesion counts decreased by 35.9% *versus* 11.8% on the control side. Sebum measurements were also decreased, 28.1% *versus* 6.4%.

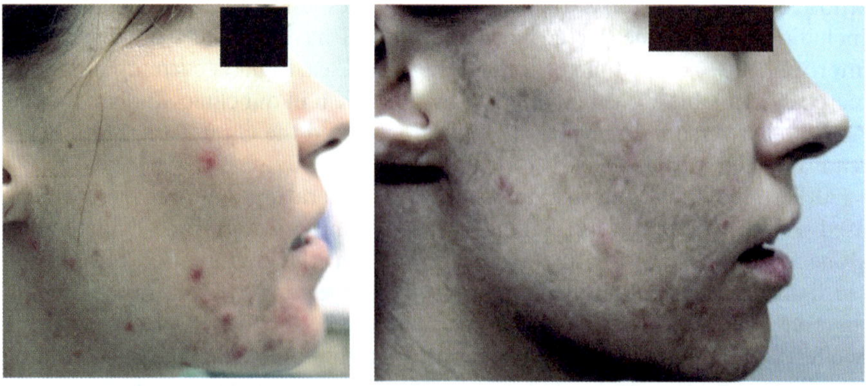

Figure 10.5 Clinical example of ALA-Blu-U for treatment of inflammatory acne vulgaris.

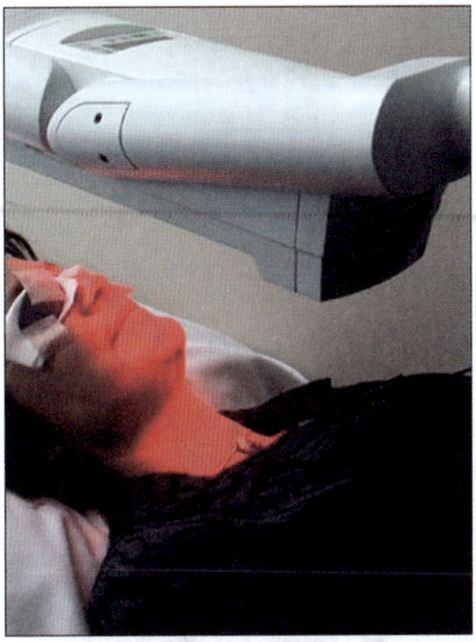

Figure 10.6 Aktilite red light source (Galderma, Ft. Worth, TX).

Yellow light pulsed dye lasers (585–595 nm) also treat inflammatory acne vulgaris. Several very well-designed PDLs exist in the US market place; one even multiplexes the pulse of light of the 585 nm with a 1064 nm laser system to make the treatments more effective. The first published clinical trial with the PDL and inflammatory acne[19] evaluated 41 individuals and showed the effectiveness of a low-fluence PDL. It was a double-blind, randomized clinical trial where 31 individuals received PDL and 10 patients received a sham treatment. One treatment was given to these patients. Acne severity decreased from 3.8 to 1.9 with the PDL group *versus* 3.6 to 3.5 with the sham treatment (using a modified Leeds grading system). A second clinical trial, utilizing the same low-fluence PDL, did not confirm these results[20] and, in fact, showed no improvement using this PDL *versus* the sham device. Many of us feel the PDLs work quite well in acne vulgaris – more research is need to look at whether there is sufficient evidence to support its use.

A variety of intense pulsed light (IPL) sources are available for commercial use and they have been used to treat a number of cutaneous entities over their years in existence, including inflammatory acne vulgaris lesions. The IPLs first developed for the treatment of leg veins, unwanted hair and pigmentary disorders soon became the mainstay for treating the signs of photodamage, known as photorejuvenation. Several of the currently available IPLs are shown in Figures 10.7–10.10. The first IPL acne clinical study used the ClearTouch (Radiancy Inc, Orangeburg, NY), known today as the SkinStation. It uses light and heat energy, known as "LHE technology" to trigger the destruction of the *P. acnes* bacteria. Elman *et al.*[21] treated 19 patients with the IPL device and showed that 85% of the individuals had a greater than 50% improvement in their acne vulgaris lesions following therapy. The patients had twice weekly therapy for four weeks.

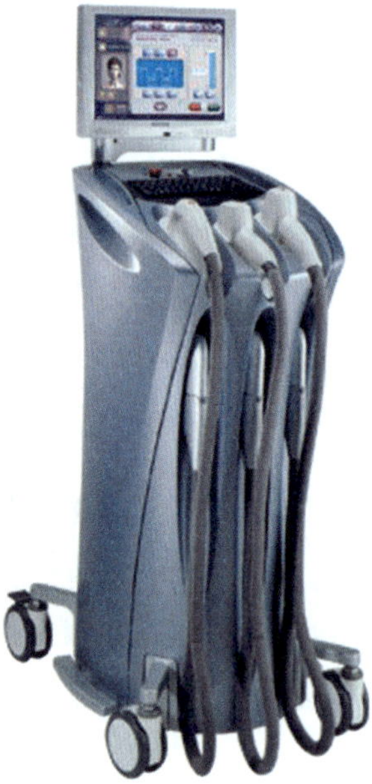

Figure 10.7 Lumenis One IPL (Lumenis, Santa Clara, CA).

Dierickx[22] reported on the use of the Lux V hand piece on the Palomar Medical Technologies (Burlington, MA) IPL systems (EsteLux, MediLux and StarLux Systems). Fourteen patients with mild to moderate inflammatory acne vulgaris lesions each received 5 treatments, given every 2–4 weeks. Two to three

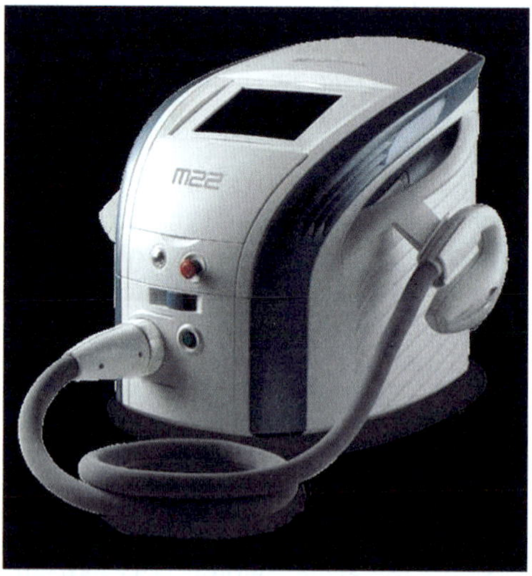

Figure 10.8 M22 IPL (Lumenis, Santa Clara, CA).

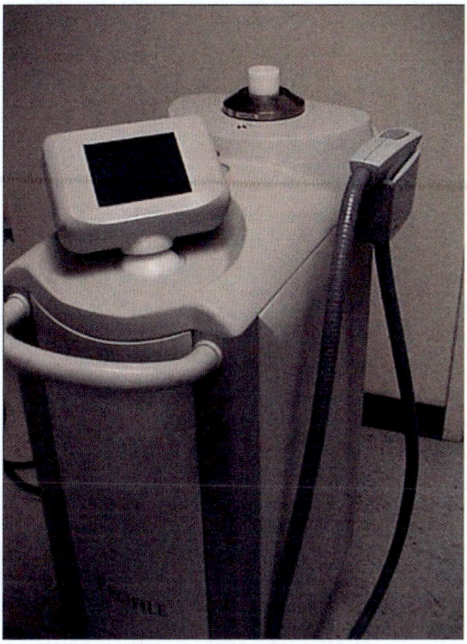

Figure 10.9 BBL IPL (Sciton, Palo Alto, CA).

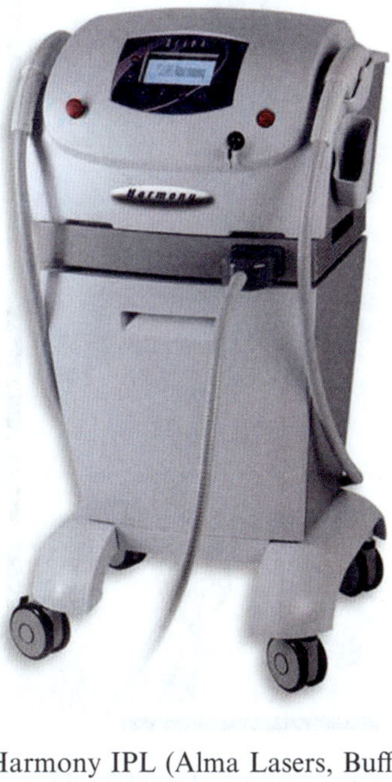

Figure 10.10 Harmony IPL (Alma Lasers, Buffalo Grove, IL).

passes were given with an average fluence of $10\,\mathrm{J\,cm^{-2}}$. At six months post-therapy, clearance rates of 72% for non-inflammatory acne vulgaris lesions were seen and 73% were demonstrated for inflammatory acne vulgaris lesions.

The newest of the IPL technologies is known commercially as the Isolaz (Solta Medical, Hayward, CA) which uses IPL technology and a vacuum to bring the tissue closer to the light source. This is known as photopneumatic therapy (PPx). It is shown in Figure 10.11. Weiss *et al.*[23] treated ten individuals with inflammatory acne vulgaris. Patients received four treatments at one-week intervals. They were given, on average, 2–4 passes with the device per treatment session. The treatment times were, on average, no more than 15 minutes; no topical analgesia was required. Patients were followed for one month past their last IPL treatment. They noted the median time to patient improvement was 5 days, with some noting improvement within 2 days. Average lesion count reduction for papules was between 50 and 60%, pustules 70 and 80% and reduction in comedones 40–50%. No adverse events were reported. They concluded this therapy provided pain-free, fast acne clearing. Shamban *et al.*[24] reported their experiences with the PPx device in a retrospective, multi-centre clinical evaluation of 56 individuals. They found a 50% reduction after the first treatment which increased to 90% after four treatments.

Gold *et al.*[25] also reported their findings with the PPx. Eleven patients were treated with four treatments at three-week intervals. The patients were followed for three months post their last treatment and found a 78% reduction in

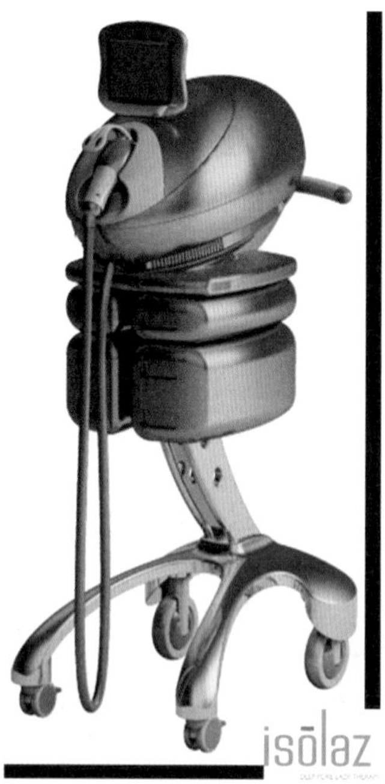

Figure 10.11 Isolaz PPx IPL (Solta Medical, Hayward, CA).

inflammatory acne lesions (p = 0.0137) and a 57.8% reduction in non-inflammatory acne lesions (p = 0.0383). There were no significant adverse events, and downtime for patients was minimal. The Isolaz is FDA cleared for both inflammatory and non-inflammatory acne lesions. Clinical examples are shown in Figures 10.12 and 10.13.

In 2010, most clinicians would recommend this device be used in a two-step process. After cleansing of the face, a pore cleansing pass is conducted first with only the vacuum. This is followed by a treatment pass with the vacuum and the light together. Patients are remarkably happy and, as seen in the clinical trials involving the device, they see results fast.

10.5. Lasers that Destroy Sebaceous Glands

There are laser systems that treat inflammatory acne lesions *via* destruction of the sebaceous glands themselves. These include the near-infrared lasers and other energy-based systems, mainly radiofrequency (RF) devices. The concept of partial destruction of the sebaceous gland, the mechanism by which PDT works *via* a topical photosensitizer, will also be discussed.

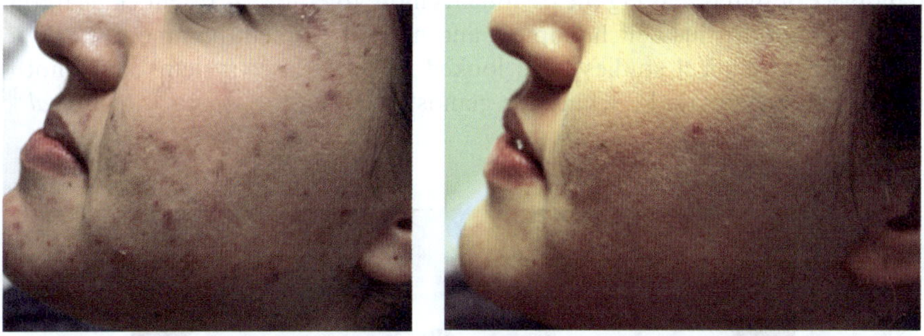

Figure 10.12 Clinical example of Isolaz PPx for the treatment of acne vulgaris.

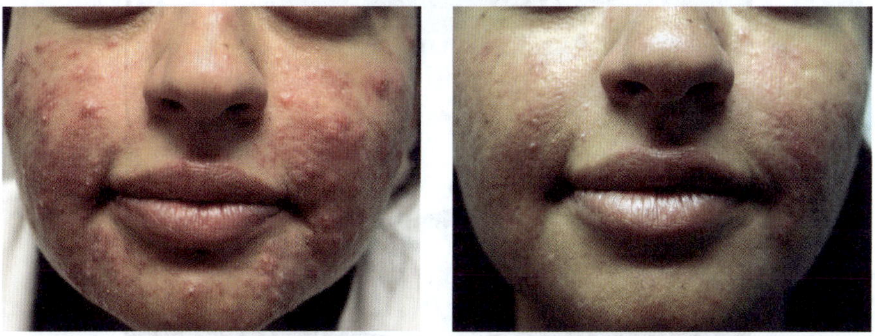

Figure 10.13 Clinical example of Isolaz PPx for the treatment of acne vulgaris.

The near-infrared lasers being studied for acne vulgaris are the 1319 nm Sciton Profile, the 1320 nm CoolTouch CT3 (CoolTouch Inc., Roseville, CA), the 1450 nm SmoothBeam (Candela, Wayland, MA) and the 1540 nm erbium glass Aramis (WaveLight-Quantel, Clermont-Ferrand, France).

Of all of the laser systems in this category the SmoothBeam has the most published literature associated with its use. Paithankar *et al.*[26] reported their experience with the 1450 nm SmoothBeam laser and its patented dynamic cooling device (Figure 10.14) to treat 27 patients with inflammatory acne on the back. Four treatments were given at three-week intervals; patients were followed for six months. They showed a 98% reduction in inflammatory acne vulgaris lesions after four treatments. At follow-up, 100% lesion clearance was seen in all but one of the participants. A second study by Friedman *et al.*[27] evaluated facial acne. Nineteen patients saw lesion counts decreased by 37% after one treatment, 58% after two treatments and 83% after three treatments. Transient erythema and edema were seen with the treatments and topical anesthetics were utilized to minimize the discomfort routinely observed with these treatments.

The SmoothBeam is an effective acne treatment, limited on occasion by pain associated with the treatment. Recent advances and protocols of lower energies and multiple passes are achieving good results. As well, a new and larger handpiece design is making the treatments more tolerable. A recent clinical trial seems to support this new handpiece and newer treatment parameters.[28]

Additional clinical studies have looked at other laser systems and photosensitizers to destroy the sebaceous glands. Tuchin *et al.*[29] and Lloyd *et al.*[30]

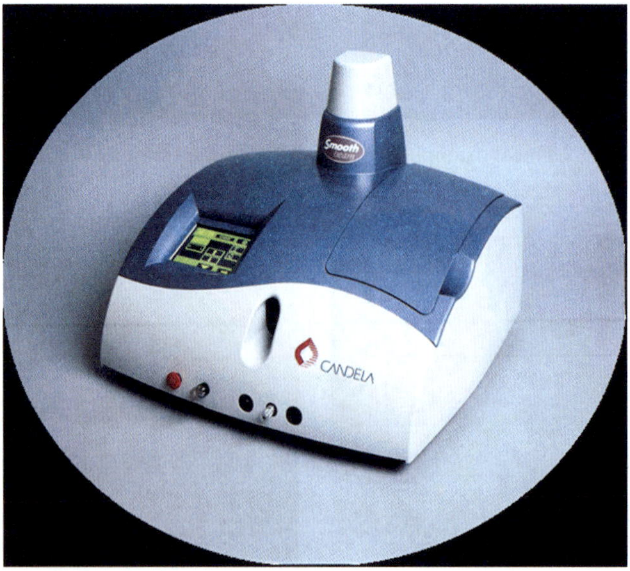

Figure 10.14 1450 nm SmoothBeam laser (Candela, Wayland, MA).

used indocyanine green (ICG) and a diode laser for the treatment of inflammatory acne *via* the destruction of sebaceous glands. Both of the clinical trials showed a positive effect. ICG is not a readily available photosensitizer and therefore, not routinely utilized in clinical practice. The 1540 nm laser has also been reported effective in the treatment of inflammatory acne vulgaris.[31]

Ruiz-Esparza *et al.*[32] reported a preliminary observation with the use of monopolar radiofrequency (RF) in the treatment of inflammatory acne vulgaris. Twenty-two patients were evaluated with the ThermaCool device (Solta Medical, Hayward, CA). Excellent responses were seen in 82%, modest responses in 9% and no response in 9%. Further clinical studies are needed to confirm RFs role in the treatment of acne vulgaris. Figure 10.15 shows an example of a patient treated with a UniPolar RF device.

10.6. ALA-PDT and MAL-PDT in the Treatment of Inflammatory Acne Vulgaris

The use of ALA-PDT and MAL-PDT is now recognized around the globe as a useful therapeutic modality for the treatment of non-melanoma skin cancers, actinic keratoses (AKs) and photorejuvenation, and for patients suffering from moderate to severe inflammatory acne vulgaris. The photosensitizers have been shown to accumulate readily in non-melanoma skin cancer cells, actinically

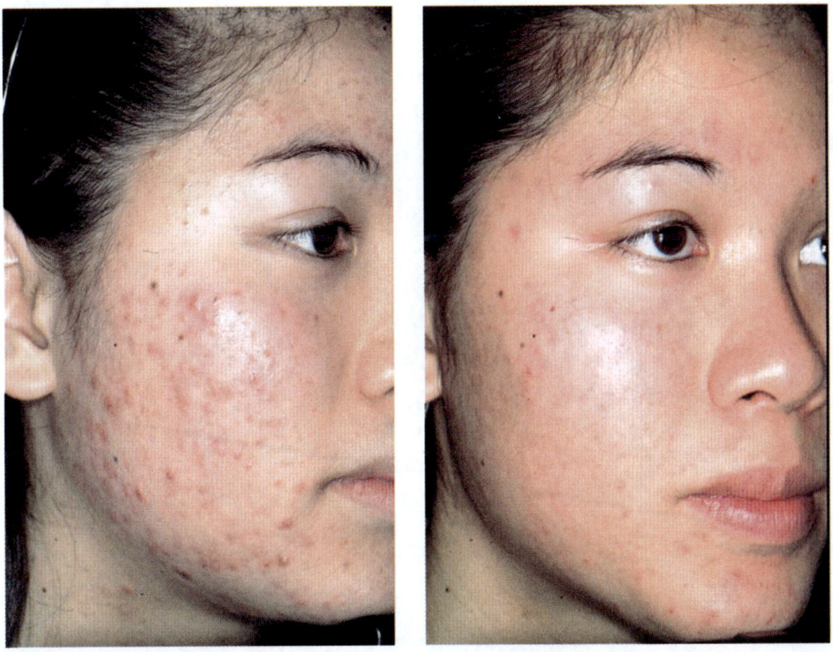

Figure 10.15 Clinical example with UniPolar RF device for the treatment of acne vulgaris. Photos courtesy of Martin Braun MD.

damaged skin cells and AKs and in the pilosebaceous unit, where acne vulgaris lesions arise. This chapter will not review ALA and MAL in detail. Various reports and references exist which are very thorough and in which one can find all the detailed pertinent information.[33,34] The remainder of this section will deal with PDT for inflammatory acne vulgaris.

ALA-PDT is FDA approved in the US for the treatment of non-hyperkeratotic AKs of the face and scalp utilizing a blue light source for 16 minutes and 40 seconds. The only currently available ALA in the US market is Levulan® Kerastick™, by DUSA Pharmaceuticals (Wilmington, MA) Figure 10.16. The methyl ester of ALA (MAL) is known as Metvix®, and is now distributed by Galderma (Fort Worth, TX). Metvix®, or Metvixia® as it is known in the US, has FDA clearance in the US for the treatment of non-hyperkeratotic AKs of the face and scalp. It is shown in Figure 10.17.

The first reported clinical trial utilizing ALA in the treatment of acne vulgaris was by Hongcharu *et al.*[35] In their report, 22 individuals were evaluated utilizing ALA and a 550–570 nm broadband light source. The ALA was incubated for three hours without occlusion. Significant clinical clearance was noted after four weekly ALA-light treatments. The patients in this trial experienced significant downtime, known as the "PDT effect". The PDT effect (downtime with healing) in this study consisted of acneform folliculitis, post-inflammatory hyperpigmentation, superficial peeling and crusting. The mechanism of action for this improvement was felt to be through partial destruction of the sebaceous glands.

Itoh *et al.*[36] reported their experience with ALA-PDT for acne vulgaris in utilizing a 635 nm pulsed excimer dye laser and a four-hour ALA drug incubation in a single patient with intractable acne vulgaris on the face. The treated

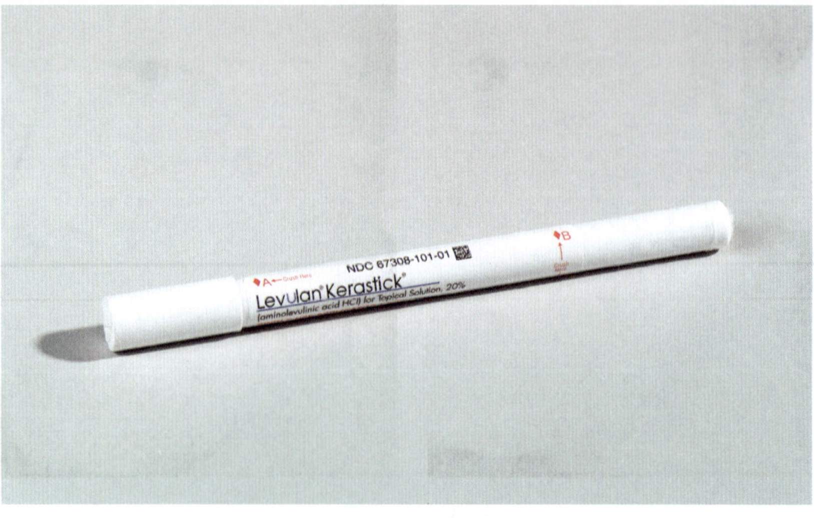

Figure 10.16 Levulan® Kerastick™ (Dusa Pharmaceuticals, Wilmington, MA).

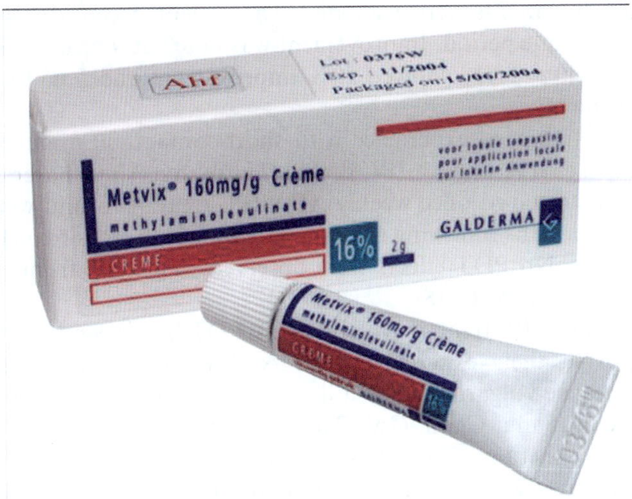

Figure 10.17 Metvixia® methyl ester of ALA (MAL) (Galderma, Ft. Worth, TX).

area remained disease free for the eight months of the follow-up time period. The patient did experience a PDT effect, with erythema, edema and crusting immediately after the therapy. In a subsequent study, Itoh *et al.*[37] reported their experience with 13 patients utilizing ALA-PDT and a 600–700 nm light from a halogen light source. All patients showed improvement in their inflammatory acne vulgaris lesions with new acne vulgaris lesions reduced at 1, 3 and 7 months following their therapy. A PDT effect once again was seen and some recurrence was noted after the following six months.

After these initial successes, many investigators in the US began looking at short-contact, full-face drug incubation for the treatment of moderate to severe inflammatory acne vulgaris, just as they did with AKs and photodamage. Goldman[38] reported his experience treating inflammatory acne vulgaris and sebaceous gland hyperplasia with a one-hour ALA incubation and therapy with either an IPL device or a blue light source. "Relative clearing" of the inflammatory acne lesions were seen after 2–4 once weekly ALA-PDT treatments. Treatments were noted to be pain free and no PDT effect was seen.

Gold[39] used a 30–60 minute ALA drug incubation and a blue light source to evaluate moderate to severe inflammatory acne vulgaris lesions. The patients received one weekly ALA-blue light treatment for four weeks and were followed at one and three months. Response rates of 60% were found. No adverse events were observed.

Goldman *et al.*[40] treated 22 patients with moderate to severe inflammatory acne vulgaris with blue light with and without the ALA. There was a greater response in the ALA-PDT blue light group than the blue light group alone. Taub[41] used short-contact, full-face therapy with blue light sources (ClearLight or Blu-U) or an IPL with radiofrequency (Aurora®, Syneron, Yokneam,

Israel). Eighteen patients were evaluated. The patients received 2–4 treatments over a 4–8 week time period. Improvement was noted at four months following the last treatment – 11/18 showed 50% improvement and 5 had greater than 75% improvement.

Gold *et al.*[42] reported experience with short-contact (30 minutes), full-face therapy utilizing ALA and an IPL device (Harmony, Alma Lasers, Caesarea, Israel). Patients received once weekly ALA-IPL treatments and were followed for up to three months following their final treatment. A 72% reduction in acne lesions was seen. No PDT effects were observed. A patient from that clinical trial is shown in Figure 10.18.

Three split-face IPL treatments with ALA-PDT have been reported. Santos *et al.*[43] reported their experience with ALA-PDT in moderate to severe inflammatory acne vulgaris lesions utilizing ALA-PDT and the Quantum IPL device (Lumenis, Santa Clara, CA). Thirteen patients were treated with short-contact, full-face therapy. The IPL was used with a 560 nm filter, double pulsed with 2.4/6.0 msec, a 25-msec pulse delay and fluences of 26–34 $J\,cm^{-2}$. In this split-face analysis, 10 of 13 patients showed a marked response in the ALA-IPL treated side *versus* the IPL side alone after a single treatment. The second split-face clinical trial, performed by Rojanamatin *et al.*[44] confirmed the results

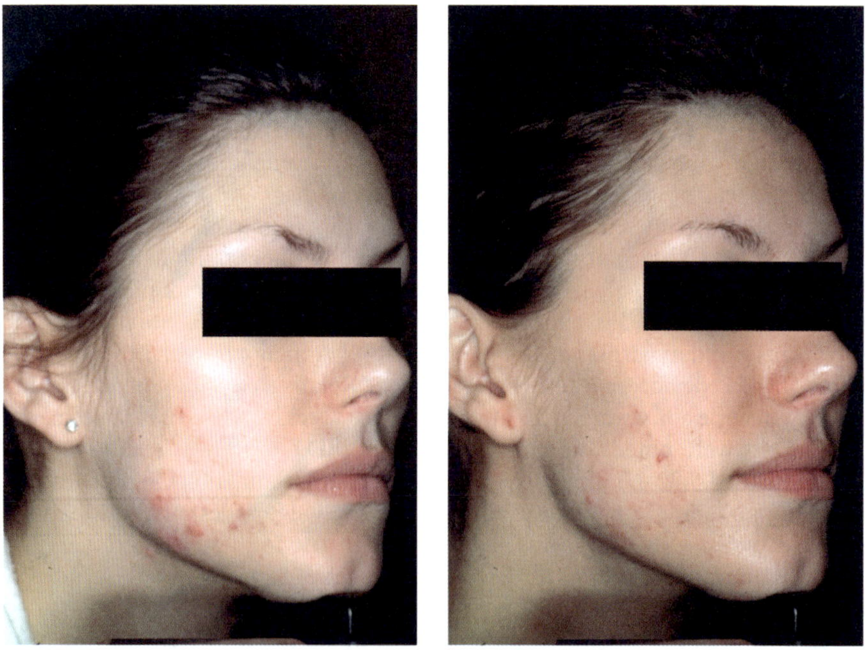

Figure 10.18 Clinical example of patient with acne lesions treated with short-contact (30 minutes), full face therapy with ALA and the Harmony IPL device. Photos courtesy of Michael H. Gold MD, Tennessee Clinical Research Center, Nashville, TN.

described by Santos *et al.* They evaluated 14 patients in a split-face fashion with an IPL and found that the ALA-IPL treated area was superior to treatment with the IPL alone. A third split-face ALA-IPL study by Oh *et al.*[45] also confirmed the use of ALA and an IPL *versus* an IPL alone. In their study, Oh and his colleagues also examined different drug incubation times in their patients. The group treated 20 individuals and found that the longer drug incubation time group had better clearance in their patient population; the ALA-IPL sides faired better than the IPL sides alone.

Alexiades-Armenakas[46] reported the PDL and ALA in inflammatory acne. A drug incubation average time of 45 minutes and three PDL sessions cleared all of the 14 patients. Miller and Van Camp[47] reported on the successful use of ALA and the KTP laser in patients with inflammatory acne vulgaris. A recent split-faced ALA-PDL clinical trial by Orringer *et al.*[48] showed that the combination of ALA-PDL was better than PDL alone. Clinical examples of inflammatory acne vulgaris treated with ALA-PDT are shown in Figures 10.19 and 10.20.

ALA-PDT is available in the US, on a limited basis in certain countries in Latin America and on a limited basis in certain Asia-Pacific nations as well, although the interest worldwide with this therapy continues to grow year in year out.

With the excitement generated by these clinical trials, it was hoped that an FDA indication for inflammatory acne and perhaps insurance reimbursement would be forthcoming. Two clinical trials undertaken by the ALA manufacturer have failed to show statistical clinical improvement of the combination of light and sensitizer *versus* light alone. We know blue light therapy works for inflammatory acne vulgaris and we know through the initial clinical trials that ALA and light works well for inflammatory acne vulgaris. Most of us would argue, however, that the combination works better than blue light alone and many of us will continue to utilize this therapy in our clinics for patients suffering from moderate to severe inflammatory acne vulgaris.

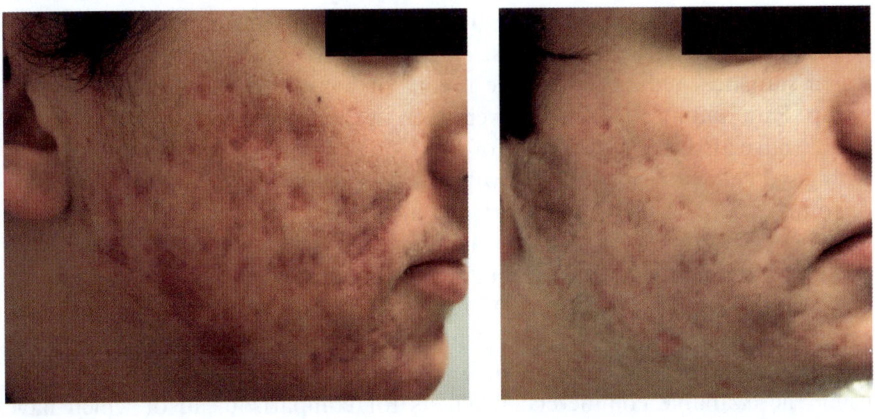

Figure 10.19 Clinical example of treatment of inflammatory acne vulgaris with ALA-PDT.

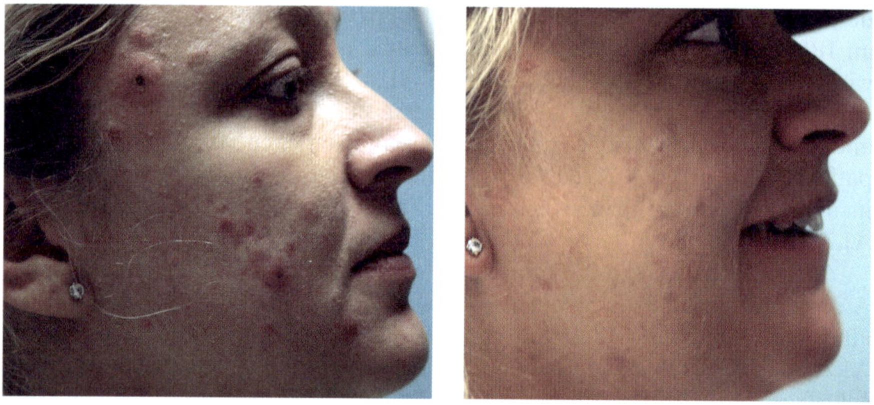

Figure 10.20 Clinical example of treatment of inflammatory acne vulgaris with ALA-PDT.

MAL is the most common PDT drug used in Europe and had been available in the US since the end of 2008, although the company rollout of this drug was not until 2009. It is used primarily in Europe as a therapy for non-melanoma skin cancers, and has been used successfully to treat non-hyperkeratotic AKs and Bowen's disease. Recently it has been shown to be effective in the treatment of photorejuvenation and has recently been studied in several small clinical trials for the treatment of inflammatory acne vulgaris in Europe. Wiegell and Wulf[49] evaluated 21 patients with moderate to severe inflammatory acne vulgaris. Two treatments were given two weeks apart. Twelve weeks after the treatments, there was a 68% reduction in inflammatory acne lesion counts, with no change in the control group. All patients in the study experienced a "PDT effect" consisting of severe erythema, pustular eruptions and exfoliation of the skin. Moderate to severe pain during the treatments was also noted. A second clinical trial by Horfelt *et al.*[50] looked at 30 individuals with moderate to severe inflammatory acne vulgaris lesions. This was a split-face analysis, with a three-hour under-occlusion drug incubation, exposure to red light, with two treatments given at two-week intervals. At 12 weeks after the last treatment, there was a statistical reduction in acne lesions of 54% *versus* 20% in the control group. Pain and a "PDT effect" were seen in the patients treated. Christensen *et al.*[51] have also reported the use of MAL-PDT in inflammatory acne vulgaris with good clinical results but still with a PDT effect. Further work in Europe is ongoing with MAL-PDT and shortened incubation times to determine the best treatment paradigms for its use in inflammatory acne vulgaris.

Several reports have looked at pain associated with PDT treatments and, in doing so, have tried to compare ALA and MAL and the pain associated with each treatment.[52,53] As has been pointed out,[54] it is difficult to compare the therapies without using the standards of care used for each therapy and by not using the available commercial products for comparison, all of which have never been performed. Short-contact, full-face therapy for the treatment of inflammatory acne vulgaris has to be pain-free and there can be no associated

PDT effect for this to be an accepted, widely used therapeutic modality, at least in the opinion of this author.

10.7. Home Devices for Acne Vulgaris

Over the past several years, much interest has turned to the consumer market and to the potential to have available safe and effective home devices which could transform the entire acne market away from the physician as the provider to the patient utilizing home devices to treat their own acne vulgaris at home. Several home devices have made their way to the market and do provide effective home use control for inflammatory acne vulgaris. These include the Tanda device (Figure 10.21) and the OmniLux home LED blue and red light sources (Figure 10.22). Clinical trial work has shown that these devices are effective and that they reduce the inflammatory component of acne in those using them.[55–57] These light sources utilize both blue and red LEDs and can produce clinical results in the home arena.

Time will tell if these home devices have a long-lasting effect on the overall treatment options for acne vulgaris. The acne market is extremely large and there seems to be room for many players. And time will also tell if other photosensitizers, even over-the-counter photosensitizers, make it to the market, allowing more and more patients to be treated in this manner.

10.8. Conclusions

Ultimately, dermatologists are the best equipped to care for patients suffering from acne vulgaris. We have all of the tools of the trade at our disposal –

Figure 10.21 Tanda (Pharos Life) home device for treatment of acne vulgaris.

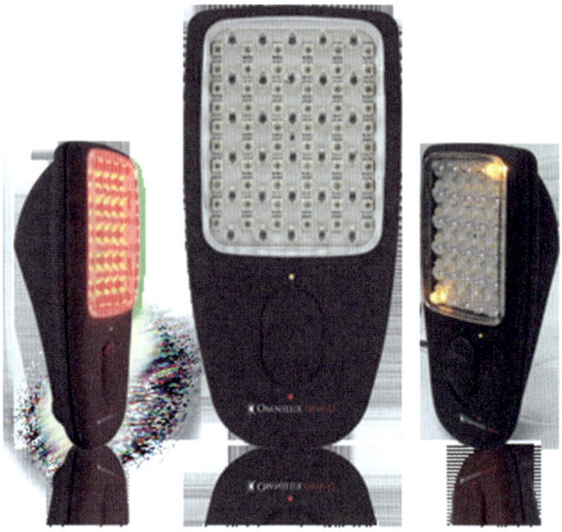

Figure 10.22 OmniLux home LED blue and red light sources for treatment of acne vulgaris.

medicines and technology. We have to determine when to add one to the other – when one medicine is not enough and which device will work for a particular patient. We live in the technology age; we should use technology in our work for the benefit of our patients, and lasers and light sources for acne vulgaris should definitely be high on the list of priorities.

PDT for acne vulgaris works, and it works well. The major limiting factor, at the time of this writing, is that this is a non-covered insurance benefit, which means that the patient is ultimately responsible for the cost of the procedure. All lasers and light therapies are out-of-pocket expenses for our patients and with PDT; we are adding an additional expense that many might not be able to afford. Therefore, it is prudent that we offer these therapies to our patients with that understanding, use the compassion that medicine has taught us and utilize these therapies when it is an appropriate thing to do.

References

1. J.J. Leyden, Therapy for acne vulgaris, *N. Engl. J. Med.*, 1997, **336**, 1156–1162.
2. J.J. Leyden, Oral isotretinoin. How can we treat difficult acne patients?, *Dermatology*, 1997, **195**(S1), 29–33.
3. M.H. Gold, Acne and PDT: new techniques with lasers and light sources, *Lasers Med. Sci.*, 2007, **22**, 67–72.
4. H. Gollnick, W. Cunliffe, D. Berson, B. Dreno, A. Finlay, J.J. Leyden, A. R. Shalita and D. Thiboutot, Global Alliance to Improve Outcomes in Acne, *Management of acne: a report from a Global Alliance to Improve Outcomes in Acne, J. Am. Acad. Dermatol.*, 2003, **49**(1S), S1–S37.
5. P. Coates, S. Vyakrnam, E.A. Eady, C.E. Jones, J.H. Cove and W.J. Cunliffe, Prevalence of antibiotic-resistant propionibacteria on the skin of acne patients:

10-year surveillance data and snapshot distribution study, *Br. J. Dermatol.*, 2002, **146**, 840–848.
6. C.M. Velicer, S.R. Heckbert, J.W. Lampe, J.D. Potter, C.A. Robertson and S.H. Taplin, Antibiotic use in relation to the risk of breast cancer, *JAMA*, 2004, **291**, 827–835.
7. C.N. Ellis and K.J. Krach, Uses and complications of isotretinoin therapy, *J. Am. Acad. Dermatol.*, 2001, **45**, S150–S157.
8. A. Kawada, Y. Aragane, H. Kameyama, Y. Sangen and T. Tezuka, Acne phototherapy with a high-intensity, enhanced, narrow-band, blue light source: an open study and in vitro investigation, *J. Dermatol. Sci.*, 2002, **30**, 129–135.
9. A.R. Shalita, Y. Harth and M. Elman, Acne photoclearing using a novel, high-intensity, enhanced, narrow-band, blue light source, *Clin. Application Notes*, 2003, **9**, 1.
10. M.H. Gold, The utilization of ALA-PDT and a new photoclearing device for the treatment of severe inflammatory acne vulgaris – results of an initial clinical trial, *J. Lasers Surg. Med.*, 2003, **15**(S), 46.
11. P. Papageorgiou, A. Katsambas and A. Chu, Phototherapy with blue (415 nm) and red (660 nm) light in the treatment of acne vulgaris, *Br. J. Dermatol.*, 2000, **142**, 973–978.
12. H. Meffert, K. Gaunitz, T. Gutewort and U.J. Amlong, Therapy of acne with visible light: decreased irradiation time by using a blue-light high energy lamp, *Dermatol. Monatsschr.*, 1990, **176**, 597–560.
13. M.P. Goldman, D. Atkin and S. Kincaid, PDT/ALA in the treatment of actinic keratoses: real world experience, *J. Lasers Surg. Med.*, 2002, **14**(S), 24.
14. M.H. Gold and M.P. Goldman, A multicenter clinical evaluation of the treatment of mild to moderate inflammatory acne vulgaris of the face with visible blue light in comparison to topical 1% clindamycin antibiotic solution, *J. Drugs Dermatol.*, 2005, **4**(1), 64–70.
15. C.C. Dierickx, Lasers, light and radiofrequency for treatment of acne, Med. Laser Appl., 2004, **19**, 196–204.
16. C. Zane, R. Capezzera, A. Pedretti, E. Facchinetti and P. Calzavara-Pinton, Non-invasive diagnostic evaluation of phototherapeutic effects of red light phototherapy of acne vulgaris, *Photodermatol. Photoimmunol. Photomed.*, 2008, **24**(5), 244–248.
17. J.I. Na and D.H. Suh, Red light phototherapy alone is effective for acne vulgaris: randomized, single-blinded clinical trial, *Dermatol. Surg.*, 2007, **33**(10), 1228–1233, discussion 1233.
18. L.E. Bowes, D. Manstein and R.R. Anderson, Effect of 532 nm KTP laser exposure on acne and sebaceous glands, *Lasers Med. Sci.*, 2003, **18**, S6–S7.
19. E.D. Seaton, A. Charakida, P.E. Mouser, I. Grace and R.M. Clement, A.C. Chu. Pulsed-dye laser treatment for inflammatory acne vulgaris: randomized controlled trial, *Lancet*, 2003, **362**, 1347–1352.
20. J.S. Orringer, S. Kang, T. Hamilton, W. Schumacher, S. Cho, C. Hammerberg, G.J. Fisher, D.J. Karimipour, T.M. Johnson and J.J. Voorhees, Treatment of acne vulgaris with a pulsed dye laser: a randomized controlled trial, *JAMA*, 2004, **291**, 2834–2839.
21. M. Elman and G. Lask, The role of pulsed light and heat energy (LHE™) in acne clearance, *J. Cosmet. Laser Ther.*, 2004, **6**, 91–95.
22. C.C. Dierickx, Treatment of acne vulgaris with a variable-filtration IPL system, *Lasers Surg. Med.*, 2004, **S16**, 66.

23. G.S. Munavalli and R.A. Weiss, Rapid acne regression using photopneumatic (PPX) therapy, *Lasers Surg. Med.*, 2007, **S19**, 22.
24. A.T. Shamban, M. Enokibori, V. Narurkar and D. Wilson, Photopneumatic technology for the treatment of acne vulgaris, *J. Drugs Dermatol.*, 2008, **7**(2), 139–145.
25. M.H. Gold and J. Biron, Efficacy of a novel combination of pneumatic energy and broadband light for the treatment of acne, *J. Drugs Dermatol.*, 2008, **7**(7), 639–642.
26. D.Y. Paithankar, E.V. Ross, B.A. Saleh, M.A. Blair and B.S. Graham, Acne treatment with a 1,450 nm wavelength laser and cryogen spray cooling, *Lasers Surg. Med.*, 2002, **31**, 106–114.
27. P.M. Friedman, M.H. Jih, A. Kimyai-Asadi and L.H. Goldberg, Treatment of inflammatory facial acne vulgaris with the 1450-nm diode laser: a pilot study, *Dermatol. Surg.*, 2004, **30**, 147–151.
28. M. Sherling, P.M. Friedman, R. Adrian, A.J. Burns, H. Conn, R. Fitzpatrick, R. Gregory, S. Kilmer, G. Lask, V. Narurkar, T.M. Katz and M. Avram, Consensus recommendations on the use of an erbium-doped 1,550-nm fractionated laser and its applications in dermatologic laser surgery, *Dermatol. Surg.*, 2010 Apr, **36**(4), 461–469.
29. V.V. Tuchin, E.A. Genina, A.N. Bashkatov, G.V. Simonenko, O.D. Odoevskaya and G.B. Altshuler, A pilot study of ICG laser therapy of acne vulgaris: photodynamic and photothermolysis treatment, *Lasers Surg. Med.*, 2003, **3**, 296–310.
30. J.R. Lloyd and M. Mirkov, Selective photothermolysis of the sebaceous glands for acne treatment, *Lasers Surg. Med.*, 2002, **31**, 115–120.
31. M. Kassir, D. Newton, M. Maris, R. Euwer and P. Servell, Er:glass (1.54 mm) laser for the treatment of facial acne vulgaris, *Lasers Surg. Med.*, 2004, **34**, S65.
32. J. Ruiz-Esparza and J.B. Gomez, Nonablative radiofrequency for active acne vulgaris: the use of deep dermal heat in the treatment of moderate to severe active acne vulgaris (thermotherapy): a report of 22 patients, *Dermatol. Surg.*, 2003, **29**, 333–339.
33. M.H. Gold, Aminolevulinic acid photodynamic therapy: medical evidence for its expanded use, *Exp. Rev. Medical Dev.*, 2006, **3**(3), 357–371.
34. M.H. Gold, Novel treatment options for severe inflammatory acne vulgaris, *Expert Rev. Dermatol.*, 2006, **1**(1), 13–23.
35. W. Hongcharu, C.R. Taylor, Y. Chang, D. Aghassi, K. Suthamjariya and R.R. Anderson, Topical ALA-photodynamic therapy for the treatment of acne vulgaris, *J. Invest. Dermatol.*, 2000, **115**(2), 183–192.
36. Y. Itoh, Y. Ninomiya, S. Tajima and A. Ishibashi, Photodynamic therapy for acne vulgaris with topical 5-aminolevulinic acid, *Arch. Dermatol.*, 2000, **136**(9), 1093–1095.
37. Y. Itoh, Y. Ninomiya, S. Tajima and A. Ishibashi, Photodynamic therapy of acne vulgaris with topical delta-aminolaevulinic acid and incoherent light in Japanese patients, *Brit. J. Dermatol.*, 2001, **144**, 575–579.
38. M.P. Goldman, Using 5-aminolevulinic acid to treat acne and sebaceous hyperplasia, *Cosmet. Dermatol.*, 2003, **16**, 57–58.
39. M.H. Gold, The utilization of ALA-PDT and a new photoclearing device for the treatment of severe inflammatory acne vulgaris – results of an initial clinical trial, *J. Lasers Surg. Med.*, 2003, **15**(S), 46.
40. M.P. Goldman and S. Boyce, A single-center study of aminolevulinic acid and 417nm photodynamic therapy in the treatment of moderate to severe acne vulgaris, *J. Drugs Dermatol.*, 2003, **2**, 393–396.

41. A.F. Taub, Photodynamic therapy for the treatment of acne: a pilot study. *J. Drugs Dermatol.*, 2004, **3**, S10–S14.
42. M.H. Gold, The use of a novel intense pulsed light and heat source and ALA-PDT in the treatment of moderate to severe inflammatory acne vulgaris, *J. Drugs Dermatol.*, 2004, **3**(6), S15–S19.
43. A.V. Santos, V.G. Belo and G. Santos, Effectiveness of photodynamic therapy with topical 5-aminolevulinic acid and intense pulsed light versus intense pulsed light alone in the treatment of acne vulgaris: comparative study, *Dermatol. Surg.*, 2005, Aug, **31**(8 Pt 1), 910–915.
44. J. Rojanamatin and P. Choawawanich, Treatment of inflammatory facial acne vulgaris with intense pulsed light and short contact of topical 5-aminolevulinc acid: a pilot study, *Dermatol. Surg.*, 2006, **32**, 991–997.
45. S.H. Oh, D.J. Ryu, E.C. Han, K.H. Lee and J.H. Lee, A comparative study of topical 5-aminolevulinic acid incubation times in photodynamic therapy with intense pulsed light for the treatment of inflammatory acne, *Dermatol. Surg.*, 2009, Dec, **35**(12), 1918–1926.
46. M.R. Alexiades-Armenakas, Long-pulsed dye laser-mediated photodynamic therapy combined with topical therapy for mild-to-severe comedonal, inflammatory, or cystic acne, *J. Drugs Dermatol.*, 2006, **5**, 45–55.
47. A. Miller and A. Van Camp, Treatment of acne vulgaris with photodynamic therapy: The use of aminolevulinic acid and green light, *Cosm. Derm.*, 2006, **19**(10), 624–627.
48. S.R. Wiegell and H.C. Wulf, Photodynamic therapy of acne vulgaris using methyl aminolaevulinate: a blinded, randomized, controlled trial, *Br. J. Dermatol.*, 2006, **154**, 969–976.
49. C. Hörfelt, J. Funk, M. Frohm-Nilsson, D. Wiegleb Edström and A.M. Wennberg, Topical methyl aminolaevulinate photodynamic therapy for treatment of facial acne vulgaris: results of a randomized, controlled study, *Br. J. Dermatol.*, 2006, **155**, 608–613.
50. S.R. Wiegell and H.C. Wulf, Topical methyl aminolaevulinate photodynamic therapy for treatment of facial acne vulgaris: results of a randomized, controlled study, *Br. J. Am. Acad. Dermatol.*, 2006, **54**(4), 647–651.
51. E. Christensen, T. Warloe, S. Kroon, J. Funk, P. Helsing, A.M. Soler, H.J. Stang, O. Vatne and C. Mørk; Norwegian Photodynamic Therapy (PDT) Group, Guidelines for practical use of MAL-PDT in non-melanoma skin cancer, *J. Eur. Acad. Dermatol. Venereol.*, 2010, May, **24**(5), 505–512, Epub 2009 Oct 6.
52. A. Kasche, S. Luderschmidt, J. Ring and R. Hein, Photodynamic therapy induces less pain in patients treated with methyl aminolevulinate compared to aminolevulinic acid, *J. Drugs Dermatol.*, 2006, Apr, **5**(4), 353–356.
53. M.H. Gold, 5-Aminolevulinic acid photodynamic therapy versus methyl aminolevulinate photodynamic therapy for inflammatory acne vulgaris, *J. Am. Acad. Dermatol.*, 2008, **58**(2S), 60–62.
54. M.H. Gold, A. Andriessen, J. Biron and H. Andriessen, Clinical efficacy of self-applied blue light therapy for mild-to-moderate facial acne, *J. Clin. Aesthet. Dermatol.*, 2009, **2**(3), 44–50.
55. M.H. Gold, A. Andriessen and J. Biron, Self-diagnosis of mild-to-moderate acne for self treatment with blue light therapy, *J. Clin. Aesthet. Dermatol.*, 2009, **2**(4), 40–44.
56. N.S. Sadick, Handheld LED array device in the treatment of acne vulgaris, *J. Drugs Dermatol.*, 2008, **7**(4), 347–350.

Chapter 11

PDT for Viral Lesions

Michael H. Gold

Medical Director, Gold Skin Care Center, Tennessee Clinical Research Center, Clinical Assistant Professor, Vanderbilt University School of Medicine, Department of Dermatology, Vanderbilt University School of Nursing, Nashville, TN, USA; Visiting Professor of Dermatology, Huashan Hospital, Fudan University, Shanghai, China; No. 1 Hospital of China Medical University, Shenyang, China

Table of Contents

Photodynamic Inactivation of Microbial Pathogens: Medical and Environmental Applications
Editors Michael R. Hamblin and Giulio Jori

Published by the Royal Society of Chemistry, www.rsc.org

11.1. Introduction

Photodynamic therapy (PDT) has been used to treat a variety of dermatologic skin concerns, many of which have been previously described in this textbook. One of the dermatologic disorders described in the medical literature with the use of PDT is its role in the treatment of recalcitrant verrucae, including the treatment of condylomata acuminata, and for the treatment of molluscum contagiosum. Various case reports and series have been published with the use of PDT for these conditions and this chapter will review these reports and show how PDT can be used for these conditions.

As is the case with many of the entities PDT is commonly used for, this remains an off-label use for the two drugs commonly used in the US for PDT, ALA and MAL. The Food and Drug Administration (FDA) approval for 20% 5-ALA in the United States (US) is for the treatment of non-hyperkeratotic AKs of the face and scalp, utilizing a blue light source and a drug incubation time of 14–18 hours.[1] This drug is commonly referred to as Levulan® Kerastick® (Dusa Pharmaceuticals, Wilmington, MA). The FDA approval for methyl ALA (MAL) is for the non-hyperkeratotic AKs of the face and scalp after a three-hour, under-occlusion drug incubation period and treatment with red light.[1] This drug is commonly referred to as Metvix outside of the US and Metvixia in the US (Galderma Laboratories, Fort Worth TX, Paris, France).

11.1.1. Verrucae, Condylomata and Molluscum

Verrucae are double-stranded DNA viruses, which are produced by the human papilloma virus, commonly known as HPV. HPV is a member of the papova viral family. HPV, or verrucae, are seen almost every day in dermatology offices and have been identified to occur on virtually all body parts. Over 70 genotypes of verrucae have been identified. Both skin and mucosal surfaces are affected by these lesions. Verrucae, commonly referred to as warts, are clinically identified as being papular to nodular structures which have a horny layer on their surface; they range in size from 1–2 mm to several centimetres and may become confluent forming even larger lesions. These lesions are usually asymptomatic; however, pain and functional abnormalities have been reported, especially with lesions which arise on the hands and on the feet. Epidemiologic studies have shown up to 22% of school-age children may develop verrucous lesions and upwards of 50–95% of renal transplant patients may develop verrucous lesions, some of which may progress towards squamous cell carcinomas and therefore require aggressive treatment strategies, of which PDT may play an important role.[2]

The histology of verrucous lesions shows epidermal hyperplasia, acanthosis, papillomatosis, hyperkeratosis and parakeratosis, as well as thrombosed capillaries within the dermal papillae. Elongated rete ridges curve, in most incidences, to the centre of the lesion. Viral replication takes place in differentiated keratinocytes in or above the stratum granulosum. Vacuolated cells, known as koilocytotic cells, are found in the mid-upper dermis.[2]

The treatments of verrucae are numerous and varied, and often discouraging. Numerous manuscripts have summarized treatment options for verrucae. Treatment modalities usually focus on the physical destruction of the individual lesion, *i.e.* the viral cell, and include cryotherapy,[3] curettage, excision, carbon dioxide laser ablation,[4] pulsed dye laser therapy,[5] liquid nitrogen, electrosurgery and the application of a variety of topical acid preparations.[6,7] Other therapies reported in the literature include the infrared coagulation,[8] alpha-interferon, topically applied 5-flurouracil,[9] intralesional bleomycin[10] and topically applied DNCB.[11] Despite these treatment modalities, some lesions remain recalcitrant to the therapies, and some of these therapies are so painful as to not be practical in everyday clinical medicine.

Condylomata acuminata are anogenital verrucous lesions caused by the HPV virus. They have been related to genital carcinomas and so treatment is essential to prevent this potential conversion into carcinoma. Treatment options are similar to what has been described for recalcitrant verrucae.[12]

Molluscum contagiosum are also caused by viruses although we know they are caused by a different family of viruses than verrucae. They are large DNA pox viruses. Molluscum is commonly seen in children and has shown a resurgence in those afflicted with HIV disease, especially when the viral load in these individuals is low.[13–18] They are clinically described as being discrete skin-coloured smooth papules with an umbilicated centre. These lesions occur on both the skin and mucous membranes. In children, lesions are reported to occur on all body surfaces; those in HIV-positive individuals are more commonly described on the head and neck area. Those associated with HIV are also reported to be larger than the typical small papular lesions seen in children and are, in general, more recalcitrant to therapeutic options.[19]

The incidence of molluscum lesions is reported to be as high as 5% in children and up to 5–18% in the HIV and immunocompromised population. Histologic examination shows a hypertrophied and hyperplastic epidermis. Above the basal layer there are lobules of enlarged epidermal cells with inclusion bodies, commonly known as molluscum bodies. The inclusion bodies contain the viral particles.[17,20–25]

Treatment options for molluscum are numerous and similar to those described for verrucous lesions. Typical treatment includes liquid nitrogen, cantharidin, tretinoin cream, podophyllin 20–25%, salicylic acid, tincture of iodine, silver nitrate, trichloroacetic acid and surgery with curettage with or without electrical desiccation. In HIV-positive individuals, antiretroviral therapy, intralesional alpha interferon and injection of streptococcal antigen OK-432 have been reported to be useful in treating molluscum in these lesions. Despite these modalities, recalcitrant lesions are also often reported.[16]

11.2. PDT and Its Use in Verrucae, Condylomata and Molluscum

11.2.1. Verrucae

PDT has been used in the treatment of recalcitrant verrucous lesions and molluscum lesions over the past ten years. The remainder of this chapter will review the various case reports and series which have been published with ALA, MAL, showing its value in treating recalcitrant disease.

In the first series published in 1995, Ammann *et al.*[26] reported findings utilizing PDT in six patients with refractory verrucae vulgaris lesions of the hands. The patients had all failed conventional therapy. The duration of the lesions varied from 2–10 years. The group utilized a 20% ALA in an oil-in-water emulsion which was applied under occlusion for 5–6 hours before being illuminated utilizing a slide projector light source for 30 minutes. The patients were followed for up to 2 months following their therapy. The treated areas in all the patients studied showed an acute inflammatory skin reaction following the treatments. In five of the individuals, there was no change in the verrucae; however, in one, complete resolution was achieved. The authors noted PDT was not a success in the treatment of verrucae, but they were encouraged by the one treatment success and encouraged further clinical evaluations.

In 1997, Smetana *et al.*[27] reported on their experiences with both an *in vivo* and an *in vitro* analysis utilizing PDT for both verrucae and molluscum patients. In their verrucae case report, a 15-year kidney transplant patient presented with recalcitrant verrucae on the hands. A 20% ALA cream containing 2% EDTA and 2% DMSO was applied to the affected areas and incubated for a period of 4 hours. Red light was used to activate the photosensitizer, at 120 J cm^{-2}. They noted dramatic improvement at the one-month follow-up period. No recurrence was noted two years later. Their second patient, an HIV-positive individual, presented with numerous molluscum lesions on the face. Utilizing a similar protocol, the authors achieved similar clearance at the one-month follow-up period. The authors concluded that ALA-PDT may be a useful modality for recalcitrant verrucae and molluscum, especially with the unique ingredients that were added to the ALA preparation.

Stender *et al.*,[28] in 2000, published findings in 232 hand and foot verrucous lesions found in 45 individuals in the first placebo-controlled clinical trial for ALA-PDT in the treatment of verrucous lesions. They randomized 117 verrucous lesions to receive ALA-PDT and 115 to receive placebo-PDT. Each lesion was treated with occlusion of the affected area for 4 hours and the lesions were irradiated with a red light source (Waldman PDT 1200, Waldmann-Medizin-Technik, Villingen-Schwenningen, Germany), with a wavelength in the range of 590–700 nm. The lesions were exposed to 50 J cm^{-2} for 23 minutes and 20 seconds, yielding a total dose of 70 J cm^{-2}. All of the verrucae were treated at baseline, at 1 and at 2 weeks, with another treatment regimen of three treatments given one month later if clearance wasn't achieved. Follow-up was at one and two months following the last treatment. Their results showed that at week 14 there was a relative reduction in wart area of 98% in the ALA group *vs.* 52% in the placebo group. At week 18, the relative reduction in the ALA group was 100% *versus* 71% in the placebo group. Pain was more evident in the ALA treated group compared to the placebo group. The authors concluded their study supported the use of ALA-PDT in the treatment of recalcitrant verrucae.

Fabbrocini *et al.*,[29] in 2001, reported their experience with recalcitrant plantar warts in a controlled placebo clinical trial. They evaluated 67 patients utilizing PDT in these recalcitrant lesions. Each lesion was pre-treated for seven days with a topical ointment containing 10% urea and 10% salicylic acid in order to remove the superficial hyperkeratotic layers commonly seen in these lesions. A gentle curettage was also used prior to the application of the ALA; the ALA was a 20% cream preparation utilizing Eucerin® cream as its base. The ALA preparation was occluded for 5 hours prior to light therapy. Sixty-four lesions received the ALA cream, while 57 verrucae received only the Eucerin® cream base. The light source used was a tungsten lamp that emitted a spectrum of light from 400–700 nm, with a peak at 630 nm. The power incidence was 50 mW cm^{-2}, at a distance of 10 cm. Patients received one treatment and, if not clear, received up to two more therapies at one-week intervals. The patients were then followed for 22 months following their last therapy. The results showed that 48 of 64 warts (75%) completely healed compared to 13 out of the 57 lesions in the placebo group (22.8%). Of interest, 47.9% of the ALA lesions cleared following one treatment, 31.3% required two treatments and only 20.8% needed three treatments for clearance to occur. The authors concluded that ALA was successful in the treatment of recalcitrant verrucae.

Stender *et al.*,[30] in 1999, reported further work on the use of ALA-PDT in the treatment of recalcitrant verrucae of the hands and feet. They evaluated 30 individuals with a total of 250 verrucous lesions and randomized patients to receive one of five treatment protocols: ALA-PDT with white light three times in ten days; ALA-PDT with white light for one treatment; ALA-PDT with red light for three treatments in ten days; ALA-PDT with blue light for three treatments in ten days; and cryotherapy, up to four treatments within a two-month period. The ALA utilized was a 20% ALA preparation; each lesion was incubated for 5 hours under occlusion. The areas treated were illuminated with light from slide projectors (halogen lamp) at a distance of 15 cm, with a total dose given of 40 J cm^{-2}. The treatments were repeated if the lesions were not completely cleared after the first course. The results from the study showed complete clearance of 73% with white light with three treatments, 71% after one white light treatment, 42% after red light therapy, 23% after blue light therapy and 20% after cryotherapy. Of the areas that cleared, no recurrences were noted after a 12-month follow-up period. The authors concluded that ALA-PDT was a useful modality for the treatment of recalcitrant verrucae of the hands and feet.

Mizuki *et al.*,[31] in 2003, reported on the use of ALA-PDT in a 13-year-old Japanese boy who had a two-year history of multiple "plane" warts of the face, unresponsive to previous therapies. They used a 20% ALA oil-in-water emulsion and a single 500-W metal halide lamp with peaks of 630 nm and 700 nm. The ALA was incubated for 6 hours under occlusion; illumination was for 20 minutes for an energy of 120 J cm^{-2}. The patient underwent two ALA-PDT sessions. Five months after the final treatment, the areas remained disease free. This suggested ALA-PDT was useful for the treatment of recalcitrant plane warts.

Smucler and Jatsova,[32] in 2005, reported on the use of a pulsed dye laser (PDL) with ALA in the treatment of verrucous lesions. They compared PDL alone *versus* PDL plus PDT as well as PDT and an LED light source. They found that all three of their protocols were effective in treating these recalcitrant lesions, but the combination of PDL plus PDT resulted in the highest cure rate with the shortest number of treatments. PDL cured 81% of the lesions treated (n = 112), with a mean number of sessions being 3.34. PDL plus PDT cured 100% (n = 86) with a mean number of sessions being 1.95. PDT plus LED cured 96% of the lesions (n = 76), with the mean number of sessions being 2.53. The authors strongly suggested this combined modality to become a treatment of choice for viral warts.

Gold and Pope[33] reported a case of fractional laser resurfacing and the use of ALA-PDT in treating a recalcitrant verrucous lesion of the foot in 2008. The patient had presented with a verrucous lesion on the heel of the foot which did not respond to conventional therapy, including the use of topical medications, liquid nitrogen, PDL treatments and conventional PDT using both short- and long-contact ALA and treatment with the PDL. The patient was successfully treated with an erbium:YAG fractional laser, which served as an enhancement for the delivery of ALA into the skin. Twenty percent 5-ALA solution was then applied and occluded for 24 hours. After two PDL treatments, the verrucous lesion had resolved. The use of the erbium:YAG fractionated laser delivered fractionated microthermal zones into the skin which, as noted, enhanced the delivery of ALA into the thick skin of the foot, making the PDT reaction work better than what had previously been performed. The results are shown in Figure 11.1.

11.2.2. Condylomata

Several reports have shown the successes of PDT in the treatment of condylomata acuminata. In 2002, Stefanaki *et al.*[34] reported on the use of ALA-PDT in the treatment of 12 male patients with condylomata acuminata. The lesions were treated with 20% ALA and irradiated with optimal doses of red light at 70 J cm^{-2} and 100 J cm^{-2}. The overall cure rate reported was 72.9% at the 12-month follow-up period. Wang *et al.*,[35] in 2004, reported on 164 patients with urethral condylomata acuminata treated with 10% ALA. Complete response rates at 6–24 months were 95.12% with recurrence rates of 5.13%. Chen *et al.*,[36] in 2007, reported their experiences with ALA-PDT *versus* conventional CO_2 laser therapy in patients with condylomata acuminata. Sixty-five patients received ALA-PDT and 21 patients CO_2 laser therapy. The complete response was 95% after one treatment, with a recurrence rate noted to be 6.3% at 12 weeks. This was better than conventional CO_2 laser therapy and adverse events were also noted to be lower in the PDT group. And Liang *et al.*,[37] in 2009, reported on 90 patients treated with ALA-PDT (n = 67) or conventional CO_2 laser therapy (n = 23). In this study, 20% ALA was used with a three-hour drug incubation. Light irradiation with red light of 100 J cm^{-2} at

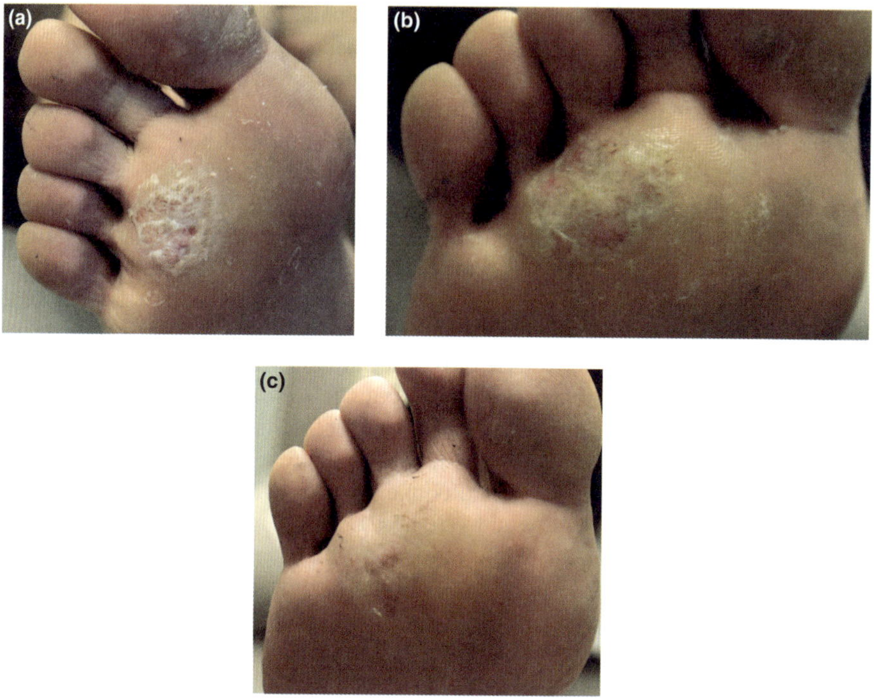

Figure 11.1 Clinical example of fractional laser resurfacing and the use of ALA-PDT in treating a recalcitrant verrucous lesion of the foot; before treatment (a) and one month following the first treatment with marked reduction in size (b); one month after second treatment with complete resolution of lesion (c).

100 mW was applied to the lesions, and up to three weekly treatments were given. Complete clearance was noted in 95.93% of the ALA group and 100% in the CO_2 laser group one week after the last treatment but recurrence rates and adverse events were much less in the ALA treated group.

11.2.3. Molluscum

Gold *et al.*[38] reported on the successful use of ALA-PDT in the treatment of recalcitrant molluscum contagiosum in an HIV patient in 2004. They treated an HIV-positive individual who had numerous large face and neck molluscum lesions with four sessions of ALA-PDT utilizing the original FDA treatment protocol. The ALA was applied to individual lesions and allowed to incubate for 16 hours before exposure to a blue light source (ClearLight, CureLight, Yokneam, Israel) for 16 minutes and 40 seconds. The therapy was reported painful during the illumination, so cool water and a fan were utilized to ease this discomfort. There was a significant inflammatory response following each treatment, which required systemic and topical corticosteroids for relief.

The treatments were performed at two-week intervals and a dramatic response was seen after the fourth treatment. Unfortunately, he succumbed to his HIV disease during a trip that he noted was made possible by the use of the PDT on his recalcitrant molluscum lesions, and his friends conveyed back to us his gratitude for clearing his skin to allow him to take this journey, which turned out to be the last travels of his life.

Moiin[39] evaluated 40 patients with molluscum lesions treated with ALA-PDT and a blue light source, the BluU® (Dusa Pharmaceuticals, Wilmington, MA). Six of the individuals had a history of HIV infection, and were the basis of his report. The ALA was applied broadly to the affected areas, incubated from 14–24 hours and treated with the blue light for 16 minutes and 40 seconds. All of the patients studied improved in their lesion count. In two patients, one treatment produced an improvement in the lesion count of 60%. Furthermore, patients who received 3–5 treatments were found to have a 75–80% reduction in the number of lesions counted. The therapy did have adverse events, including erythema, edema, vesiculation, hyperpigmentation, hypopigmentation, pain, stinging/burning and itching. The only deliberately intense side effect was a burning sensation described by one patient when placed under the blue light source. Only one child did not tolerate the procedure fully due to erythema formation. The conclusion was that PDT is a viable option for treating molluscum contagiosum in HIV-positive patients and immunocompromised children. All of the patients expressed preference of ALA-PDT over previous therapies for their molluscum.

11.3. Conclusions

The use of ALA-PDT has been shown to be successful for the treatment of recalcitrant verrucae, condylomata acuminate and molluscum contagiosum lesions. Studies have shown that ALA-PDT is a useful modality for these lesions and should be considered when confronted with a patient with these viral conditions. Further studies are warranted to further evaluate protocols used, light sources and incubation times, to determine how to make the therapy even more appealing to clinicians and patients alike.

References

1. E.W. Jeffes, J.L. McCullough, G.D. Weinstein, R. Kaplan, S.D. Glazer and J.R. Taylor, Photodynamic therapy of actinic keratoses with topical aminolevulinic acid hydrochloride and fluorescent blue light, *J. Am. Acad. Dermatol.*, 2001, **45**, 96–104.
2. I.M. Stender, Treatment of human papilloma virus, in *Photodynamic Therapy*, ed. M.P. Goldman, Elsevier, Munich, Germany, 2005, 77–88.
3. J.F. Bourke, J. Berth-Jones and P.E. Hutchinson, Cryotherapy of common viral warts at intervals of 1, 2, and 3 weeks, *Br. J. Dermatol.*, 1995, **132**, 433–436.
4. R.A. Logan and C.B. Zachary, Outcome of carbon dioxide laser therapy for persistent cutaneous viral warts, *Br. J. Dermatol.*, 1989, **121**, 99.

5. O.T. Tan, T.M. Hurwitz and T.J. Stafford, Pulsed dye laser treatment of recalcitrant verrucae, a preliminary report, *Lasers Surg. Med.*, 1993, **13**, 127–137.
6. R. Hirose, M. Hori, T. Shukuwa, M. Udono, M. Yamada, T. Koide and H. Yoshida, Topical treatment of resistant warts with glutaraldehyde, *J. Dermatol.*, 1994, **21**, 248–253.
7. M.H. Bunney, M.W. Nolan and D.A. Williams, An assessment of methods of treating viral warts by comparative treatment trials based on standard design, *Br. J. Dermatol.*, 1976, **94**, 667–680.
8. C.L.G. Halasz, Treatment of common warts using the infrared coagulator, *J. Dermatol. Surg. Oncol.*, 1994, **20**, 252–256.
9. R.T. Brodell and D.L. Breadle, The treatment of palmar and plantar warts using natural alpha interferon and a needleless injector, *Dermatol. Surg.*, 1995, **21**, 213–218.
10. M.P. James, P.M. Collier, W. Aherne, A. Hardcastle and S. Lovegrove, Histologic, pharmacologic and immunocytochemical effects of injection of bleomycin, *J. Am. Acad. Dermatol.*, 1993, **28**, 933–937.
11. K.C. Shah, R.M. Patel and D.P. Umrigar, Dinitroclorobenzene treatment of verrucae plana, *J. Dermatol.*, 1991, **18**, 639–642.
12. P. Schneede, Genital human papillomarvirus infections, *Curr. Opin. Urol.*, 2002, **12**, 57–61.
13. E.C. Siegfried, Warts and molluscum contagiosum on children: an approach to therapy, *Dermatol. Ther.*, 1997, **2**, 51–67.
14. J. Verbov, How to manage warts, *Arch. Der. Child.*, 1999, **80**, 97–99.
15. E.J. Lewis, M. Lam and C.E. Crutchfield, An update on molluscum contagiosum, *Cutis*, 1997, **60**, 29–34.
16. K. Husar and M. Skerlev, Molluscum contagiosum from infancy to maturity, *Clin. Dermatol.*, 2002, **20**, 170–172.
17. B.M. Coldiron and P.R. Bergstresser, Prevalence and clinical spectrum of skin disease in patients infected with human immunodeficiency virus, *Arch. Dermatol.*, 1989, **125**, 357–361.
18. A. Czelusta, A. Yen-Moore, M. Van der Straten, D. Carrasco and S.K. Tyring, An overview of sexually transmitted diseases. Part III. Sexually transmitted diseases in HIV-infected patients, *J. Am. Acad. Dermatol.*, 2000, **43**, 409–432.
19. J.J. Schwartz and P.L. Myskowsk, Molluscum contagiosum in patients with human immunodeficiency virus infection: a review of twenty-seven patients, *J. Am. Acad. Dermatol.*, 1992, **27**, 583–588.
20. D.R. Lowy, Molluscum contagiosum, in *Dermatology in General Medicine*, ed. I.M. Freedberg, A.Z. Eisen, K. Wolf, K.F. Austen, L.A. Goldsmith and S.I. Katz, The McGraw-Hill Companies, Inc., New York, 1999, 2478–2481.
21. W.L. Matis, A. Triana, R. Shapiro, L. Eldred, B.F. Polk and A.F. Hood, Dermatologic findings associated with human immunodeficiency virus infection, *J. Am. Acad. Dermatol.*, 1987, **17**(5 Pt 1), 746–751.
22. D.S. Goodman, E.D. Teplitz and A. Wishner *et al.*, Prevalence of cutaneous disease in patients with acquired immunodeficiency syndrome (AIDS) or AIDS-related complex, *J. Am. Acad. Dermatol.*, 1987, **17**(2 Pt 1), 210–220.
23. J.J. Schwartz and P.L. Myskowski, Molluscum contagiosum in patients with human immunodeficiency virus infection. A review of twenty-seven patients, *J. Am. Acad. Dermatol.*, 1992, **27**(4), 583–588.
24. W.L. Epstein, I. Senecal and H. Krasnobrod, Viral antigens in human epidermal tumors: localization of an antigen to molluscum contagiosum, *J. Invest. Dermatol.*, 1963, **40**, 51.

25. J. Kwittken, Molluscum contagiosum: some new histologic observations, *Mt. Sina. J. Med., NY*, 1980, **47**, 583.
26. R. Ammann, T. Hunziker and L.R. Braathen, Topical photodynamic therapy in verrucae, *Dermatol.*, 1995, **191**, 346–347.
27. Z. Smetana, Z. Malik, A. Orenstein, E. Mendelson and E. Ben-Hur, Treatment of viral infections with 5-aminolevulinic acid and light, *Lasers Surg. Med.*, 1997, **21**, 351–358.
28. I.M. Stender, R. Na, H. Fogh, C. Gluud and H.C. Wulf, Photodynamic therapy with 5-aminolevulinic acid or placebo for recalcitrant foot and hand warts: randomized double-blind trial, *Lancet*, 2000, **March**, 355.
29. G. Fabbrocini, M.P. Di Costanzo, A.M. Riccardo, M. Quarto, A. Colasanti, G. Roberti and G. Monfrecola, Photodynamic therapy with topical 5-aminolevulinic acid for the treatment of plantar warts. *J. Photochem. Photobiol. Bio.*, 2001, Aug 15, **61**(1–2), 30–34.
30. I.M. Stender, J. Lock-Andersen and H.C. Wulf, Recalcitrant hand and foot warts successfully treated with photodynamic therapy with topical 5-aminolaevulinic acid: a pilot study, *Clin. Exp. Dermatol.*, 1999, May, **24**(3), 154–159.
31. D. Mizuki, T. Kaneko and K. Hanada, Successful treatment of topical photodynamic therapy using 5-aminolevulinic acid for plane warts, *Br. J. Dermatol.*, 2003, Nov, **149**(5), 1087–1088.
32. R. Smucler and E. Jatsova, Comparative study of aminolevulinic acid photodynamic therapy plus pulsed dye laser versus pulsed dye laser alone in treatment of viral warts, *Photomed. Laser Surg.*, 2005, Apr, **23**(2), 202–205.
33. M.H. Gold and A. Pope, Fractional resurfacing aiding photodynamic therapy of a recalcitrant plantar verrucae – a case report & review of the literature, *J. Clin. Aesthet. Dermatol.*, 2008, **1**(1), 30–33.
34. I.M. Stefanaki, S. Georgiou, G.C. Themelis, E.M. Vazgiouraki and A.D. Tosca, In vivo fluorescence kinetics and photodynamic therapy in condylomata acuminate, *Br. J. Dermatol.*, 2003, **149**, 972–976.
35. X.L. Wang, H.W. Wang, H.S. Wang, S.Z. Xu, K.H. Liao and P. Hillemanns, Topical 5-aminovaevulinic acid-Photodynamic therapy for the treatment of urethralcondylomata acuminate, *Br. J. Dermatol.*, 2004, **151**, 880–885.
36. K. Chen, B.Z. Chang, M. Ju, X.H. Zhang and H. Gu, Comparative study of photodynamic therapy vs CO2 laser vaporization in treatment of condylomata acuminate, a randomized clinical trial, *Br. J. Dermatol.*, 2007, **156**, 516–520.
37. J. Liang, X.N. Lu, H. Tang Z. Zhang, J. Fan and J.H. Xu, Evaluation of photodynamic therapy using topical aminolevulinic acid hydrochloride in the treatment of condylomata acuminata: a comparative, randomized clinical trial. *Photodermatol. Photoimmunol. Photomed.*, 2009, Dec, **25**(6), 293–297.
38. M.H. Gold, The use of ALA-PDT in the treatment of recalcitrant molluscum contagiosum in HIV/AIDS affected individuals, *J. Laser Surg. Med.*, 2003, **15**(supp), 40.
39. A. Moiin and M.H. Gold, Treatment of verrucae vulgaris and molluscum contagiosum with photodynamic therapy, *Dermatologic Clinics*, 2007, **25**, 75–80.

Chapter 12

PDT for Cutaneous Leishmaniasis

Oleg E. Akilov,[a,*] Ulysses W. Sallum[b,*] and Tayyaba Hasan[c]

[a] Instructor, Department of Dermatology, University of Pittsburgh, 200 Lothrop Street, Presbyterian South Tower, Suite 3880, Pittsburgh, PA 15213, USA; [b] Research Fellow, Department of Dermatology, Harvard Medical School, Wellman Center for Photomedicine, Massachusetts General Hospital, 40 Blossom Street, Boston, MA 02114, USA; [c] Professor, Department of Dermatology, Harvard Medical School, Wellman Center for Photomedicine, Massachusetts General Hospital, 40 Blossom Street, Boston, MA 02114, USA

Table of Contents

* These authors contributed equally to this work

Photodynamic Inactivation of Microbial Pathogens: Medical and Environmental Applications
Editors Michael R. Hamblin and Giulio Jori

Published by the Royal Society of Chemistry, www.rsc.org

12.1. Introduction

Leishmania is an intracellular protozoan parasite of the family *Trypanosomatidae*.[1,2] Transmitted by the sandfly, *Leishmania* spp. parasitize in tissue macrophages clinically manifesting as skin lesions, mucocutanous ulceration or hepatosplenomegaly (Figure 12.1). The geographic distribution of *Leishmania* is dependent on its insect vector (*Diptera: Psychodidae*), which is endemic to tropical and subtropical regions of the Western Hemisphere, Asia, Africa and South America. In recent decades, migratory trends among people due to urbanization, political conflicts, *etc.* have resulted in an increase in the incidence of cutaneous leishmaniasis (CL). Current WHO reports estimate the annual global incidence of CL cases to be 1 to 1.5 million.[3,4] In endemic zones (*e.g.* Afghanistan, India, Pakistan, Iran, Iraq, Sudan, Somalia, Kenya) incidence of CL is estimated to be approximately 2% to 3% of the population.[5–8] Often the CL is self-healing but, where it is not, treatment becomes a significant problem.

For cutaneous CL, a primary lesion in CL can be limited by inflammatory infiltrate in the papillary dermis, or can evolve into granuloma clinically presenting as nodules, depending on the concentration of parasites and the host immune response. Generally, the cutaneous infection caused by dermotropic species of *Leishmania* manifest as mild to severe ulceration. The ulcerate lesion on the exposed skin caused by *L. major* can persist for 18 weeks. In contrast,

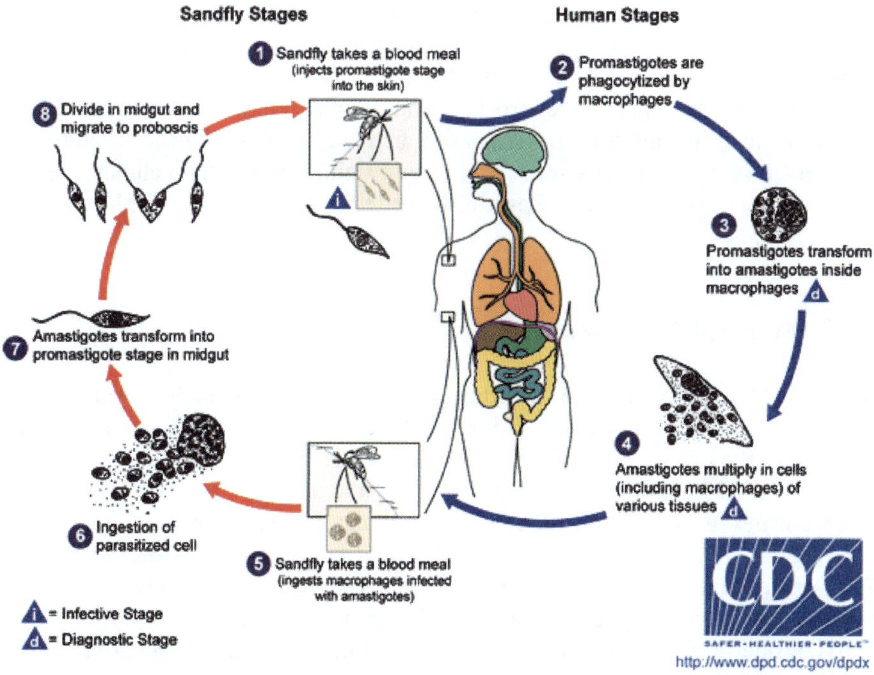

Figure 12.1 Life cycle of *Leishmania*.

L. tropica efflorescence typically lasts for one or two years, but may present for decades.[9] Resolution of the granulomata leads to fibrosis and scarring. The extended duration of the skin lesions and subsequent scarring also cause social consequences for patients.

While numerous treatment modalities exist, there is no ideal therapy for CL.[9] Currently employed treatment options include highly popular physical methods and relatively expensive chemotherapeutic agents. The variety of physical treatments – such as thermotherapy, cryotherapy and electrotherapy – cannot, in most cases, assure adequate parasite destruction and often require additional treatments. Curettage and surgical or laser removal may cause disfiguring scars that are aesthetically undesirable. Thermotherapy does not demonstrate a high curative rate in CL when compared to intramuscular meglumine antimoniate.[10] The procedure is painful, often accompanied by secondary infection with modest cosmetic consequences, and does not have any advantages over other topical therapies. Current systemic therapeutics are limited by adverse pleiotropic effects. Recurrent infection is common, most likely owing to increased drug resistance of the parasite.[11] Thus, a search for an effective, simple and low-cost treatment for CL that can be administered conveniently is still an active topic in the field of biomedical research in applied parasitology.

PDT, an emerging photochemistry-based technology, appears to be applicable to this infection, is also safe and can be made affordable. Topical procedures avoid systemic toxic effects and combine the advantages of both physical treatments and chemotherapy. Various photosensitizers (PSs) have demonstrated light-induced toxicities for different cell types. However, in dermatologic settings, topical formulations of aminolevulinic acid (ALA) or its derivates (methyl aminolevulinate hydrochloride) have been the most commonly investigated.[12] ALA and its derivatives are pro-PSs and require enzymes along the haem biosynthetic pathway for conversion to activatable PSs. The availability of different PSs, with various spectra of activity and preferential cellular sites of accumulation, prompted us to establish strategies for the optimal photokilling of *Leishmania*.

12.2. General Considerations

While PDT is a relatively new indication for CL, efforts made towards optimizing this technology, could bring PDT to the forefront of Leishmaniasis therapy in the near future. Six clinical studies have been completed on the effect of ALA-based PDT for Old World CL in 39 patients with a total of 77 lesions.[13] Mechanisms of action and methods of optimization have been proposed and investigated.[14–17] An ideal PDT regimen would facilitate resorption of pathological tissue and promote self-healing. With respect to minimizing host tissue damage and improving the efficacy of PDT, two general strategies can be taken into consideration: 1) improve the selectivity of the treatment by destroying only the diseased cells, using a PDT regimen directed towards specific cell targets and 2) stimulate the immune system through the application of subcurative PDT, which releases immunogenic antigens for induction of a specific immune response. Most likely, a combination of the two processes will provide the most benefit.

A compact, organized collection of mature mononuclear phagocytes presents difficulties for drug access. A consequence of granuloma formation is the physical restriction of the parasite within the infection site. As a function of this limitation, the granuloma is resistant to passive drug diffusion. The development of therapeutics that are actively transported into the infection could enable improved treatment response. Previously, we showed the importance of the charge interaction between the anionic parasitic membrane and the cationic charge of the PS.[16] Bristow *et al.* also confirmed that positively charged PS yields selective killing of parasites, without significant damage of macrophages or keratinocytes.[18] Nanotechnology-based delivery systems that exploit active transport systems and harbour a cationic charge may present a promising strategy for improving PS delivery into leishmanial granulomata.

An advantage with regards to drug delivery for CL treatment is the presence of an ulcerative defect, which breaches one of the barriers for PS penetration into the lesion, the epidermis, thus providing a portal for drug delivery into the granuloma. Indeed, the level of penetration into granulomatous tissue was 8 times higher through the ulcer floor than through the intact epidermis in our studies.[19] Serous crusts and scabs can delay the drug penetration, thus their removal should be recommended before topical application. Light curettage can also be helpful in providing an improved condition for the drug delivery.

12.3. PDT for CL in Clinical Settings

The first attempt to treat patients with CL with PDT was documented by a group of dermatologists from the Hadassah-Germany Skin Center in 2003.[20] A 34-year-old man from Libya presented with a 5-month history of multiple CL lesions caused by *L. donovani*. Metvix was applied under occlusion for 5 h with a margin of 5 mm. Treatment with red light 57–670 nm with a fluence of 75 J cm^{-2} (150 mW cm^{-2}) was performed 28 times for the first 3 months twice weekly and thereafter once a week for 1 month. The lesions improved after four to six PDT sessions. After 20 sessions, all CL lesions disappeared. A clinical study conducted by Gardlo *et al.* concluded that PDT, in contrast to systemic drug treatments, showed no sign of systemic toxicity and had an excellent cosmetic outcome.[20,21] Asilian *et al.* compared the effectiveness of the topical drug, paromomycin, with PDT and concluded that PDT was more effective.[22] In this study, PDT was found to cure the lesions rapidly as demonstrated by the flattening, reduction in size and ramollissement of the lesions. In none of these studies were the parasite counts reported and therapeutic success was based on the physical dissipation of the lesions. These impressive treatment outcomes in the clinic with ALA-PDT are intriguing, as the efficacy of this PDT strategy requires that the target organism convert ALA, a pro-PS, to protoporhyrin IX, the actual PS, *via* a series of enzymes along the haem biosynthesis pathway.

Seven of the eight enzymes of the haem biosynthetic pathway are missing in most of the *Leishmania* spp.[23] This fact stimulated us to investigate the mechanism of action of ALA-PDT.[15,24] We hoped that such studies might provide mechanistic insights to the efficacy of ALA-PDT, point the way toward enhanced treatment strategies and provide a quantitative basis for the treatment protocol by establishing not only the clinical outcome but also the level of parasite destruction.

12.4. Mechanism of Parasite Destruction in CL by ALA-PDT

Our investigation revealed a great difference between the level of parasite eradication by ALA-PDT achieved using a *L. major* infected Mφ cell culture model as compared to *in vivo*. Although effective reduction of parasite burden was observed *in vivo* there was no significant change observed *in vitro* (Figure 12.2

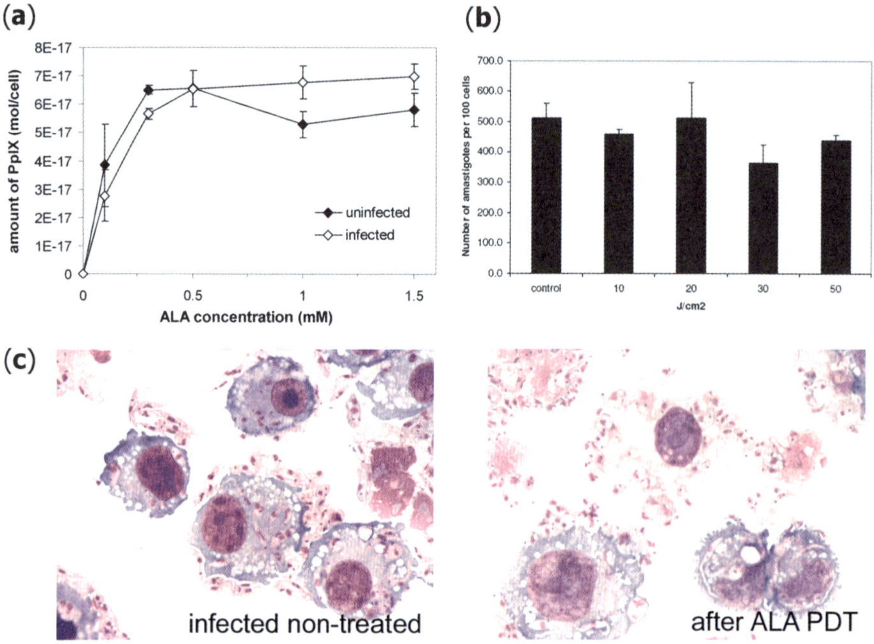

Figure 12.2 ALA-PDT does not decrease the number of parasites in the *in vitro* model of infected Mφ. (a) The quantity of ALA-derived PpIX in *L. major*-infected and non-infected J774.2 cells was similar. J774.2 cells (5×10^5 cells/fish) were co-incubated with 0–1.5 mm of ALA for 4 h at 37 °C. ALA-derived PpIX was quantified as described in Methods. Each point represents the mean ± SD from duplicates. (b) Number of amastigotes per 100 J774.2 cells after ALA-PDT. Infected J774.2 cells were incubated with 1 mm ALA for 4 h, and treated by 635 nm laser at 0–50 J cm^{-2}; 18 hours after irradiation, cytospine samples were prepared, and the number of amastigotes per 100 cells was counted under light microscope. Each bar represents the mean ± SD from triplicates. (c) Infected J774.2 cells 18 h after ALA-PDT.

and Figure 12.3). Perhaps factors not present in the cell culture system were responsible for mediating parasite cell death following ALA-PDT *in vivo*.

Photodynamic therapy was performed following the topical application of ALA to CL ear lesions of BALB/c mice after three weeks of *L. major* infection.[15] Inflammation and tissue necrosis were evaluated seven days after ALA-PDT.[15] Inflammation caused by the ALA-PDT and some necrosis was observed on the ears of the ALA-PDT group in contrast to the lesions of the control group or the ALA-only group (Figure 12.3a,b).[15] The wedge-shaped necroses directly in the vascular area of *Auricularis anterior* in all of the ears of the ALA-PDT group were indicative of PDT-induced vascular damage

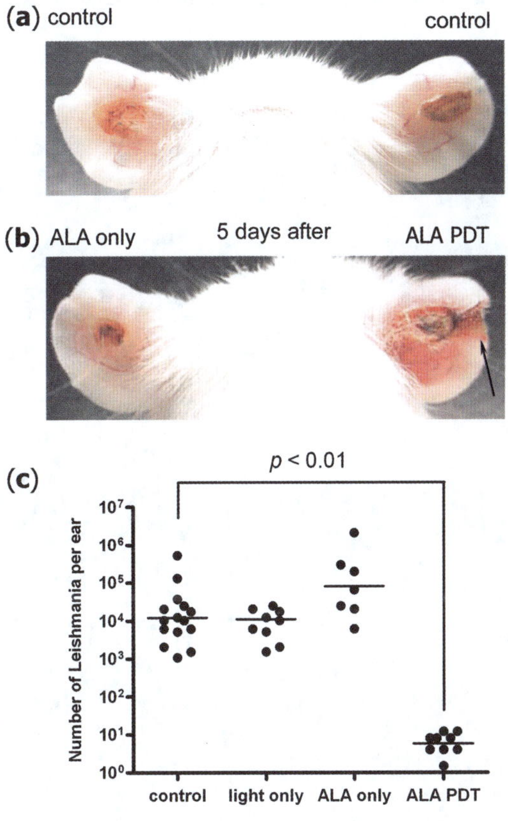

Figure 12.3 Topical ALA-PDT is efficient in parasite eradication. (a) Ears with no treatment compiled a control group ($n = 7$). (b) Topical ALA-PDT was performed on CL lesions 3 weeks after *L. major* infection. For the ALA group, 20% of ALA solution was applied; incubation lasted 4 h under dark conditions. Subsequently ears were irradiated by a 635 nm diode-laser at 50 J cm^{-2} (66 mW cm^{-2}) ($n = 5$). Ears without irradiation but after ALA exposure served as the ALA-only group ($n = 5$). Photographs of the lesions were taken 7 days after ALA-PDT. (c) Quantification of *L. major* parasites was performed as described in Methods. Each point represents the quantity of parasites per one sample, and each bar represents the mean value. *$P < 0.05$.

(Figure 12.3b, arrows).[15] Furthermore, there was an observable inflammatory response in the ears of the ALA-PDT mice as compared to the control groups (Figure 12.3a,b). The parasite burden was quantified for each group and a 24.5-fold reduction in *L. major* burden was observed for the ALA-PDT group as compared to the ALA alone group (Figure 12.3c).[15] These findings indicated to us that ALA-PDT was effective in reducing parasite burden, but what of the mechanism of action?

In search of answers we looked more closely at the inflammatory response associated with ALA-PDT. Figure 12.4a,b represents typical histological findings in the group of infected non-treated mice.[15] The observed dense infiltrate of histiocytes and mixture of lymphocytes are characteristic of acute CL.[15] Leishman-Donovan bodies (amastigotes within histiocytes) are the pale blue-grey

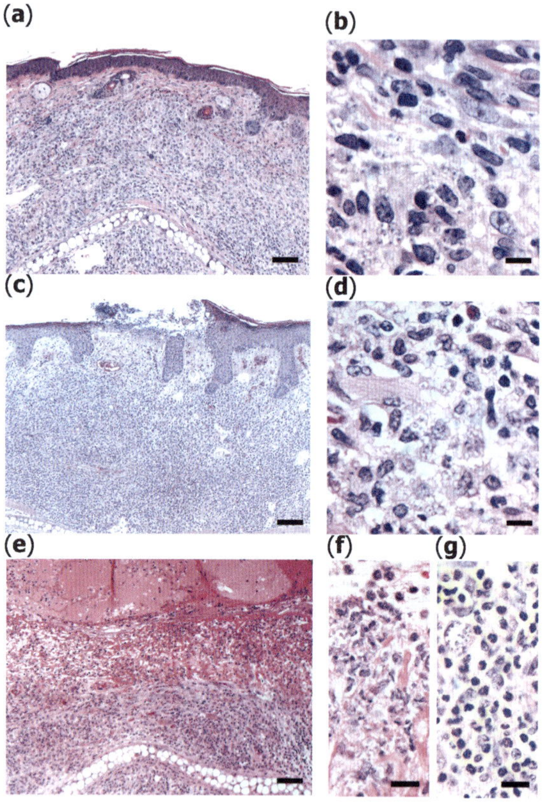

Figure 12.4 ALA-PDT initiated a rigorous inflammatory response with the release of *L. major* amastigotes from histiocytes. Topical ALA-PDT was performed 3 weeks after *L. major* infection as described in Methods. Skin samples obtained from the ears of infected BALB/c mice without treatment (a, b), samples from 7 days after ALA only (c, d), and samples from 7 days after ALA-PDT (e–g) were stained with haematoxylin and eosin. Scale bar = 200 μm (a, c, e), scale bar = 5 μm (b, d, f, g).

round or oval bodies with a centrally located nucleus and a lateral kinetoplast (Figure 12.4b).[15] The epidermal changes consist of slight hyperkeratosis and epidermal hyperplasia. The findings in skin samples from the mice of the ALA-only group demonstrated epidermal spongiosis (Figure 12.4c).[15] A strong edematous response was present throughout the dermis, accompanied by extracellular amastigotes released from histiocytes (Figure 12.4d).[15] Furthermore, blood stasis in the vessels of the papillary dermis was observed. Twenty-four hours after PDT observable effects included (figures not shown): parakeratosis, spongiosis, hydropic degeneration of basal cells of the epidermis, dermal oedema, local destruction of collagen fibres, blood congestion in distended vessels and diapedesis of red blood cells. Seven days after PDT (Figure 12.4e–g), the dermis remained locally necrotic with neutrophil infiltration in the upper part (Figure 12.4f).[15] The remaining ulcer was covered by a scab and extravasation of red blood cells was noted in the subpapillary layer of the dermis. Observation at increased magnification revealed that the remaining cell infiltrate in the deeper part of the dermis consisted of primarily neutrophils and lymphocytes, and a small number of histiocytes (Figure 12.4g).[15]

In order to define more clearly our histological findings we sought to quantify the relative proportion of leukocytes infiltrated within the lesions of the experimental groups *via* flow cytometry. The percentage of neutrophils (Ly-6G/Ly-6C), macrophages (F4/80), T- (CD3) and B- (CD19) cells were determined at seven days after ALA-PDT. Results of these experiments are presented in Figure 12.5a.[15] The percentage of macrophages dramatically decreased from 31.4% (in non-treated mice) to 2.7% seven days after ALA-PDT. The percentage of neutrophils stayed within the range of 15.1% seven days after PDT.[15] The percentage of T cells did not undergo statistically different changes. However, a dramatic 8.7-fold increase in the percentage of CD19-positive B cells indicated recruitment to the site of inflammation (Figure 12.5a).[15]

These results prompted our evaluation of the level of cytokines (IL-6, IL-10, MCP-1, IFN-γ, TNF-α and IL-12p70) within the lesions. ALA-PDT treatment significantly increased the amount of MCP-1 (Figure 12.5).[15] The amount of MCP-1 after ALA-PDT was 4.3-fold greater compared with the infected non-treated group of mice (412.8 ± 96.0 pg/ear *vs.* 133.0 ± 31.5 pg/ear, $P < 0.05$) or ALA-only group mice (85.3 ± 31.9 pg/ear, $P < 0.05$) (Figure 12.5). This change could be explained by the observed reduction in the number of monocytes within the lesions of mice at seven days post ALA-PDT. A dramatic reduction in the macrophage population could lead to an upregulation of MCP-1, which increases the recruitment of monocytes to sites of inflammation. IL-6 was also significantly increased after ALA-PDT compared with ALA-only group (163.6 ± 94.8 pg/ear *vs.* 37.5 ± 31.5 pg/ear, $P < 0.05$) (Figure 12.4b).[15] IL-6 is a mediator of the acute phase response and an activator of T- and B-cells. ALA-PDT of mouse CL lesions caused a decrease in Mφ count and an increase in B-cell numbers at seven days post treatment.

Through these studies we determined that while effective parasite destruction could be achieved with ALA-PDT, it was the result of the killing of infected and uninfected macrophages (host cells for parasite replication) and perhaps

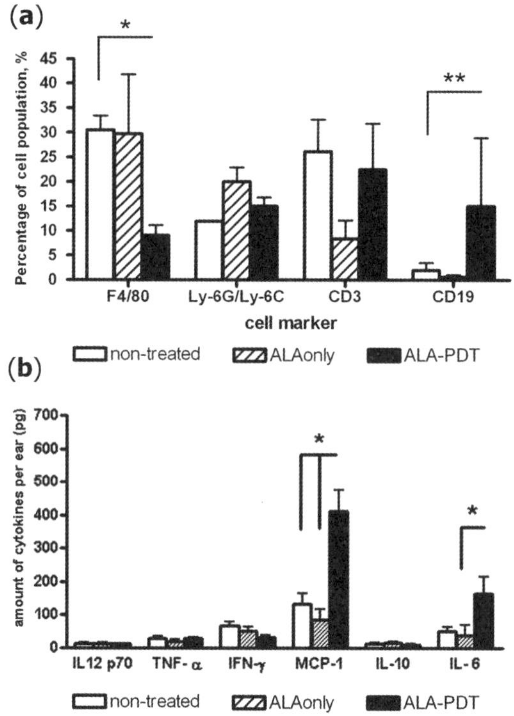

Figure 12.5 Phenotypic profile of cells in ears with CL lesions after ALA-PDT. The amount of neutrophils (Ly-6G/Ly-6C), macrophages (F4/80), T lymphocytes (CD3) and B lymphocytes (CDI9) was determined by flow cytometry as described in Methods. (a) Column plots for expression of the different cellular markers are shown for infected non-treated mice, treated mice with ALA only and treated mice 7 days after ALA-PDT. Each bar represents the mean $\pm$ SD values from five to seven samples. *P $<$ 0.05 *vs.* ALA-PDT, **P $<$ 0.001 *vs.* ALA-PDT. (b) The cytokine profile in skin lesions after ALA-PDT signified an acute inflammatory reaction. The level of cytokines IL-6, IL-10, MCP-1, IFN-γ and TNF-α and IL-12p70 was measured in each ear 7 days after ALA-PDT. We used a BD CBA Mouse Inflammation Kit and the assay was performed according to manufacturer's instructions, using supernatant of the ear tissue homogenate. Each bar represents the mean $\pm$ SD values from five to seven samples. *P $<$ 0.05 *vs.* ALA-PDT.

associated immune mechanisms rather than the result of direct parasiticidal effects.[15]

For some cutaneous conditions, PDT acts *via* a combination of direct cytotoxicity and immunomodulatory effects. However, the PDT effect on the immune system seems to be different from that with other methods of tissue destruction (as noted in our example describing outcomes of cryotherapy). It was reported that PDT induced a release of TNF-α by macrophages, an increase of the functional expression of adhesion molecules, a production of IL-6 and an enhancement of phagocytosis.[25,26]

CL is a "spectral" disease directly depending on the immune status of the host organism. Some authors have even tried to draw clinical parallels between CL and leprosy, due to the fact that the various morphological types of leprosy are determined by different types of immune responses.[27–37] A correlation of clinical type and the potency of the patient's immune response undoubtedly exists. With heightened immunity parasite replication is limited to the top layer of dermis as observed in the eczematoid, erysipeloid and psoriasiform forms of CL under normal conditions, a compromised immune system fails to limit parasite propagation, and the spread of *Leishmania* spp. leading to infiltration of the deeper layer of the skin as observed during sporotrichoid CL.[40,28–30]

Pathogenesis of *Leishmania* infection is immune-mediated. The imbalance of T-lymphocyte populations has been thought to govern some, if not all, of the chronic and sustained pathological mechanisms. The balance of Th1 and Th2 cytokines produced in response to *Leishmania* infection has provided the main conceptual framework for understanding healing *vs.* non-healing forms of CL manifestations.[31] The Th1 host immune response, and in particular IL-12p70 production, is considered to be an essential prerequisite for protection against such intracellular organisms as *Leishmania*.[32,33] After infection, IL-12p70 stimulates NK and T cells to produce IFN-γ, which activates Mφ to produce leishmanicidal nitric oxide.[34,35] In contrast to resistant mice, BALB mice develop aberrant Th2 responses and suffer progressive disease after infection with *L. major*. These outcomes depend on the production of IL-4 and IL-13, during the first two days of infection, by $CD4^+$ T cells that express the Vβ4-Vα8 T cell receptors specific for a dominant I-A(d) restricted epitope of the LACK antigen from *L. major*.[36–38] Increased production of IL-4 inhibits leishmanicidal activity of infected Mφ by decreasing nitric oxide generation.[39]

During CL, immune cells are found in the granulomatous stroma, separated from infected cells and amastigotes by extracellular matrix and the basal membrane of host cells. PDT destroys the structure of the granuloma and enables direct interaction between immune cells and amastigotes, thus resulting in a systemic anti-parasitic immune response. Our studies on the effects of PDT during CL confirmed a significant involvement of the immune mechanisms in the response to therapy.[15,40] We have shown that the activation of a Th1 immune response contributes to parasite elimination in CL lesions after PDT with phenothiazines.[40] Interestingly, ALA-PDT for CL did not involve Th1 skewing (no increase of Th-1 induced IL-12p70 and IFN-γ was observed).[15] Possibly, the high level of IL-6 and neutrophils infiltration may signify Th17 immune response. This could be the subject of future studies.

We suspect that not only the inflammatory response of tissues to PDT may facilitate parasite eradication, but also that the mechanism death induced by PDT may produce a specific immunogenic signal.[41] The involvement of immune mechanism in parasite clearance by PDT may yield differential treatment responses among immuno-supressed patients (*e.g.* AIDS, advance lymphoma/leukaemia, post-transplant) with CL.[13]

Although ALA-PDT proved unable to directly destroy amastigotes our comparative analysis of the outcome of ALA-PDT to that of cryotherapy revealed several positive features, including a comparable efficacy of parasite eradication and the formation of a dry scab, while healing after cryotherapy was accompanied by severe burn reaction with obligatory bulla formation. This aesthetically desirable outcome is especially important when the CL lesions appear on the face and other open skin areas.[42] When ALA-PDT and cryotherapy was used for treatment of multiple CL lesions on the same patients, both methods led to complete healing of all treated lesions. However, the lesions treated with PDT improved after four sessions with compete healing after six sessions, while the lesions treated with cryotherapy healed after five applications with minor scar formation in the centre of the lesions. Visible differences in cosmetic effects were observed as well. Cryotherapy led to more intensive peripheral hyperpigmentation and visible atrophic scarring in the central area of the lesions.[42]

Analyses of the phylogenetic relationships among *Leishmania* species indicate a divergence between the *Leishmania* complexes (*i.e. L. major*, *L. mexicana*, *L. donovani* and *Leishmania* (*Viannia*) subgenus) approximately 40–80 million years ago, which is in the range of the divergence of mammalian orders (*i.e.* mice and humans).[43–45] Thus, the cutaneous disease-causing species (New World *vs.* Old World) are as different from one another as they are from those causing visceral disease. Although, there were reports that *L. chagasi*, *L. panamensis*, *L. amazonensis* promastigotes were sensitive to photodynamic therapy *in vitro*, it is possible that New World species may respond to PDT differently *in vivo*.[46,47] In our work we studied Old World species exclusively.

12.5. The Search for More Effective PSs for CL

We pursued the identification of a PS with maximal phototoxicity towards *Leishmania*. We found that benzophenoxazine analogues, 5-ethylamino-9-diethyl-aminobenzo[*a*]phenoselenazinium (EtNBSe) in particular, exhibited the maximal therapeutic effect ($83.8 \pm 0.3\%$ of killing at fluence $10\,\mathrm{J\,cm^{-2}}$) at the minimal concentration (15 nmol per 2×10^7 *L. major* promastigotes) as compared to protoporphyrin IX, δ-aminolevulinic acid, a benzoporphyrin derivative monoacid (BPD), and 5-ethylamino-9-diethylaminobenzo[*a*]phenothiazinium chloride (EtNBS) (Figure 12.6).[16] Hydrophilic cationic benzophenoxazine analogues are most effective due to their strong attraction to the negatively charged parasitic membrane, their aqueous solubility, their high singlet oxygen quantum yield and their efficacy in targeting intracellular organelles.[16]

Translation of our *in vitro* results to an animal model yielded insufficient parasite elimination from CL lesions with a single treatment prompting us to pursue a multiple treatment regimen. We based our interval of treatment on the four days required to reach the lowest parasite count following a single PDT.[40] We hypothesized that several consecutive PDTs with an interval of four days might be more effective in reducing parasite burden than a single treatment. To test this hypothesis we performed three PDTs using 3,7-bis-(*N*,*N*-dibutylamino) phenothiazinium bromide (PPA904) provided by Photopharmica (UK) every

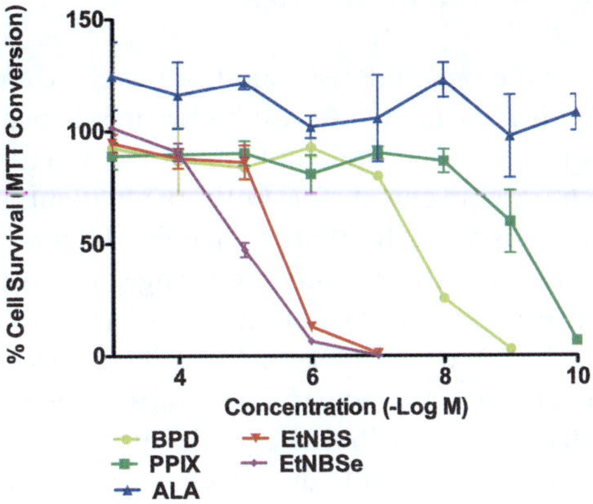

Figure 12.6 Effect of PDT on *L. major* (VI) Pro-mastigotes *in vitro* cytotoxic photo-activities of the different PSs toward *L. major* (VI) (18 hr after irradiation). Promastigotes (2 3 107 ml^{-1}) were incubated with various concentrations (M) of PSs for 2 hr and then light irradiated as described in Experimental Procedures. Cell viability was determined as the percentage of MTT conversion activity.

fifth day in mice four weeks following infection with *L. major*. The number of parasites in the CL lesions progressively decreased after every single treatment from 1×10^6 at inoculation reaching 34 ± 24 parasites four days after the third treatment, which was a 5.2-log reduction in comparison with the control (Figure 12.7) ($p<0.001$). Thus, consecutive PDT seems to be a more effective strategy for eradication of Leishmania parasites from the CL lesions after topical PS administration.[19]

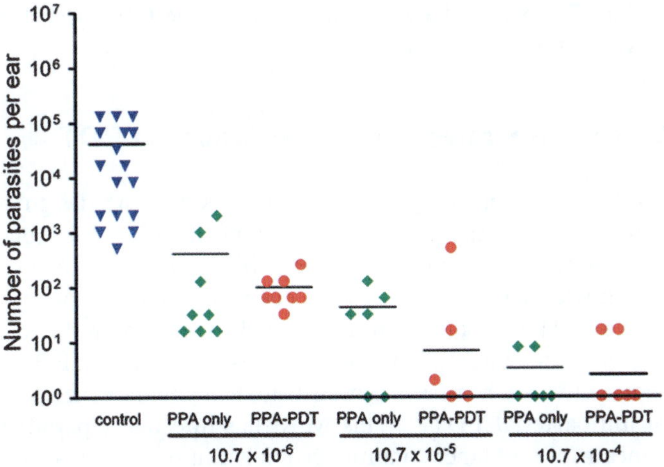

Figure 12.7 Parasite quantification 5 days after PPA904-PDT.

12.6. Considerations for Light Dose

Irradiance is another important determinant of the clinical efficacy of PDT for CL.[48] At first sight, it is attractive to use higher irradiances when performing PDT in clinical settings, since the total time of irradiation can be reduced significantly, which is convenient for both the patient and the physician. Although complex, it is thought that higher irradiances may lead to rapid oxygen consumption or photobleaching rendering some of the "later photons" ineffective for PDT as either the PS concentration or the oxygen concentration or both are reduced.[49–51] We have observed a high incidence of thermal damage when using high irradiances. Our experimental observation does not justify the use of more than 50 mW cm^{-2} in BALB/c model of CL. In C57BL/6 mice, even a lower irradiance maybe preferable due to high absorbance by melanin.

12.7. From Bench to Bedside

PDT has emerged as a very promising and effective treatment for CL but, despite the positive results obtained, current PDT treatments decrease, but do not eradicate, parasitic load.[19,20,22,52] Optimizing the outcome of PDT in murine models of CL is a key step towards the clinical translation of this therapy. The enhancement of drug delivery and penetration into infectious granulomata can provide better drug accumulation and distribution. However, unequal drug diffusion throughout the deeper layers of the granuloma may prevent equal exposure of Leishmania parasites to PS. Systemic administration of PSs may be advantageous under such circumstances. Alternatively, PSs specifically targeted to *Leishmania* could enable rapid healing and reduced tissue remodelling with prospective use for the treatment of the generalized form of leishmaniasis. In this regard recent advances in nanotechnology-based drug delivery systems hold promise where the potential exists for not only increasing the PS payload delivery but also the penetration depth by manipulating delivery using nanoconstructs.

12.8. Prospective Strategies for Antileishmanial PDT

There is perhaps no field of greater opportunity for the design of actively targeted PS constructs than that of infectious disease. The abundance of pathogen associated molecular patterns (PAMPs) provides an avenue to design therapies of greater specificity for use at deep, sensitive anatomical sites that afford reduced collateral damage to uninfected tissues and reduced toxicity to the host. Recently, the targeting of the drug resistance enzyme, ß-lactamase, for the release of active PS has been demonstrated to enhance the specific phototoxicity of antibacterial PDT.[53] This work has affected a paradigm shift in which the means of resistance to antimicrobials can be used against pathogens to promote their destruction and where the use of agents made up of moieties with distinct antibiotic properties that apply selective pressure in opposite directions with respect to their molecular target can effectively turn resistance mechanisms upside down.[53] Just as the active agent remains the PS the ability of PDT to destroy multiple cellular targets remains a great strength for its individual use as well as in combination with other therapies. Such strategies could help to overcome *Leishmania* resistance to conventional therapies.

Photodynamic therapy has shown promise in the killing of *Leishmania*, where, at least in the short term, ulcers associated with the disease disappeared with good cosmetic outcomes.[20] However multiple treatments per week were required over the course of months.[20] Treatment response could be improved through increasing the selectivity of the PDT as well as through the rational combination of PDT with other therapies. PDT selectivity is derived from the preferential localization of PS in targets of interest and from the spatial confinement of light delivery. For PDT to be used effectively in the treatment of leishmaniasis, a greater level of selectivity for the parasite or the components of the disease-associated lesion is necessary to enhance the long-term outcome.

The complete genomic DNA sequences of multiple *Leishmania* species are now available enabling our capacity to better understand the biology of *Leishmania* within its vector and its host. The molecular biology of antimicrobial resistance, host-parasite interactions and the pathways that support the nutritional requirements of the parasites are potential targets for the development of PDTs with enhanced specificity.

12.8.1. Major Surface Metalloprotease GP63

New approaches to increase the specificity of PDT involve photosensitizer targeting. Improved targeting can be achieved *via* conjugation of the PS to moieties, such as monoclonal antibodies directed against target-associated antigens, polymers and peptide scaffolds that may aid in overcoming the present difficulties.[54–58] Therefore, there are two significant components in targeted PDT: proper target identification and the method of conjugation.

The *Leishmania* major metalloproteinase of 63 kDa (GP63) is the major protein ($\sim 5\times10^5$ molecules/promastigote cell) expressed on the surface of *Leishmania* promastigotes.[59–67] Due to its numerous functions, GP63 has been postulated as an important virulence factor.[68–70] The targeting of GP63 for PS delivery could improve eradication of both extracellular and intracellular promastigotes (Figure 12.8). The proteinase activity of GP63 has been demonstrated to be a defence mechanism against the complement system. This results in enhanced resistance to complement-mediated lysis by cleavage of C3.[68,71,72] GP63 acts as a primary acceptor for iC3b deposition, resulting in facilitated phagocytosis and survival within the phagolysosome through its cleavage of lysosomal enzymes.[71,73–76] GP63 may play a further role during interaction with immune cells of the host as it has been shown to diminish T cell response by the cleavage of CD4, inhibit Mφ chemotaxis and enhance the migrating capacity of *Leishmania* by degradation of collagen IV and fibronectin.[77–79]

12.8.2. Purine Salvage

All genera of protozoan parasites must acquire purines from the host in order to survive and grow.[80] Thus purine acquisition is a potential target pathway for the design of novel therapies (Figure 12.8b). Interestingly thiopurines can be

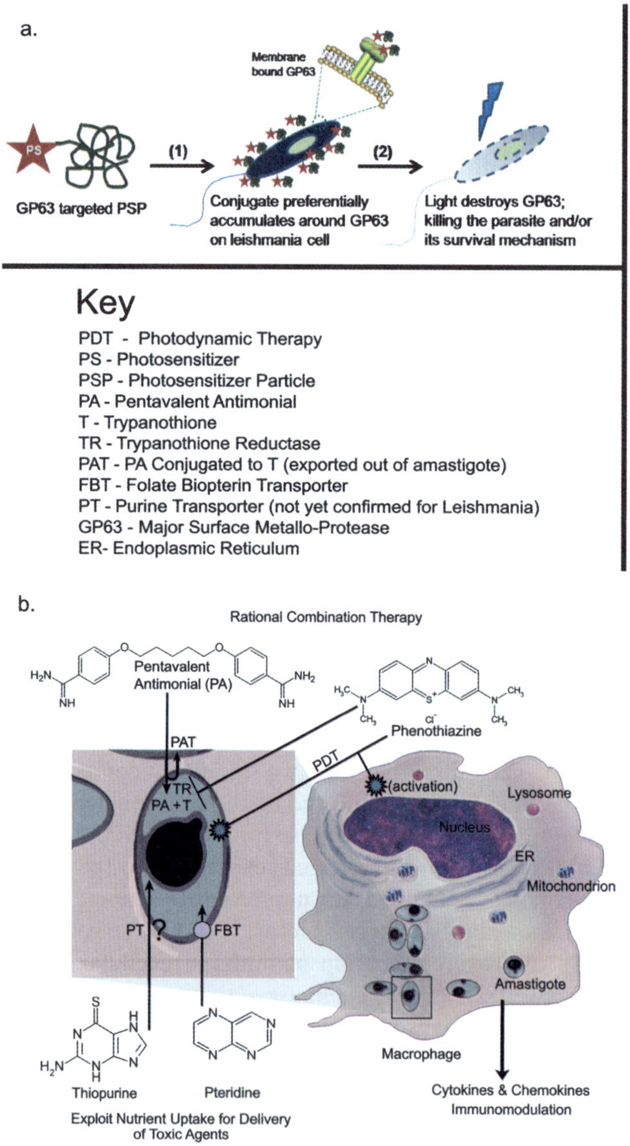

Figure 12.8 Prospective strategies for targeted antileishmanial PDT. a. GP63 targeting strategy for promastigotes, b. targeting strategies and proposed combination therapy for amastigotes.

incorporated into the DNA of replicating cells and increase their sensitivity to UV-A-induced cell death.[81] There is potential for the use of topical thiopurines and UV-A phototherapy for the treatment of cutaneous leishmaniasis.

12.8.3. Pentavalent Antimonial Resistance

A mainstay for antileishmanial therapy has been the use of pentavalent antimonials. The resistance of *Leishmania* to therapies with pentavalent antimonials relies, in part, on the activity of enzymes integral to thiol metabolism.[82] Antimonial resistance in *Leishmania* is accomplished through detoxification and sequestration of the drugs. The inactivation of pentavalent antimonials by the parasite is accomplished through the addition of a thiol group to the drug. Typanothione reductase (TR) has been shown to be important in this regard and has been designated as a target for inhibitors in the rationale design of combinatorial therapies for Leishmaniasis (Figure 12.8). The phenothiazine PS methylene blue is a potent competitive-inhibitor of TR.[83,84] Methylene blue is a substrate for TR in a reaction that uses NADPH as a cofactor and ultimately results in the formation of Leocomethylene blue.[83] This reaction is spontaneously reversible and results in the recycling of methylene blue and the concurrent production of a hydrogen peroxide molecule in the absence of light.[83,85] Through using PSs that not only generate reactive oxygen species *via* PDT, but also inhibit the activity of TR, the specific toxicity of PDT and combination therapies with antimonials could be enhanced. Methylene blue could competitively inhibit the activity of TR and thereby reduce leishmanial resistance to pentavalent antimonial therapy while also serving as a PS. The possibility of this combination in improving treatment response merits further study.

12.8.4. Pteridine Uptake

Like other parasitic members of the family *Trypanosomatidae*, Leishmania are pteridine auxotrophs. Folate and biopterin uptake is regulated with growth phase in Leishmania, with heightened uptake in actively dividing cells.[86] Once transported into the cells *via* their respective transporters they must be reduced to their respective tertrahydro forms in order to serve as cofactors for thymdilate synthesis.[86] The absolute dependence on the host for these precursor molecules makes their respective uptake pathways particularly attractive for targeted therapies including PDT (Figure 12.8). Such strategies have been explored for the targeting of PSs with promising results.[87] There has been considerable work towards the development of vitamin B as a targeting moiety for anti-cancer drugs and nanoparticles.[88–93]

The use of pteridine conjugates as well as other vitamin conjugates deserves exploration for targeted therapies against leishmaniasis.

12.8.5. Metacaspases as Therapeutic Targets and Indicators of PDT Response

Metacaspases (MCAs) are orthologues of caspases that are produced by plants, fungi and trypanosomes and have a role in DNA replication and programmed cell death (PCD).[94] Interestingly, the MCAs of Leishmania *donovani* are unable to cleave caspase-specific substrates, but do cleave trypsin-specific substrates.[85] The expression of MCAs in *L. donovani* is induced by oxidative stress and the overexpression of MCAs in *L. donovani* results in an increase in apoptosis and an increased sensitivity to peroxide-induced apoptosis.[85] Overexpression of MCA A in *Leishmania major* leads to growth retardation and changes in ploidy resulting from errors in mitotic spindle segregation.[95] In *L. major*, MCA A has been shown to possess trypsin-like activity and to be involved in PCD.[94] The complete genome of *L. major* has been shown to possess a single copy of the MCA A gene and both alleles were shown to be essential for the replication of the parasite.[95] These findings are significant indicators that MCA expression could serve as a temporal marker of sensitivity to a second treatment when increased, and also a marker of the cytostatic effect of therapy on the parasite when decreased. The gene sequences of *L. major* MCA A and *L. donovani* MCA 2 share complete identity. The gene sequence of *L. donovani* metacaspase 1 shares near complete identity to both. Photodynamic therapy generates reactive oxygen species (ROS) that are directly responsible for its antimicrobial effect.[96] In assaying the effect of PDT on the temporal expression of MCA genes one could obtain information that indicates not only the viability of the parasite, but also points in time where the parasite would be most vulnerable to further oxidative stress following an initial PDT treatment. The potential use of metacaspase expression as an indicator for the timing of consecutive and/or combinatorial therapies in the treatment of Leishmaniasis warrants future study.

12.9. Conclusions

In the next decade PDT could become a viable therapy for CL. As for the visceral form of the disease, the development of disease-specific PS targeting strategies will facilitate the success of PDT as a therapy. There is an obvious need for more effective topical drug delivery systems in improving the depth of PS penetration. The dually selective nature of PDT is one of its great advantages and the advent of nanoscale delivery strategies combined with our increased understanding of *Leishmania* molecular biology will enable a heightening of disease-specific PS localization. All of these avenues are under active investigation. Progress along these lines will afford PDT a broad role in the localized treatment of systemic disease.

Acknowledgements

Supported by funding from the United States Department of Defense (FA955-04-1-0079); the Bill and Melinda Gates Foundation (51813); and the National Institutes of Health (AR007098-35).

References

1. D.H. Connor and R.C. Neafie, Cutaneous leishmaniasis, in *Pathology of Tropical and Extraordinary Diseases*, ed. C.H. Bunford and D.H. Connor, Armed Forces Institute of Pathology, Washington, vol. 1, pp. 258–264.
2. J.A. Huaman and M.C. Castillo, Mucocutaneous leishmaniasis. Histologic classification and anatomoclinical correlations. Apropos of 102 cases, *Arch. Anat. Cytol. Pathol.*, 1990, **38**, 135–142.
3. WHO, The leishmaniases. Report of a WHO Expert Committee, *World Health Organ. Tech. Rep. Ser.*, 1984, **701**, 1–140.
4. P. Desjeux, Leishmaniasis: current situation and new perspectives, *Comp. Immunol. Microbiol. Infect. Dis.*, 2004, **27**, 305–318.
5. R. Reithinger, M. Mohsen, K. Aadil, M. Sidiqi, P. Erasmus and P.G. Coleman, Anthroponotic cutaneous leishmaniasis, Kabul, Afghanistan, *Emerg. Inf. Dis.*, 2003, **9**, 727–729.
6. H.W. Murray, J.D. Berman, C.R. Davies and N.G. Saravia, Advances in leishmaniasis, *Lancet*, 2005, **366**, 1561–1577.
7. S. Brooker, N. Mohammed, K. Adil, S. Agha, R. Reithinger, M. Rowland, I. Ali and J. Kolaczinski, Leishmaniasis in refugee and local Pakistani populations, *Emerg. Infect. Dis.*, 2004, **10**, 1681–1684.
8. J. Kolaczinski, S. Brooker, H. Reyburn and M. Rowland, Epidemiology of anthroponotic cutaneous leishmaniasis in Afghan refugee camps in northwest Pakistan, *Trans. R. Soc. Trop. Med. Hyg.*, 2004, **98**, 373–378.
9. A.A. Alrajhi, Cutaneous leishmaniasis of the Old World, *Skin Therapy Lett.*, 2003, **8**, 1–4.
10. T.R. Navin, B.A. Arana, F.E. Arana, A.M. de Merida, A.L. Castillo and J.L. Pozuelos, Placebo-controlled clinical trial of meglumine antimonate (glucantime) vs. localized controlled heat in the treatment of cutaneous leishmaniasis in Guatemala, *Am. J. Trop. Med. Hyg.*, 1990, **42**, 43–50.
11. M. Ouellette, J. Drummelsmith and B. Papadopoulou, Leishmaniasis: drugs in the clinic, resistance and new developments, *Drug Resist. Updates*, 2004, **7**, 257–266.
12. C.J. Kelty, N.J. Brown, M.W. Reed and R. Ackroyd, The use of 5-aminolaevulinic acid as a photosensitiser in photodynamic therapy and photodiagnosis, *Photochem. Photobiol. Sci.*, 2002, **1**, 158–168.
13. E.M. van der Snoek, D.J. Robinson, J.J. van Hellemond and H.A. Neumann, A review of photodynamic therapy in cutaneous leishmaniasis, *J. Eur. Acad. Dermatol. Venereol.*, 2008, **22**, 918–922.
14. O.E. Akilov, S. Kosaka, J.W. Foley and T. Hasan, Photodynamic killing of Leishmania major with cationic dyes, 11th Congress of the European Society for Photobiology, Aix-Les-Bains, France, 2005, 101.
15. O.E. Akilov, S. Kosaka, K. O'Riordan and T. Hasan, Parasiticidal effect of delta-aminolevulinic acid-based photodynamic therapy for cutaneous leishmaniasis is indirect and mediated through the killing of the host cells, *Exp. Dermatol.*, 2007, **16**, 651–660.
16. O.E. Akilov, S. Kosaka, K. O'Riordan, X. Song, M. Sherwood, T.J. Flotte, J.W. Foley and T. Hasan, The role of photosensitizer molecular charge and structure on the efficacy of photodynamic therapy against leishmania parasites, *Chem. Biol.*, 2006, **13**, 839–847.
17. H.S. de Bruijn, W. Sluiter, A. van der Ploeg-van den Heuvel, H.J. Sterenborg and D.J. Robinson, Evidence for a bystander role of neutrophils in the response to

systemic 5-aminolevulinic acid-based photodynamic therapy, *Photodermatol. Photoimmunol. Photomed.*, 2006, **22**, 238–246.

18. C.-A. Bristow, R. Hudson, T.A. Paget and R.W. Boyle, Potential of cationic porphyrins for photodynamic treatment of cutaneous Leishmaniasis, *Photodiagn. Photodyn. Ther.*, 2006, **3**, 162–167.
19. O.E. Akilov, W. Yousaf, S.X. Lukjan, S. Verma and T. Hasan, Optimization of topical photodynamic therapy with 3,7-bis(di-n-butylamino)phenothiazin-5-ium bromide for cutaneous leishmaniasis, *Lasers Surg. Med.*, 2009, **41**, 358–365.
20. K. Gardlo, Z. Horska, C.D. Enk, L. Rauch, M. Megahed, T. Ruzicka and C. Fritsch, Treatment of cutaneous leishmaniasis by photodynamic therapy, *J. Am. Acad. Dermatol.*, 2003, **48**, 893–896.
21. K. Gardlo, S. Hanneken, T. Ruzicka and N.J. Neumann, Photodynamic therapy of cutaneous leishmaniasis. A promising new therapeutic modality, *Hautarzt*, 2004, **55**, 381–383.
22. A. Asilian and M. Davami, Comparison between the efficacy of photodynamic therapy and topical paromomycin in the treatment of Old World cutaneous leishmaniasis: a placebo-controlled, *randomized clinical trial, Clin. Exp. Dermatol.*, 2006, **31**, 634–637.
23. J.F. Sah, H. Ito, B.K. Kolli, D.A. Peterson, S. Sassa and K.P. Chang, Genetic rescue of Leishmania deficiency in porphyrin biosynthesis creates mutants suitable for analysis of cellular events in uroporphyria and for photodynamic therapy, *J. Biol. Chem.*, 2002, **277**, 14902–14909.
24. S. Kosaka, O.E. Akilov, K. O'Riordan and T. Hasan, A mechanistic study of delta-aminolevulinic acid-based photodynamic therapy for cutaneous leishmaniasis, *J. Invest. Dermatol.*, 2007, **127**, 1546–1549.
25. S. Evans, W. Matthews, R. Perry, D. Fraker, J. Norton and H.I. Pass, Effect of photodynamic therapy on tumor necrosis factor production by murine macrophages, *J. Natl. Cancer Inst.*, 1990, **82**, 34–39.
26. N. Yamamoto, S. Homma, T.W. Sery, L.A. Donoso and J.K. Hoober, Photodynamic immunopotentiation: in vitro activation of macrophages by treatment of mouse peritoneal cells with haematoporphyrin derivative and light, *Eur. J. Cancer*, 1991, **27**, 467–471.
27. R.L. Modlin, Learning from leprosy: insights into contemporary immunology from an ancient disease, *Skin Pharmacol. Appl. Skin Physiol.*, 2002, **15**, 1–6.
28. N. Iftikhar, I. Bari and A. Ejaz, Rare variants of Cutaneous Leishmaniasis: whitlow, paronychia, and sporotrichoid, *Int. J. Dermatol.*, 2003, **42**, 807–809.
29. A.G. Kibbi, P.G. Karam and A.K. Kurban, Sporotrichoid leishmaniasis in patients from Saudi Arabia: clinical and histologic features, *J. Am. Acad. Dermatol.*, 1987, **17**, 759–764.
30. S. Spier, M. Medenica, S. McMillan and C. Virtue, Sporotrichoid leishmaniasis, *Arch. Dermatol.*, 1977, **113**, 1104–1105.
31. N. Peters and D. Sacks, Immune privilege in sites of chronic infection: Leishmania and regulatory T cells, *Immunol. Rev.*, 2006, **213**, 159–179.
32. B. Raupach and S.H. Kaufmann, Immune responses to intracellular bacteria, *Curr. Opin. Immunol.*, 2001, **13**, 417–428.
33. G. Trinchieri and P. Scott, Interleukin-12: a proinflammatory cytokine with immunoregulatory functions, *Res. Immunol.*, 1995, **146**, 423–431.
34. G. Trinchieri, Interleukin-12: a cytokine produced by antigen-presenting cells with immunoregulatory functions in the generation of T-helper cells type 1 and cytotoxic lymphocytes, *Blood*, 1994, **84**, 4008–4027.

35. J. Assreuy, F.Q. Cunha, M. Epperlein, A. Noronha-Dutra, C.A. O'Donnell, F.Y. Liew and S. Moncada, Production of nitric oxide and superoxide by activated macrophages and killing of Leishmania major, *Eur. J. Immunol.*, 1994, **24**, 672–676.
36. J.A. Louis, A. Gumy, H. Voigt, P. Launois and M. Rocken, The use of the murine model of infection with Leishmania major to reveal the antagonistic effects that IL-4 can exert on T helper cell development and demonstrate that these opposite effects depend upon the nature of the cells targeted for IL-4 signaling, *Pathol. Biol. (Paris)*, 2003, **51**, 71–73.
37. R.M. Locksley and J.A. Louis, Immunology of leishmaniasis, *Curr. Opin. Immunol.*, 1992, **4**, 413–418.
38. D. Sacks and N. Noben-Trauth, The immunology of susceptibility and resistance to Leishmania major in mice, *Nat. Rev. Immunol.*, 2002, **2**, 845–858.
39. I. Vouldoukis, P.A. Becherel, V. Riveros-Moreno, M. Arock, O. da Silva, P. Debre, D. Mazier and M.D. Mossalayi, Interleukin-10 and interleukin-4 inhibit intracellular killing of Leishmania infantum and Leishmania major by human macrophages by decreasing nitric oxide generation, *Eur. J. Immunol.*, 1997, **27**, 860–865.
40. O.E. Akilov, S. Kosaka, K. O'Riordan and T. Hasan, Photodynamic therapy for cutaneous leishmaniasis: the effectiveness of topical phenothiaziniums in parasite eradication and Th1 immune response stimulation, *Photochem. Photobiol. Sci.*, 2007, **6**, 1067–1075.
41. A. Tesniere, T. Panaretakis, O. Kepp, L. Apetoh, F. Ghiringhelli, L. Zitvogel and G. Kroemer, Molecular characteristics of immunogenic cancer cell death, *Cell Death Differ.*, 2008, **15**, 3–12.
42. K. Pizinger, P. Cetkovska, D. Kacerovska and M. Kumpova, Successful treatment of cutaneous leishmaniasis by photodynamic therapy and cryotherapy, *Eur. J. Dermatol.*, 2009, **19**, 172–173.
43. D.G. Croan, D.A. Morrison and J.T. Ellis, Evolution of the genus Leishmania revealed by comparison of DNA and RNA polymerase gene sequences, *Mol. Biochem. Parasitol.*, 1997, **89**, 149–159.
44. A.P. Fernandes, K. Nelson and S.M. Beverley, Evolution of nuclear ribosomal RNAs in kinetoplastid protozoa: perspectives on the age and origins of parasitism, *Proc. Natl. Acad. Sci. USA*, 1993, **90**, 11608–11612.
45. D. McMahon-Pratt and J. Alexander, Does the Leishmania major paradigm of pathogenesis and protection hold for New World cutaneous leishmaniases or the visceral disease? *Immunol. Rev.*, 2004, **201**, 206–224.
46. P. Escobar, I.P. Hernandez, C. M. Rueda, F. Martinez and E. Paez, Photodynamic activity of aluminium (III) and zinc (II) phthalocyanines in Leishmania promastigotes, *Biomedica*, 2006, **26**(1), 49–56.
47. S. Dutta, D. Ray, B.K. Kolli and K.P. Chang, Photodynamic sensitization of Leishmania amazonensis in both extracellular and intracellular stages with aluminum phthalocyanine chloride for photolysis in vitro, *Antimicrob. Agents Chemother.*, 2005, **49**, 4474–4484.
48. E. Latorre-Esteves, O.E. Akilov, P. Rai, S.M. Beverley and T. Hasan, Monitoring the efficacy of antimicrobial photodynamic therapy in a murine model of cutaneous leish maniasis using L. major expressing GFP, *Journal of Biophotonics*, 2010, **3**, 328–335.
49. N. Solban, B. Ortel, B. Pogue and T. Hasan, Targeted optical imaging and photodynamic therapy, *Ernst Schering Res. Found. Workshop*, 2005, 229–258.
50. T.M. Sitnik, J.A. Hampton and B.W. Henderson, Reduction of tumour oxygenation during and after photodynamic therapy in vivo: effects of fluence rate, *Br. J. Cancer*, 1998, **77**, 1386–1394.

51. T.M. Sitnik and B.W. Henderson, The effect of fluence rate on tumor and normal tissue responses to photodynamic therapy, *Photochem. Photobiol.*, 1998, **67**, 462–466.
52. C.D. Enk, C. Fritsch, F. Jonas, A. Nasereddin, A. Ingber, C.L. Jaffe and T. Ruzicka, Treatment of cutaneous leishmaniasis with photodynamic therapy, *Arch. Dermatol.*, 2003, **139**, 432–434.
53. X. Zheng, U.W. Sallum, S. Verma, H. Athar, C.L. Evans and T. Hasan, Exploiting a bacterial drug-resistance mechanism: a light-activated construct for the destruction of MRSA, *Angew. Chem. Int. Ed. Engl.*, 2009, **48**, 2148–2151.
54. M. Carcenac, C. Larroque, R. Langlois, J.E. van Lier, J.C. Artus and A. Pelegrin, Preparation, phototoxicity and biodistribution studies of anti-carcinoembryonic antigen monoclonal antibody-phthalocyanine conjugates, *Photochem. Photobiol.*, 1999, **70**, 930–936.
55. M.R. Hamblin, M. Del Governatore, I. Rizvi and T. Hasan, Biodistribution of charged 17.1A photoimmunoconjugates in a murine model of hepatic metastasis of colorectal cancer, *Br. J. Cancer*, 2000, **83**, 1544–1551.
56. N. Brasseur, R. Ouellet, C. La Madeleine and J.E. van Lier, Water-soluble aluminium phthalocyanine-polymer conjugates for PDT: photodynamic activities and pharmacokinetics in tumour-bearing mice, *Br. J. Cancer*, 1999, **80**, 1533–1541.
57. R.D. Brokx, S.K. Bisland and J. Gariepy, Designing peptide-based scaffolds as drug delivery vehicles, *J. Contr. Release*, 2002, **78**, 115–123.
58. Y.N. Konan, R. Gurny and E. Allemann, State of the art in the delivery of photosensitizers for photodynamic therapy, *J. Photochem. Photobiol. B*, 2002, **66**, 89–106.
59. C. Bordier, The promastigote surface protease of Leishmania, *Parasitol. Today*, 1987, **3**, 151–153.
60. L.L. Button and W.R. McMaster, Molecular cloning of the major surface antigen of leishmania, *J. Exp. Med.*, 1988, **167**, 724–729.
61. L.L. Button, D.G. Russell, H.L. Klein, E. Medina-Acosta, R.E. Karess and W.R. McMaster, Genes encoding the major surface glycoprotein in Leishmania are tandemly linked at a single chromosomal locus and are constitutively transcribed, *Mol. Biochem. Parasitol.*, 1989, **32**, 271–283.
62. D. Fong and K.P. Chang, Surface antigenic change during differentiation of a parasitic protozoan, Leishmania mexicana: Identification by monoclonal antibodies, *Proc. Natl. Acad. Sci. USA*, 1982, **79**, 7366–7370.
63. E. Medina-Acosta, R.E. Karess and D.G. Russell, Structurally distinct genes for the surface protease of Leishmania mexicana are developmentally regulated, *Mol. Biochem. Parasitol.*, 1993, **57**, 31–45.
64. H.B. Steinkraus, J.M. Greer, D.C. Stephenson and P.J. Langer, Sequence heterogeneity and polymorphic gene arrangements of the Leishmania guyanensis gp63 genes, *Mol. Biochem. Parasitol.*, 1993, **62**, 173–185.
65. J.R. Webb, L.L. Button and W.R. McMaster, Heterogeneity of the genes encoding the major surface glycoprotein of Leishmania donovani, *Mol. Biochem. Parasitol.*, 1991, **48**, 173–184.
66. M.E. Wilson, K.K. Hardin and J.E. Donelson, Expression of the major surface glycoprotein of Leishmania donovani chagasi in virulent and attenuated promastigotes, *J. Immunol.*, 1989, **143**, 678–684.
67. C. Yao, K.G. Leidal, A. Brittingham, D.E. Tarr, J.E. Donelson and M.E. Wilson, Biosynthesis of the major surface protease GP63 of Leishmania chagasi, *Mol. Biochem. Parasitol.*, 2002, **121**, 119–128.

68. P.B. Joshi, D.L. Sacks, G. Modi and W.R. McMaster, Targeted gene deletion of Leishmania major genes encoding developmental stage-specific leishmanolysin (GP63), *Mol. Microbiol.*, 1998, **27**, 519–530.
69. J. Bouvier, C. Bordier, H. Vogel, R. Reichelt and R. Etges, Characterization of the promastigote surface protease of Leishmania as a membrane-bound zinc endopeptidase, *Mol. Biochem. Parasitol.*, 1989, **37**, 235–245.
70. R. Etges, J. Bouvier and C. Bordier, The major surface protein of Leishmania promastigotes is a protease, *J. Biol. Chem.*, 1986, **261**, 9098–9101.
71. A. Brittingham, C.J. Morrison, W.R. McMaster, B.S. McGwire, K.P. Chang and D.M. Mosser, Role of the Leishmania surface protease gp63 in complement fixation, cell adhesion, and resistance to complement-mediated lysis, *J. Immunol.*, 1995, **155**, 3102–3111.
72. B. McGwire and K.P. Chang, Genetic rescue of surface metalloproteinase (gp63)-deficiency in Leishmania amazonensis variants increases their infection of macrophages at the early phase, *Mol. Biochem. Parasitol.*, 1994, **66**, 345–347.
73. M.B. Seay, P.L. Heard and G. Chaudhuri, Surface Zn-proteinase as a molecule for defense of Leishmania mexicana amazonensis promastigotes against cytolysis inside macrophage phagolysosomes, *Infect. Immun.*, 1996, **64**, 5129–5137.
74. D.G. Russell, The macrophage-attachment glycoprotein gp63 is the predominant C3-acceptor site on Leishmania mexicana promastigotes, *Eur. J. Biochem.*, 1987, **164**, 213–221.
75. G. Chaudhuri, M. Chaudhuri, A. Pan and K.P. Chang, Surface acid proteinase (gp63) of Leishmania mexicana. A metalloenzyme capable of protecting liposome-encapsulated proteins from phagolysosomal degradation by macrophages, *J. Biol. Chem.*, 1989, **264**, 7483–7489.
76. D.Q. Chen, B.K. Kolli, N. Yadava, H.G. Lu, A. Gilman-Sachs, D.A. Peterson and K.P. Chang, Episomal expression of specific sense and antisense mRNAs in Leishmania amazonensis: modulation of gp63 level in promastigotes and their infection of macrophages in vitro, *Infect. Immun.*, 2000, **68**, 80–86.
77. A.S. Hey, T.G. Theander, L. Hviid, S.M. Hazrati, M. Kemp and A. Kharazmi, The major surface glycoprotein (gp63) from Leishmania major and Leishmania donovani cleaves CD4 molecules on human T cells, *J. Immunol.*, 1994, **152**, 4542–4548.
78. A.A. Ahmed, A. Wahbi, K. Nordlind, A. Kharazmi, K.G. Sundqvist, V. Mutt and S. Liden, In vitro Leishmania major promastigote-induced macrophage migration is modulated by sensory and autonomic neuropeptides, *Scand. J. Immunol.*, 1998, **48**, 79–85.
79. B.S. McGwire, K.P. Chang and D.M. Engman, Migration through the extracellular matrix by the parasitic protozoan Leishmania is enhanced by surface metalloprotease gp63, *Infect. Immun.*, 2003, **71**, 1008–1010.
80. B. Ullman and D. Carter, Molecular and biochemical studies on the hypoxanthine-guanine phosphoribosyltransferases of the pathogenic haemoflagellates, *Int. J. Parasitol.*, 1997, **27**, 203–213.
81. A. Massey, Y.Z. Xu and P. Karran, Photoactivation of DNA thiobases as a potential novel therapeutic option, *Curr. Biol.*, 2001, **11**, 1142–1146.
82. M.K. Mittal, S. Rai, A. Ravinder, S. Gupta, S. Sundar and N. Goyal, Characterization of natural antimony resistance in Leishmania donovani isolates, *Am. J. Trop. Med. Hyg.*, 2007, **76**, 681–688.
83. K. Buchholz, M.A. Comini, D. Wissenbach, R.H. Schirmer, R.L. Krauth-Siegel and S. Gromer, Cytotoxic interactions of methylene blue with trypanosomatid-specific

disulfide reductases and their dithiol products, *Mol. Biochem. Parasitol.*, 2008, **160**, 65–69.

84. C. Chan, H. Yin, J. Garforth, J.H. McKie, R. Jaouhari, P. Speers, K.T. Douglas, P.J. Rock, V. Yardley, S.L. Croft and A.H. Fairlamb, Phenothiazine inhibitors of trypanothione reductase as potential antitrypanosomal and antileishmanial drugs, *J. Med. Chem.*, 1998, **41**, 148–156.
85. N. Lee, S. Gannavaram, A. Selvapandiyan and A. Debrabant, Characterization of metacaspases with trypsin-like activity and their putative role in programmed cell death in the protozoan parasite Leishmania, *Eukaryot. Cell*, 2007, **6**, 1745–1757.
86. M.L. Cunningham and S.M. Beverley, Pteridine salvage throughout the Leishmania infectious cycle: implications for antifolate chemotherapy, *Mol. Biochem. Parasitol.*, 2001, **113**, 199–213.
87. R. Schneider, F. Schmitt, C. Frochot, Y. Fort, N. Lourette, F. Guillemin, J.F. Muller and M. Barberi-Heyob, Design, synthesis, and biological evaluation of folic acid targeted tetraphenylporphyrin as novel photosensitizers for selective photodynamic therapy, *Bioorg. Med. Chem.*, 2005, **13**, 2799–2808.
88. C.P. Leamon and P.S. Low, Folate-mediated targeting: from diagnostics to drug and gene delivery, *Drug Discov. Today*, 2001, **6**, 44–51.
89. C.M. Ward, N. Acheson and L.W. Seymour, Folic acid targeting of protein conjugates into ascites tumour cells from ovarian cancer patients, *J. Drug Target.*, 2000, **8**, 119–123.
90. B. Stella, S. Arpicco, M.T. Peracchia, D. Desmaele, J. Hoebeke, M. Renoir, J. D'Angelo, L. Cattel and P. Couvreur, Design of folic acid-conjugated nanoparticles for drug targeting, *J. Pharm. Sci.*, 2000, **89**, 1452–1464.
91. S.R. Holladay, Z. Yang, M.D. Kennedy, C.P. Leamon, R.J. Lee, M. Jayamani, T. Mason and P.S. Low, Riboflavin-mediated delivery of a macromolecule into cultured human cells, *Biochim. Biophys. Acta*, 1999, **1426**, 195–204.
92. J.E. Schroeder, I. Shweky, H. Shmeeda, U. Banin and A. Gabizon, Folate-mediated tumor cell uptake of quantum dots entrapped in lipid nanoparticles, *J. Contr. Release*, 2007, **124**, 28–34.
93. F. Esmaeili, M.H. Ghahremani, S.N. Ostad, F. Atyabi, M. Seyedabadi, M.R. Malekshahi, M. Amini and R. Dinarvand, Folate-receptor-targeted delivery of docetaxel nanoparticles prepared by PLGA-PEG-folate conjugate, *J. Drug Target.*, 2008, **16**, 415–423.
94. I.J. Gonzalez, C. Desponds, C. Schaff, J.C. Mottram and N. Fasel, Leishmania major metacaspase can replace yeast metacaspase in programmed cell death and has arginine-specific cysteine peptidase activity, *Int. J. Parasitol.*, 2007, **37**, 161–172.
95. A. Ambit, N. Fasel, G.H. Coombs and J.C. Mottram, An essential role for the Leishmania major metacaspase in cell cycle progression, *Cell Death Differ.*, 2008, **15**, 113–122.
96. M.R. Hamblin and T. Hasan, Photodynamic therapy: a new antimicrobial approach to infectious disease?, *Photochem. Photobiol. Sci.*, 2004, **3**, 436–450.

Chapter 13

Antimicrobial Photodynamic Therapy (aPDT) for Oral Infections

Jonathan Pratten,[a] Veronique Benhamou[b] and Cale Street[c]

[a] Microbial Diseases Research Department, UCL Eastman Dental Institute, 256 Gray's Inn Road, London, WC1X 8LD, UK; [b] McGill University Dental School, Montreal, Quebec, Canada; [c] Ondine Biopharma Corporation, 19017-120th Ave NE, Bothell, WA 98011, USA

Table of Contents

Photodynamic Inactivation of Microbial Pathogens: Medical and Environmental Applications
Editors Michael R. Hamblin and Giulio Jori

Published by the Royal Society of Chemistry, www.rsc.org

13.1. Introduction

The mouth provides a large number of diverse surfaces, such as soft shedding tissues of the buccal mucosa and hard non-shedding surfaces of the teeth, on which a wide variety of complex biofilms are able to form. Dental plaque is the term commonly used for the biofilm formed on teeth; however, the term plaque has now been extended to encompass biofilms on all the oral surfaces. These biofilms consist of a complex microbial community embedded in a matrix of polymers of bacterial and salivary origin. They are inherently more resistant to antibiotics, antimicrobials and antifungal agents which make the use of novel antimicrobials, such as antimicrobial Photodynamic Therapy (aPDT) to be discussed in this chapter, all the more important.

Small alterations in an environment can lead to ecological shifts and subsequent population changes in these complex biofilms and in certain specific cases this may predispose to a more "pathogenic" microbial community. This is the basis of the "ecological plaque hypothesis",[1] which can be used to understand the microbial aspects of a range of oral infections.

13.2. Caries

Dental caries can be defined as the localized demineralization of the tooth tissue by various acids produced by bacterial fermentation of dietary carbohydrates. Treatment of the disease involves removal of the damaged tooth tissue and its replacement with a restorative material. This disease is arguably the most common, chronic infectious disease in humans. Caries can be split into two categories: coronal (crown) caries and root surface caries.

13.2.1. Coronal Caries

Coronal carries can occur on all surfaces of the crown where the supragingival plaque biofilm is allowed to develop and mature. However, it is most commonly associated with pits and fissures and approximal sites. A major question has been which specific bacteria, if any, are involved in the progression of disease. The "ecological plaque hypothesis" suggests that a shift in the microbiota is brought about by increases in the amount and frequency of dietary fermentable carbohydrates. These substrates are fermented by the bacteria in the supragingival plaque biofilm leading to the production of acid end products like lactic acid. This acid production serves to lower the local pH and favour a shift in the biofilm population to acid-tolerant bacteria, for example, mutans streptococci. In addition to being aciduric, mutans streptococci are also highly acidogenic and can produce enough acid to lower the pH to levels that are inhibitory to many other bacteria within the biofilm (pH 4.5). The production of lactic acid is fundamental to the pathogenesis of mutans streptococci. It has been shown that mutant strains of *Streptococcus rattus* (a caries causing species in rats and hamsters) that lack lactate dehydrogenase activity fail to demineralize teeth in an animal model despite colonization and plaque formation.[2] Lactobacilli are usually found in supragingival plaque biofilms in low numbers; conversely, however, they have been shown to be present in elevated proportions in established caries lesions and have therefore been associated with the progression of the lesion rather than its initiation.[3–5] Non-mutans streptococci are thought rarely to be involved with the disease progression of carious lesions although they usually outnumber mutans streptococci. However, they have been shown to produce acid in an acidic environment.[6]

13.2.2. Root Surface Caries

Root surface caries, as the name implies, occur on root cementum or dentine and again are caused by a microbial biofilm. The disease is secondary to gingival recession since in a healthy mouth cementum and dentine are not exposed to the microbiota and are therefore unavailable for colonization. Gingival recession can be caused by a number of factors, for example old age, mechanical injury or periodontal treatment regimens. In industrialized countries the proportion of the population over 65 years of age is increasing, additionally the percentage of these remaining dentate is also increasing. A survey carried out by Steele *et al.*[7] showed, amongst other things, that in southern England 67% of patients over 60 years old were dentate compared to an equivalent cohort in 1962 which showed only 15% to be dentate.

The microbiological nature of the associated plaque biofilm is different from that associated with crown caries even though it is technically still supragingival plaque. The lesion has been shown to have a definite progression since changes in its clinical appearance are observed over time. Briefly, as the gingiva recedes new cementum/dentine is exposed and, in susceptible hosts, a lesion may start to form. The appearance of the lesion is described as "soft" and consists of a highly demineralized lesion replete with bacteria. Surrounding this, a further demineralized area is apparent and infiltrating bacteria can be observed; this lesion is actively carious. The progression of the lesion leads to a change in appearance and is categorized as "leathery". This is an intermediate stage and consists of a re-mineralized surface overlaying a heterogeneous mix of bacteria, de-mineralized tissue and re-mineralized tissue. A further progression is to a "hard" lesion; these are fully re-mineralized and inactive with respect to caries. The microbiology of this biofilm has been the subject of numerous investigations over the years; however, only within the last 10–20 years have the problems associated with sampling of the infected underlying dentine, and not the overlying biofilm, been identified and addressed.[8,9] In their study Beighton and Lynch[8] showed that the bacterial composition of the carious dentine biofilm associated with "soft" lesions consisted of significantly more lactobacilli and Gram-positive pleomorphic rods but, conversely, significantly fewer streptococci compared to the overlying plaque biofilm. Additionally, an increased number and/or proportion of *S. mutans* in "soft" lesions was observed compared to "hard" lesions or sound surfaces. *Actinomyces* spp. have historically been associated with root surface caries; however, studies by Brailsford *et al.*[10,11] showed *A. naeslundii* was not associated with active carious lesions and that *A. israelii* and *A. gerencseriae* predominated.

It is thought that the lesion occurs as a function of accumulation and subsequent stagnation of a plaque biofilm at the gingival margin. The nature of the microbiota is poorly understood but there seems to be no one species responsible for disease initiation and progression. What may be important is the presence of particular strains of a defined but heterogeneous group of bacteria that are particularly suited to that environment. Indeed, Sansone *et al.*[6] showed that the acidogenic and aciduric microbiota associated with a carious lesion is far more diverse (approx. 25 taxa) than the corresponding acidogenic and aciduric microbiota associated with sound root surfaces (approx. 8 taxa).

13.3. Periodontal Diseases

Periodontal disease defines a broad group of diseases affecting the periodontal tissues, the commonest of which are inflammatory processes of the gingiva and tissues attaching the tooth. These diseases are usually associated with microbial infection due to accumulation of a plaque biofilm and calculus.

13.3.1. Gingivitis

The bacteria and their extracellular products present within the plaque biofilm on the surfaces of teeth at the gum margin can cause inflammation. This is termed gingivitis and is the most common of the periodontal diseases usually brought about by poor oral hygiene. It can be defined as "a non-specific inflammatory process of the gingiva (gum) without destruction of the supporting tissues". A complex range of gingival diseases are recognized and have been classified into two main groups with 12 headings and numerous subheadings.[12] This disease is usually reversible and on removal of the biofilm (*i.e.* a return to good oral hygiene) the tissues revert to a healthy clinical state. It is likely that the entire human population suffers to some extent from this disease. The microbiota associated with dental plaque-induced gingivitis is different from that associated with health. In addition to an increase in biofilm mass (for example due to poor oral hygiene) the composition shifts from one dominated by streptococci[13] to one where *Actinomyces* spp. are dominant.[14,15]

13.3.2. Periodontitis

Periodontitis refers to a group of more advanced and related diseases within the broad heading of periodontal disease. It can be defined as "an apical extension of gingival inflammation to involve the tissues supporting the tooth (periodontal ligament and bone)". The destruction of the fibre attachment results in a periodontal pocket. There are 48 specific periodontitis disease categories now recognized.[12]

By far the most common is chronic periodontitis, which is the major cause of tooth loss in the adult population. The disease is mediated by the microbiota forming the plaque biofilm on the tooth surface. Additionally, and as a consequence of the immune response elicited by the bacteria, further destruction may occur due to the host inflammatory response.

The biofilm present in the gingival crevice and later in the periodontal pocket is extremely diverse with up to 100 culturable species from a single pocket.[16] Since such a diverse microbiota is present, trying to identify the particular species responsible for disease initiation and progression is a very complex and difficult undertaking. The most plausible explanation for the aetiology is again supplied by the "ecological plaque hypothesis". Briefly, the primary inflammation events due to a large biofilm mass at the gingival margin increase the flow of gingival crevicular fluid (GCF; a tissue exudate that bathes the periodontal tissues) thus changing the local environment and allowing proteolytic and anaerobic species to predominate. GCF is present in small quantities in health but in much larger volumes in diseased sites and contains a wide range of complex molecules derived from a number of sources *e.g.* serum, connective tissue and epithelium.[17] In addition, polymorphonuclear neutropathic granulocytes and monocytes are present in GCF in the pocket.

The World Workshop on Clinical Periodontology (American Academy of Periodontology Consensus report 1996) has designated three species as etiologic agents of periodontitis in a susceptible host, namely *Actinobacillus actinomycetemcomitans*, *Porphyromonas gingivalis* and *Tannerella forsythensis*.[18] The findings from the majority of other microbiology studies are based on data derived from the cultivable microbiota. However, it has been estimated that only 50% of the oral microbiota is cultivable.[19,20] More recently molecular techniques have been used to detect and identify the uncultivable portion of this highly diverse biofilm.[21,22] The aetiology of periodontal disease is further complicated by a range of predisposing factors. The susceptibility of the host has been shown to be important. A specific genotype of the polymorphic interleukin-1 (a pro-inflammatory cytokine and key regulator of the host responses to microbial infection and a major modulator of extracellular matrix catabolism and bone resorption) gene cluster is associated with severity of periodontitis.[23] There is also evidence to suggest that smoking is a significant risk factor for periodontal disease-associated tooth loss.[24] Additionally diabetes mellitus is a major risk factor for chronic periodontitis and the more severe and rapidly progressing forms.[25,26]

As discussed previously, because of the relationship between the presence of bacterial biofilms and caries and periodontal diseases the treatment regimes are essentially based on mechanical removal of these biofilms. However, many individuals are unable or unmotivated to perform tooth brushing with the regularity and efficiency necessary, so that in recent decades there has been an increased interest in the possibility to replace or complement mechanical therapy with the use of antiseptic agents or antibiotics. However, given the significant increase in the emergence of species resistant to traditional antimicrobial agents and the inherent resistance of biofilms, it has increased the scientific interest in developing alternative antimicrobial therapies that can act as adjuncts in the control of plaque, thus assisting in maintaining the oral health of individuals.

13.3.3. Phototherapy

The therapeutic use of light phototherapy is a term that includes all treatments that use non-ionizing radiation for therapeutic purposes inducing reactions in the human body that may cause benefits to patients. This process can occur when we are exposed to natural light, UVA or UVB. Historically, this treatment has been recommended for several purposes, with applications in the treatment of winter depression in the regulation of circadian rhythm and sleep disorders. Sometimes it is necessary to use a photosensitive drug or photosensitizer with the use of light to achieve specific effects. This process is known as photochemotherapy. In such cases, the drug is activated by light reacting with specific cells that may be causing the disease. A good example of this therapy includes the use of a UV lamp to treat newborns with jaundice where light triggers the release of a chemical reaction that allows the body to eliminate the yellow pigment (photosensitizer) characteristic of the disease.[27]

Since the advent of laser lights in the past century, the possibilities of the therapeutic use of light have expanded in all areas of science. According to their interaction with the irradiated tissue, lasers can be classified into two major groups: surgical lasers, or lasers operating at high power, and non-surgical lasers, also known as therapeutic or lasers that operate at low power.[28] Recently, light emitting diodes (LEDs) have been used in medicine as an alternative to the use of conventional lasers. LEDs are semiconductor devices that emit light when biased properly in the visible or invisible range. Although both lasers and LEDs produce monochromatic lights, LEDs do not show good control and coherence, resulting in emission bands that are wider and reduced cost of therapy.

Regarding the effect of the lights in the ability to kill bacteria, we know that this effect is highly influenced by the type of light used. The antimicrobial action of lasers operating at high power is well established[29] in periodontal therapy, root canal treatment[30,31] and sterilization of instruments and implants of titanium.[32] In general these lasers require high doses of energy and reach the surface irradiated with high temperatures promoting sterilization by local thermal action.

The low-power lasers, in turn, are meant to restore the biological balance by improving the conditions of cellular tissue vitality, for example, by modulating anti-inflammatory responses. These effects are generated by photochemical and photoelectric properties of aPDT rather than a thermal action. In the specific case of aPDT, the antimicrobial effect can be achieved once the photosensitizing drug is activated by a low-intensity light (laser or LED) passing to an excited state and generates substances that can damage and ultimately kill the treated cell.

13.4. Antimicrobial Photodynamic Therapy

aPDT is a technique that can potentially reach harmful cells without affecting normal tissues of the host. This therapy was initially designed for the treatment of cancer, based on the observation that some non-toxic molecules, such as photosensitizers, accumulate primarily in malignant cells. Using this technique, the photosensitizer is initially administered to the patient systemically, and by itself does not cause damage to healthy tissues. However, when a light (usually a laser or LED) is applied to tissues containing the drug, the drug is activated and tissues are destroyed rapidly, in precisely the places where the radiation is beamed. Thus, careful application of the light beam can guarantee that the therapy reaches the diseased tissues selectively.

Thus, photosensitizers are molecules that have the special property of absorbing light energy and use that energy to promote chemical reactions in cells and tissues undergoing photodynamic therapy. To this end, the photosensitizer should be able to absorb light at a wavelength of light emitted by the equipment used. In this process, there is only an adverse effect since some drugs may produce skin sensitivity, which involves the removal of the patient from any light source during the period in which the drug is active in the human body. The effectiveness of therapy combines two principles: the preferential accumulation of photosensitizer in the target cells with a precise irradiation of light, which allows localized and selective use of aPDT.

13.4.1. Antimicrobial Action of Photodynamic Therapy

The rapid emergence of resistance among pathogenic bacteria has led to a growing interest in the scientific community in finding alternatives to antimicrobial therapy which make the emergence of resistant species unlikely. Examples of such therapies include the use of bacteriophages, antimicrobial peptides and photodynamic therapy.[33] Studies evaluating the susceptibility of photodynamic therapy on normal and resistant strains found that both strains are susceptible to photodynamic therapy when the appropriate combination of photosensitizer and light is used.[34]

The ability of a component to absorb incident light does not necessarily mean that it can act as a photosensitizer. To be clinically effective, the photosensitizer must have absorption peaks near the wavelength of light used and must not show toxic effects to the host.[35]

In such cases, photodynamic therapy involves the localized application of drug and light on the infected area as opposed to its application for cancer treatment, where the drug is usually administered systemically. aPDT aims to achieve a sufficiently high antimicrobial effect on the causative agent of infection by selective action of the photosensitizer and light on the target organisms, avoiding damage to healthy tissues of the host. During this process, cellular photosensitive components pass to an excited state when exposed to light of a wavelength that is further characterized by the passage of electrons to higher energy levels. In this excited state, the photosensitizer can interact with molecular oxygen initiating the formation of highly reactive singlet oxygen (photo processes Type II) or interact with other molecules as electron acceptors resulting in the production of hydroxyl radicals and other organic molecules (photo processes Type I).[36] The products of these photochemical reactions can then damage the essential components of cells or irreversibly alter their metabolic activity, resulting in bacterial death.[37]

Among the factors that may influence the success of the antimicrobial effect by aPDT are those derived from photobiological factors and those derived from irradiated bacterial cells. Among the factors are photosensitizer concentration, the time of pre-irradiation, the wavelength of light, the power of the device, the irradiation time and the density of energy used. Among the bacterial factors that may interfere with the process are those related to the physiological characteristics of the bacterial cell, such as the growth phase, whether they are growing as biofilms and whether multiple species are involved.

13.5. aPDT Against Cariogenic Bacteria

A large number of studies have shown that photodynamic therapy is effective in reducing the numbers of bacteria in broth culture.[38,39] However, the bacteria that cause caries are more often found in the oral cavity organized in biofilms on the surface of the teeth and surrounding tissues. This intrinsic resistance shown by biofilms appears to be related to the presence of matrix polysaccharides, changes in cell wall composition, slower growth rate as well as the expression of genes for resistance to antimicrobial agents.[40]

Currently, there are relatively few groups that evaluate the antimicrobial potential of aPDT on biofilms. Previously, Wilson *et al.*[41] showed that the combination of aluminium disulfonated phthalocyanine (AlPcS2) and a laser diode emitting at a wavelength of 660 nm was effective in significantly reducing the numbers of *S. sanguinis* grown as biofilms. In a study involving multi-species biofilms, Wood *et al.*[42] showed a high mortality rate (results were obtained by transmission electron microscopy and confocal laser microscopy and no quantitative analysis was performed) when bacterial biofilms *in situ* were treated with a cationic photosensitizer and exposed to white light of 400 W for 30 min. In 2002, O'Neill *et al.*[43] showed a significant antimicrobial effect was obtained in oral biofilms grown on nitrocellulose discs when they were sensitized with toluidine blue and then irradiated with red light with an energy density of 81.9 $J\,cm^{-2}$. In 2005, Zanin *et al.*[44] found that photodynamic therapy was effective to reduce up to 99% of the viable population of *S. mutans* grown *in vitro* and sampled at various times (3, 7 and 10 days).

Comparative studies evaluating the antimicrobial effect of aPDT using conventional lasers and LED lights have been performed on biofilms of *S. mutans* grown in an "artificial mouth" biofilm model (Constant Depth Film Fermentor[45]). The model enabled the simulation of the formation of acquired pellicle on hydroxyapatite discs, while the biofilms of *S. mutans* were grown in an environment continually bathed in artificial saliva. Furthermore, the choice of this model enabled precise control of the frequency distribution of sucrose, temperature, atmosphere and thickness of biofilm. The results demonstrated a significant reduction in the counts of bacteria in biofilms of *S. mutans* although there was no significant difference in antimicrobial effect promoted when a Helium-Neon (HeNe) laser or an LED light were used. Interestingly, confocal laser microscopy showed that most bacterial killing by aPDT occurred predominantly in the outer layers of the biofilm, which may have been caused by the difficulty of the photosensitizer to penetrate.[44] Such issues could be overcome by the choice of photosensitizers with greater ability to penetrate cationic biofilms as well as by irradiation of the biofilms internally through the use of a fibre optic.

Since the photodynamic therapy can be effectively employed to reduce the microorganisms in the biofilm, thereby helping to prevent the onset of caries lesions, another very interesting application is the use of photodynamic therapy directly on dental caries. Although further research must be performed, some studies have observed that *S. mutans* can be killed in a position very similar to those found in a carious lesion. Burns *et al.*[46] observed reductions in the number of viable *S. mutans* when they were embedded in a collagen matrix with varying degrees of demineralization.

The use of photodynamic therapy for the disinfection of caries could be beneficial for reducing the amount of dental tissue removed during cavity preparation and promoting an effective decontamination of the treated area before the sealing of the lesion. The treatment might become less traumatic for the patient and faster for the dentist besides improving the prognosis by decreasing the need for dental tissue during cavity preparation.

13.6. aPDT and Periodontal Disease – *in vitro* Studies

Several *in vitro* studies have demonstrated that aPDT using different light sources and photosensitizers can eradicate periopathogenic bacterial strains *in vitro*. Wilson was the first to demonstrate effective killing of pathogens including *P. gingivalis*, *Fusobacterium nucleatum* and *A. actinomycetemcomitans* in homogenous planktonic cultures[35] and planktonic human subgingival plaque samples[47] using low-power laser light and dyes such as toluidine blue, methylene blue and crystal violet. Importantly, these studies showed that irradiation in the absence of photosensitizer had no detectable effect, nor did the photosensitizer by itself have a bactericidal effect. Furthermore, two later studies published by this group[48,49] examined aPDT dosimetry and mechanism of action during lethal photosensitization of *P. gingivalis*. In particular, they tested photosensitizer concentration, light dose (varied by changing exposure time), pre-incubation time with photosensitizer, presence or absence of serum, pH and bacterial growth phase in an attempt to optimize the antibacterial effect. An energy dose of 4.4 joules with 25 μg ml^{-1} photosensitizer provided the greatest effect, and efficacy was found to decrease with higher photosensitizer concentrations. Wilson's group has also demonstrated that these and other strains are susceptible when they comprise a biofilm.[50]

Subsequent to the original published work by the Wilson group, several others have demonstrated the efficacy of aPDT against oral microbes. Indeed, when the susceptibility of the three "key" periodontal pathogens grown as biofilms *in vitro* has been investigated the reductions have been in the order of 99.9%.[51] Pfitzner *et al.*[52] used several experimental photosensitizers and light doses in excess of 20 J cm^{-2} to completely eradicate cultures of *P. gingivalis* and *F. nucleatum*. Furthermore, Chan and Lai[53] showed that Gram-negative anaerobes were susceptible to methylene blue mediated aPDT, with *F. nucleatum* and *A. actinomycetemcomitans* being reduced by about 95% compared to the non-treated control. Another group,[54] using a porphyrin-derivative photosensitizer, reported a light dose-dependent $>6\ \log_{10}$ reduction in viable count of *P. gingivalis*. Prates *et al.*[55] were able to achieve 2–3 $\log_{10}$ reductions from control in planktonic cultures of *A. actinomycetemcomitans* with malachite green and energy doses of 6–9 J cm^{-2}. Another group[56] evaluated the effectiveness of aPDT *in vitro* on biofilms comprised primarily of the oral bacteria *Actinomyces israelii*, *F. nucleatum*, *P. gingivalis* and *Prevotella intermedia*. In this study extracted human teeth were inoculated for biofilm growth, and it was found that up to 80% eradication could be achieved after treatment with methylene blue and activation with 30 J cm^{-2} laser light. Matevski *et al.*[57] observed a light dose dependent killing of *P. gingivalis* and several other periopathogens in the presence of serum and blood using a helium-neon laser and the photosensitizer toluidine blue O. Studies such as this, which incorporate serum and proteins into an *in vitro* test model, are important for predicting the overall efficacy of aPDT in an oral *in vivo* environment. Other *in vitro* antimicrobial PDT models have utilized sub- or supragingival mixed plaque samples isolated from human subjects, as opposed to heterogeneous cultures grown in the laboratory. For example, Qin *et al.*[58] used such a model to demonstrate that aPDT consisting of 1 mg ml^{-1} toluidine blue O (TBO) and an energy dose of 12 J cm^{-2} was effective in eliminating supragingival plaque cultures. Furthermore, they showed that efficacy increased as power density and energy dose were increased, but decreased as photosensitizer concentration was increased (due to interstitial absorbance of light by bulk sensitizer molecules).

Many studies evaluating the effects of photodynamic therapy on periodontal bacteria have focused on the therapeutic potential of therapy as well as reducing the viability of the microorganisms. However, studies have shown that aPDT, in addition to having an antimicrobial effect, seems to have some effect on microbial virulence factors. Lipopolysaccharides are one of the most important virulence factors of the bacteria which cause periodontal disease, having an important role by inducing the host inflammatory response and stimulating the release of cytokines. Komerik *et al.*[59] demonstrated that the ability of Gram-negative bacteria to induce cytokine production was reduced after exposure to a HeNe laser in the presence of TBO. Furthermore, aPDT is able to inactivate destructive host cytokines tumour necrosis factor-alpha (TNF-alpha) and interleukin (IL)-1 beta.[60] Other important virulence factors for the development of periodontal disease are the bacterial enzymes capable of degrading the host tissue. *P. gingivalis* produce proteolytic enzymes that are important mediators of cellular destruction during periodontitis. Packer *et al.*[61] demonstrated that exposure of *P. gingivalis* photodynamic therapy using the combination of TBO and HeNe laser reduced the proteolytic activity of these enzymes by 50%. Furthermore, aPDT is able to inactivate destructive host cytokines tumour necrosis factor-alpha (TNF-alpha) and interleukin (IL)-1 beta.[60]

If we consider the clinical use of aPDT in periodontal disease we need to take into account the removal of viable pathogenic microorganisms and reduce the expression of some of its virulence factors. This represents a considerable advantage compared to the use of antibiotics and conventional antiseptics since the latter are capable of decreasing the viability of the bacteria but have no effect on their virulence factors. Therefore, with all of these factors combined, it would be anticipated that by using aPDT there would be a more favourable healing environment.

The possibility of the occurrence of adverse reactions with the use of aPDT has been studied. Animal models have shown that concentrations of photosensitizer and energy densities used to cause damage to microorganisms have minimal effects on host tissues.[27] Furthermore, Soukos *et al.*[62] evaluated the effect of chlorin e6 conjugated to pentalysine associated with red light to kill *P. gingivalis* in the presence of epithelial cells. The study showed that it was possible to selectively eliminate *P. gingivalis* without causing significant damage to skin cells.

The use of traditional antimicrobial therapies for the treatment of periodontitis has two obvious disadvantages compared with the use of aPDT. Initially, there is a difficulty in maintaining the antibiotic or antiseptic in high enough concentrations for a period of time sufficient to cause a significant reduction in the viability of bacteria and, secondly, this strategy favours the emergence of resistant bacterial strains.

13.7. Animal Models for Oral aPDT

Animal models have been used extensively by researchers in the field of aPDT to demonstrate both safety (*i.e.* absence of damage to host tissue) and efficacy in the mouth. As discussed previously in the chapter, the aetiology of periodontal disease is very complex and likely is driven by a combination of subgingival microbiota changes and resulting host immunological response.[16] While *in vitro* studies showing antimicrobial efficacy against implicated periopathogens are useful to demonstrate proof-of-principle for oral aPDT applications, only studies performed *in vivo* can accurately predict clinical response. This is because the presence of blood, saliva, interstitial fluid and other factors in the periodontal microenvironment has an impact on aPDT efficacy that would not be observed in less complex *in vitro* efficacy models.

The use of rodent models for aPDT in the oral cavity provided early evidence that the *in vitro* efficacy observed against periopathogens carries over to the more complex oral microenvironment. In one of the earlier *in vivo* reports, investigators evaluated the efficacy of aPDT against the periopathogen *Porphyromonas gingivalis* using a rat oral model.[63] In this study, maxillary molars of rats were inoculated with *P. gingivalis* and exposed to up to 48 joules of 630 nm laser light in the presence of the phenothiazinium photosensitizer toluidine blue O (TBO). Immediately following treatment, microbiological specimens were collected using paper point sampling. At three days post treatment animals were sacrificed and jaw samples were excised for histological sectioning and evaluation. In addition, other animals were sacrificed at 90 days post treatment and alveolar bone loss was measured using morphometric and radiographic methods. Results showed that no viable bacteria were detected after application of 1 $mg\,ml^{-1}$ TBO in combination with any of the light doses evaluated. At lower concentrations of TBO (0.01 and 0.1 $mg\,ml^{-1}$) and light doses of 6 joules, less bacterial killing (1–2 $\log_{10}$ reduction in viability from control) was observed. Histological examination of tissue samples showed no adverse effects on cell/tissue morphology in the aPDT treated samples.

Several more recent studies using rodent oral aPDT models have been conducted by a Brazilian group. Almeida *et al.*[64] used a ligature-induced model of periodontal disease in rats, and aPDT treatment in that study consisted of application of the photosensitizer methylene blue followed by activation with laser light. The results of this study showed significantly less bone loss in the aPDT treatment group compared to the control groups at both 5 and 15 days post-treatment. In a subsequent study published by the same group,[65] the investigators focused on evaluating furcation defects in rats resulting from ligature induced periodontal disease. In that study, the aPDT treatment group again exhibited less bone loss compared to the control groups at day 7 and day 15. A third report from the same group[66] used ligatures and chemical induction of diabetes to evaluate the effect of the diabetic state on improvement of periodontal disease *in vivo*. The results confirmed that aPDT using methylene blue significantly reduced bone loss at all time points in both diabetic and non-diabetic animals. Finally, a recent study by the same group[67] used a rodent model to evaluate the potential to use aPDT for periodontal treatment in immunocompromised individuals. In that study, experimentally immunosuppressed rats were induced to develop periodontal disease and treated with scaling a root planing prior to aPDT consisting of TBO and laser illumination. Results of the study showed significantly less bone loss in animals treated with the full aPDT protocol, as opposed to those treated with SRP or TBO alone. In addition, histological analysis showed healthier periodontal ligament tissue with less evidence of bone resorption in aPDT treated samples. Results were similar regardless of whether the animal had been previously immunocompromised by chemical injection. While this series of Brazilian studies is important in showing tissue/bone effects after periodontal treatment with aPDT, it should be noted that none directly examined microbiological reductions of periopathogens in response to treatment in an *in vivo* model of periodontal disease.

In an *in vivo* study that did examine microbiological end points after aPDT, Qin *et al.*[68] compared the therapeutic efficacy of aPDT to conventional scaling and root planing in a rat model of ligature-induced periodontal disease. Treatment in the aPDT group was administered using 12 J cm^{-2} of 635 nm laser light and 1 mg ml^{-1} TBO, while treatment in the control group consisted of manual debridement followed by cleaning with a topical 0.5% chlorhexidine solution. The authors concluded that aPDT produced similar results to scaling and root planing with approximately 4% bacterial survival in both conditions. In addition, they reported that clinical signs of inflammation were reduced by both treatments with no signs of injury to host tissue. Lin *et al.*[69] used a different model in which aPDT was used to treat bacterially infected excision wounds in the oral cavities of rats. Again, the photosensitizer used in this study was TBO and illumination was performed using a 635 nm diode laser system. The authors observed a light dose dependent killing effect whereby 48 joules cm^{-2} produced a 97% kill of human supragingival plaque bacteria inoculated into the wound site. Furthermore, progression of the lesions over time was reduced in the aPDT treatment group *vs.* non-treated control. The authors concluded that aPDT could serve as a valuable alternative to other chemical disinfectants such as chlorhexidine gluconate for the treatment of oral infections such as periodontal disease, peri-implantitis and endodontic infection.

Several groups have also used large animal models to investigate the effects of antimicrobial PDT in the oral cavity. In a beagle model, in which healthy mouths were inoculated and infected with *P. gingivalis* and *F. nucleatum* and subsequently treated with aPDT,[70] it was observed that clinical measures of periodontal disease such as inflammation and bleeding on probing were significantly decreased from control. In addition, the number of positively colonized *P. gingivalis* sites was greatly reduced from that of non-treatment control animals. In a related model to ligature-induced periodontal disease, Hayek *et al.*[71] compared the effects of aPDT and conventional treatment on microbial reduction in ligature-induced peri-implantitis in dogs. Implants were placed in 18 third premolars in dogs and, following osseointegration and ligature induced peri-implantitis, animals were divided into two treatment groups. One group received conventional treatment of mucoperiosteal flaps to scale implant surfaces followed by irrigation with 0.12% chlorhexidine, while the other received aPDT consisting of mucoperiosteal scaling followed by placement of azulene (final concentration 0.01% w/w) paste photosensitizer and activation with low-power 660 nm laser light. Results showed that *Prevotella* spp., *Fusobacterium* spp. and *S. Beta-haemolyticus* were significantly reduced in both groups, and that the photosensitizer did not stain the implant or the surrounding tissue.

The authors concluded that aPDT treatment was an effective non-invasive alternative to conventional therapy for peri-implantitis. Finally, Shibli *et al.*[72] also used a dog model of ligature-induced peri-implantitis and exposed infected sites to aPDT using TBO (100 μg ml) and laser light (685 nm for 80 seconds, energy density of 200 J cm^{-2}). In block biopsies performed at 5 months post-treatment, this group observed new bone formation in all samples. Unfortunately, as all groups (including controls) in this study underwent aPDT treatment it was difficult to ascertain the relative contribution of aPDT to bone formation and reosseointegration in the mouth.

In addition to the *in vivo* models designed to show the efficacy of antimicrobial PDT in the oral cavity, there have also been several published reports focusing on safety in animal models. In a 2002 study by Komerik *et al.*,[59] rat buccal mucosal tissue was subjected to varying concentrations of the photosensitizer TBO with subsequent exposure to various power densities of 633 nm laser light. Additionally, control groups consisting of light-alone, photosensitizer-alone and no treatment were performed. After treatment, mucosal tissue was removed and examined histologically while TBO biodistribution was assayed using digital fluorescence imaging. The authors observed that aPDT resulted in no observable necrotic or inflammatory changes in the buccal mucosa. Furthermore, fluorescence imaging showed that TBO penetrated only to the depth of the epithelium, with virtually no fluorescence observed in the underlying connective tissue. These results suggested that TBO-mediated aPDT does not adversely affect buccal mucosa. Another study[73] subjected oral mouse gingival tissue to aPDT using TBO with 635 nm laser light at an energy dose of 60 J cm^{-2}. Additionally, control groups of light only (140 J cm^{-2}), photosensitizer only (2.5 mg ml^{-1}) and no treatment were also performed. Subsequent histological examination showed that aPDT did not alter the gingival tissue structure or gross cellular morphology as compared to untreated controls.

Taken together, the *in vivo* studies summarized above build upon the body of *in vitro* evidence supporting safety and efficacy of aPDT in the oral cavity. While several of the large animal studies examined clinically relevant end points such as bleeding and inflammation, none specifically addressed soft tissue reattachment – a necessary outcome for resolution of periodontal disease in humans. The next step was therefore to advance aPDT into controlled clinical studies in order to demonstrate that preclinical results translated to clinical outcome.

13.8. Human Clinical Applications for Oral aPDT

Scaling and root planing (SRP) is currently known as the gold standard for treatment of periodontal disease. This technique has been practised for many years and involves manually removing plaque/biofilm deposits on the tooth below the gum line followed by refinishing of the surface of the root to promote soft issue reattachment. SRP is well known to result in modest clinical improvement in most patients suffering from mild to severe periodontal disease.[74–76] However, a recent meta-analysis found that while SRP alone was somewhat effective in improving clinical outcome, the most effective overall approach was SRP combined with a local antibiotic therapy.[77] Furthermore, Bonito *et al.*[78] conducted a review of over 40 published studies using local adjuncts to SRP and concluded that adding an antibacterial modality to the treatment of periodontal disease had a measurable positive impact on probing depth (PD) and clinical attachment level (CAL). These reviews suggest that manual removal of bacterial biofilms using SRP does not completely remove all pathogenic organisms, and that combining this with an approach to eradicate the small number of remaining organisms is the best way to create a healthy microenvironment for healing and reattachment. As a result, multiple adjunctive approaches have been evaluated including systemic antibiotics,[79] locally applied antibiotics[80] and topical disinfectants such as chlorhexidine gluconate.[81] In practice, however, growing concerns over antibiotic resistance have made the widespread use of systemic antibiotics impractical. Furthermore, the difficulty in applying topical agents sub-gingivally combined with the propensity for rapid flushing due to excessive secretion of gingival crevicular fluid has limited the effectiveness of most topical adjuncts to SRP.

In recent years lasers have been developed for the thermal removal of sub-gingival plaque and biofilm[82,83] with varying degrees of reported success. More recently, clinical studies using non-thermal laser systems in the mouth for aPDT have also emerged in the literature. As noted previously in this chapter, but worth reiterating in this context, aPDT eradicates biofilms *via* an oxygen-mediated photoreaction as opposed to traditional lasers for plaque removal which rely on thermal effects. Importantly, clinical studies of periodontal therapies can focus on microbiological end points (absolute reductions in periopathogenic or total sub-gingival bacterial counts), clinical end points (bleeding on probing, pocket probing depth, attachment level, *etc.*) or a combination of both. While clinical end points are relatively easily measured by a trained dental professional, studies involving microbiological end points require carefully constructed protocols and assays and can be prone to error during sampling. For these and other reasons relating to regulatory authority requirements, most of the published clinical work to date using aPDT to treat periodontal disease has focused on clinical end points as success indicators.

In one of the first published reports of aPDT for the treatment of periodontal disease in humans, ten patients with aggressive periodontal disease were treated using either methylene blue and laser light or conventional scaling and root planing.[84] This small study utilized a split-mouth design and patients were measured at pre-treatment baseline and 3 months post-treatment for various clinical parameters including bleeding on probing, probing depth, gingival recession and clinical attachment level. The investigators found significant improvements in all measures from baseline to 3 months post-treatment, and no complications or adverse events were noted. The conclusion of the study was that aPDT produced clinical improvements in periodontal disease similar to that using the gold standard of SRP. It should be noted that the split-mouth study design has been criticized in the past because treatment in one quadrant can cause effects on the other side of the mouth (*i.e.* whole-mouth effect).[85]

In 2009, the same group reported on further results from the same ten-patient study,[86] in which samples of gingival crevicular fluid (GCF) were collected from aPDT and SRP treated periodontal pockets and subjected to protein analysis (ELISA) to quantify the cytokines TNF-alpha and RANKL (receptor activator of nuclear factor kappa-B ligand). The results of the study showed that both aPDT and SRP treatments led to significant reductions in TNF-alpha and RANKL in GCF at a period of 90 days post-treatment. The cytokine reductions observed were similar across the two treatment groups. These results show that aPDT using methylene blue and laser activation can lead to significant reductions in host-derived cytokines in the periodontal microenvironment. Since it is known that chronic sub-gingival bacterial colonization leads to immune activation and production of inflammatory cytokines, and that the presence of these cytokines over sustained periods plays a part in the tissue/bone destruction characteristic of periodontal disease, it can be concluded that inactivation of cytokines in the sub-gingival space would at least delay the progression of periodontal disease. The correlation in clinical measures and cytokine profiles across the two studies would seem to support this conclusion.

In another study using the Helbo Photodynamic Systems aPDT system, consisting of methylene blue with red laser light activation, 20 subjects were treated again using a split-mouth study design.[87] In this design all patients received conventional SRP before randomization into groups receiving aPDT treatment in two of four quadrants. The clinical parameters measured at baseline, 1 week and 3 months post-treatment were sulcular fluid flow rate and bleeding on probing. In addition, measurements of probing depth, gingival recession and relative attachment level, as well as degree of tooth mobility and furcation involvement, were taken at baseline and 3 months post-treatment. The results of this study showed that at 3 months post-treatment, relative attachment level, probing depth, sulcular fluid flow rate and bleeding on probing improved significantly more in the aPDT group as compared to the SRP-alone group. As a result, the investigators concluded that adjunctive aPDT can improve outcomes seen with SRP in patients with chronic periodontal disease.

In 2007, Andersen *et al.*[88] reported the results of a pilot human study to evaluate oral aPDT using the Periowave™ system, consisting of methylene blue in liquid formulation activated using 670 nm laser light. This randomized controlled trial involved 33 adult subjects presenting with chronic adult periodontal disease. Initially, probing depth measurements were obtained from all teeth (six sites per tooth) and candidate treatment sites with ≥ 6 mm depth were identified. In addition, bleeding on probing and clinical attachment level were measured before treatment. Subjects were randomly assigned to three groups: 1) SRP alone, 2) aPDT alone and 3) SRP with aPDT administered adjunctively. Measurements of probing depth, bleeding on probing and clinical attachment level were taken at 3, 6 and 12 weeks following treatment for each group. A total of 622 individual sub-gingival sites were treated as part of this study, and no treatment-related adverse events were reported. In the SRP with aPDT group, a significant improvement in attachment level was seen at 6 and

12 weeks compared to the SRP alone group (approximately 3-fold difference). Bleeding on probing was significantly improved in all three groups; however, the largest improvement was seen in the aPDT-alone group, in which bleeding decreased 71% at 6 weeks and 73% at 12 weeks. In the group treated with SRP and aPDT, there were also large improvements in probing depth, with mean reductions of 1.16 ± 0.39 mm and 1.11 ± 0.53 mm at 6 and 12 weeks, respectively ($p < 0.06$ for 6 weeks and $p < 0.05$ for 12 weeks compared to SRP alone). The results of this study showed that the Periowave™ aPDT system was effective when used adjunctively to the current standard therapy for periodontal disease. While this study suggested that aPDT alone showed clinical improvements similar to that of the SRP-alone group, eliminating the manual plaque/biofilm removal step in practice may not be advisable as presumably bacterial recolonization could occur faster on rough surfaces containing unremoved, previously treated, plaque deposits.

In a 2009 study, Chondros *et al.*[89] examined the clinical and microbiological effects of aPDT adjunctively to SRP in patients already receiving supportive periodontal therapy. In that study, 24 patients received sub-gingival SRP alone or SRP followed by a single treatment of aPDT. Multiple clinical parameters (bleeding on probing, pocket depth, clinical attachment level, full mouth plaque score, gingival recession) as well as microbiological assessments were taken at baseline, 3 and 6 months. It was found that patients receiving adjunctive aPDT exhibited a statistically significant improvement in bleeding on probing compared to SRP-alone at 3 and 6 months post-treatment, although no differences were observed in pocket depth, clinical attachment level and full mouth plaque score. Microbiological results for this study were variable. While both SRP-alone and SRP with adjunctive aPDT groups showed significant reductions from baseline in several common periopathogens, the aPDT group also had significantly lower levels of the bacterial species *Fusobacterium nucleatum* and *Eikenella corrodens* at 3 months as compared to the SRP-alone treatment group.

In the largest clinical study of aPDT in the oral cavity conducted to date, 121 Canadian patients were enrolled in a randomized, prospective, examiner-blinded, parallel-group study evaluating aPDT for the treatment of chronic adult periodontal disease. The results of this study, taking place across three major dental schools and one private clinical practice, have been submitted in full elsewhere for publication but will be summarized as part of this chapter. Subjects diagnosed with chronic periodontal disease, with at least 18 or more fully erupted teeth and at least 4 measurement sites exhibiting probing depth of 6–9 mm in at least two quadrants of the mouth were eligible to participate in this study. Subjects were randomized into two protocol groups: (1) Control Group receiving SRP-alone and (2) Test Group receiving SRP followed by aPDT treatment in qualifying sites. aPDT was carried out using the Periowave™ system, consisting of sub-gingival placement of a methylene blue based photosensitizer formulation followed by 60 second (per site) illumination with 670 nm wavelength laser light. Two different energy doses were used over the course of the study (15 patients received 9 joules per aPDT site and 48 patients received 13 joules per aPDT site), and the population treated with the higher dose was analysed along with the total data set as well as separately (post-hoc) to determine whether an increase in energy dose influenced clinical efficacy. Two clinical measurement indices (clinical attachment level and probing depth) as well as a scoring evaluation (bleeding on probing) were used as end points, with no microbiological assessment performed as part of the study. For the treatment protocol, SRP treatment was administered over two separate sessions (half mouth per session), and aPDT was carried out during the second visit in the test arm. Clinical outcome was assessed at 6 and 12 weeks after the final treatment visit. In all, measurements were taken from more than 4000 treated periodontal sites over the course of the study. Taking all aPDT treated pockets together (*i.e.* combining subjects that received 9 and 13 joule energy doses), with a single treatment of aPDT adjunctively to SRP the improvements achieved for clinical attachment level and probing depth at 6 and 12 weeks, while clinically significant, fell just short of statistical significance in terms of superiority to SRP alone ($p = 0.07$ and $p = 0.10$, respectively). However, in the subset of patients ($n = 48$) receiving the higher energy dose (13 joules per site), Periowave™ adjunctive to SRP resulted in highly statistically significant improvements in both attachment level and probing depth compared to SRP alone ($p < 0.0001$ *vs.* SRP alone for attachment level and probing depth). These results are summarized in Table 13.1. It should also be noted that clinical attachment level improvement for all measured pockets (both treated and untreated) in the aPDT treatment arm showed a statistically significant improvement ($p < 0.01$) over that in the SRP-alone arm. This observation implies a derived benefit from aPDT, most likely *via* a "whole-mouth effect", which is a derived benefit throughout the mouth in response to treatment of specific infected sites. This whole-mouth effect extended to probing depth reduction as well; analysis of probing depth reduction for all measured pockets in the aPDT arm demonstrated a statistically significant response ($p < 0.05$) compared to the SRP-alone arm. The results of this controlled multicentre clinical study clearly demonstrated the utility of aPDT administered adjunctively to the current standard of care in a North American population. In post-hoc analysis it was also shown that energy dose is a key variable, and there appears to be a clear relationship between energy applied and magnitude of

Table 13.1 Clinical outcomes in patients ($n = 106$) receiving SRP alone or SRP with adjunctive aPDT using a 13 joule energy dose. Values are reported as change in clinical attachment level or probing depth in millimetres from a pre-treatment baseline measurement.

	Clinical attachment level		Probing depth	
	SRP-Alone	SRP + aPDT	SRP-Alone	SRP + aPDT
Number of pockets	2349	2463	2352	2460
Week 6 change (mm)	−0.50 + 0.05	−0.69 + 0.05[a]	−0.69 + 0.03	−0.82 + 0.04[a]
Week 12 change (mm)	−0.54 + 0.05	−0.71 + 0.05[a]	−0.68 + 0.03	−0.85 + 0.04[a]

[a] $p < 0.0001$ *vs.* SRP alone for global treatment effect (week 6 and 12 results combined).

response that should be considered carefully in the application of aPDT in a clinical setting.

Since the concept of using aPDT to treat oral infections was first contemplated by Wilson and others in the early 1990s, several reviews have been published worldwide detailing the potential of this application.[90–95] The majority of these reviews have concluded that aPDT will emerge as an important treatment modality in oral/dental applications, with the caveat that further clinical evidence is needed to optimize the technology and drive adoption by clinicians. As noted in a more recent systematic review by Azarpazhooh *et al.*,[96] to date the success of aPDT in the clinic for the adjunctive treatment of periodontal disease has been variable. Reasons for this variability, while likely in part due to small sample sizes of studies published to date, can be theorized from preclinical studies that preceded this clinical work. As demonstrated in several preclinical aPDT studies,[51,97,98] the importance of delivering an optimal activating light energy dose cannot be understated. While it is difficult to pinpoint a precise power density or energy dose threshold for clinical use using preclinical results, it is clear that antibacterial efficacy drops off dramatically when insufficient light is delivered to the site of treatment. Importantly, insufficient light dose can also occur with the use of non-optimized photosensitizer concentrations, as very high concentrations of dyes such as methylene blue and TBO lead to extreme interstitial light absorption, effectively "shielding" the bacterial biofilm from the photoreaction. These considerations further accentuate the need to optimize not only the light source/energy dose, but also the photosensitizer formulation used in clinical application.

13.9. Case Study 1 – Adjunctive aPDT for Severe Aggressive Adult Periodontitis

13.9.1. Background

Patient #1 was a healthy 28-year-old male with no familial history of systemic diseases. The patient was not taking any medications at the time of presentation, and did not have a history of smoking. A thorough medical check-up including extensive blood work revealed no significant findings. The patient did report a history of early tooth loss in his immediate family, and was subsequently diagnosed with severe generalized aggressive periodontitis. Because of the severity and presentation of the condition, this patient was deemed to be a suitable candidate for aPDT treatment in addition to standard non-surgical debridement (SRP).

13.9.2. Treatment

All examinations and treatments were performed at the Department of Periodontics at McGill University in Montreal, Canada. After a complete initial periodontal examination, non-surgical SRP was performed under local anaesthesia over the course of three appointments. aPDT was administered after SRP at each appointment using the Periowave™ system, consisting of sub-gingival application of a liquid methylene blue-based photosensitizer formulation followed by 60 second illumination from a non-thermal 670 nm diode laser. The full treatment schedule was as follows: at appointment #1 SRP and aPDT were performed in quadrants 1 and 4, at appointment #2 SRP and aPDT were performed in quadrants 2 and 3 with aPDT repeated in quadrants 1 and 4 and at appointment #3 aPDT was repeated in quadrants 2 and 3. The time between each of these appointments was 2 to 3 weeks. Teeth 34 and 35 were extracted during appointment #2, and tooth 11 exfoliated spontaneously prior to therapy. The patient was seen every three months for maintenance recall therapy where supra- and sub-gingival scaling was performed and oral hygiene instructions re-enforced. Photographs and panoramic radiographs were taken prior to therapy and at 1 year post-treatment. The patient did not take any local or systemic antibiotics over the course of the treatment and follow-up periods.

13.9.3. Results

Figure 13.1 shows a photograph of the patient's upper and lower dentition prior to therapy. Note the inflammation of the gingival tissues and the heavy calculus deposits. Probing depths >6 mm and bleeding on probing were generalized. All the teeth had a class II or higher mobility. Figure 13.2 shows the panoramic radiographic presentation of the case pre-operatively. Note the extensive generalized alveolar bone loss. Figure 13.3 shows the same photographic view taken at a 1-year post-treatment follow-up visit. Figure 13.4 depicts the corresponding panoramic radiograph at 1 year post-treatment. The probing depths were reduced to <4 mm, and minimal bleeding on probing was

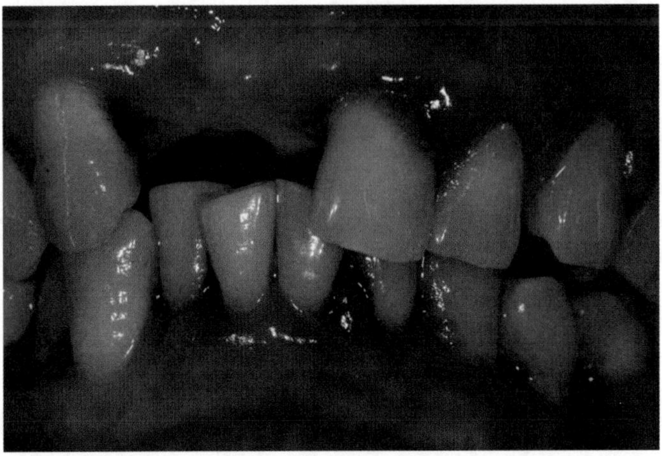

Figure 13.1 Dental photograph of Patient #1 taken prior to debridement and aPDT.

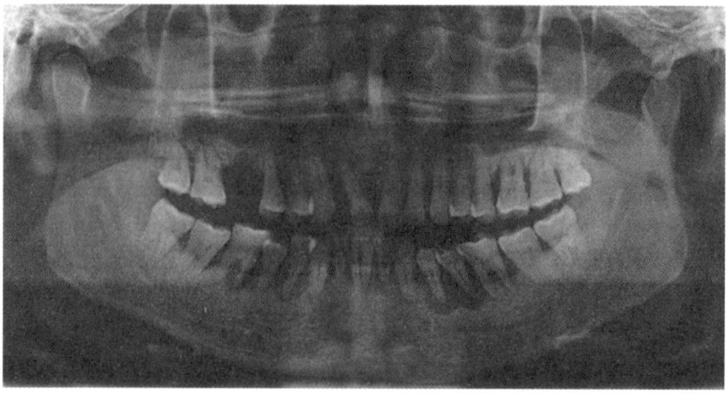

Figure 13.2 Panoramic radiograph of Patient #1 taken prior to debridement and aPDT.

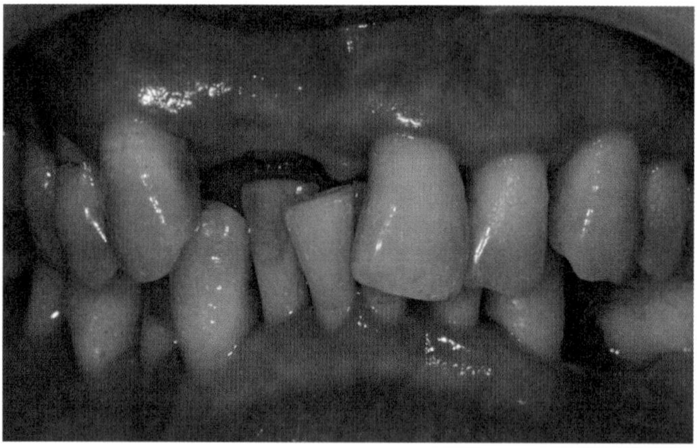

Figure 13.3 Dental photograph of Patient #1 taken at one year post-treatment. Treatment consisted of three appointments to conduct full mouth debridement with two rounds of adjunctive aPDT.

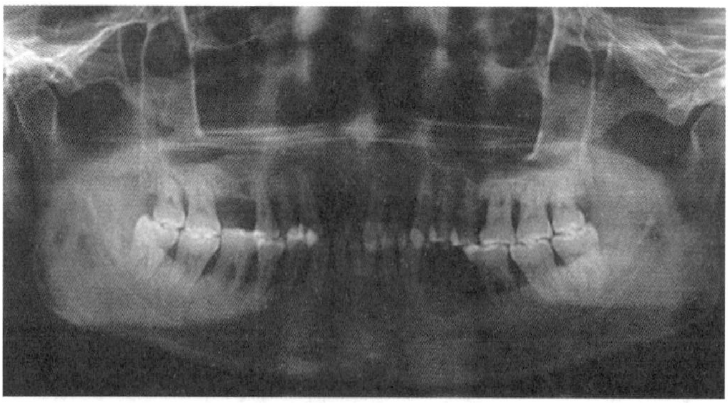

Figure 13.4 Panoramic radiograph of Patient #1 taken at one year post-treatment.

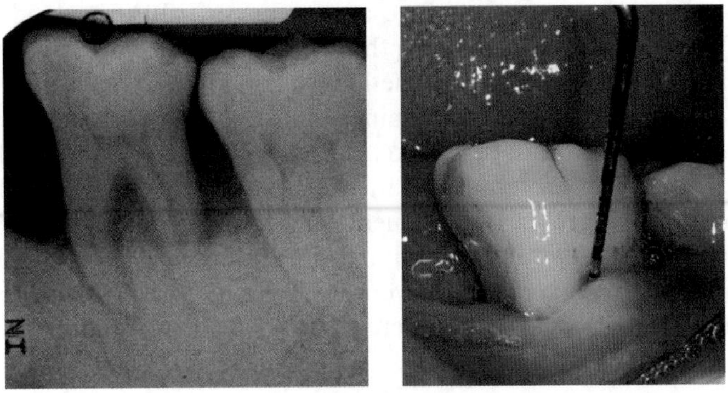

Figure 13.5 Dental radiograph and photograph of tooth 3.6 of patient #1 taken at one year post-treatment. Note the bone fill in the furcation area and the minimal probing depth.

observed. In addition, the tooth mobility was significantly reduced. Of special note was the minimal amount of gingival recession observed. Figure 13.5 illustrates the probing depth and osseous improvements seen in one of Patient #1's treated molar sites at 1 year follow-up post-initiation of treatment.

13.9.4. Conclusions

The results seen in this patient suggest that the combination of standard SRP and adjunctive aPDT are effective in restoring healthy gingival tissue, reducing probing depth and promoting bone regeneration in adult individuals with aggressive periodontitis. The clinical improvements seen in this patient following treatment are well in excess of what would normally be expected with standard SRP/debridement techniques.

13.10. Case Study 2 – Osseous Changes after Adjunctive aPDT Treatment for Aggressive Periodontitis

13.10.1. Background

Patient #2 was a 12-year-old boy presenting to the clinic with severe aggressive periodontitis. The patient was a non-smoker, was not on any medications and had no significant medical history otherwise.

13.10.2. Treatment

All examinations and treatments were performed at the Department of Periodontics at McGill University in Montreal, Canada. After a complete initial periodontal examination, non-surgical SRP was performed under local anaesthesia over the course of three appointments. aPDT was administered after SRP at each appointment using the Periowave™ system, consisting of subgingival application of a liquid methylene blue-based photosensitizer formulation followed by 60 second illumination from a non-thermal 670 nm diode laser. The teeth exhibiting deep pockets (>7 mm) were treated at each probing site for 60 seconds or more bucally and lingually. Figures 13.6 and 13.7 show representative photographs of placement of photosensitizer and illumination in a single site, respectively. The full treatment schedule was as follows: at appointment #1 SRP and aPDT were performed in quadrants 1 and 4, at appointment #2 SRP and aPDT were performed in quadrants 2 and 3 with

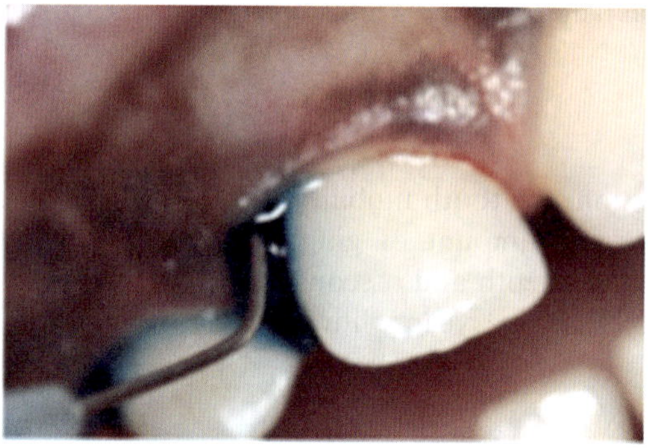

Figure 13.6 Photograph of methylene blue photosensitizer formulation being placed subgingivally during aPDT treatment.

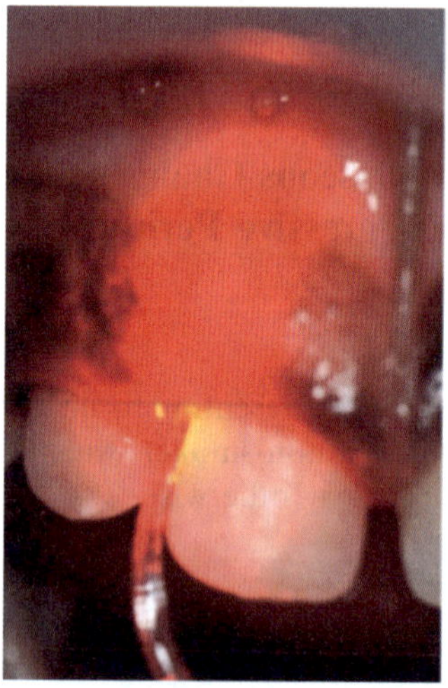

Figure 13.7 Photograph of subgingival illumination using the Periowave™ aPDT system during aPDT treatment.

aPDT repeated in quadrants 1 and 4 and at appointment #3 aPDT was repeated in quadrants 2 and 3. After treatment, the patient was seen every three months for maintenance recall therapy where supra- and sub-gingival scaling was performed and oral hygiene instructions re-enforced. Radiographs were taken prior to initiation of treatment and at 6 months post-treatment. The patient did not take any local or systemic antibiotics over the course of the treatment and follow-up periods.

13.10.3. Results

In Figure 13.8, the radiographs taken from Patient #2 prior to therapy show extensive bone loss around all first molars. Figure 13.9 shows the radiographs from the same angle taken at 6 months post-SRP with adjunctive aPDT treatment. Note the regeneration of the bone within the defects.

13.10.4. Conclusions

Patient #2 exhibited severe periodontitis with extensive bone loss, and was at risk to lose several teeth at a young age. However, with the combination of non-surgical debridement therapy and adjunctive aPDT significant long-term clinical improvements were seen in both soft and hard tissue. Importantly, bone regeneration as shown by radiographs before and after treatment was pronounced, and certainly in excess of what would normally be anticipated after

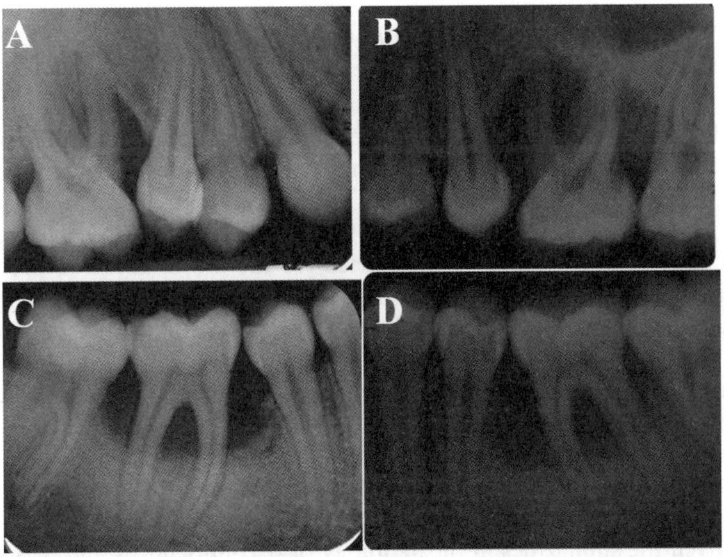

Figure 13.8 Radiographs from Patient #2 taken prior to debridement and adjunctive aPDT treatment.

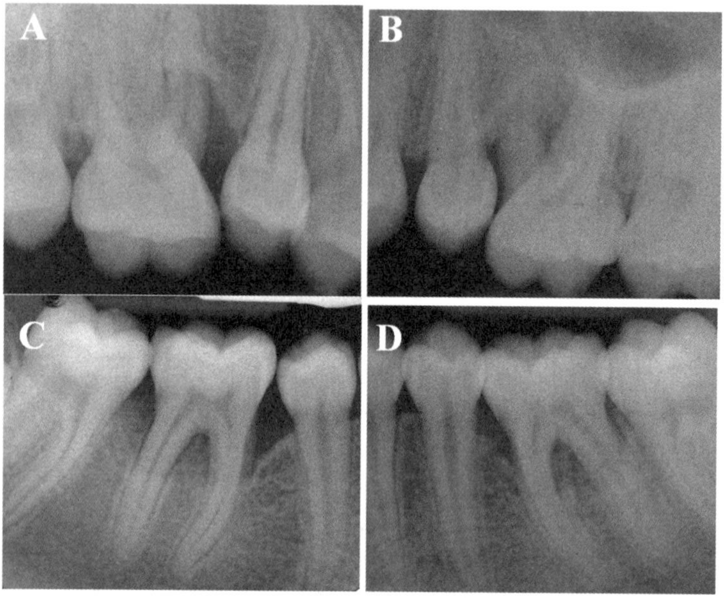

Figure 13.9 Radiographs from Patient #2 taken at six months post-treatment consisting of debridement and adjunctive aPDT.

standard SRP therapy. This result supports the use of aPDT as an adjunct to non-surgical debridement for the treatment of aggressive periodontitis.

13.11. Other Clinical Applications for Oral aPDT

While the majority of preclinical and clinical oral aPDT work to date has focused on periodontal disease, there are several other microbiologically mediated oral conditions that aPDT may be useful for. As mentioned previously in the chapter, root caries are localized hard tissue demineralization caused by supragingival bacterial biofilm processes. However, to date no human clinical studies have been reported for the use of aPDT to prevent or treat carious lesions. Similarly, peri-implantitis is a condition occurring when bacterial pathogens infect the soft tissue around dental implants. This can rapidly lead to tissue/bone breakdown and eventually will result in loss of the implant itself. Peri-implantitis is known to be caused primarily by the same Gram-negative pathogens that mediate periodontal disease subgingivally,[99] so it is reasonable to conclude that aPDT could be used for this application. While large animal (dog) studies have been conducted using aPDT to treat ligature-induced peri-implantitis, no clinical trials have been published to date in this area. Given the similarities in the aetiology of peri-implantitis and periodontal disease, it seems reasonable to predict that aPDT could be successfully used for this application.

Endodontic infections lead to necrosis of soft tissue within the tooth structure, and elimination of this infection is another potential application for aPDT. This level of infection almost always requires mechanical removal of infected tissue followed by sealing of the empty space created within the root; however, reinfection can result due to inadequate removal of persistent microorganisms. The major species of bacteria that cause endodontic infections are often different from those that are commonly associated with periodontal disease, and include *Enterococcus faecalis* along with various species of *Firmicutes*, *Bacteroiditetes*, *Actinobacteria* and *Proteobacteria*.[100] Several groups have published reports of photodynamic eradication of these bacterial strains in extracted tooth models. Soukos *et al.*[101] demonstrated that methylene blue activated by high doses of red light eliminated 97% of *E. faecalis* biofilm inside experimentally infected root canals of extracted teeth. Several years later the same group measured the effectiveness of aPDT *in vitro* on root canal biofilms comprised of the oral bacteria *A. israelii*, *F. nucleatum*, *P. gingivalis* and *P. intermedia*.[56] In that study, scanning electron microscopy (SEM) and viability assessments post-treatment showed 80% reductions in bacterial counts after treatment with methylene blue and red light. Garcez *et al.*[102] also used an extracted tooth model infected with engineered bioluminescent strains of *P. mirabalis* and *P. aeruginosa* to evaluate the potential for aPDT to treat endodontic infections. In that study, the combination of conventional endodontic treatment and aPDT led to >98% reductions in bacterial viability and delayed re-growth within the root canal. Finally, another study[103] showed that the optimization of excipients/solvents used to formulate methylene blue led to enhanced penetration of dentinal tubules and greater antibacterial efficacy in infected extracted tooth root canals. To the author's knowledge, no clinical studies evaluating aPDT for endodontic infections have been undertaken to date. The relatively rare incidence of endodontic therapy failures, coupled with the long follow-up periods required, complicate the design of a clinical study in this area, although the promising preclinical results reported suggest utility in this application.

Finally, oral candidiasis represents the most common fungal infection in humans.[104] Pathogenesis of this condition is not fully understood, but it is believed sometimes to result from a bacterial disturbance/imbalance allowing unchecked growth of the fungal species on previously occupied oral tissues. Several preclinical studies have shown that aPDT eradicates *Candida sp.* in lab models,[105–107] and that this likely occurs *via* membrane damage and permeabilization.[108] Specific methods of photosensitizer delivery in the mouth have even been proposed for this application,[109] but no clinical studies have been reported to date using aPDT to treat oral candidiasis.

13.12. Conclusions and Future Directions

The phenothiazinium photosensitizers methylene blue and TBO have been the most widely used to date for oral antimicrobial applications. While there has been recent work examining the mechanism of antibacterial action for activated phenothiaziniums,[110,111] future studies should address the contribution of type I and type II photoreactions at a biochemical level with attention to variability between pathogenic strains. Moreover, other phenotypic strain-specific characteristics such as matrix secretion, production of singlet oxygen scavengers and propensity for biofilm formation may influence the efficacy of aPDT *in vivo*. In addition, the significance of differences in the mechanism of action of aPDT acting on Gram-positive and Gram-negative organisms should be examined *in vivo* with an eye towards selectivity to periodontal pathogens. Finally, light dosimetry is another parameter that has been addressed in several preclinical studies, but needs to be studied more closely in future clinical work. So far, clinical evidence suggests that there is a threshold energy dose required to achieve significant clinical outcomes in the treatment of periodontal disease. Most importantly, as a rule any aPDT systems under development should be extensively optimized in *in vitro* and animal models with respect to matching of photosensitizer (type, concentration, formulation) and light source (type, power output, wavelength) before being subjected to the rigours of clinical trials in humans. As with any emerging technology, premature human testing and/or commercialization of non-optimized systems poses a serious threat to the acceptance of aPDT for oral indications and to eventual adoption of the technology into clinical practice.

Taken together, the preclinical and clinical data reported to date convincingly support the use of aPDT in the oral cavity from both a safety and an efficacy standpoint. The vast majority of published work to date focuses on aPDT for the treatment of periodontal disease; however, there are several other indications for which the technology may be useful. In addition to a growing body of literature supporting aPDT efficacy in the treatment of periodontal diseases, the impressive safety profile for this technology should also be noted. Specifically, none of the published clinical reports to date have noted significant adverse events or safety concerns related to oral aPDT treatment. This is to be expected, as the treatment is topical in nature and the photosensitizers used clinically to date have well-known properties and safety profiles. It is anticipated that continued development of aPDT applications for the oral cavity will underscore both the clinical utility and "patient-friendly" nature of the technology.

References

1. P.D. Marsh, Microbial ecology of dental plaque and its significance in health and disease, *Adv. Dent. Res.*, 1994, **8**, 263.
2. K.P. Stashenko and J.D. Hillman, Microflora of plaque in rats following infection with and LDH deficient mutant of *Streptococcus rattus*, *Caries Res.*, 1989, **23**, 375.
3. W.J. Loesche, Role of *Streptococcus mutans* in human decay, *Microbiol. Res.*, 1986, **50**, 353.
4. G.H.W. Bowden, in *Risk Markers for Oral Diseases*, ed. N.W. Johnson, Cambridge University Press, Cambridge, 1991, p. 266.

5. J. van Houte, Role of micro-organisms in caries etiology, *J. Dent. Res.*, 1994, **73**, 672.
6. C. Sansone, J. Van Houte, K. Joshipura, R. Kent and H.C. Margolis, The association of mutans streptococci and non-mutans streptococci capable of acidigenesis at a low pH with dental caries on enamel and root surfaces, *J. Dent. Res.*, 1993, **72**, 508.
7. J.G. Steele, A.W. Walls, S.M. Ayatollahi and J.J. Murray, Major clinical findings from a dental survey of elderly people in three different English communities, *Br. Dent. J.*, 1996, **180**, 17.
8. D. Beighton and E. Lynch, Comparison of selected microflora of plaque and underlying carious dentine associated with primary root caries lesions, *Caries Res.*, 1995, **29**, 154.
9. P. Schupbach, V. Osterwalder and B. Guggenheim, Human root caries: microbiota of a limited number of root caries lesions, *Caries Res.*, 1996, **30**, 52.
10. S.R. Brailsford, E. Lynch and D. Beighton, The isolation of *Actinomyces naeslundii* from sound root surfaces and root carious lesions, *Caries Res.*, 1998, **32**, 100.
11. S.R. Brailsford, R.B. Tregaskis, H.S. Leftwich and D. Beighton, The predominant *Actinomyces* spp. isolated from infected dentin of active root caries lesions, *J. Dent. Res.*, 1999, **78**, 1525.
12. G.C. Armitage, Development of a classification system for periodontal diseases and conditions, *Ann. Periodontol.*, 1999, **4**, 1.
13. J. Slots, Microflora in the healthy gingival sulcus in man, *Scand. J. Dent. Res.*, 1977, **85**, 247.
14. L.V. Moore, W.E. Moore, E.P. Cato, R.M. Smibert, J.A. Burmeister, A.M. Best and R.R. Ranney, Bacteriology of human gingivitis, *J. Dent. Res.*, 1987, **66**, 989.
15. F. Dalwai, D.A. Spratt and J. Pratten, The use of quantitative PCR and culture methods to characterise ecological flux's in bacterial biofilms, *J. Clin. Microbiol.*, 2007, **45**, 3072.
16. A.D. Haffajee and S.S. Socransky, Microbial etiological agents of destructive periodontal diseases, *Periodontol. 2000*, 1994, **5**, 78.
17. I.B. Lamster, Evaluation of components of gingival crevicular fluid as diagnostic tests, *Ann. Periodontol.*, 1997, **2**, 123.
18. D. Ready, F. D'Aiuto, D.A. Spratt, J. Suvan, M.S. Tonetti and M. Wilson, Disease severity associated with presence in subgingival plaque of *Porphyromonas gingivalis*, *Aggregatibacter actinomycetemcomitans*, and *Tannerella forsythia*, singly or in combination, as detected by nested multiplex PCR, *J. Clin. Microbiol.*, 2008, **46**, 3380.
19. S.S. Socransky, R.J. Gibbons, A.C. Dale, L. Bortnik, E. Rosenthal and J.B. MacDonald, The microbiota of the gingival crevice in man. 1 Total microscopic and viable counts and counts of specific organisms, *Arch. Oral Biol.*, 1963, **8**, 275.
20. A. Tanner, M.F. Maiden, B.J. Paster and F.E. Dewhirst, The impact of 16S ribosomal RNA-based phylogeny on the taxonomy of oral bacteria, *Periodontol. 2000*, 1994, **5**, 26.
21. D. Dymock, A.J. Weightman, C. Scully and W.G Wade, Molecular analysis of microflora associated with dentoalveolar abscesses, *J. Clin. Microbiol.*, 1996, **34**, 537.
22. D.A. Spratt, A.J. Weightman and W.G. Wade, Diversity of oral asaccharolytic *Eubacterium* species in periodontitis – identification of novel phylotypes representing uncultivated taxa. *Oral Microbiol. Immunol.*, 1999, **14**, 56.
23. K.S. Kornman, A. Crane, H.Y. Wang, F.S. di Giovine, M.G. Newman, F.W. Pirk, T.G. Wilson, F.L. Higginbottom and G.W. Duff, The interleukin-1 genotype as a severity factor in adult periodontal disease, *J. Clin. Periodontol.*, 1997, **24**, 72.

24. G. Holm, Smoking as an additional risk for tooth loss, *J. Periodontol.*, 1994, **65**, 996.
25. R.C. Oliver and T. Tervonen, Diabetes – a risk factor for periodontitis in adults?, *J. Periodontol.*, 1994, **65**, 530.
26. L.J. Cianciola, B.H. Park and E. Bruck, Prevalence of periodontal disease in insulin-dependent diabetes mellitus (juvenile diabetes), *J. Am. Dent. Assoc.*, 1982, **104**, 653.
27. E. Ledo and A. Ledo, Phototherapy, photochemotherapy, and photodynamic therapy: unapproved uses or indications, *Clin. Dermatol.*, 2000, **18**, 77.
28. J. Tunér and L. Hode in, *Low Level Laser Therapy – Clinical Practice and Scientific Background*, Prima Books, Grängesberg, Sweden, 1999, p. 233.
29. Y. Ando, A. Aoki, H. Watanabe and I. Ishikawa, Bactericidal effect of erbium YAG laser on periodontopathic bacteria, *Lasers Surg. Med.*, 1996, **19**, 190.
30. A. Brugnera, F. Zanin, E.L. Barbin, J.C. Spano, R. Santana and J. D Pecora, Effects of Er:YAG and Nd:YAG laser irradiation on radicular dentine permeability using different irrigating solutions, *Lasers Surg. Med.*, 2003, **33**, 256.
31. A. Moritz, U. Schoop, K. Goharkhay, S. Jakolitsch, W. Kluger, J. Wernisch and W. Sperr, The bactericidal effect of Nd:YAG, Ho:YAG, and Er:YAG laser irradiation in the root canal: an in vitro comparison, *J. Clin. Laser Med. Surg.*, 1999, **17**, 161.
32. T. Kato, H. Kusakari and E. Hoshino, Bactericidal efficacy of carbon dioxide laser against bacteria-contaminated titanium implant and subsequent cellular adhesion to irradiated area, *Lasers Surg. Med.*, 1998, **23**, 299.
33. M.R. Hamblin and T. Hasan, Photodynamic therapy: a new antimicrobial approach to infectious disease?, *Photochem. Photobiol. Sci.*, 2004, **3**, 436.
34. M. Wilson and C. Yianni, Killing of methicillin-resistant *Staphylococcus aureus* by low-power laser light, *J. Med. Microbiol.*, 1995, **42**, 62.
35. M. Wilson, J. Dobson and W. Harvey, Sensitization of oral bacteria to killing by low-power laser radiation, *Curr. Microbiol.*, 1992, **25**, 77.
36. A.J. MacRobert, S.G. Bown and D. Phillips, What are the ideal photoproperties for a sensitizer?, *Ciba. Found. Symp.*, 1989, **146**, 4.
37. Z. Malik, J. Hanania and Z. Nitzan, Bactericidal effects of photoactivated porphyrins – an alternative approach to antimicrobial drugs, *J. Photochem. Photobiol. B.*, 1990, **5**, 281.
38. M. Wilson, T. Burns, J. Pratten and G.J. Pearson, Killing of *Streptococcus sanguis* in biofilms using a light-activated antimicrobial agent, *J. Appl. Bacteriol.*, 1995, **78**, 569.
39. J.A. Williams, G.J. Pearson, M.J. Colles and M. Wilson, The effect of variable energy input from a novel light source on the photoactivated bactericidal action of toluidine blue O on *Streptococcus mutans*, *Caries Res.*, 2003, **37**, 190.
40. J.W. Costerton, P.S. Stewart and E.P. Greenberg, Bacterial biofilms: a common cause of persistent infections, *Science*, 1999, **284**, 1318.
41. M. Wilson, T. Burns and J. Pratten, Killing of *Streptococcus sanguis* in biofilms using a light-activated antimicrobial agent, *J. Antimicrob. Chemother.*, 1996, **37**, 377.
42. S. Wood, B. Nattress, J. Kirkham, R. Shore, S. Brookes and J. Griffiths, An *in vitro* study of the use of photodynamic therapy for the treatment of natural oral plaque biofilms formed *in vivo*, *Photochem. Photobiol. B*, 1999, **50**, 1.
43. J.F. O'Neill, C. Hope and M. Wilson, Oral bacteria in multi-species biofilms can be killed by red light in the presence of toluidine blue, *Lasers Surg. Med.*, 2002, **3**, 86.
44. I.C. Zanin, R.B. Goncalves, A.B. Junior, C.K. Hope and J. Pratten, *In vitro* study of bactericidal effect of low level laser therapy in the presence of photosensitizer on cariogenic bacteria, *J. Antimicrob. Chemother.*, 2005, **56**, 324.

45. J. Pratten, Growing oral biofilms in a Constant Depth Film Fermentor (CDFF), *Curr. Protoc. Microbiol.*, 2007, **1**, 1B.5.
46. T. Burns, M. Wilson and G.J. Pearson, Effect of dentine and collagen on the lethal photosensitization of *Streptococcus mutans*, *Caries Res.*, 1995, **29**, 192.
47. S. Sarkar and M. Wilson, Lethal photosensitization of bacteria in subgingival plaque from patients with chronic periodontitis, *J. Periodontal. Res.*, 1993, **28**, 204.
48. B. Manpreet, A. MacRobert, S. Meghji, B. Henderson and M. Wilson, Effect of dosimetric and physiological factors on the lethal photosensitization of *Porphyromonas gingivalis in vitro*, *Photochem. Photobiol.*, 1997, **65**, 1026.
49. B. Manpreet, A. MacRobert, S. Meghji, B. Henderson and M. Wilson, A study of the uptake of toluidine blue O by Porphyromonas gingivalis and the mechanism of lethal photosensitization, *Photochem. Photobiol.*, 1998, **68**, 370.
50. J. Dobson and M. Wilson, Sensitization of oral bacteria in biofilms to killing by light from a low-power laser, *Arch. Oral Biol.*, 1992, **37**, 883.
51. C.N. Street, L. Pedigo and N.G. Loebel, Energy dose parameters affect antimicrobial photodynamic therapy-mediated eradication of periopathogenic biofilm and planktonic cultures, *Photomed. Laser Surg.*, 2010, **28**, S61–66.
52. A. Pfitzner, B. Sigusch, V. Albrecht and E. Glockmann, Killing of periodontopathogenic bacteria by photodynamic therapy, *J. Periodontol.*, 2004, **75**, 1343.
53. Y. Chan and C. Lai, Bactericidal effects of different laser wavelengths on periodontopathic germs in photodynamic therapy, *Lasers Med. Sci.*, 2003, **18**, 51.
54. C.R. Rovaldi, A. Pievsky, N.A. Sole, P.M. Friden, D.M. Rothstein and P. Spacciapoli, Photoactive porphyrin derivative with broad-spectrum activity against oral pathogens in vitro, *Antimicrob. Agents Chemother.*, 2000, **44**, 3364.
55. R. Prates, A. Yamada, L. Suzuki, M. Hashimoto, S. Cai, S. Gouw-Soares, L. Gomes and M. Ribeiro, The irradiation parameters investigation of photodynamic therapy on yeast cells, *Photochem. Photobiol. B: Biol.*, 2006, **86**, 70.
56. J. Fimple, C. Fontana, F. Foschi, K. Ruggiero, X. Song, T. Pagonis, A. Tanner, R. Kent, A. Doukas, P. Stashenko and N. Soukos, Photodynamic treatment of endodontic polymicrobial infection *in vitro*, *J. Endod.*, 2008, **34**, 728.
57. D. Matevski, R. Weersink, H. Tenenbaum, B. Wilson, R. Ellen and G. Lepine, Lethal photosensitization of periodontal pathogens by a red-filtered Xenon lamp in vitro, *J. Periodontal Res.*, 2003, **38**, 428.
58. Y. Qin, X. Luan, L. Bi, G. He, X. Bai, C. Zhou and Z. Zhang, Toluidine blue-mediated photoinactivation of periodontal pathogens from supragingival plaques, *Lasers Med. Sci.*, 2007, **23**, 49.
59. N. Komerik, A. Curnow, A.J. MacRobert, C. Hopper, P.M. Speight and M. Wilson, Fluorescence biodistribution and photosensitising activity of toluidine blue O on rat buccal mucosa, *Lasers Med. Sci.*, 2002, **17**, 86.
60. P. Braham, C. Herron, C. Street and R. Darveau, Antimicrobial photodynamic therapy may promote periodontal healing through multiple mechanisms, *J. Periodontol.*, 2009, **80**, 1790.
61. S. Packer, M. Bhatti, T. Burns and M. Wilson, Inactivation of proteolytic enzymes from *Porphyromonas gingivalis* using light activated agents, *Lasers Med. Sci.*, 2000, **15**, 24.
62. N.S. Soukos, L.A. Ximenez-Fyvie, M.R. Hamblin, S.S. Socransky and T. Hasan, Targeted antimicrobial photochemotherapy, *Antimicrob. Agents Chemother.*, 1998, **42**, 2595.
63. N. Kömerik, H. Nakanishi, A. MacRobert, B. Henderson, P. Speight and M. Wilson, In vivo killing of *Porphyromonas gingivalis* by toluidine blue-mediated

photosensitization in an animal model, *Antimicrob. Agents Chemother.*, 2003, **47**, 932.

64. J. de Almeida, L. Theodoro, A. Bosco, M. Nagata, M. Oshiiwa and V. Garcia, Influence of photodynamic therapy on the development of ligature-induced periodontitis in rats, *J. Periodontol.*, 2007, **78**, 566.
65. J. de Almeida, L. Theodoro, A. Bosco, M. Nagata, M. Oshiiwa and V. Garcia, In vivo effect of photodynamic therapy on periodontal bone loss in dental furcations, *J. Periodontol.*, 2008, **79**, 1081.
66. J. de Almeida, L. Theodoro, A. Bosco, M. Nagata, S. Bonfante and V. Garcia, Treatment of experimental periodontal disease by photodynamic therapy in rats with diabetes, *J. Periodontol.*, 2008, **79**, 2156.
67. L.A. Fernandes, J. Almeida, L.H. Theodoro, A.F. Bosco, M.J. Nagata, T.M. Martins, T. Okamoto and V.G. Garcia, Treatment of experimental periodontal disease by photodynamic therapy in immunosuppressed rats, *J. Clin. Periodontol.*, 2009, **36**, 219.
68. Y.L. Qin, X. Luan, L. Bi, Y. Sheng, C. Zhou and Z. Zhang, Comparison of toluidine blue-mediated photodynamic therapy and conventional scaling treatment for periodontitis in rats, *J. Periodontal Res.*, 2008, **43**, 162.
69. J. Lin, L.J. Bi, G. Zhang, Y.M. Fu and T.T. Dong, Toluidine blue-mediated photodynamic therapy of oral wound infections in rats, *Lasers Med. Sci.*, 2010, **25**, 233.
70. B. Sigusch, A. Pfitzner, V. Albrecht and E. Glockmann, Efficacy of photodynamic therapy on inflammatory signs and two selected periodontopathogenic species in a beagle dog model, *J. Perodontol.*, 2005, **76**, 1100.
71. R. Hayek, N. Araújo, M. Gioso, J. Ferreira, C. Baptista-Sobrinho, A. Yamada and M. Ribeiro, Comparative study between the effects of photodynamic therapy and conventional therapy on microbial reduction in ligature-induced peri-implantitis in dogs, *J. Periodontol.*, 2005, **76**, 1275.
72. J. Shibli, M. Martins, F. Nociti, V. Garcia and E. Marcantonio, Treatment of ligature-induced peri-implantitis by lethal photosensitization and guided bone regeneration: a preliminary histologic study in dogs, *J. Periodontol.*, 2003, **74**, 338.
73. X. Luan, Y. Qin, L. Bi, C. Hu, Z. Zhang, J. Lin and C. Zhou, Histological evaluation of the safety of toluidine blue-mediated photosensitization to periodontal tissues in mice, *Lasers Med. Sci.*, 2009, **24**, 162.
74. M. Farman and R.I. Joshi, Full-mouth treatment versus quadrant root surface debridement in the treatment of chronic periodontitis: a systematic review, *Br. Dent. J.*, 2008, **8**, 205.
75. N.P. Lang, W.C. Tan, M.A. Krahenmann and M. Zwahlen, A systematic review of the effects of full-mouth debridement with and without antiseptics in patients with chronic periodontitis, *J. Clin. Periodontol.*, 2008, **35**, 8.
76. C.H. Drisko, Nonsurgical periodontal therapy, *Periodontol. 2000*, **25**, 2001, 77.
77. H. Hung and C. Douglass, Meta-analysis of the effect of scaling and root planing, surgical treatment and antibiotic therapies on periodontal probing depth and attachment loss, *J. Clin. Periodontol.*, 2002, **29**, 975.
78. A.J. Bonito, L. Lux and K.N. Lohr, Impact of local adjuncts to scaling and root planing in periodontal disease therapy: a systematic review, *J. Periodontol.*, 2005, **76**, 1227.
79. A.D. Haffajee, S.S. Socransky and J.C. Gunsolley, *Ann. Periodontol.*, 2003, **8**, 115.

80. A.J. Bonito, L. Lux and K.N. Lohr, Systemic anti-infective periodontal therapy. A systematic review, *J. Periodontol.*, 2005, **76**, 1227.
81. J. Cosyn and I. Wyn, A systematic review on the effects of the chlorhexidine chip when used as an adjunct to scaling and root planing in the treatment of chronic periodontitis, *J. Periodontol.*, 2006, **77**, 257.
82. E. Bornstein, Method and dosimetry for thermolysis and removal of biofilm in the periodontal pocket with near-infrared diode lasers: a case report, *Dent. Today*, 2005, **24**, 1.
83. I. Ishikawa, A. Aoki and A.A. Takasaki, Clinical application of erbium: YAG laser in periodontology, *J. Int. Acad. Periodontol.*, 2008, **10**, 22.
84. R. de Oliveira, H.O. Schwartz-Filho, A.B. Novaes and M. Taba, Antimicrobial photodynamic therapy in the non-surgical treatment of aggressive periodontitis: a preliminary randomized controlled clinical study, *J. Periodontol.*, 2007, **78**, 965.
85. A.A. Antczak-Bouckoms, J.F. Tulloch and C.S. Berkey, Split-mouth and cross-over designs in dental research, *J. Clin. Periodontol.*, 1990, **17**, 446.
86. R. de Oliveira, H.O. Schwartz-Filho, A.B. Novaes, G.P. Garlet, R. Freitas de Souza, M. Taba, S.L. Scombatti de Souza and F.J. Ribeiro, Antimicrobial photodynamic therapy in the non-surgical treatment of aggressive periodontitis: cytokine profile in gingival crevicular fluid, preliminary results, *J. Periodontol.*, 2009, **80**, 98.
87. A. Braun, C. Dehn, F. Krause and S. Jepsen, Short-term clinical effects of adjunctive antimicrobial photodynamic therapy in periodontal treatment: a randomized clinical trial, *J. Clin. Periodontol.*, 2008, **35**, 877.
88. R. Andersen, N.G. Loebel, D. Hammond and M. Wilson, Treatment of periodontal disease by photodisinfection compared to scaling and root planing, *J. Clin. Dent.*, 2007, **18**, 34.
89. P. Chondros, D. Nikolidakis, N. Christodoulides and R. Rössler, Comparison of the effectiveness of the pulsed dye laser 585 nm versus 595 nm in the treatment of new surgical scars, *Lasers Med. Sci.*, 2009, **24**, 681.
90. M. Wilson, Lethal photosensitization of oral bacteria and its potential application in the photodynamic therapy of oral infections, *Photochem. Photobiol. Sci.*, 2004, **3**, 412.
91. M. Wainwright, Photodynamic antimicrobial chemotherapy (PACT), *J. Antimicrob. Chemother.*, 1998, **42**, 12.
92. J. Tardivo, A. Giglio, C. Oliveira, D. Gabrielli, H. Junqueira, D. Tada, D. Severino, R. Turchiello and M. Baptista, Methylene blue in photodynamic therapy: From basic mechanisms to clinical applications, *Photodiagnosis Photodyn. Ther.*, 2005, **2**, 175.
93. P. Meisel and T. Kocher, Photodynamic therapy for periodontal diseases: state of the art, *J. Photochem. Photobiol.*, 2005, **79**, 159.
94. K. Konopka and T. Goslinski, Photodynamic therapy in dentistry, *J. Dent. Res.*, 2007, **86**, 694.
95. N. Kömerik and A. MacRobert, Photodynamic therapy as an alternative antimicrobial modality for oral infections, *J. Environ. Pathol. Toxicol. Oncol.*, 2006, **25**, 487.
96. A. Azarpazhooh, P.S. Shah, H.C. Tenenbaum and M.B. Goldberg, The effect of photodynamic therapy for periodontitis: a systematic review and meta-analysis, *J. Periodontol.*, 2010, **81**, 4.
97. R.A. Prates, A.M. Yamada, L.C. Suzuki, M.C. Hashimoto, S. Cai, S. Gouw-Soares, L. Gomes and M.S. Ribeiro, Bactericidal effect of malachite green and red

laser on *Actinobacillus actinomycetemcomitans*, *J. Photochem. Photobiol. B: Biol.*, 2007, **86**, 70.

98. C.N. Street, A. Gibbs, L. Pedigo, D. Andersen and N.G. Loebel, *In vitro* photodynamic eradication of *Pseudomonas aeruginosa* in planktonic and biofilm culture, *Photochem. Photobiol.*, 2009, **85**, 137.
99. A.D. Pye, D.E. Lockhart, M.P. Dawson, C.A. Murray and A.J. Smith, A review of dental implants and infection, *J. Hosp. Infect.*, 2009, **72**, 104.
100. J.F. Siqueira and I.N. Rocas, The microbiota of acute apical abscesses, *J. Dent. Res.*, 2009, **88**, 969.
101. N.S. Soukos, P.S. Chen, J.T. Morris, K. Ruggiero, A.D. Abernethy, S. Som, F. Foschi and S. Doucette, Photodynamic therapy for endodontic disinfection, *J. Endod.*, 2006, **32**, 979.
102. A.S. Garcez, M.S. Ribeiro, G.P. Tegos, S.C. Nunez, A. Jorge and M.R. Hamblin, Antimicrobial photodynamic therapy combined with conventional endodontic treatment to combat root canal infection, *Lasers Surg. Med.*, 2007, **39**, 59.
103. S. George and A. Kishen, Photophysical, photochemical, and photobiological characterization of methylene blue formulations for light-activated root canal disinfection, *J. Biomed. Optics*, 2007, **12**, 1.
104. N.S. Soysa, L.P. Samaranayake and A.N. Ellepola, Antimicrobials as a contributory factor in oral candidosis – a brief overview, *Oral Dis.*, 2008, **14**, 138.
105. J.M. Bliss, C.E. Bigelow, T.H. Foster and C.G. Haidaris, Susceptibility of *Candida* species to photodynamic effects of Photofrin, *Antimicrob. Agents Chemother.*, 2004, **48**, 2000.
106. E. Munin, L.M. Giroldo, L.P. Alves and M.S. Costa, Study of germ tube formation by Candida albicans after photodynamic antimicrobial chemotherapy (PACT), *J. Photochem. Photobiol. B: Biol.*, 2007, **88**, 16.
107. B.M. Soares, D.L. da Silva, G.R. Sousa, J.C. Amorim, M.A. de Resende, M. Pinotti and P.S. Cisalpino, In vitro photodynamic inactivation of Candida spp. growth and adhesion to buccal epithelial cells, *J. Photochem. Photobiol. B: Biol.*, 2009, **94**, 65.
108. L.M. Giroldo, M.P. Felipe, M.A. de Oliveira, E. Munin, L.P. Alves and M.S. Costa, Photodynamic antimicrobial chemotherapy (PACT) with methylene blue increases membrane permeability in *Candida albicans*, *Lasers Med. Sci.*, 2009, **24**, 109.
109. R.F. Donnelly, P.A. McCarron, M.M. Tunney and A.D. Woolfson, Antifungal photodynamic therapy, *J. Photochem. Photobiol. B: Biol.*, 2007, **86**, 59.
110. K. Sahu, H. Bansal, C. Mukhurjee, M. Sharma and P.K. Gupta, Atomic force microscopic study on morphological alterations induced by photodynamic action of Toluidine Blue O in *Staphylococcus aureus* and *Escherichia coli*, *J. Photochem. Photobiol. B: Biol.*, 2009, **96**, 9.
111. W. Caetano, P.S. Haddad, R. Itri, D. Severino, V.C. Vieira, M.S. Baptista, A.P. Schroder and C.M. Marques, Photo-induced destruction of giant vesicles in methylene blue solutions, *Langmuir*, 2007, **23**, 1307.

Chapter 14

Photodynamic Antifungal Chemotherapy

P. G. Calzavara-Pinton,[a] M. T. Rossi[a] and B. Ortel[b]

[a] Department of Dermatology, Spedali Civili Hospital, P. le Spedali Civili 1, 25123 Brescia, Italy; [b] Section of Dermatology, University of Chicago, 5841 S. Maryland, Chicago, IL 60615, USA

Table of Contents

Photodynamic Inactivation of Microbial Pathogens: Medical and Environmental Applications
Editors Michael R. Hamblin and Giulio Jori

Published by the Royal Society of Chemistry, www.rsc.org

14.1. General Principles of Photosensitization of Fungi

Several exogenous photosensitizers have been investigated, mainly belonging to three types of molecules: phenothiazine dyes, porphyrins and phthalocyanines (Figure 14.1). As a general rule, photosensitizers that are of interest for therapeutic application in humans are those without dark toxicity, *i.e.* those that ideally have no pharmacological or toxic effect in the absence of light activation. In addition, upon irradiation the photosensitizers should not be genotoxic or mutagenic.[1] PDT in itself has been shown to have a low risk of DNA damage; in eukaryotic fungi this hazard is further reduced by the presence of a membrane that envelopes the nucleus and acts as an additional barrier to penetration of sensitizing dyes or their high-energy photoproducts.[2] Mechanisms and efficiency of cell inactivation are critically dependent on chemical properties, such as the lipophilicity/hydrophilicity balance (logP), the degree of ionization (pKa) and the presence of charged groups. These properties determine the mechanisms and the efficacy of cellular uptake of the photosensitizer. They are also critical for the pattern of relative photosensitizer distribution among the various sub-cellular compartments.[1] Once uptake is established, the antimicrobial photodynamic effects have been found strictly to depend on photophysical parameters, including absorption peak (λ_{max}), intensity of absorption (ε) and quantum yield (Φ) for singlet oxygen. Unlike mammalian cells, fungi are surrounded by a rigid cell wall that is composed principally of soluble and insoluble polysaccharides, including chitin, beta-glucans and

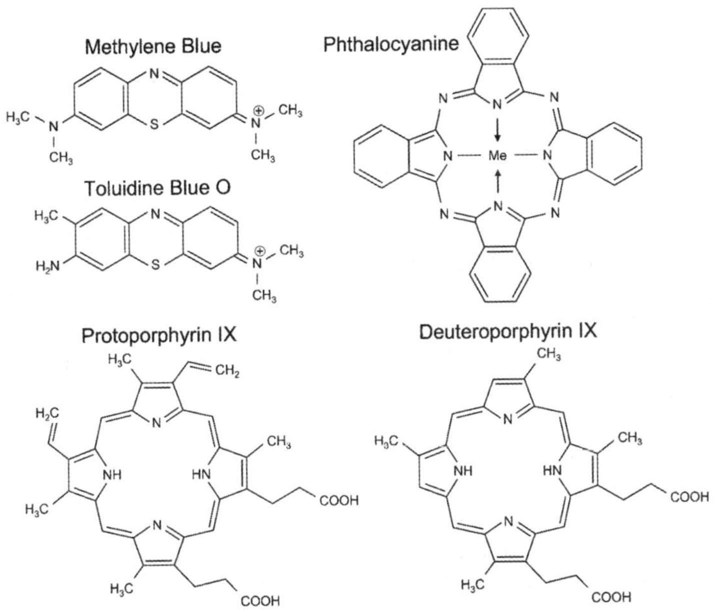

Figure 14.1 Chemical structures of a few photosensitizers for anti-fungal PDT.

glycoproteins. As a general rule, cellular uptake is negatively impacted by the lipophilicity of photosensitizers and enhanced by hydrophilicity and presence of charged moieties. Following uptake, sensitizers are distributed to subcellular targets. The pattern of intracellular localization is very relevant for photosensitization effects, because structures closest to the sensitizer have the greatest probability of being targeted in the photochemical process. This is due to the high reactivity and short intrinsic lifetime (= short diffusion radius) of reactive oxygen species (ROS).[3] The biochemical and functional effects of photosensitization include inactivation of enzymes and other proteins and peroxidation of lipids leading to the lysis of cell membranes, lysosomes and mitochondria.[4] The multiplicity of cellular targets in fungi likely reduces the risk of selection of mutant strains resistant to PDT and this risk should be further minimized by the absence of mutagenic effects of PDT.

The spectrum of the light used for photoactivation is another critical issue. For PDT of malignancies, photoensitizers and matching light sources in the far red or near infrared are selected for deep tissue penetration. Light in the red and far red wavelength region can penetrate deeply into mammalian skin. Sensitizers with an absorption peak in these wavebands are suitable for the treatment of dermatophytes as these infect the interfollicular epidermis, hair follicles and nails. Photosensitizers absorbing in the blue region may be better suited for the treatment of *Candida* species that invade only the stratum corneum.

14.2. Photosensitizers

14.2.1. Phenothiazine Dyes

Oxidized phenothiazines have a simple planar, tricyclic skeleton and are normally encountered as cations. The most widely used compounds are methylene blue (MB) and toluidine blue O (TBO). Both are efficient producers of singlet oxygen[5] and their maximum wavelengths of absorption (λ_{max}) in water are 656 nm for MB and 625 nm for TBO.[6] The phenothiazines are known to localize to the plasma membrane of yeasts. Consequently, this is the cellular structure primarily damaged upon illumination. It has been proposed that the increased permeability resulting from such damage is the reason for cell death.[7] The fungicidal effect of low concentrations of MB (0.1 mg ml^{-1}) followed by irradiation with 28 J cm^{-2} red light (685 nm from a diode laser) has been demonstrated in several *Candida* species: *C. albicans*, *C. dubliniensis*, *C. krusei* and *C. tropicalis*.[8]

In vitro studies show that Candida species are efficiently killed by PDT using TBO, whether grown planktonically[9] (in suspension) or in biofilms,[10] which is the growth pattern most relevant for human disease. However, they are still considerably less susceptible to photodynamic killing than a number of prokaryotic bacteria, including *Staphylococcus aureus* and *Propionibacterium acnes*.[11] In fact, TBO doses as high as 2 mg ml^{-1} have been necessary to induce satisfactory levels (>99%) of lethal cytotoxicity in planktonic *C. albicans* upon illumination. Similarly, *C. albicans* biofilms required TBO doses of 5 mg ml^{-1} to achieve highly efficient killing. Light dose requirements (red light at 635 nm) in photosensitized killing of *C. albicans* are also high (200 J cm^{-2}).[10]

The reduced sensitivity of yeasts relative to bacteria may be explained according to the target theory: the killing effect of photoactivated MB in prokaryotic cells appears to be a single-hit process, where all the putative targets in the cell are equally susceptible and – if damaged – can lead to cell death. In contrast, in the eukaryotic cells a multi-hit process is needed, where damage to more than one molecular target is required before cell death occurs. These differences in susceptibility may be due to or amplified by differences in the cell size. *Candida* species are approximately 25–50 times larger than bacteria and therefore contain a greater number of targets per cell.[12] Under the same experimental conditions, photosensitized killing rates for fungi were 18–200-fold higher than those determined for keratinocytes, suggesting that MB-PDT of fungal infections of the skin may be not only effective but also highly selective.[11] PDT with identical drug and light doses did not cause any detectable genotoxic and mutagenic effects on either fungi or keratinocytes.[12]

The interest that has been raised by the described *in vitro* findings has prompted *in vivo* investigations. Oral azole-resistant candidiasis of severe combined immunodeficiency (SCID) beige nude mice, a murine immunodeficiency model, was treated with the application of an aqueous solution of MB followed by irradiation with 100 J cm^{-2} of laser light with an emission peak at 664 nm delivered through a cylindrical diffuser. The *Candida* infection was responsive to PDT and treatment effects were dependent on drug and light doses.[13] A drawback of this study is that results of follow-up examinations were not reported. These findings still suggest that MB-PDT may have the potential to become an effective, non-toxic, simple, inexpensive and repeatable therapy of oral candidiasis, specifically in immunodeficient HIV-positive patients, which often select for multi-drug resistant fungal strains. It will be necessary to develop convenient light delivery systems for the treatment of oesophageal infection, which often accompanies oropharyngeal candidiasis in immunosuppressed patients.

A potential unwanted effect of the administration of these highly coloured photosensitizers as mouthwash to humans is an undesirable staining of teeth, lips and oral mucosa. Therefore targeted delivery of the photosensitizer directly to the site of infection is highly desirable. Donnelly *et al.* have formulated patches containing TBO that were capable of resisting dissolution when immersed in artificial saliva, thus indicating clinical applicability.[10] Patches containing TBO at concentrations between 10 and 100‘mg cm^{-2} were studied for photosensitizer release into an aqueous receiver solution. The results suggest that short application times of TBO-containing mucoadhesive patches should allow treatment of recently acquired oropharyngeal candidiasis caused solely by planktonic cells whereas longer patch application (longer release) times may be required for persistent disease where biofilms are implicated.

Although yeasts of the genera *Candida* and *Cryptococcus* continue to be the main opportunistic fungi responsible for invasive mycoses in humans, serious infections caused by *Aspergillus* spp. and by other genera of filamentous fungi have emerged all over the world. The primary reason is the increased numbers of immunocompromised individuals.[14] Additionally, the continuous and indiscriminate administration of fungicides has favoured the selection of fungal strains resistant to currently used fungicides.[15] Metarhizium anisopliae is a worldwide entomopathogenic (causing disease in insects) deuteromycete that has been used for decades in programs of agricultural pest and disease-vector control in various countries. *M. anisopliae* is not considered to be pathogenic for humans or domestic animals. However, starting in the late 1990s some rare cases of infections caused by this fungus in both immunocompromised and immunocompetent individuals have been reported.[16] Gonzales and colleagues[17] compared conidial photosensitization in two fungal species (*M. anisopliae* and *Aspergillus nidulans*) with MB and TBO under different drug incubation and light exposure conditions. Conidial inactivation reached up to 99.7%, but none of the treatments achieved a 100% fungicidal effect although photosensitization delayed the germination of those conidia that survived PDT.

Photodynamic inactivation of dermatophytes has been studied as well. Eight strains of dermatophytes (Trichophyton mentagrophytes, Trichophyton rubrum, Trichophyton tonsurans, Microsporum cookei, Microsporum canis, Microsporum gypseum, Epidermophyton floccosum and Nannizia cajetani) were exposed to UVA irradiation after sensitization with two thiophenes (2.20 : 50.200-terthienyl and 5-(4-OH-1-butinyl) 2,2′-bithienyl). A strong and dose-dependent inhibition of the growth of all tested strains was found although a complete inactivation was never obtained.[18]

14.2.2. Porphyrins

The biological role of photosensitized reactions mediated by naturally occurring endogenous porphyrins, which are metabolic by-products of haem synthesis, is presently being debated. Irradiation with 20–50 J cm^{-2} of broadband visible light without the addition of exogenous sensitizers was found to produce an oxygen-dependent lethal effect on plasma membranes and mitochondria of *C. guilliermondii*.[19] However, in another study[9] the viability of suspensions of *C. albicans* was not affected by irradiation with 66 J cm^{-2} of red laser light (632.8 nm).

Haematoporphyrin derivative (HpD) is a complex mixture of mono- and oligomeric porphyrins derived from blood. It has various absorption peaks in the 400–650 nm region and its antifungal properties have been investigated since the early 1980s.[20] Within a broad range of concentrations, HpD is not toxic for cells in the dark,[21] but with light exposure it can effectively kill fungal cells. The mechanisms of photosensitization are quite different from those described for MB and other phenothiazine dyes. HpD is not taken up by *Saccharomyces cerevisiae* (*S. cerevisiae*) and *C. albicans*[20,22] and the photocytotoxic activity is mediated by unbound dye molecules in the surrounding aqueous medium. With irradiation, the photoactivated HpD molecules cause an initial limited alteration of the cytoplasmic membrane that then allows for the penetration of the photosensitizer into the cell, the translocation to the inner membrane and the consequent photodamage of intracellular targets. Freeze-fracture electron microscopy studies of *C. albicans* cells, which had been photosensitized by a meso-substituted cationic porphyrin, clearly indicate that the photosensitized damage progresses from the outer leaflet of the plasma membrane to the inner leaflet.[23] At a biochemical level, the phototoxic process involves mainly the peroxidation of lipids and, only to a minor extent, the inactivation of proteins of the cell wall[23] and is inhibited in the presence of albumin.[24]

In order to determine the importance of sterol oxidation, sterol auxotrophic strains of *S. cerevisiae* were grown on media containing sterols with different levels of saturation before photosensitization.[25] Cells grown on a completely saturated sterol, cholestanol, were markedly resistant to the photosensitizing process using HpD or protoporphyrin IX. This demonstrated that photo-oxidation of the native unsaturated yeast sterol, ergosterol, played a critical role in PDT-mediated cytotoxicity in native yeast cells.

The more highly purified components derived from haematoporphyrin (HpD), Photofrin® (QLT Vancouver, BC) and Photosan III® (Seehof Laboratorium, GmbH, Munich, Germany) are also effective in the photo-inactivation of bacteria and fungi upon red light exposures. Recently, also Photogem® (Timtec Co. Newark, DE), another purified HpD derivative activated by blue light irradiation from an LED source, was found effective in the inactivation of *C. albicans*, *C. dubliniensis*, *C. tropicalis* and *C. krusei*.[26] However, the experiments of this study were conducted with planktonic cultures of the microorganisms, which is not an optimal experimental model for investigating *in vivo* conditions.

Recently, several new synthetic porphyrin-derived compounds have been evaluated for their possible antifungal activity. These drugs have been synthesized in order to obtain chemically pure dyes with absorption peaks in the 650–700 nm range, with high molar extinction coefficients and high quantum yields for singlet oxygen. Polarity of the molecule is a critical feature. The hydrophilicity of porphyrins strongly influences their ability to efficiently photosensitize yeast cells, with more highly water soluble porphyrins being much more effective than their more hydrophobic counterparts.[27,28]

Neutral amphiphilic meso-arylglycosylporphyrins[27] and water-soluble diamino-acid derivatives of porphyrins[28] are characterized by a better uptake and, consequently, a greater activity against the yeast *S. cerevisiae* in comparison to hydrophobic molecules such as HpD, chlorins and benzoporphyrin derivatives. Photoactivated hydrophilic porphyrins, such as deuteroporphyrin, deuteroporphyrin monomethylesther (DP mme) and 5,10,15-tris(4-methylpyridinium)-20-phenyl-[21H, 23H]-porphine trichloride (Sylsens B) consistently showed a high activity against the dermatophyte *T. rubrum*. Again, their activity was found greater than that of lipophilic phthalocyanines (ZnPc, PcS4 and AlPcS4) and Photofrin® that displayed nothing more than a fungistatic effect for about one week.[29]

In an *ex vivo* model using human stratum corneum the application of a formulation containing 10 μM Sylsens B followed by exposure to 18 J cm^{-2} UVA-1 radiation (340–400 nm) proved to be effective against *T. rubrum*.[30] These results clearly indicate the potential of synthetic hydrophilic porphyrins for clinical applications in human fungal diseases.

14.2.3. 5-Aminolevulinic Acid

5-Aminolevulinic acid (ALA) is not a photosensitizer by itself but most eukaryotic cells metabolize it into an active endogenous sensitizer, protoporphyrin IX (PpIX), through the haem synthetic pathway.[31] The main difference between exogenously added PpIX and ALA-induced PpIX in mammalian cells is the difference in sub-cellular distribution and the resulting increased photosensitivity with ALA-PpIX.[32] In comparison to mammalian cells, fungi display differences in the enzymatic machinery: coproporphyrinogen oxidase is found in the cytoplasm, and both ALA-synthase and ALA-dehydratase-mediated processes are rate-limiting steps. Their activity is controlled by the intracellular free haem pool.[33] Consequently, the ALA-induced accumulation of PpIX may be increased by the addition of iron chelators, such as desferrioxamine, EDTA, 1,2-diethyl-3-hydroxypyridin-4-one hydrochloride and 2,2′-dipyridyl. These agents inhibit the conversion of PpIX to haem by restricting iron availability, thus preventing haem-mediated negative feedback on the rate limiting enzymes. PpIX may be photoactivated by wavelengths ranging from UVA to the visible wavebands with maximum absorption in the Soret band region at 375–405 nm. There are several smaller peaks (Q-bands) including one at 630–633 nm that is often targeted *in vivo* for better tissue penetration. As described for exogenous photosensitizers, upon light exposure PpIX induces cytotoxic effects through oxygen-dependent mechanisms. These photochemical reactions damage mainly the mitochondria, where PpIX is synthesized, and plasma membranes.[34] Prolonged irradiation causes additional alteration of other cytoplasmic structures and the inhibition of the synthesis of DNA and RNA,[35] but genotoxic and mutagenic effects were never detected in yeasts.[33]

ALA-based PDT seems suitable for the treatment of human mycosis because of its institutional approval status for human skin disease. In fact, two agents for PpIX induction are commercially available, ALA (Levulan®) and its methyl ester (Metvix®). *In vitro* studies have demonstrated that fungi can be effectively photosensitized after topical delivery of ALA and killed at dose rates much lower than those that kill keratinocytes.[36] The clinical efficacy of ALA-PDT has already been evaluated in several investigations. One open pilot study enrolled nine patients with interdigital mycosis of the feet. Before therapy, skin scrapings of lesional skin were inoculated on Sabouraud's dextrose agar containing antibacterial antibiotics. Colonies of *C. albicans* grew in three cases, *Trichophyton dermagrophytes* in four and *T. rubrum* in two. After incubation in a 20% aqueous ALA solution all colonies showed strong red fluorescence under illumination with a UVA-emitting lamp (Fluolight, Saalman GmbH,

Germany) demonstrating efficient poprhyrin synthesis (Figure 14.2). The treatment protocol consisted of the application of a 20% ALA cream under an occlusive dressing to affected skin, followed four hours later by irradiation with 75 J cm^{-2} of broadband red light. Clinical and microbiological recovery was documented in six out of nine patients after one (four cases) or four (two cases) treatments. Overall tolerability was good in all patients. However, only two patients (one with *C. albicans* and one with T. mentagrophytes) had persistent remission at a follow-up visit after four weeks. It is not clear why the good initial therapeutic effect was followed by quick recurrence in almost all patients. A first possible explanation is that *in vivo* environmental conditions, such as temperature, humidity and pH of the interdigital skin, could result in poor cellular ALA uptake and insufficient biosynthesis of PpIX. In addition the uneven delivery of ALA cream and light must be taken into account because of the irregular three-dimensional shape of this peculiar anatomical area.[37]

PDT is an attractive treatment option for fungal nail infections, given the several-months-long duration of conventional treatment with oral and/or topical antifungal drugs. Donnelly *et al.* developed a bioadhesive patch containing 50 mg cm^{-2} ALA that allowed a high ALA concentration on the ventral side of excised human nail to be achieved.[10] There are multiple opportunities to enhance ALA permeation across the nail, including penetration enhancers and

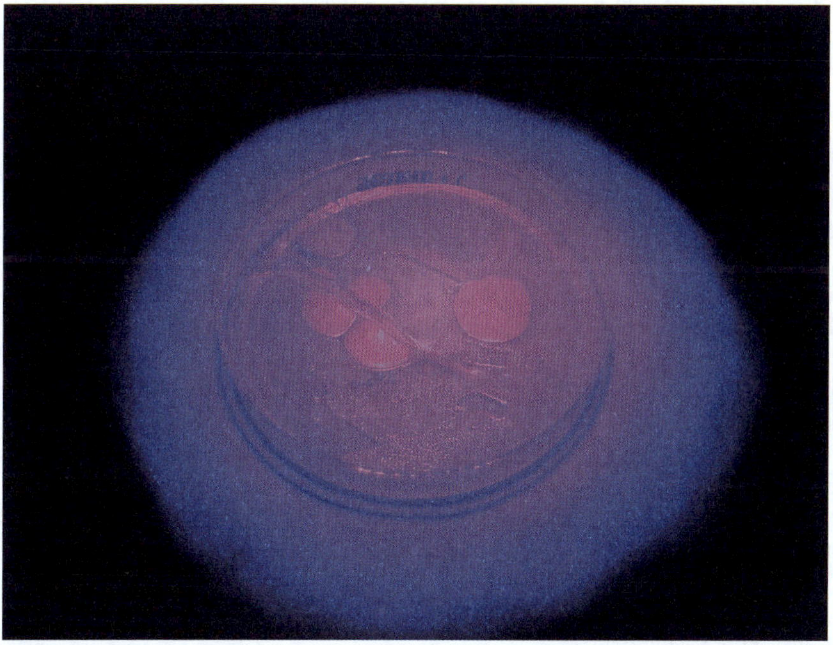

Figure 14.2 Strong red fluorescence is detected after incubation of colonies of *Candida albicans* with an aqueous 20% ALA solution and irradiation with a UVA lamp.

ablation of the dorsal nail surface. In addition, inclusion of iron chelators in the patch may enhance ALA efficacy.[38]

Two patients with onychomycosis of the toenail received 20% urea ointment under occlusion for keratolytic treatment. Subsequently a 20% ALA methyl ester cream was applied for 5 hours under occlusion and light protection.[39] The nail was then irradiated with 100 J cm^{-2} of 630 nm pulsed light from an excimer-pumped dye laser. Complete clearing was obtained in both patients after seven and six treatments, respectively. No recurrence was observed clinically at 3- and 6-month follow-up visits.

Another successful strategy was demonstrated by Piraccini *et al.* Onychomycosis of the toe caused by *T. rubrum* was treated with PDT after removal of the infected nail plate and nail bed hyperkeratosis using softening with 40% urea and mechanical debridement. Sixteen per cent ALA methyl ester cream (Metvix®) was applied under occlusion for 3 hours. Three PDT sessions using 37 J cm^{-2} at 630 nm were delivered over a period of 45 days. They were well tolerated and at the 12 and 24 months follow-up visits, KOH examinations and cultures were still negative and the toenails were considered recovered with residual mild traumatic onycholysis.[40]

An open trial conducted by Sotirou and colleagues reported quite different results. Thirty patients with distal and lateral subungual toenail onychomycosis were treated with the application of a 20% ALA cream after removal of the nail plate and subungual hyperkeratosis. ALA exposure was for 3 hours and 40 J cm^{-2} of 570–670 nm red light were used. The treatment was performed three times at two-week intervals. Cure rates of 43.3% and 36.6% were seen 12 and 18 months after treatment, respectively.[41] There are several possible explanations for the unsatisfactory outcome in many of these patients. *In vitro* studies demonstrated that formation of PpIX induced by ALA in the growth media was restricted to selected parts of the fungal mycelium, leading to a significant but not higher than 50% growth inhibition of *T. rubrum in vitro*. Thus, according to the investigators, the growth of *T. rubrum* seems to be only time-delayed by PDT.[42] In view of these *in vitro* findings, achievement of any clinical cure appears almost surprising. Smijs *et al.* hypothesize that, although PDT inactivates both hyphae and spores, a better fungicidal effect could be achieved by the application of a second treatment within 24–48 hours, a time interval in which hyphal tip morphology could not change into more resistant globular structures.[30]

This means that repeated treatments are crucial and that modifications of the described clinical regimens will potentially enhance efficiency. The inherent versatility of PDT offers many opportunities to develop improved clinical strategies.

14.2.4. Phthalocyanines

A high number of synthetic phthalocyanines (Pc) are currently available. All these compounds have in common a porphyrin-like chemical structure characterized by the condensation of benzene rings with pyrrole moieties. They share high singlet oxygen quantum yields and high extinction coefficient in the far red (680–720 nm) spectral region,[1] which are both desirable for clinical use. However, they differ from each other in physico-chemical properties that may be modified by changing the central ligand atom and altering the number and type of side chains. Central metal ligands play a crucial role in the photobiological activity and influence triplet yield and triplet lifetime of the compound. The introduction of polar substituents on the side chains modifies the overall balance of polarity. The hydrophilicity/hydrophobicity relationship is an important chemical property with respect to photosensitizer uptake and its localization into microorganisms. The Pc macrocycle is essentially hydrophobic and hydrophobic compounds have been found poorly effective for PDT, likely due to their tendency to form aggregates that have very little photosensitizing activity.[43] This view is supported by a study that demonstrated that the photodynamic effect of the non-sulfonated Al-Pcs is mediated by the monomeric form of the dye.[35] Several hydrophilic mono- and tetra-sulfonated AlPc were not taken up efficiently and did not inactivate the yeast *Kluyveromyces marxianus*[43] whereas an appreciable amount of another water-soluble compound, the mono-sulfonated ZnPcS, was tightly bound to intracellular targets and showed a high photosensitizing activity[1] in *S. cerevisiae*. This discrepancy may be explained by recent findings showing that the presence of cationic charges is necessary for inactivation of *C. albicans*, but an excess of charges, especially if homogeneously distributed, causes a decrease in activity.[44] On the basis of these findings, a tetra-cationic ZnPc bearing four aminoalkylated peripheral substituents proved highly capable of inactivating even strains of *C. albicans* that are multi-drug resistant.[1] An additional advantage of Pc-PDT is that no resistant fungal strains developed even after multiple treatments. The clinical applicability is supported by good selectivity as demonstrated by a much greater photosensitizing activity in *C. albicans* than in murine and human keratinocytes and fibroblasts.[45] Some authors have investigated the incorporation of Pc into liposomes as a strategy for enhancing cell uptake by yeasts.[1] However, results were disappointing but understandably so, because the size of liposomes ranged from 20 to 100 nm[46] and it appears that only globular structures with a diameter of less than 5.8 nm can pass through the cell membrane of yeasts.[47] Mutagenic effects of photodynamic treatment with ClAl-Pc and RPL068 were not found in *Kluyveromyces marxianus* (*K. marxianus*)[35] and *C. albicans*,[45] confirming findings in other species and with other photosensitizers.

14.3. Light Delivery

By definition, PDT requires a source of radiation that supplies the energy required for singlet oxygen production *in situ*. The optimal wavelength for excitation is determined by the molecular structure of the photosensitizer and, thus, a different wavelength range is best for the phenothiaziniums (*ca.* 600–660 nm) and for the phthalocyanines (*ca.* 630–690 nm). The porphyrins can be excited by light in the red region of the spectrum, but they absorb blue light around 400 nm much more efficiently. If many possible wavelengths are available for activation, the choice will be determined by other conditions, such as desired depth of tissue penetration. Light fluence is exponentially reduced with tissue thickness. The decrease is determined by absorption, particularly by haemoglobin and melanin, and by scattering. The respective parameters for absorption and scattering vary between different tissue types.[48] As a general rule, longer (red) wavelengths in the visible spectrum penetrate deeper into tissue than the shorter ones. However, a high degree of tissue penetration is only required to kill fungi residing below the surface of the skin, in the matrix or the thickened plate of a nail, or in the deeper portion of the hair follicle.

The ideal light source should provide a high output at the requisite wavelength for activation of the photosensitizer. Lasers, light emitting diodes (LEDs) and filtered incoherent fluorescent or metal halide lamps are most commonly employed in PDT today. Use of unfiltered light sources with emission in the ultraviolet range should be minimized, as that portion of the spectrum poses a risk of erythemogenicity and mutagenesis. Similarly, emission in the infrared range is also undesirable, in order to avoid uncontrolled heating of tissue or cell cultures. Typical power levels of light sources used in antifungal PDT are in the range of 10 to 100 mW cm^{-2}.[10]

References

1. G. Bertoloni, F. Rossi, G. Valduga, G. Jori, H. Ali and J.E. Van Lier, Photosensitizing activity of water- and lipid-soluble phthalocyanines on prokaryotic and eukaryotic microbial cells, *Microbios*, 1992, **71**, 33–46.
2. B. Zeina, J. Greenman, W.M. Purcell and B. Das, Killing of cutaneous microbial species by photodynamic therapy, *Br. J. Dermatol.*, 2001, **144**, 274–278.
3. C. Zhou, Mechanisms of tumour necrosis induced by photodynamic therapy, *J. Photochem. Photobiol. B*, 1989, **3**, 299–318.
4. G. Bertoloni, F. Zambotto, L. Conventi, E. Reddi and G. Jori, Role of specific cellular targets in the hematoporphyrin-sensitized photoinactivation of microbial cells, *Photochem. Photobiol.*, 1987, **46**, 695–698.
5. M. Wainwright, Methylene blue derivatives: suitable photoantimicrobials for blood product disinfection?, *Int. J. Antimicr. Agents*, 2000, **16**, 381–394.
6. T. Ito, Toluidine blue: the mode of photodynamic action in yeast cells, *Photochem. Photobiol.*, 1977, **25**, 47–53.
7. L.M. Giroldo, M.P. Felipe, M.A. de Oliveira, E. Munin, L.P. Alves and M.S. Costa, Photodynamic antimicrobial chemotherapy (PACT) with methylene blue increases membrane permeability in Candida albicans, *Lasers Med. Sci.*, 2009, **24**, 109–112.
8. S.C. de Souza, J.C. Junqueira, I. Balducci, C.Y. Koga-Ito, E. Munin and A.O. Jorge, Photosensitization of different Candida species by low power laser light, *J. Photochem. Photobiol. B.*, 2006, **83**(1), 34–38.
9. M. Wilson and N. Mia, Sensitisation of Candida albicans to killing by low-power laser light, *J. Oral Pathol. Med.*, 1993, **22**(8), 354–357.

10. R.F. Donnelly, P.A. McCarron, M.M. Tunney and A.D. Woolfson, Potential of photodynamic therapy in treatment of fungal infections of the mouth. Design and characterization of a mucoadhesive patch containing toluidine blue O, *J. Photochem. Photobiol. B: Biol.*, 2007, **86**, 59–69.
11. B. Zeina, J. Greenman, D. Corry and W.M. Purcell, Cytotoxic effects of antimicrobial photodynamic therapy on keratinocytes in vitro, *Br. J. Dermatol.*, 2002, **146**, 568–573.
12. B. Zeina, J. Greenman, D. Corry and W.M. Purcell, Antimicrobial photodynamic therapy: assessment of genotoxic effects on keratinocytes in vitro, *Br. J. Dermatol.*, 2003, **148**, 229–232.
13. M.C. Teichert, J.W. Jones, M.N. Usacheva and M.A. Biel, Treatment of oral candidiasis with methylene blue-mediated photodynamic therapy in an immunodeficient murine model, *Oral. Surg. Oral. Med. Oral. Pathol. Oral. Radiol. Endod.*, 2002, **93**, 155–160.
14. A. Espinel-Ingroff, A. Fothergill, M. Ghannoum, E. Manavathu, L. Ostrosky-Zeichner, M. Pfaller, M. Rinaldi, W. Schell and T. Walsh, Quality control and reference guidelines for CLSI broth microdilution susceptibility method (M38 document) for amphotericin B, itraconazole, posaconazole, and voriconazole, *J. Clin. Microbiol.*, 2005, **43**, 5243–5246.
15. J.B. Anderson, Evolution of antifungal-drug resistance: mechanisms and pathogen fitness, *Nat. Rev. Microbiol.*, 2005, **3**, 547–556.
16. S. Osorio, R. de la Cámara, M.C. Monteserin, R. Granados, F. Oña, J.L. Rodriguez-Tudela and M. Cuenca-Estrella, Recurrent disseminated skin lesions due to Metarhizium anisopliae in an adult patient with acute myelogenous leukemia, *J. Clin. Microbiol.*, 2007, **45**, 651–655.
17. F.P. Gonzales, S.H. da Silva, D.W. Roberts and G.U. Braga, Photodynamic inactivation of conidia of the fungi Metarhizium anisopliae and Aspergillus nidulans with methylene blue and toluidine blue, *Photochem Photobiol.*, 2010, **22**, [Epub ahead of print].
18. C. Romagnoli, D. Mares, G. Sacchetti and A. Bruni, The photodynamic effect of 5-(4-hydroxy-1-butinyl)-2,2-bithienyl on dermatophytes, *Mycol. Res.*, 1998, **102**, 1519–1524.
19. G.Y. Fraikin, M.G. Strakhovskaya and A.B. Rubin, The role of membrane-bound porphyrin-type compound as endogenous sensitizer in photodynamic damage to yeast plasma membranes, *J. Photochem. Photobiol. B*, 1996, **34**, 129–135.
20. T. Ito, Photodynamic action of haematorphyrin on yeast cell: a kinetic approach, *Photochem. Photobiol.*, 1981, **34**, 521–524.
21. G. Bertoloni, F. Zambotto, L. Conventi, E. Reddi and G. Jori, Role of specific cellular targets in the hematoporphyrin-sensitized photoinactivation of microbial cells, *Photochem. Photobiol.*, 1987, **46**, 695–698.
22. G. Bertoloni, R. Sacchetto, G. Jori, D.J. Vernon and S.B. Brown, Protoporphyrin photosensitization of Enterococcus hirae and Candida albicans cells, *Las. Life Sci.*, 1993, **5**, 5267–5275.
23. S.A. Lambrechts, M.C. Aalders and J. van Marle, Mechanistic study of the photodynamic inactivation of Candida albicans by a cationic porphyrin, *Antimicrob. Agents Chemother.*, 2005, **49**(5), 2026–2034.
24. S.A. Lambrechts, M.C. Aalders, F.D. Verbraak, J.W. Lagerberg, J.B. Dankert and J.J. Schuitmaker, Effect of albumin on the photodynamic inactivation of microorganisms by a cationic porphyrin, *J. Photochem. Photobiol. B*, 2005, **79**(1), 51–57.

25. W.M. Casey and L.W. Parks, A role for sterols in the porphyrin mediated photosensitization of yeast, *Photochem. Photobiol.*, 1989, **50**, 553–556.
26. L.N. Dovigo, A.C. Pavarina, D.G. Ribeiro, C.S. Adriano and V.S. Bagnato, Photodynamic inactivation of four Candida species induced by Photogem®, *Braz. J. Microbiol.*, 2010, **41**, 42–49.
27. V. Carré, O. Gaud, I. Sylvain, O. Bourdon, M. Spiro, J. Blais, R. Granet, P. Krausz and M. Guilloton, Fungicidal properties of meso-arylglycosylporphyrins: influence of sugar substituents on photoinduced damage in the yeast Saccharomyces cerevisiae, *J. Photochem. Photobiol. B*, 1999, **48**, 57–62.
28. T. Zoladek, B.N. Nguyen, I. Jagiello, A. Graczyk and J. Rytka, Diamino acid derivatives of porphyrins penetrate into yeast cellsinduce photodamage, but have no mutagenic effect, *Photochem. Photobiol.*, 1997, **66**, 253–259.
29. T.G. M. Smijs and H.J. Schuitmaker, Photodynamic inactivation of the dermatophyte Trichophyton rubrum, *Photochem. Photobiol.*, 2003, **77**, 556–560.
30. T.G. Smijs, S. Pavel, M. Talebi and J.A. Bouwstra, Preclinical studies with 5,10,15-Tris(4-methylpyridinium)-20-phenyl-(21H,23H)-porphine trichloride for the photodynamic treatment of superficial mycoses caused by Trichophyton rubrum, *Photochem. Photobiol.*, 2009, **85**, 733–739.
31. N. Brouillet, G. Arselin-De Chateaubodeau and C. Volland, Studies on protoporphyrin biosynthetic pathway in Saccaromyces cerevisiae: characterization of the tetrapyrrole intermediates, *Biochimie*, 1975, **57**, 647–655.
32. Z. Ji, G. Yang, V. Vasovic, B. Cunderlikova, Z. Suo, J.M. Nesland and Q. Peng, Subcellular localization pattern of protoporphyrin IX is an important determinant for its photodynamic efficiency of human carcinoma and normal cell lines, *J. Photochem. Photobiol. B.*, 2006, **84**(3), 213–220.
33. S. Bermudez Moretti, S. Correa García and A. Batle, Porphyrin biosynthesis intermediates are not regulating delta-aminlevulinic acid transport in Saccharomyces cerevisiae, *Biochem. Biophys. Res. Commun.*, 2000, **272**, 946–950.
34. M.G. Strakhovskaya, A.O. Shumarina, G.Y. Fraikin and A.B. Rubin, Synthesis of PPIX induced by 5-aminolevulinic acid in yeast cells in the presence of 2,20-dipyridyl, *Biochemistry (Mosc.)*, 1998, **63**, 725–728.
35. M. Paardekooper, A.E. Van Gompel, H.J. G.M. De Boont, J.F. Nagelkerke, J. Van Steveninck and P.J. A. Van der Broek, Photodynamic treatment of yeast with chloroaluminium-phthalocyanine: role of the monomeric form of the dye, *Photochem. Photobiol.*, 1994, **59**, 161–166.
36. G. Monfrecola, E.M. Procaccini, M. Bevilacqua, A. Manco, G. Calabro and P. Santoianni, In vitro effect of 5-aminolaevulinic acid plus visible light on Candida albicans, *Photochem. Photobiol. Sci.*, 2004, **3**, 419–422.
37. P.G. Calzavara-Pinton, M. Venturini, R. Capezzera, R. Sala and C. Zane, Photodynamic therapy of interdigital mycoses of the feet with topical application of 5-aminolevulinic acid, *Photodermatol. Photoimmunol. Photomed.*, 2004, **20**, 144–147.
38. C.M. Cassidy, M.M. Tunney, P.A. McCarron and R.F. Donnelly, Drug delivery strategies for photodynamic antimicrobial chemotherapy: From benchtop to clinical practice, *J. Photochem. Photobiol. B: Biology*, 2009, **95**, 71–80.
39. D. Watanabe, C. Kawamura, Y. Masuda, Y. Akita, Y. Tamada and Y. Matsumoto, Successful treatment of toenail onychomycosis with photodynamic therapy, *Arch. Dermatol.*, 2008, **144**, 19–21.
40. B.M. Piraccini, G. Rech and A. Tosti, Photodynamic therapy of onychomycosis caused by Trichophyton rubrum, *J. Am. Acad. Dermatol.*, 2008, **59**, S75–S76.

41. E. Sotiriou, T. Koussidou-Eremonti, G. Chaidemenos, Z. Apalla and D. Ioannides, Photodynamic therapy for distal and lateral subungual toenail onychomycosis caused by Trichophyton rubrum: Preliminary results of a single-centre open trial, *Acta Derm. Venereol.*, 2010, **90**(2), 216–217.
42. H. Kamp, H.J. Tietz, M. Lutz, H. Piazena, P. Sowyrda and J. Lademann, Antifungal effect of 5-amilolevulinic acid PDT in Trichophyton rubrum, *Mycoses*, 2005, **48**, 101–107.
43. K. Berg, J.C. Bommer and J. Moan, Evaluation of sulfonated aluminium phthalocyanines for use in photochemotherapy. A study on the relative efficiencies of photoinactivation, *Photochem. Photobiol.*, 1989, **49**, 587–594.
44. A. Segalla, C.D. Borsarelli, S.E. Braslavsky, J.D. Spikes, G. Roncucci, D. Dei, G. Chiti, G. Jori and E. Reddi, Photophysical, photochemical and antibacterial photosensitizing properties of a novel octacationic Zn(II)-phtalocyanine, *Photochem. Photobiol. Sci.*, 2002, **1**, 641–648.
45. D. Dei, A. Busetti, G. Chiti, M.P. De Filippis, L. Fantetti, F. Giuntini, M. Soncin, G. Jori and G. Roncucci, Phthalocyanines as photodynamic agents for the inactivation of microbial pathogens, *J. Porph. Phtaloc.*, 2006, **10**, 147–159.
46. D.D. Lasic, The mechanism of vesicle formation, *Biochem. J.*, 1988, **256**, 1–11.
47. J.G. De Nobel, C. Dijkers, E. Hooijberg and F.M. Klis, Increased cell wall porosity in Saccharomyces cerevisiae after treatment with dithiothreitol or EDTA, *J. Gen. Microbiol.*, 1989, **135**, 2077–2084.
48. L. Brancaleon and H. Moseley, Laser and non-laser light sources for photodynamic therapy, *Lasers Med. Sci.*, 2002, **17**, 173–186.

Chapter 15

Photodynamic Therapy for Wound Infections

Stanley B. Brown[a,b] and Clare L. Conway[b]

[a] Institute of Molecular and Cell Biology, Faculty of Biological Sciences, University of Leeds, Leeds, LS2 9JT, UK; [b] Photopharmica Ltd, Leeds Innovation Centre, 103 Clarendon Road, Leeds, LS2 9DF, UK

Table of Contents

Photodynamic Inactivation of Microbial Pathogens: Medical and Environmental Applications
Editors Michael R. Hamblin and Giulio Jori

Published by the Royal Society of Chemistry, www.rsc.org

15.1. Introduction

The skin is the main barrier between the internal organs and tissues and the external environment. Key among its many functions is its role in protection against external microorganisms. When the integrity of the skin breaks down locally, a wound may be created and this may occur for a number of reasons. Acute wounds are often caused by trauma, for example as in a wound caused by a burn, an accident or during surgery. However, wounds may also be caused by underlying pathophysiological conditions, as in leg ulcers or diabetic foot ulcers. Often such wounds are difficult to heal and become chronic. In any of these types of wounds, entry of microorganisms into the body becomes facilitated with negative consequences, which may result in life-threatening infections. In chronic wounds especially, the interaction between the natural healing process, the microbiological flora present and the underlying physiology is highly complex, which goes a long way to explain why scientific understanding of these processes remains relatively poor and why adequate treatment is challenging. Chronic wounds include leg and foot ulcers and pressure sores. In this brief account, we explore the complexity of chronic wounds and the potential of photodynamic therapy (PDT) for their treatment, especially from the perspective of antimicrobial therapy.

15.2. Wounds, Healing and Microorganisms

Normal, acute wound healing is a complex and highly regulated process, which has been much studied and which is reasonably well understood in terms of the molecular and physiological factors that control it. Healing progresses through several stages including initial blood clot formation, recruitment of inflammatory mediators and proliferation of the required vascular and epithelial cells. Healing relies on a fragile physiological process, vulnerable to disruption, and can be compromised by both endogenous (pathophysiological) factors and exogenous factors (*e.g.* microorganisms).[1]

Under normal physiological conditions, a mutually beneficial relationship exists within and on the human body between the host and a range of microbiological species, primarily bacteria.[1] However, this is a sensitive equilibrium, vulnerable to disturbance. Many bacteria, which are harmless or beneficial in the healthy individual, may demonstrate highly opportunistic behaviour and become pathogenic when the host is weakened either systemically or physically, as in the cutaneous wound. In the latter case, wounds become contaminated by microorganisms and new microbial ecosystems may develop.[1] Nevertheless, in the absence of underlying disease, natural healing and defence processes are normally adequate to deal with such microbial contamination and result in healing over a period of days to weeks. It is important to note therefore that the presence of microorganisms alone is not sufficient to account for slow or non-healing wounds which become chronic and therefore antimicrobial treatments alone are unlikely to achieve healing.

15.3. Chronic Leg and Diabetic Foot Ulcers – Prevalence and Impact

These wounds are a very substantial cause of morbidity and, in the case of diabetic foot ulcers, mortality, with major effects on quality of life and they represent a very large health economic burden. The prevalence of these ulcers is not easy to establish accurately, partly because their management is disseminated across different clinical specialties (including dermatology, vascular surgery, endocrinology and general medicine) and also within different types of health care providers, *e.g.* hospitals, community nurses and elderly care homes. In the UK, the prevalence of venous leg ulcers at any one time has been quoted as around 1.7% for the population over 65 years,[2,3] although this may well be an underestimate.

The figures for diabetic foot ulcers are even higher, with cross-sectional community surveys in the UK showing that 7.4% of patients with Type 2 or Type 1 diabetes had a history of active or previous foot ulcer.[4,5] The three-year cumulative incidence in the USA from hospital discharge data has been reported as 5.8%.[4] The lifetime risk for any diabetic patient is quoted as between 15 and 25%.[4,6,7] Diabetic foot ulceration has a poor prognosis, with amputation being a common sequel. In a review, Jeffcoate and Harding[4] report that between 15% and 27% of all foot ulcers result in amputation involving surgical removal of bone, but this varies between countries and is very dependent upon the varying standards of care and different management practices. Mortality is high following amputation, with figures quoted around 10–15%.[8,9] In addition to the quality-of-life issues for patients, diabetic foot ulcers have substantial economic consequences.[10] Costs are affected by preventative measures, management strategies to shorten wound healing time and prevent amputation and for the care for disability following amputation.[11] In 2001, diabetic foot ulcers (and subsequent amputations) were reported to have cost $10.9 billion in the USA[12] and the total annual cost of diabetic foot complications in the UK was estimated at £252 million.[13]

15.4. Chronic Wounds and Infection

With the exception of pressure sores, the great majority of chronic wounds occur on either the legs or the feet and include venous and arterial leg ulcers and diabetic foot ulcers. Chronic leg ulcers and pressure ulcers may be large in area, but are usually relatively superficial, whereas diabetic foot ulcers are generally smaller in area, but may be much deeper and microbiological invasion is therefore much more likely to cause serious sequelae. Much confusion may arise because of different uses of the term "wound infection". Most authors use this term in the sense that a wound contains infective agents including bacteria or other microbial species which cause or have the potential to cause damage to normal health. This includes a narrower use of the term infection (often termed "clinical infection") when referring to the advanced stages of infection associated with the classical signs such as oedema, erythema and purulent exudates. For chronic wounds, the following sub-divisions of infection may be helpful.

Contaminated wounds: Here, the wound contains bacteria which have gained access from surrounding skin or during exposure during dressing changes. The bacteria are in a quiescent state *i.e.* not invading the tissue or even replicating significantly. In this state, it is unlikely that the bacteria will interfere with wound healing.

Colonised wounds: Here, the bacteria are able to find nutrition in the wound (*e.g.* from necrotic tissue and other wound products) and are therefore able to multiply. The density that the bacteria are able to attain in the wound depends on factors including the amount of nutrients available and the species of bacteria present. This is the commonest situation in chronic leg ulcers, where it is likely that a competition exists between the host healing response and bacterial replication, resulting in a pseudo-equilibrium, in which the host defence mechanisms prevent the further bacterial multiplication which would signal clinical infection, but the bacterial load and demand for nutrients is sufficient to prevent significant healing. Importantly, in such colonized wounds, there is no significant damage to the underlying tissue. This is the situation in most chronic leg ulcers, although there is always a risk that the balance can be tipped towards clinical infection.

Clinically infected wounds: Here, the bacteria "win" the competition and invade and degrade normal tissue. This induces the characteristic inflammatory responses of erythema and purulent exudates. Such clinically significant infections will certainly impede healing and may become systemic and life-threatening, especially in the case of diabetic foot ulcers.

Although there is increasing acceptance of the importance of microbial colonization in chronic wound healing, there is a surprising lack of detailed understanding of the relationship between the microbiological load of a chronic wound and the healing rate. Some practitioners have used a general rule of thumb that an excess of 10^5 colony-forming units (cfu) per cm^2 is inhibitory to healing (and presumably therefore that one aim of therapy should be to reduce the load below that level). However, it is now recognized that the interaction between wound and microorganisms is much more subtle and is dependent on many factors, including the identity of the bacterial species present.[14] Nevertheless, rigorous studies of the relationship between the quantitative microbiological load (including the identity of bacterial species) and healing remain sparse. One challenge is the sampling of chronic wounds for microbiological assessment. Surface swabbing is relatively simple and allows for sampling from the whole of the wound surface. However, swabbing may not accurately report on bacteria deeper in the wound. This can be better achieved through biopsy, but this can be painful and only reports on the precise area sampled whereas, in large wounds, the bacterial load may not be uniformly distributed (in either density or distribution of species) across the entire surface. However, a recent study has demonstrated good correlation between swabbing and biopsy[15] and, in clinical studies, appropriately controlled swabbing has been acceptable to regulatory agencies in Phase I/II studies, though some biopsy confirmatory data may be required for Phase III regulatory studies.

A recent study by Moore and coworkers[16] has investigated the relationship between the surface bacteriology of chronic venous leg ulcers and healing outcome. In ulcers from 178 patients, 153 individual bacterial species were identified, but there was no correlation between healing and any single species, the presence of anaerobes or the diversity of bacterial species, previously thought to be predictive of impaired healing. Another significant component of wound microbiology may be the existence of biofilm, although it is only recently that the studies have focused on the potential role of biofilm in wound healing.[17] In spite of the challenges, there is a growing consensus that antimicrobial therapy is an appropriate and important component of wound management and, moreover, that new antimicrobial treatments are needed. It is in this context that there is a niche that PDT may well be able to fill.

15.5. Chronic Ulcers and Current Treatments

15.5.1. Leg Ulcers

The ultimate aim of any treatment of wounds is to achieve healing and restoration of the overlying skin functions. Chronic leg ulcers may exist for many months and sometimes for several years. Even if healing can be achieved, recurrence is a significant threat. The underlying pathophysiology of chronic leg ulcers is associated with reduced blood perfusion and includes impaired arterial supply (peripheral vascular disease), leading to arterial leg ulcers and impaired venous drainage (venous hypertension), leading to venous leg ulcers. The degree to which a leg ulcer is venous or arterial is routinely measured quantitatively by a parameter known as the ankle brachial pressure index (ABPI), which is a measure of the ratio of the blood pressure in the lower limbs to that in the arms. Significantly lower blood pressure in the legs is an indication of arterial disease. For patients with chronic leg ulcers, an ABPI >0.8 is taken to signify underlying venous disease, whilst an ABPI <0.8 indicates at least some arterial disease. Although venous disease predominates in up to 80% of leg ulcers, in practice there is often mixed arterial and venous disease, yet another complicating factor in the understanding and treatment of chronic leg ulcers.

A detailed discussion of the clinical management of chronic leg ulcers is beyond the scope of this brief review. The key aims are to achieve healing of the wound as rapidly as possible (though even in favourable cases this may take many weeks) and to prevent recurrence. Current management of venous leg ulcers is largely through four-layer compression bandaging and simple physical techniques such as washing, along with removal of necrotic tissue through debridement. However, there is a growing consensus that an antimicrobial component needs to be added to management to achieve accelerated healing. Topical antibiotics are relatively ineffective and, although silver dressings[18] and honey are in use, there is a clear need for new and more effective antimicrobial treatments. Whilst antimicrobial therapy may improve healing by removal of competing bacteria, there is also interest in biological therapies which positively promote healing, for example through administration of growth factors.[19]

15.5.2. Diabetic Foot Ulcers

Many overlapping factors may lead to foot ulceration but the key factors are the peripheral neuropathy which renders skin damage more likely because of lack of sensation, coupled with diabetic ischaemia.[4] Bacterial contamination invariably follows wound formation and may be superficial and local, may spread into soft tissue (cellulitis) or may penetrate into bone (osteomyelitis). Jeffcoate and Harding[4] have described four principles in treating chronic diabetic foot ulcers. The first is to treat any infection, for which antibiotics remain the front line, although increased resistance is a major concern. This is an area in which PDT may have a valuable role. Other treatments are to establish whether any associated ischaemia is amenable to revascularization, to ensure that forces applied to the affected area are minimized and to improve wound bed condition through debridement, topical applications and removal of callus.

15.6. Potential of PDT in Treatment of Chronic Wounds

The sections above demonstrate the complexity of chronic leg and diabetic foot ulcers and their underlying pathophysiology. It would be naive to imagine that a single treatment regime such as PDT is likely to be successful, in isolation, in achieving wound healing. However, given the wide scale of the morbidity and mortality associated with these conditions, and the low success rate in the healing of chronic wounds, it is also clear that new treatments are urgently needed. How then might PDT be developed to help to solve these major clinical problems? There are two broad responses to this question which give PDT advantages over other modalities. Firstly, PDT has a unique antimicrobial action, which can be accurately targeted, which works equally well with antibiotic-sensitive and antibiotic-resistant organisms and which is not, itself, susceptible to development of resistance. This benefit should be directed to reducing the inhibition of wound healing by microorganisms. Secondly, PDT application to chronic wounds offers the potential for induction of a host response, either through the immune system or through stimulation of biological effects such as production of growth factors, which may positively contribute to wound healing. If PDT treatment regimes can be developed to treat chronic wounds harnessing one or both of these mechanisms, in association with standard therapies such as compression bandaging, there are good prospects for new product development in which PDT can contribute to the management of these debilitating conditions.

15.7. Pre-clinical Models

There is now a large and expanding literature on the antimicrobial effects of PDT *in vitro*, under laboratory culture conditions. Almost any photosensitizer and light will inactivate Gram-positive bacteria, but negatively charged and neutral photosensitizers have little effect on Gram-negative bacteria, unless used in conjunction with a disrupting agent such as polymixin.[20] In order to obtain broad spectrum antimicrobial activity, photosensitizers need to be cationic[20,21] and there is a large variation in PDT efficacy depending upon the chemical structure of the cationic molecule. For example, methylene blue, whilst being an excellent photosensitizer, has a relatively poor efficacy in bacterial inactivation, whilst its tetrabutyl analogue (3,7-bis[*N*,*N*-dibutylamino] phenothiazin-5-ium bromide) demonstrates very high antimicrobial activity against both Gram-positive and Gram-negative organisms, as well as fungal species. Using such high-activity photosensitizers *in vitro*, it is possible to achieve typically 5–7 logs of cell kill with comparatively very low photosensitizer and light doses.

A number of experimental animal model systems for tissue infection and for acute wounds have now been developed and are considered in detail in Chapter 8 of this book. Perhaps the most immediate impact of these models is that the high log kills seen by PDT regimes *in vitro* are not easily reproduced *in vivo*, where much higher photosensitizer and light doses are needed and lower log kills are typically achieved. Nevertheless, such models are valuable in establishing the most appropriate parameters to be used in clinical studies.

15.8. Clinical Studies

To date there appears to be only one case report in the literature of a clinical study using PDT for treatment of chronic wounds (see below). However, extensive clinical studies have been performed by Photopharmica since 2006, with completion of an initial feasibility study, a placebo-controlled Phase IIa study and with an ongoing Phase IIb study. It is not possible to present detailed results of these studies since they have not yet been published in the primary literature but, in this section, we consider implications of some of the clinical studies and how such studies may impact on wound healing.

15.8.1. Case Report

In a case study reported in 2006, Clayton and Harrison[22] treated a 72-year-old woman with chronic recalcitrant venous ulceration of her right lower leg, persisting for more than 12 months and complicated by cellulitis and colonization with methicillin-resistant *Staphylococcus aureus* (MRSA). Conventional treatments including long-term four-layer bandaging, topical antiseptics and a variety of bacteriostatic dressings and application of larval therapy had proved unsuccessful. PDT was delivered twice weekly over four weeks to the ulcer at different 5-cm diameter sites each time using topical 5-aminolaevulinic acid (5-ALA) and irradiation at 633 nm. The PDT was well tolerated with minimal discomfort to the patient, even without the use of topical local anaesthesia (unlike the use of 5-ALA in treatment of non-melanoma skin cancer, where pain is frequently experienced) and significant improvement was subsequently observed in the ulcer.

15.8.2. Clinical Feasibility Study

Following extensive preclinical safety and toxicology testing, Photopharmica Ltd are developing the photosensitizer PPA904 (3,7-bis[*N*,*N*-dibutylamino] phenothiazin-5-ium bromide) (Figure 15.1) for treatment of chronic leg ulcers. This photosensitizer has several key advantages for antimicrobial PDT as follows:

- Very high, broad spectrum antimicrobial activity compared with many other photosensitizers (*e.g.* methylene blue)
- Strong light absorption at 680 nm
- Easily manufactured and purified as GMP material (no isomers)
- Small molecular size compared with porphyrins and phthalocyanines
- Excellent pre-clinical safety profile.

In a small antimicrobial PDT pilot study using broad spectrum light, eight patients with bacterially colonized chronic ulcers, four with diabetic foot ulcers and four with venous leg ulcers were given a single PDT treatment with PPA 904. The PDT was well tolerated by all patients and no patient experienced pain, which is a common problem in other topical PDT treatments. In each of the patients there was a decrease in total bacterial load immediately after treatment compared to pre-treatment (median 2.5 $\log_{10}$ reduction) and this was statistically significant ($p = 0.0078$) as determined by the Wilcoxon Rank non-parametric test. No pain or other adverse events were recorded and no evidence of systemic uptake of the drug was found following blood analysis in three patients. This was the first demonstration of a clinically significant reduction in

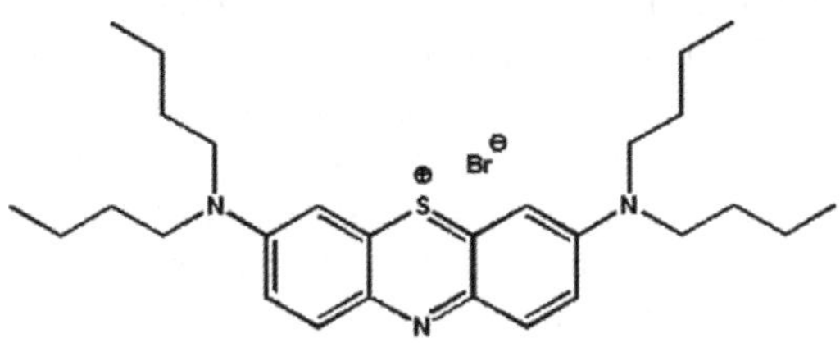

PPA 904

Figure 15.1 Structure of PPA904 (3,7-bis[*N*,*N*-dibutylamino]phenothiazin-5-ium bromide).

bacterial load by a PDT treatment in this indication and both Gram-positive and Gram-negative microorganisms were susceptible to the treatment.

15.8.3. Single Treatment, Randomized, Placebo-controlled Study

In the first study of its kind, a Phase IIa, randomized, blinded, placebo-controlled trial in 32 patients was carried out with PPA 904 PDT. The study population consisted of 16 patients with chronic venous leg ulcers (8 received PPA 904 plus light and 8 received placebo plus light) and 16 patients with chronic diabetic foot ulcers (8 received PPA 904 plus light and 8 received placebo plus light). All patients received standard wound therapy (compression bandaging for venous leg ulcers and standard dressing for diabetic foot ulcers) as well as a single treatment of either PPA 904 plus light or placebo plus light. The data showed a strongly significant reduction in bacterial load following PDT and a trend towards accelerated healing in the PPA 904 group compared with the placebo group. The study showed the therapy to be safe and without treatment-induced pain.

15.8.4. Repeat Treatment, Randomized Placebo-controlled Study

Based on the encouraging results from the early clinical data, a repeat dose study was initiated early in 2010, whereby 69 patients with chronic leg ulcers receive weekly doses of PPA 904 PDT plus light or placebo plus light. The trial is powered to detect significance in reduction of microbiological load, but may also reveal accelerated healing. The study is blinded and no efficacy results will be available until completion, which is expected late in 2011.

15.9. Conclusions

The complexity of chronic wounds represents a challenge for development of new therapies, but these are urgently needed because of the extent of morbidity and mortality and the inadequacy of existing approaches. PDT represents a novel and promising treatment option, provided that this complexity is recognized in the design of clinical trials. Such studies require delivery, not just of PDT alone, but PDT superimposed on existing standard wound management, as for example in the four-layer compression bandaging in chronic venous leg ulcers. An appropriate control arm is essential for such trials, not least because the rigorous application of standard treatment during hospital-based clinical trials may produce a significant placebo effect against which the active treatment must be judged.

It is likely that frequent reduction of bacterial load (*e.g.* through once weekly PDT) may achieve accelerated healing by repeatedly shifting the balance against bacterial colonization and in favour of the healing processes. Also important may be the positive effect of PDT in stimulation of growth factors to assist wound healing. The clinical study results obtained so far are encouraging in supporting the hypothesis that PDT may add real benefit to the treatment of chronic wounds and the results from the ongoing Phase IIb study should further increase our knowledge in this important area of medicine.

References

1. P.G. Bowler, Wound pathophysiology, infection and therapeutic options, *Ann. Med.*, 2002, **34**, 419.
2. R.S. Howell-Jones, M.J. Wilson, K.E. Hill, A.J. Howard, P.E. Price and D.W. Thomas, A review of the microbiology, antibiotic usage and resistance in chronic skin wounds, *J. Antimicrob. Chemother.*, 2005, **55**, 143.
3. D.J. Margolis, W. Bilker, J. Santanna and M. Baumgarten, Venous leg ulcer: incidence and prevalence in the elderly, *Am. Acad. Dermatol.*, 2002, **46**, 381.
4. W.J. Jeffcoate and J.G. Harding, Diabetic foot ulcers, *Lancet*, 2003, **361**, 1545.
5. D.P. Walters, W. Gatling, A. C Houston, M.A. Mullee, S.A. Julious and R.D. Hull, Mortality in diabetic subjects: an eleven-year follow-up of a community-based population, *Diabet. Med.*, 1992, **9**, 354.
6. G.E. Rieber, B.A. Lipsky and G.W. Gibbons, The burden of diabetic foot ulcers, *Am. J. Surg.*, 1998, **176**, 5S.
7. N. Singh, D.G. Armstrong and B.A. Lipsky, Preventing foot ulcers in patients with diabetes, *JAMA*, 2005, **293**, 217.
8. L.A. Lavery, W.H. van Houtum and L.B. Harkless, In-hospital mortality and disposition of diabetic amputees in The Netherlands, *Diabet. Med.*, 1996, **13**, 192.
9. A.F. Da Silva, P. Desgranges, J. Holdsworth, P.L. Harris, P. McCollum, S.M. Jones, J. Beard and M. Callam, In-hospital mortality and disposition of diabetic amputees in The Netherlands, *Diabet. Med.*, 1996, **13**, 726.
10. A.J.M. Boulton, L. Vileikyte, T.G. Ragnorson and J. Apelqvist, The global burden of diabetic foot disease, *Lancet*, 2005, **366**, 1719.
11. T.G. Ragnarson and J. Apelqvist, Health-economic consequences of diabetic foot lesions, *Clin. Infect. Dis.*, 2004, **39**(2), 132.
12. A. Shearer, P. Scuffham, A. Gordois and A. Oglesby, Predicted costs and outcomes from reduced vibration detection in people with diabetes in the U.S, *Diabetes Care*, 2003, **26**, 2305.
13. A. Gordois, P. Scuffham, A. Shearer, A. Oglesby and J.A. Tobian, The health care costs of diabetic peripheral neuropathy in the US, *Diabetes Care*, 2003, **26**, 1790.
14. P.G. Bowler, The 10(5) bacterial growth guideline: reassessing its clinical relevance in wound healing, *Ostomy Wound Manage.*, 2003, **49**, 44.
15. C.E. Davies, K.E. Hill, R.G. Newcombe, P. Stephens, M.J. Wilson, K.G. Harding and D.W. Thomas, A prospective study of the microbiology of chronic venous leg ulcers to reevaluate the clinical predictive value of tissue biopsies and swabs, *Wound Repair Regen.*, 2007, **15**, 17.
16. K. Moore, V. Hall, A. Paull, T. Morris, S. Brown, D. McCulloch, M.C. Richardson and K.G. Harding, Surface bacteriology of venous leg ulcers and healing outcome, *J. Clin. Pathol.,* 2010, Jul 29 [Epub ahead of print].
17. J. Hurlow and P.G. Bowler, Clinical experience with wound biofilm and management: a case series, *Ostomy Wound Manage.*, 2009, **55**, 38.
18. S.L. Percival, P.G. Bowler and D. Russell, Bacterial resistance to silver in wound care, *J. Hosp. Infect.*, 2005, **60**, 1.

19. R. Braund, S. Hook and N.J. Medlicott, The role of topical growth factors in chronic wounds, *Curr. Drug Deliv.*, 2007, **4**, 195.
20. M. Merchat, G. Bertoloni, P. Giacomini, A. Villanueva and G. Jori, Meso-substituted cationic porphyrins as efficient photosensitizers of gram-positive and gram-negative bacteria, *J. Photochem. Photobiol. B*, 1996, **32**, 153.
21. A. Minnock, D.I. Vernon, J. Schofield, J. Griffiths, J.H. Parish and S.B. Brown, Photoinactivation of bacteria. Use of a cationic water-soluble zinc phthalocyanine to photoinactivate both gram-negative and gram-positive bacteria, *J. Photochem. Photobiol. B*, 1996, **32**, 159.
22. T.H. Clayton and P.V. Harrison, Photodynamic therapy for infected leg ulcers, *Br. J. Dermatol.*, 2007, **156**, 384.

Chapter 16

Photodynamic Therapy for *Helicobacter pylori* Infections

Javad T. Hashmi[a,b] and Michael R. Hamblin[a,b,c]

[a] Wellman Center for Photomedicine, Massachusetts General Hospital, Boston, MA 02114, USA; [b] Department of Dermatology, Harvard Medical School, Boston, MA 02114, USA; [c] Harvard-MIT Division of Health Sciences and Technology, Cambridge, MA 02139, USA

Table of Contents

Photodynamic Inactivation of Microbial Pathogens: Medical and Environmental Applications
Editors Michael R. Hamblin and Giulio Jori

Published by the Royal Society of Chemistry, www.rsc.org

16.1. Introduction

Mucosa forms the innermost layer of the human gastrointestinal tract (GIT) and surrounds the gastrointestinal lumen. This membranous lining comes in direct contact with ingested food and is responsible for important aspects of digestion, including absorption and secretion. Parietal cells of the stomach epithelium produce gastric acid (HCl), which assists in the decomposition of food molecules. The GIT's mucosa is normally protected against this acidic solution by the protective lining; however, excessive secretion of gastric acid, and subsequent breakdown of the GIT's protective lining, can lead to inflammation or ulceration. Peptic ulcer disease (PUD) is a common condition in which the mucosal layer of the GIT ulcerates, causing exquisite levels of epigastric pain or discomfort.[1] Other symptoms of PUD include nausea, vomiting, waterbrash, haematochezia, melena, fatigue, weight loss, anorexia, bloating, abdominal fullness and chest pain (often confused for myocardial infarction). PUD has several significant complications, including bleeding (which can cause anaemia or even death), penetration or perforation of the GIT tract (and subsequent peritonitis) and swelling or scarring that can cause GIT obstruction. Patients afflicted with the disease are also at increased risk of developing gastric cancer.

PUD arises from the stomach (gastric ulcer) or more commonly from the duodenum (duodenal ulcer). A majority of cases of PUD (70–90%) are associated with infection by the Gram-negative microaerophilic bacterium, *Helicobacter pylori*, which selectively colonizes the mucosal layer of the GIT.[2] In normal circumstances, the immune system is unable to eliminate *H. pylori* infection, thus allowing for chronic and "steady state" illness, persisting for life. *H. pylori* infection causes a defect in the GIT's protective lining, partly induced by the host inflammatory response towards the bacteria, resulting in mucosal erosions and PUD.[3] *H. pylori* is a common pathogen in humans with an adult prevalence rate of 20–50% in the developed world and up to 80% in developing countries.[4] As such, it can be said that *H. pylori* is the most common infectious organism in the world.[5] It should be noted that not all patients with *H. pylori* develop PUD; nonetheless, the majority of those with PUD are infected with *H. pylori*. The burden of this disease is great, and the Centers for Disease Control and Prevention (CDC) estimates that one of out of ten Americans suffers from PUD at least once in their lives. In the United States alone, over one million hospitalizations and 6500 deaths per year can be attributed to PUD, and the annual health care costs are estimated at nearly $6 billion (http://www.cdc.gov/ulcer/economic.htm).

Conventional treatment of PUD involves the eradication of *H. pylori* using a combination of antibiotics and proton pump inhibitors (PPI), usually administered for a duration of one to two weeks. The favoured regimens consist of metronidazole combined with a PPI, bismuth and tetracycline, or clarithromycin combined with a PPI and amoxicillin.[6] Because these antibiotic combinations can be complex and associated with unpleasant side effects, patient compliance is often suboptimal. Eradication rates with such regimens vary between 72% and 95% and the CDC states that "in more than 80 percent of patients the ulcer is cured and does not recur" (http://www.cdc.gov/ulcer/economic.htm).[7] Yet, it is clear that a significant number of patients are not afforded relief by pharmacological means alone (the remaining 20 or so percent), and an alternative modality of treatment is warranted.[7] The need for novel therapy is underscored by the well-documented emergence of resistant strains of *H. pylori*.[8] The widespread use of antibiotics has resulted in the rapid emergence of bacterial resistance, necessitating the search for alternative means of treatment. Resistance to metronidazole has been identified in 13–39% of isolates, and to clarithromycin in 8–12% of isolates. Resistant strains have a lower cure rate than susceptible strains, and can result in failure in excess of 20% for metronidazole-based regimens. For clarithromycin-resistant strains, the treatment failure rate of clarithromycin-based regimens can be as high as 77%.[9] Moreover, there is also evidence that eradication rates are also influenced by the virulence factors of the patient's individual strain of *H. pylori*.[10] For these aforementioned reasons it is of interest to ask whether PDT (or indeed phototherapy alone) could be used to kill *H. pylori* and possibly eradicate the infection within the human stomach.

16.2. PDT *in vitro* and *ex vivo*

The first report of the use of PDT to kill *H. pylori* was described in a letter to *The Lancet* in 1990.[11] Bedwell and colleagues isolated *H. pylori* from biopsy tissue taken from the antrum of six patients with gastritis and sub-cultured them onto brain/heart infusion agar containing 10 μg ml^{-1} amphotericin and incubated under microaerobic conditions. All six strains had similar sensitivity to PDT so just one was used subsequently. The photosensitizer used was aluminium sulfonated phthalocyanine (AlSPc) (Ciba-Geigy). A three-day agar culture of *H. pylori* was incubated for four hours with PS in saline at concentrations ranging from 25 nM to 2.5 mM. Fifty-μl aliquots were then plated onto agar and exposed to red light at 675 nm from a copper vapour pumped dye laser (Oxford Lasers) to give light doses of 1.5 and 5.5 J cm^{-2}. Plates were incubated for five days and colonies were counted. Independent confirmation of uptake of AlSPc was achieved by fluorescence microscopy on *H. pylori* cultures exposed to AlSPc. Red laser light at 1.5 J cm^{-2} killed 100% of *H. pylori* at AlSPc concentrations as low as 25 μM and with 5.5 J cm^{-2} there was 100% kill down to 2.5 μM.

Millson and coworkers investigated which PS were able to kill *H. pylori in vitro* in a similar procedure to that described above.[12] Crystal violet and thionine were ineffective as PS, but zones of inhibition appeared with methylene blue (MB), protoporphyrin IX (PPIX), haematoporphyrin derivative (HPD), toluidine blue O (TBO) and disulfonated aluminium phthalocyanine (S2). Laser light or PS alone did not affect bacterial viability. S2 (100 μg ml^{-1}) with a laser light energy density of 16 J cm^{-2}, HPD (100 μg ml^{-1}) with 160 J cm^{-2}, MB

(100 μg ml^{-1}) with 21 J cm^{-2}, PPIX (150 μg ml^{-1}) with 320 J cm^{-2} and TBO (50 μg ml^{-1}) with 160 J cm^{-2} all reduced bacterial viability by >99%.

Millson *et al.* went on to take advantage of an interesting animal model of gastric helicobacter infection.[13] Ferrets in many parts of the world are naturally infected with *Helicobacter mustelae*, a spiral microaerophilic bacterium very similar to *H. pylori*. They treated explanted ferret gastric mucosa using one of five PS (methylene blue (MB), toluidine blue O (TBO), phthalocyanine, haematoporphyrin derivative and 5-aminolavulinic acid), followed by irradiation with an appropriately tuned copper vapour pumped dye laser. A 90% reduction in counts of bacteria sensitized with 0.75 mg kg^{-1} TBO was seen after irradiation with 200 J cm^{-2}. Concentrations of MB of 0.75 mg kg^{-1} and 7.5 mg kg^{-1} were not toxic to *H. mustelae*, but the further addition of 20 J cm^{-2} laser light reduced colony counts by more than 99%. The remaining three compounds were ineffective. They studied the microscopic fluorescence distribution of MB (7.5 mg kg^{-1}) on ferret gastric mucosa after topical administration. Fluorescence was greatest in the superficial mucosal layer, upon which lie the bacteria. However, from experiments on rats, the energy required to kill the sensitized bacteria was insufficient to damage the underlying mucosa.

16.3. PDT Clinical Trial

Wilder-Smith and colleagues investigated the photoeradication of *H. pylori* in patients using oral 5-aminolevulinic acid (5-ALA) and intra-gastric laser and white light.[14] In 13 *H. pylori*-positive volunteers, a zone of gastric antrum was irradiated with laser (410 nm, 50 J cm^{-2}) or endoscopic white light (10 J cm^{-2}) 45 minutes after 20 mg kg^{-1} oral ALA. They used biopsy urease testing and culture from irradiated and control zones taken 5 minutes, 4 and 48 hours post-irradiation. They found a maximum eradication effect at 4 hours post-irradiation when 85% of biopsies in the 410-nm laser and 66% in the white-light exposed zones were negative compared to 58% and 33% in the respective control zones.

16.4. Phototherapy *in vitro*

It has been known for many years that a few selected species of bacteria naturally accumulate sufficient porphyrins to make them susceptible to illumination (particularly if violet light around 405 nm is used). This phenomenon is best studied in *Propionibacterium acnes*, a Gram-positive bacterium recognized as a major cause of acne. Recently, it was found that blue and red light killed *P. acnes* (in the absence of photosensitizing dyes), and the US Food and Drug Administration (FDA) has approved a high-intensity narrow-band blue light therapy for the treatment of acne vulgaris.[15] The effectiveness of light in killing *P. acnes* in the absence of a PS was better understood after sensitive fluorescence spectroscopy determined that *P. acnes* naturally synthesizes its own PS, protoporphyrin IX (together with lesser amounts of coproporphyrin and uroporphyrin), and that these endogenous porphyrins mediate the killing effect of visible light.[16,17] Some oral bacterial species that cause periodontitis such as Actinomyces, Odontolyticus and *Porphyromonas gingivalis* also naturally synthesize or accumulate PPIX (and coproporphyrin), and are similarly killed by red or blue light.[18]

In 2005, Hamblin and colleagues discovered that *H. pylori*, when cultured in liquid medium, accumulates significant quantities of coproporphyrin and protoporphyrin IX, both in the cells and secreted into the medium.[19] The porphyrins were identified by fluorescence emission spectroscopy (Figure 16.1A) and by capillary electrophoresis compared with authentic porphyrins standards (Figure 16.1B). These photoactive porphyrins in the *H. pylori* cells led to cell death (up to 5 logs) by photodynamic action upon illumination with low doses of visible light, with blue/violet light (375 to 425 nm) being most efficient in eliminating *H. pylori*, followed by the range of 425 to 475 nm (blue light) (Figure 16.2). The degree of killing increased with the age of the culture and was greater than that found with *P. acnes* (another bacterium known to be photosensitive due to porphyrin accumulation) (Figure 16.3). Both virulent and drug-resistant strains were killed. The data suggested that phototherapy might be used to treat *H. pylori* infection in the human stomach.

16.5. Phototherapy Clinical Trials

Using this information, in 2005 Ganz *et al.* reported a proof-of-principle clinical study comparing quantitative cultures from biopsies taken before and

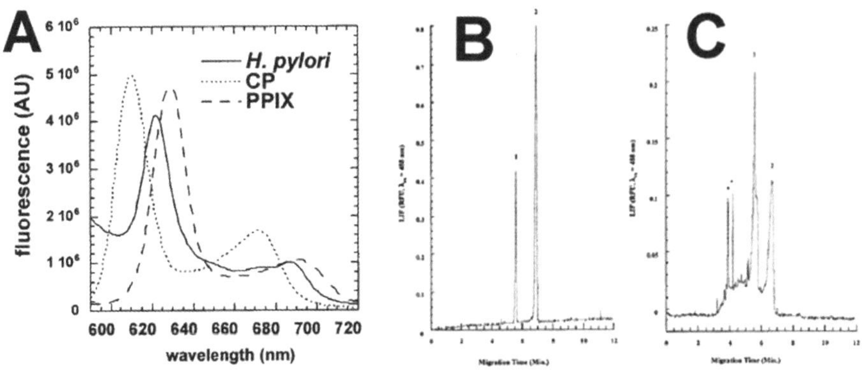

Figure 16.1 (A) Fluorescence spectroscopy (excitation 405 nm) of 4-day-old culture medium from *H. pylori* ATCC49503 compared to CP and PPIX standards. (B) Capillary electrophoresis of standard mixture of CP (peak 1) and PPIX (peak 2). (C) Capillary electrophoresis of extract from culture supernatant of 4-day-old *H. pylori* ATCC 49503. Peaks marked * may be artefacts or very low concentrations of porphyrins with more than four carboxyl groups. Reprinted from *Antimicrobial Agents and Chemotherapy*, 2005, **49**(7), doi:10.1128/AAC.49.7, 2822–2827. Copyright © 2005, American Society for Microbiology, with permission.

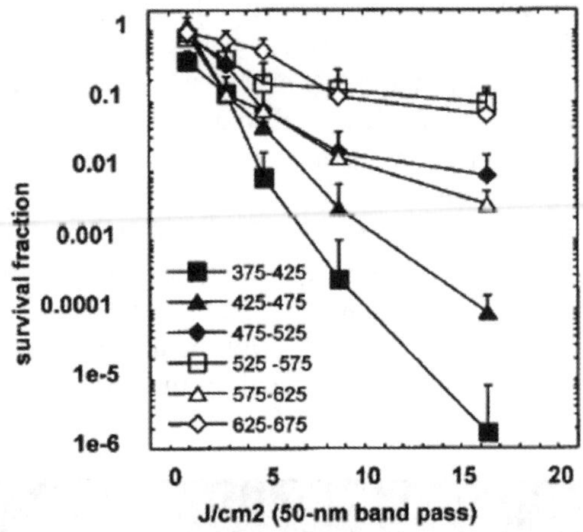

Figure 16.2 Fluence-dependent killing of *H. pylori* ATCC49503 after exposure to light produced by 50-nm band-pass filters. Values are means of three independent experiments and bars are standard errors. Reprinted from *Antimicrobial Agents and Chemotherapy*, 2005, **49**(7), doi:10.1128/AAC.49.7, 2822–2827. Copyright © 2005, American Society for Microbiology, with permission.

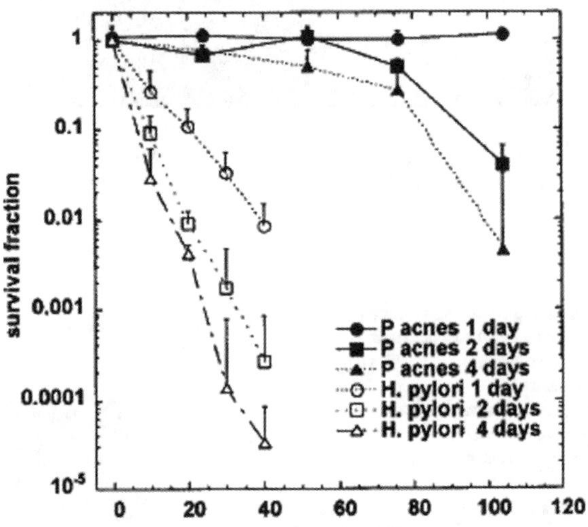

Figure 16.3 Comparison of photoinactivation using 405-nm laser light of *H. pylori* ATCC 49503 and *P. acnes* grown in liquid medium for various lengths of time. Values are means of three independent experiments and bars are standard errors. Reprinted from *Antimicrobial Agents and Chemotherapy*, 2005, **49**(7), doi:10.1128/AAC.49.7, 2822–2827. Copyright © 2005, American Society for Microbiology, with permission.

after endoscopically delivered violet light to a focal region of the mucosa in the gastric antrum.[20] This study was performed in ten chronically infected symptomatic patients between the ages of 21 and 80 who tested positive for *H. pylori*. Light (405 nm) (40 J cm^{-2}) was delivered to a 1-cm diameter spot in the gastric antrum *via* optical fibre passed through the endoscope (see Figure 16.4) and weighed biopsies were taken from treated and control spots and colonies quantitatively cultured. Blue light killed 5 logs of bacteria *in vitro*. The mean reduction in *H. pylori* colonies per gram of tissue between treated and control spots was 91% ($7.4 \pm 4.8 \times 10^6$ *vs.* $8.1 \pm 1.9 \times 10^7$, two-tailed $P < 0.0001$; Figure 16.5A). Some patients had reductions approaching 99% (Figure 16.5B). No differences were observed on histological examination of light-treated and control gastric tissue.

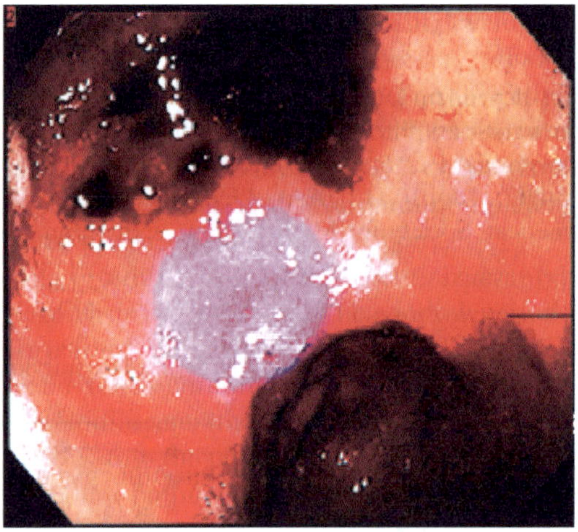

Figure 16.4 Endoscopic photograph of a control site (post-biopsy with some visible blood), and the adjacent treatment site with the spot of laser light visible as a blue circle.

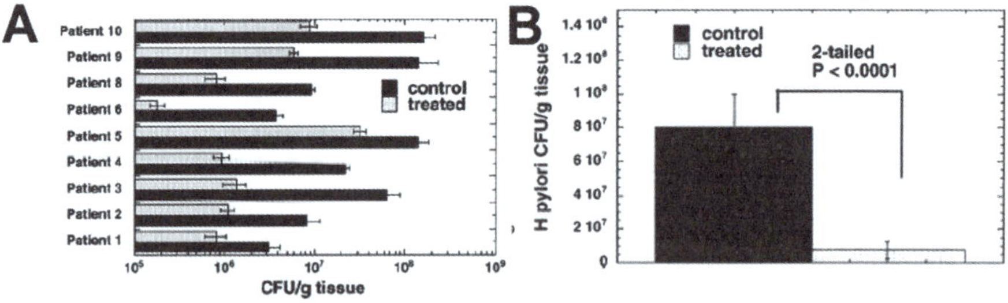

Figure 16.5 (A) Mean CFU/g tissue (+ SEM) from all cultures from all nine patients. (B) Mean CFU per gram tissue (± SEM) from all cultures from all nine patients. *P* value determined by two-tailed unpaired Student's t-test. Reprinted from *Lasers in Surgery and Medicine*, 2005, **36**, 260–265. Copyright © 2005 Wiley-Liss, Inc. with permission.

In 2009, Lembo *et al.* conducted a second clinical study using whole-stomach illumination with 12 W of violet light (408 nm) delivered endoscopically to 18 symptomatic patients suffering from *H. pylori*-induced peptic ulcer disease.[21] A suitable light source, or phototherapy system, was constructed using three functional subsystems: a power source and control system, a light wand and a catheter sheath (see Figure 16.6). The proximal portion of the light wand consisted of InGaN 408 ± 2-nm laser diodes (Nichia Corp., Torrance, CA) coupled to a fibre optic bundle. The catheter sheath consisted of a multi-segmented balloon to be inflated in the stomach to assist in positioning of the light wand. Patients were sedated with a combination of midazolam, fentanyl and topical benzocaine oral spray. The light wand was inserted into the catheter

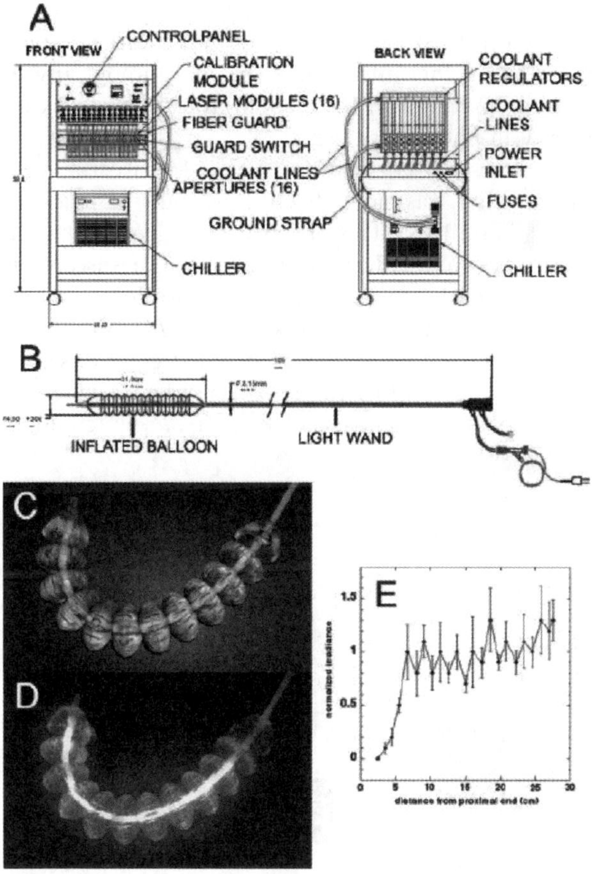

Figure 16.6 Phototherapy system. (A): The power and control system (front and back); (B) the light wand, catheter sheath and balloon; (C) light wand balloon before illumination; (D) light wand balloon illuminated; and (E) spatial distribution of irradiance along the length of the illuminated balloon. Reprinted from *Lasers in Surgery and Medicine*, 2009, **41**, 337–344. Copyright © 2009 Wiley-Liss, Inc. with permission.

sheath and then passed into the stomach. Once the device was properly placed inside the stomach (confirmed by direct visualization using upper endoscopy), the balloon was inflated and light therapy was performed. Upon completion, the balloon was deflated and the device was removed. Only one PDT procedure was administered to each patient, with treatment times ranging from 15 to 60 minutes. Results were promising, as PDT significantly (and safely) decreased bacterial load, as measured 8 hours post-treatment.

The greatest antibacterial effect was observed in the antrum as compared to the other regions of the stomach (see Figure 16.7). The most likely explanation for this phenomenon is that the endoscopic light source employed in this study delivered more light to the distal antrum as compared to the more proximal body and fundus of the stomach. On the contrary, these results may be explained by anatomical differences between the proximal and distal regions of the stomach, making it such that bacteria are more accessible to light in the antrum as compared to the body and fundus. Yet another explanation is that the *H. pylori* inhabiting the antrum may be morphologically different from those located more proximally, and could for example express increased amounts of porphyrin photosensitizers.

A number of interesting correlations were also ascertained through this study. For example, PDT was more effective in killing *H. pylori* when bacterial load was high (see Figure 16.8). This observation suggested that densely packed bacteria in the mucosa were either more accessible or susceptible to the light than sparsely distributed bacteria. Surprisingly, no dose-response correlation between increased illumination time and killing of *H. pylori* was observed. This was in contrast to our previous studies that noted good dose-response killing curves for blue light and several strains of *H. pylori in vitro*.[19,20] This unexpected finding of no dose-response correlation *in vivo* could potentially be

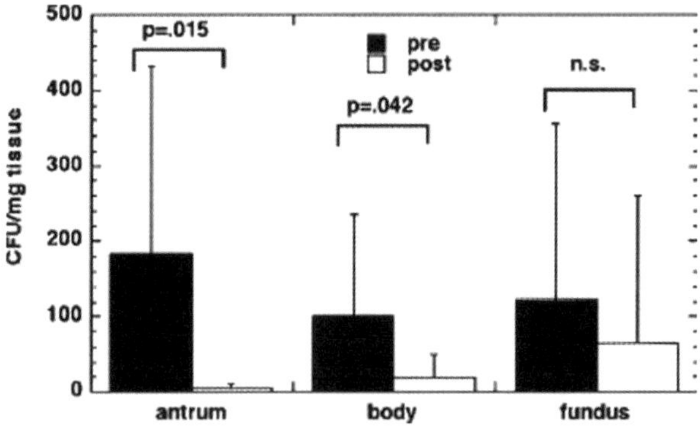

Figure 16.7 *H. pylori* culture results pre- and post-phototherapy. There was a significant difference in the mean CFU/mg cultured from biopsies removed pre- and immediately post-illumination for the antrum and body but not for the fundus. Reprinted from *Lasers in Surgery and Medicine*, 2009, **41**, 337–344. Copyright © 2009 Wiley-Liss, Inc. with permission.

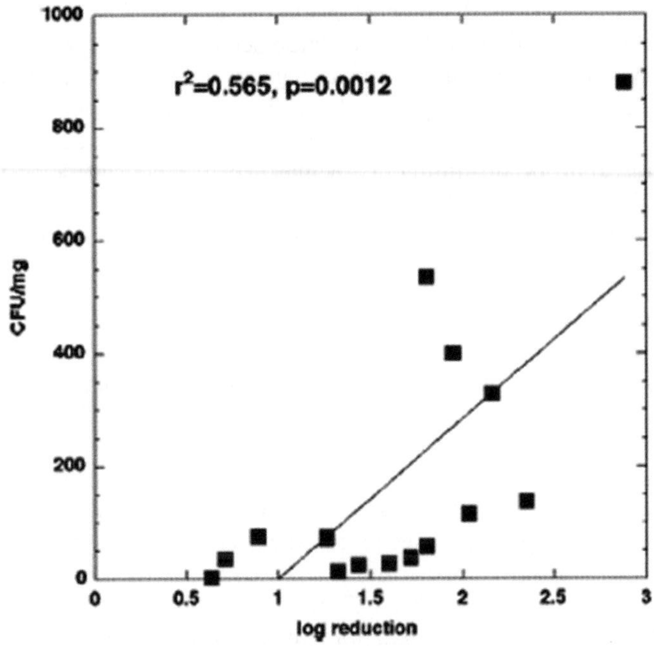

Figure 16.8 Initial *H. pylori* load (CFU/mg) in the antrum correlates with log reduction following phototherapy. Shown is the log reduction in the antrum and the initial bacterial load in CFU/mg for each patient. The linear correlation coefficient, r^2 was 0.565, $P = 0.0012$. Correlation coefficients were not significant for the stomach or fundus. Reprinted from *Lasers in Surgery and Medicine*, 2009, **41**, 337–344. Copyright © 2009 Wiley-Liss, Inc. with permission.

attributed to the existence of two (or more) populations of *H. pylori*, one easier to kill with violet light than the other. The susceptible population of bacteria may exist when the overall density of bacteria is high and may be located more superficially (and therefore be more accessible to the light). The other population may be located deeper in the mucosal ruggae or crypts (perhaps adhering to gastric epithelial cells), and thereby be concealed from the violet light. Uneven light distribution for whatever reason may be a major obstacle in achieving adequate eradication of *H. pylori* in the stomach. Presumably, this problem can be solved by a modified construction of the light source, one that allows for uniform light distribution by emitting more light in harder-to-reach places (such as the fundus) and less in the easier-to-reach areas (such as the antrum).

Even though a major decrease in bacterial load was appreciated, none of the patients achieved complete and sustained eradication of bacteria. Urease breath testing was used to estimate levels of *H. pylori*, and was used both pre- and post-treatment and significantly decreased bacterial loads were noted 8 hours post-treatment. Conversely, no significant difference was appreciated between pre-treatment levels and post-treatment levels on day 5 and beyond. It is reasonable to assume that PDT killed a large percentage of the *H. pylori* (as seen at 8 hours post-treatment), but the remaining bacteria were able to undergo rapid re-growth in order to achieve pre-treatment levels by day 5 (where even a small population of bacteria is enough to re-establish colonization). Similarly, patients treated with appropriate antibiotic regimens can experience reoccurrence of disease, a phenomenon usually attributed not to re-infection with a new bacterial strain but to recrudescence or bacterial re-growth of surviving organisms.[21]

16.6. Conclusion

The failure of PDT or phototherapy to completely eradicate *H. pylori* does not necessarily mean that there is no place for this modality of treatment in the clinical realm. It may be possible to employ PDT/phototherapy to kill approximately 99% of the initial bacterial load, and then prevent re-growth by administering antibiotics. It is reasonable to assume that a far lower dose of antibiotics (with fewer side-effects) could be employed as compared to the conventional treatment. Furthermore, PDT could be used in antibiotic-resistant strains of *H. pylori* to "kick start" treatment. Lastly, PDT may be modified to be made more effective. For example, the problem of light-resistant bacteria in patients with low bacterial loads may be overcome by the addition of a suitable "adjuvant" therapy, such as a mucolytic agent that causes deep-seated bacteria to become more accessible to the light.[22]

Notwithstanding the limitations, studies involving the role of PDT and phototherapy in the treatment of *H. pylori* infection have produced some promising results, and further research is warranted.

Acknowledgements

Research in the Hamblin laboratory is supported by the NIH (grant RO1AI050875), US Air Force MFEL program (contract FA9550-04-1-0079), Center for Integration of Medicine and Innovative Technology (DAMD17-02-2-0006), Congressionally Directed Medical Research Program (W81XWH-09-1-0514).

References

1. H. Kashiwagi, Ulcers and gastritis, *Endoscopy*, 2005, **37**, 110–115.
2. A. Kandulski, M. Selgrad and P. Malfertheiner, Helicobacter pylori infection: a clinical overview, *Dig. Liver Dis.*, 2008, **40**, 619–626.
3. J.C. Atherton, The pathogenesis of Helicobacter pylori-induced gastro-duodenal diseases, *Annu. Rev. Pathol.*, 2006, **1**, 63–96.
4. L.M. Brown, Helicobacter pylori: epidemiology and routes of transmission, *Epidemiol. Rev.*, 2000, **22**, 283–297.
5. B.E. Lacy and J. Rosemore, Helicobacter pylori: ulcers and more: the beginning of an era, *J. Nutr.*, 2001, **131**, 2789S–2793S.

6. H. Suzuki, T. Nishizawa and T. Hibi, Helicobacter pylori eradication therapy, *Future Microbiol.*, 2010, **5**, 639–648.
7. F. Parente, C. Cucino and G. Bianchi Porro, Treatment options for patients with Helicobacter pylori infection resistant to one or more eradication attempts, *Dig. Liver Dis.*, 2003, **35**, 523–528.
8. M.M. Gerrits, A.H. van Vliet, E.J. Kuipers and J.G. Kusters, Helicobacter pylori and antimicrobial resistance: molecular mechanisms and clinical implications, *Lancet Infect. Dis.*, 2006, **6**, 699–709.
9. P. Malfertheiner, U. Peitz and G. Treiber, What constitutes failure for Helicobacter pylori eradication therapy?, *Can. J. Gastroenterol.*, 2003, **17**(B), 53B–57B.
10. M. Sugimoto and Y. Yamaoka, Virulence factor genotypes of Helicobacter pylori affect cure rates of eradication therapy, *Arch. Immunol. Ther. Exp. (Warsz.)*, 2009, **57**, 45–56.
11. J. Bedwell, J. Holton, D. Vaira, A.J. MacRobert and S.G. Bown, In vitro killing of Helicobacter pylori with photodynamic therapy, *Lancet*, 1990, **335**, 1287.
12. C.E. Millson, M. Wilson, A.J. Macrobert, J. Bedwell and S.G. Bown, The killing of Helicobacter pylori by low-power laser light in the presence of a photosensitiser, *J. Med. Microbiol.*, 1996, **44**, 245–252.
13. C.E. Millson, M. Wilson, A.J. MacRobert and S.G. Bown, Ex-vivo treatment of gastric Helicobacter infection by photodynamic therapy, *J. Photochem. Photobiol. B*, 1996, **32**, 59–65.
14. C.H. Wilder-Smith, P. Wilder-Smith, P. Grosjean, H. van den Bergh, A. Woodtli, P. Monnier, G. Dorta, F. Meister and G. Wagnieres, Photoeradication of Helicobacter pylori using 5-aminolevulinic acid: preliminary human studies, *Lasers Surg. Med.*, 2002, **31**, 18–22.
15. S. Ammad, M. Gonzales, C. Edwards, A.Y. Finlay and C. Mills, An assessment of the efficacy of blue light phototherapy in the treatment of acne vulgaris, *J. Cosmet. Dermatol.*, 2008, **7**, 180–188.
16. T.B. Melo and M. Johnsson, In vivo porphyrin fluorescence for Propionibacterium acnes. A characterization of the fluorescing pigments, *Dermatologica*, 1982, **164**, 167–174.
17. T.B. Melo and G. Reisaeter, Photodestruction of endogenous porphyrins in relation to cellular inactivation of Propionibacterium acnes, *Z. Naturforsch. [C]*, 1986, **41**, 867–872.
18. C.A. Henry, M. Judy, B. Dyer, M. Wagner and J.L. Matthews, Sensitivity of Porphyromonas and Prevotella species in liquid media to argon laser, *Photochem. Photobiol.*, 1995, **61**, 410–413.
19. M.R. Hamblin, J. Viveiros, C. Yang, A. Ahmadi, R.A. Ganz and M.J. Tolkoff, Helicobacter pylori accumulates photoactive porphyrins and is killed by visible light, *Antimicrob. Agents Chemother.*, 2005, **49**, 2822–2827.
20. R.A. Ganz, J. Viveiros, A. Ahmad, A. Ahmadi, A. Khalil, M.J. Tolkoff, N.S. Nishioka and M.R. Hamblin, Helicobacter pylori in patients can be killed by visible light, *Lasers Surg. Med.*, 2005, **36**, 260–265.
21. A.J. Lembo, R.A. Ganz, S. Sheth, D. Cave, C. Kelly, P. Levin, P.T. Kazlas, P.C. Baldwin 3rd, W.R. Lindmark, J.R. McGrath and M.R. Hamblin, Treatment of Helicobacter pylori infection with intra-gastric violet light phototherapy: a pilot clinical trial, *Lasers Surg. Med.*, 2009, **41**, 337–344.
22. J. Saczko, W. Skrzypek, A. Chwîkowska, A. Choromañska, A.A. Po, A. Gamian and J. Kulbacka, Photo-oxidative action in cervix carcinoma cells induced by hpd – mediated photodynamic therapy, *Exp. Oncol.*, 2009, **31**, 195–199.

Chapter 17

Inactivation of Microbial Pathogens by Photosensitized Processes: Environmental Applications

Michela Magaraggia, Olimpia Coppellotti, Clara Fabris, Laura Guidolin and Giulio Jori

Department of Biology, University of Padova, Italy

Table of Contents

Photodynamic Inactivation of Microbial Pathogens: Medical and Environmental Applications
Editors Michael R. Hamblin and Giulio Jori

Published by the Royal Society of Chemistry, www.rsc.org

17.1. Introduction

The efficient and often selective inactivation of microbial pathogens which can be achieved by means of photosensitized processes[1] is opening favourable perspectives for the treatment of a variety of infectious diseases and is presently at an experimental stage for selected clinical applications.[2–4] On the other hand, recent findings suggest that this technique can be exploited for addressing a number of environmental problems of high scientific and social impact: in actual fact, the apparently inexorable diffusion of antibiotic-resistant microbial strains[5] is creating favourable conditions for the spreading of epidemic diseases of viral, bacterial, fungal or parasitic origin. This situation can be observed in several countries worldwide and is being steadily aggravated by the combined effect of a multiplicity of factors, such as the excessive or inappropriate prescription of antibiotics, the more and more frequent transmission of micro-organisms due to global travelling, the frequent use of antibiotics in livestock feedstuff and the expansion of poverty areas where it is hard to adequately apply prophylactic protocols. As a consequence, in addition to human patients, several constituents of the various ecosystems are at serious risk of being affected by infectious diseases, while the truly large variety of mechanisms adopted by microbial cells to enhance their resistance toward external aggressions decreases the efficacy of the majority of the currently adopted therapeutic or preventive approaches.[6] Such mechanisms include a thickening of the outer wall, encoding of new proteins which prevent the penetration of drugs, onset of mutants deficient in those porin channels allowing the influx of externally added chemicals and inter-species transfer of plasmids encoding efflux pumps and resistance enzymes. The transfer of such resistance to human pathogens represents a potentially serious threat to public health.[7]

Photosensitized processes of photodynamic type, *i.e.* involving the generation of activated oxygen species as the cytotoxic intermediates (*e.g.* singlet oxygen, hydrogen peroxide, hydroxyl radicals or the superoxide anion),[8] are endowed with specific features which make such processes a novel, convenient, low-cost and environmentally friendly methodology for an efficient and safe antimicrobial treatment in natural settings.[7,9] Thus, an extensive (3 log or larger) photodynamically induced decrease in the microbial population of a system generally occurs at photosensitizer concentrations that are by far lower than those found to be toxic for host cells or tissues, as well as for non-target organisms. Most importantly, all the available evidences point out that even the repeated use of photodynamic inactivation on a given microbial cell line does not lead to the onset of mutagenic effects or the selection of photoresistant strains. Moreover, a suitable choice of the photosensitizer leads to an extensive inactivation of a broad spectrum of pathogens, including Gram-positive and Gram-negative bacteria (both wild and multi-drug resistant strains), yeasts and fungi, mycoplasms and parasitic protozoa; this property is especially important in natural systems, which are often characterized by the presence of a heterogeneous microbial flora. Lastly, photodynamic sensitizers can be activated by inexpensive and long-lasting non-coherent visible light sources (*e.g.* filament tungsten lamps, fluorescent lamps),[10] which induce no toxic effects in the absence of the photosensitizer and do not require any special protection measures for the normal constituents of the ecosystems or the operators.

A particularly important field, where the application of photodynamic microbial inactivation for the protection of the environment and improvement of its quality can play a major role, is represented by the disinfection of microbiologically polluted water. This includes both the prevention and cure of water-borne diseases and the control of the density of harmful microorganisms in aqueous media. It is well known that a significant number of man-made activities cause the deterioration of the chemical and physical properties of the environment, promoting the proliferation of potentially dangerous microbial strains which often find a favourable habitat in humid sites, and this can spread epidemic diseases over large distances. An aqueous medium is particularly favourable for the inactivation of microbial cells by photodynamic methods since the driving force for the photoprocess is represented by an electrostatic real-time interaction between the array of negatively charged functional groups which are present at the surface of such cells (mostly, carboxylate groups) and the positively charged moieties typical of efficient antimicrobial photosensitisers.[11] As a consequence, the photoinactivation of the microbial pathogens can be performed by irradiation after very short (a few minutes) incubation times, when no appreciable accumulation of the photosensitizer in other cell types has occurred.

17.2. Porphyrins as Optimal Antimicrobial Photosensitizing Agents for Environmental Applications

Several visible light-activated photosensitizing compounds have been shown to act as antimicrobial agents; in general, those photosensitizers that bear positively charged functional groups exhibit the highest microbicidal activity.[11] However, porphyrins appear to be especially suitable for the environmental application of antimicrobial photoprocesses. This statement is based on the following considerations:

a) Porphyrins are characterized by a multi-band absorption spectrum featuring a vibrational progression in the π-π^* electronic excitation; the envelope of absorption bands ranges from the near-UV (360–400 nm) to the red (600–700 nm) spectral region. Therefore, porphyrins can be very efficiently photoexcited by essentially all the visible light wavelengths, opening the way to the use of intrinsically inexpensive lamps (*e.g.* fluorescent or filament tungsten lamps) emitting full spectrum visible light. This allows the use of low fluence-rates to achieve an important degree of microbial inactivation.
b) The most intense absorption band of porphyrins, the so-called Soret band, peaks in the 410–430 nm spectral region. Consequently, there is a very efficient interaction of the porphyrin with blue light wavelengths,

which are endowed with maximum penetration power into natural waters out of the wavelengths which are present in the solar spectrum.[12] On the basis of the Beer–Lambert law, one can estimate that for a porphyrin concentration of 0.50 μM, and a typical molar extinction coefficient ε of about 150 000 M^{-1} cm^{-1} the absorption A of the incident light will be around 0.09 per cm; since total light absorption by a system corresponds with A = 2.0–2.2, the illumination can be assumed to be sufficiently uniform to a water depth of about 25–30 cm. Less efficiently absorbed wavelengths (*e.g.* those in the green and red spectral intervals) will allow the illumination of still larger water volumes, even though the smaller probability of porphyrin photoexcitation by such wavelengths may require longer light exposure times.

c) The photosensitization mechanism typical of porphyrins involves the almost simultaneous damage of different targets located in cell membranes, including aromatic and sulfur-containing amino acids in proteins, unsaturated lipids and steroids; hence, it is very unlikely that photoresistant strains of bacterial and fungal cells are selected.[13,14] Recent investigations showed that up to ten consecutive generations of bacteria have been exposed to the photosensitizing action of cationic porphyrins with no detectable modification of their degree of photosensitivity.[15,16]

d) At the same time, the presence of the flat aromatic macrocycle favours the partitioning of porphyrin derivatives in the lipid domains of cell membranes or their stable association with a variety of proteins;[17,18] as a consequence, exogenously administered porphyrins have a high probability to be "intercepted" by specific sub-cellular sites before significant concentrations of the photosensitizing agent reach the genetic material. All the evidences available so far indicate that photoprocesses promoted by porphyrins have a very low mutagenic potential. This feature is important for the safety of the procedure in the case of repetitive treatments.

e) The overall treatment time is often reduced by the short incubation which is required in order to achieve a significantly large association of the porphyrin photosensitizer with the microbial cells. Such a fast binding kinetics is determined by the fact that the primary interaction is of electrostatic nature, namely it is promoted by the real-time interaction between the positively charged peripheral substituents in the porphyrin molecule and the large number of negatively charged functional groups (*e.g.* lipopolysaccharides, lipoteichoic and teichuronic acids) that are present in the outer wall of many microbial cells (see previous paragraph).[7,19] Thus, a number of investigations showed that prolonging the incubation of porphyrin-type derivatives with microbial cells from a few minutes to 1–2 h brings about no measurable increase in the amount of cell-bound antimicrobial agent.[20] The close proximity between the porphyrin and the potential targets of the photosensitized process represents an important factor for enhancing the efficacy of the treatment, since several photosensitive sites will be within the normal endocellular diffusion range (<0.1 μ)[21] of the photogenerated reactive oxygen species.

f) The water disinfection can be successfully obtained by using a simple and inexpensive technology. In most cases, an extensive decrease in microbial population is achieved by using low light intensities, of the order of 50 mW cm^{-2} which can be easily reached by irradiation with halogen or incandescent filament lamps, that is light sources of low cost, which have a long life span (a few thousand hours). Moreover, the low fluence-rate causes no detectable thermal effect, thus avoiding any problems related to the possible increase in water temperature. Under favourable climatic and geographical conditions, sunlight can be also chosen as a light source for porphyrin activation, thereby further reducing the total cost of the approach and opening new possibilities for the adoption of this technology in many developing countries.

g) Lastly, the proposed methodology appears to be environmentally friendly, since it is based on the use of sunlight or sunlight simulators in combination with antimicrobial agents of natural origin. In any case, the prolonged accumulation of porphyrins in the various ecosystems is unlikely owing to their gradual photobleaching under the action of ambient light: in general, the photobleaching quantum yield of porphyrins in the 10^{-4} to 10^{-5} range.[22] Moreover, the ready water solubility of cationic porphyrins should guarantee their fast dilution in the ground to levels which are below any risk threshold. In any case, the photobleaching products of these porphyrins show no major toxic effects.[23]

It is also important to underline that the chemical structure of porphyrins can be modulated by using a variety of synthetic strategies, including the insertion of substituents in the peripheral positions of the pyrrole rings or at least some of the four *meso* positions, the coordination of metal ions at the centre of the tetrapyrrolic macrocycle and the introduction of axial ligands in the fifth and sixth coordination positions of the metal ion. Thus, the possibility is open for the development of porphyrin derivatives whose physical and chemical properties, such as water-solubility, amphiphilicity, photo- and dark-stability, fine structure of the absorption spectrum, are tailored to the specific properties of the system to be sterilized.

In general, for water disinfection, *meso*-substituted porphyrins are preferentially adopted; the substituents are positively charged owing to the presence of nitrogen atoms which are quaternarized by alkylation. Typical examples are represented by *meso*-tetra(*N*-alkyl-pyridyl)porphines, whose antibacterial photoproperties have been studied in detail, even though optimal photosensitizing effects on microbial cells have been also obtained by replacement of the pyridine ring with piperidine or morpholine moieties.[24,25] One member of this class of porphyrinoid compounds is represented by *meso*-tri(*N*-methyl-pyridyl), mono(*N*-dodecyl-pyridyl)porphine (known also under the commercial name of AquaFrin), as well as by its *N*-tetradecyl analogue, which have been recently proposed as particularly efficacious photosensitizers for water disinfection:[26] the molecule of these porphyrins has been carefully engineered so that the multi-carbon dodecyl or tetradecyl chain penetrates into the lipid domains of cell membranes in order to anchor the porphyrin to the membrane and alter the three-dimensional organization of the outer wall typical of many types of microbial cells with a consequent enhancement of the membrane permeability; at the same time, the four positive charges interact electrostatically with the anionic surface of microbial cells.[27] The affinity of porphyrins for microbial cells can be also enhanced by the introduction of amphiphilic properties into their molecule,[28] as well as by the use of glycosylated derivatives.[29,30]

17.3. Photosensitized Disinfection of Water

17.3.1. The Problem of Water Contamination and Currently Used Disinfection Approaches

The question of water supply and distribution is assuming an increasingly greater significance worldwide. As of now, a growing number of countries from different areas of the planet, especially developing countries but also several Mediterranean countries, face very problematic situations concerning the availability of irrigation, industrial and potable water from both surface and ground water stocks. Many water sources are heavily polluted by hazardous chemicals, which are known to be harmful to humans and a variety of animal and vegetal ecosystems, as well as by several kinds of pathogenic micro-organisms, such as bacteria, fungi, moulds, viruses and parasitic protozoa. Potential contaminants include agricultural runoff, landfill leachates and exhausts of industrial activities.[31,32] Therefore, there is an urgent need for the definition of efficacious, reliable and environmentally safe techniques yielding an efficient water purification and disinfection. Currently, the most commonly adopted approaches are based on physical methods (*e.g.* filtration, heating or irradiation with ultraviolet light or X-rays), even though reliance is mainly placed on chemical treatments with oxidizing species, such as chlorine, chlorine dioxide or ozone.[33] However, the latter methods often exhibit a number of negative aspects, in particular the formation of endoperoxides or trihalomethanes as by-products of the reaction between the chemical disinfectant and organic matter, or the selection of multi-drug resistant microbial strains upon repeated treatment of waters characterized by a large microbial flora.[34,35] Moreover, chlorine is often phytotoxic,[36] while the use of ozone proved to be moderately expensive and involves a high consumption of energy, thereby making its implementation on a large scale a not very attractive endeavour. Therefore, it is by now apparent how difficult it is to identify a comprehensive strategy.

The use of photochemical techniques for the treatment of chemically or microbiologically contaminated waters was been proposed many years ago with so far contrasting results. Thus, the effectiveness of ultraviolet light is often limited by the reduced penetration power (less than 1 cm) of wavelengths shorter than 350 nm into natural water;[37] moreover, the direct absorption of UV-C and UV-B wavelengths by nucleic acids is a main cause of the onset of mutagenic processes leading to the selection of possibly dangerous new kinds of microbial strains.[38] To address these problems, novel photochemical pathways for water purification have been devised, including (a) the direct irradiation with full sunlight spectrum using specifically designed apparatuses in order to take advantage of the significantly deeper penetration of visible light wavelengths even into water samples characterized by relatively high levels of turbidity;[39] and (b) the introduction of photoassisted reactions *via* semiconductors (*e.g.* TiO_2 or ZnO), which can be electronically excited by light wavelengths in the UV-A spectral range giving rise to the formation of strongly oxidizing and highly cytotoxic species, *in primis* the OH radical and hydrogen peroxide.[40]

In actual fact, the disinfection of drinking water affected by an unusually high population of the Gram-negative bacterium *Escherichia coli* by means of solar reactors equipped with parabolic collectors has been obtained, even though the efficiency of the overall process was markedly enhanced by the presence of TiO_2 as a photocatalyst.[41] The addition of the semiconductor also yields an extensive destruction of a variety of organic micro-pollutants.[42] However, in spite of the intrinsic and experimentally demonstrated high efficiency of such approaches, their usefulness is presently confined to small-scale water treatment, mainly owing to the reduced water volumes which can be illuminated to a homogeneous degree by using light wavelengths in the UV-A spectral range, as well as to the relatively high costs associated with the operation of this technology. Furthermore, the oxidizing intermediates normally involved in such photoprocesses display an appreciable reactivity against most nucleotides, hence there is again a high probability of promoting mutagenic effects.[43]

Recent studies carried out at both a laboratory and a pilot plant level convincingly indicate that the photodynamic sensitization approach based on porphyrin derivatives has a high potential for the disinfection of waste waters and fish-farming waters.

17.3.2. The Photosensitized Decontamination of Microbiologically Polluted Wastewater

The possibility of reducing the population of helminth eggs and faecal coliforms in wastewater through the combined action of the tetracationic porphyrin meso-(*N*,*N*,*N*,*N*-tetramethyl-pyridyl)porphine and irradiation with visible light or direct exposure to sunlight was described by Alouini and Jemli.[44] By using micromolar porphyrin amounts and overall illumination times of about 4 hours, the coliform counts decreased by at least 5 logs. The antimicrobial photosensitizing action of the porphyrin was found to be markedly superior to that brought about under identical experimental conditions by other types of widely utilized photosensitizing agents, such as the phenothiazine derivative methylene blue or the xanthene derivative Rose Bengal.[45] Field studies showed that the efficiency of the photoprocess is controlled at least in part by the quality of the water, since the presence of significant amounts of particulate material with the consequent increase in turbidity was accompanied by a decrease in the number of faecal coliforms killed for a given totally administered light dose. The reduced photoactivity was partially due to the optical filtering carried out by the suspended particles, and partially by the occurrence of a predominantly physical quenching of the photogenerated singlet oxygen intermediate by organic or inorganic constituents of the particulate material.

Faecal coliforms and faecal enterococci also underwent an extensive mortality in wastewaters obtained from a sewage plant when the system was illuminated with full spectrum visible light in the presence of tetracationic porphyrins as the photosensitizing agents.[46] About two logs of the usually very resistant Gram-negative faecal coliforms were inactivated again upon addition of micromolar porphyrin concentrations and irradiation of the wastewater sample at a low fluence-rate, such as less than 10 mW cm^{-2} This procedure is also effective for the inactivation of viral contaminants of wastewater, as shown by experiments performed on a water sample obtained from a secondary-treated sewage plant: visible light irradiation of the water in the presence of 5 micromolar concentrations of a tetra- or tri-cationic *meso*-substituted porphyrin caused the inactivation of a T4-like bacteriophage below the limit of detection.[47] Finally, the photosensitizing action of porphyrins can also be utilized for obtaining an algicidal effect, as regards both the prevention of algae growth and the control of their proliferation; the latter process can be very important in some types of wastewater. The high activity of AquaFrin against the overgrowth of algae has been successfully tested in a variety of aquaria;[48] moreover, Schrader *et al.*[49] recently reported that both the planktonic cyanobacterium *Planktothrix perornata* (an odour-producing blue-green alga), which had been isolated from a water sample collected from a catfish pond, and a selected representative of green algae, namely *Selenastrum capricornutum*, underwent about a 50% growth inhibition upon illumination for 96 h with fluorescent light bulbs in the presence of 0.07 mg l^{-1} and, respectively, 0.2 mg l^{-1} AquaFrin.

The efficacy of a treatment protocol characterized by such mild values for some critical experimental parameters suggests that the photodynamic antimicrobial treatment can be practically applied for wastewater disinfection by using sunlight, whose spectrum optimally matches the absorption spectrum of porphyrins, even on relatively cloudy days. It appears that the presence of four positive charges in the tetrapyrrolic macrocycle of porphyrins[50] or six to eight positive charges in the tetraazaisoindole macrocycle of phthalocyanines[51] yields optimal photoinactivation results against typical faecal bacteria, such as the Gram-positive strain *Enterococcus faecalis* and the Gram-negative strain *Escherichia coli*. These *meso*-substituted porphyrins exhibit a low tendency to undergo photobleaching, especially if the irradiation is performed at low fluence rates.[44,47] Therefore, the efficiency of the photosensitized process is not impaired by a too fast decrease in the concentration of the active principle, and this feature adds to the low overall cost of this technology.

Finally, it is worth emphasizing that the tetracationic *meso*-substituted porphyrin derivatives, especially if the positive charge is localized on a quaternarized nitrogen atom, lack readily ionizable functional groups, hence they are not subjected to a typical acid-base equilibrium over a broad range of pH values. As a consequence, the photoexcitation efficiency of these porphyrins is essentially independent of the water pH unless very extreme acid and basic conditions are reached. Therefore, one would expect that the photodecontamination of wastewater from industrial sources, whose pH is often lower than 2–3 because of their significant acidity, can be readily obtained by the above-described photodynamic approach.

The safety of the photodynamic disinfection technique is further demonstrated by studies on target and non-target organisms which have their natural habitat in a water environment. Thus some free-living soil and freshwater protozoa, such as the opportunistic and highly pathogenic amoeba *Acanthamoeba palestinensis* and the ciliate *Colpoda inflata*, have been shown to be inactivated if exposed to the action of visible light-excited cationic porphyrins, in both the cystic and trophozoitic stages.[7] A detailed investigation of their degree of sensitivity indicates that in *A. palestinensis*, after incubation with 3 and 5 μM tetracationic Zn(II)pthalocyanine and irradiation, the cysts lost their excystment ability, while an almost complete drop in survival was observed with the same Zn(II)pthalocyanine in the vegetative cells at a 2 μM concentration. *C. inflata* cultures were fairly sensitive to the photocidal action of AquaFrin, showing an LD_{50} as low as 0.3 μM even in the cystic stage.[52] Other assays performed with two Branchiopod Crustacea, *Artemia franciscana* and *Daphnia magna*, show a phototoxic damage for AquaFrin concentrations higher than 0.6 μM.[53] A modulation of the response to the phototreatment depending on the photosensitizer concentration and the irradiation protocol could be elicited in all these organisms. Thus, through a careful selection of the treatment protocol, the photodynamic technique for water disinfection can be made to be selective for different microbial pathogens.

17.3.3. The Photosensitized Decontamination of Microbiologically Polluted Water from Fish-farming Plants

Waters from aquaculture ponds are often characterized by the presence of a large population of microbial pathogens owing to the generally high density of fish in the farming structures, as well as to the large concentration of food residues, organic debris and defecation products. Actually, *Vibrio anguillarum*[20,27] and *Vibrio vulnificus*,[54] two Gram-negative bacteria which occasionally colonize aquaculture plants, were shown to be rapidly inactivated by the photosensitizing action of cationic porphyrin derivatives. Similar results were obtained upon porphyrin-photosensitized killing of *Bacillus* endospores[55] and a number of bacterial species isolated from fish-farming plant waters, including *Photobacterium* sp., *Enterobacter* sp. and *Pseudomonas* sp.[56] Moreover, in *in vitro* studies the AquaFrin porphyrin has been found[49] to act as a very efficient phototherapeutant against two Gram-negative bacteria, such as *Edwardsiella ictaluri* and *Flavobacterium columnare*, which are the etiological agents of enteric septicaemia and columnaris disease; these disorders are among the leading causes of mortality in pond-raised channel catfish.[57,58] Lastly, studies performed in our laboratory using aquaria as model systems mimicking a variety of fish micro-environments clearly pointed out that the combined action of tetracationic porphyrins, such as AquaFrin or meso-tetra(*N*-methyl-pyridyl)porphine, and visible light emitted by the aquaria-incorporated fluorescent lamps induces the regression or cure of several bacterial or fungal infections in different fish without any parallel (photo)toxic effect on the fish, as well as on other constituents of the aquaria ecosystem, such as invertebrates and green plants.[48]

These findings promoted specific investigations aimed at ascertaining the possibility to actually apply the photodynamic processes for the disinfection of waters from fish-farming systems through field studies. This development was tested in pilot aquaculture plants involving rainbow trout (*Oncorhynchus mykiss*), which had been either spontaneously or artificially infected by *Saprolegnia parasitica*.[59] This choice was motivated by the consideration that infections caused by members of *Saprolegnia* species, as well as by their close water mould relatives, represent a major challenge to a large number of freshwater fish and their eggs. Prevention and control of saprolegniasis are especially difficult owing to the ubiquitous nature and the rapid spreading of such fungi. The problem is aggravated by various concomitant factors, such as (a) different stages of the fish life cycle are susceptible to saprolegniasis and in particular a marked increase in infection risk is observed during sexual maturation; (b) several forms of environmental stress, leading to a decreased water quality (*e.g.* increased average temperatures, high organic loads), act as predisposing agents for the outbreak of the disease, often by weakening the innate defence systems of the fish; (c) spores of the pathogen can successfully locate in the fish epidermis and, whilst the majority are removed or inactivated during quarantine in clear water, a significant number of viable fungal units persist on the body surface.[60] *Saprolegnia*-induced mycotic infections have been traditionally treated by the application of malachite green, a triarylmethane dye. However, recent concerns about the safety of this aromatic derivative led to the banning of its use as a fisheries chemotherapeutant in many parts of the world. Even formalin, a 24% aqueous solution of formaldehyde, which has been introduced to replace malachite green, does not eliminate the risk for acute impact on ecosystems and potential harmful effects on human health.[61]

The exposure of adult trout to AquaFrin or its tetradecyl analogue (C14) at doses of up to 1 mg l^{-1} under dim light for up to 15 consecutive days caused no significant toxic effects on the fish, as shown by the absence of mortality or development of abnormal behaviour. The porphyrins were not accumulated in detectable amounts by the various fish organs with the exception of the gills, where a significant retention of both porphyrins was observed even though a gradual clearance of the photosensitizing agents occurred when the exposure of the fish to the porphyrins was interrupted. Only in the presence of AquaFrin doses larger than 1 mg l^{-1} was some "dark" toxicity noticed, eventually leading to fish death. On the other hand, the irradiation of the porphyrin-exposed infected trout with visible light (400–800 nm) gave very encouraging results. Two protocols were developed: a *preventive treatment* based on the irradiation of artificially infected trout (about 4×10^{11} zoospores deposited on a 2 cm^2 area of scraped skin) for 1 h day^{-1} and for 10 consecutive days after addition of 0.3 mg l^{-1} porphyrin; and a *curative treatment* where trout that had already developed a marked infection by *Saprolegnia*, as shown by the presence of cottony-like patches of the mycelium, were exposed to visible light (1 h day^{-1} for 10 consecutive days) in the presence of 0.6 mg l^{-1} porphyrin. The results unequivocally showed that the outlined procedures were very successful in both inhibiting the formation of the fungal infection and causing a complete disappearance of the pre-existing infection with no detectable damage in correspondence of the perilesional tissue districts. No recurrence of the saprolegniasis in the *Saprolegnia* infected sites or in others sites of the fish was observed. Quite interestingly, such important effects were obtained by using fluence-rates of the order of 50 mW cm^{-2}, which again suggests that even sunlight on cloudy days can be used to disinfect waters in open-air fish-farming systems, thus further reducing the overall cost of the procedure and meeting the recommendations to increase the use of natural resources for energy-saving reasons.

Recent investigations from our laboratory also showed that tetracationic porphyrins at doses as low as 0.3–0.6 mg l^{-1} efficiently protect trout eggs against infections by *Saprolegnia*. In this case, the irradiation by visible light needs to be carried out on the potentially polluted water before it flows through the container of trout eggs. The direct irradiation of eggs by full spectrum visible light provokes irreversible toxic effects on the trout eggs, since a broad number of visible wavelengths are directly absorbed by carotenoid pigments, which are present in the eggs, and these once photoexcited in turn promote photochemical damage in their surroundings leading to egg death (Magaraggia and Jori, unpublished results). The preventive efficiency of specific cationic porphyrins against saprolegniasis of trout eggs is comparable with that found for some commercially used chemical disinfectants, such as formalin or peroxyacetic acid; as specified above, the latter compounds represent a serious threat for fish and environment. Hence the porphyrin/visible light combination should be preferable in terms of both safety and compliance with environmental protection.

17.3.4. Future Prospects

In spite of the encouraging results so far obtained in the field of photodynamic disinfection of wastewater and water from fish-farming plants, suitable strategies for improving the efficacy, selectivity and reliability of the technique need to be adequately explored. Thus, several lines of evidence point out that the overall photophysical properties and photosensitizing efficiency of porphyrins are enhanced by the introduction of fluorine atoms,[62] or conjugation with saccharides[63] and poly-peptides, especially those that are positively charged under physiological pH values, such as poly-lysine or poly-ornithine.[64,65] In particular, positively charged poly-peptides act as a destabilizing agent for the compact three-dimensional organization of the outer wall which surrounds several types of microbial cells, thereby enhancing their degree of photosensitivity. Very exciting perspectives have been recently opened by the preparation of water-soluble nanoparticle-photosensitizer conjugates which can efficiently generate singlet oxygen and at the same time increase the potency of light-activated antimicrobial agents. Thus, the covalent conjugation of the phenothiazine derivative toluidine blue with tiopronin-gold nanoparticles induced a four-fold enhancement in the rate and efficacy of the photoinduced killing of *Staphylococcus aureus*.[66] The improved performance of the nanoconstruct was partially a consequence of the increased extinction coefficient of the conjugate as compared with free toluidine blue.

One major step forward in the exploitation of photosensitization for water disinfection would be made possible by the utilization of photosensitizing agents that are immobilized on water-insoluble inert supports. This would allow the removal of the photosensitizer from the aqueous medium at the end of the phototreatment procedure with consequent important advantages:

- The irradiation of water prior to its introduction into (for example) the fish-farming ponds, thus avoiding any direct contact between the photoreactive species and the fish or other non-target organisms.
- The recovery of the photosensitizer from the irradiated water at the end of the treatment and its subsequent recycling, thereby further contributing to a reduction in the overall costs of the procedure.
- The complexes between the photosensitizing agent and the supporting matrix have often been found to be endowed with a significantly long shelf life, which facilitates their storage and re-use.
- Minimization of the possibility that the photosensitizer undergoes some diffusion into the environment, thus minimizing the risk of a generalized pollution by chemical contaminants.

This possibility is especially interesting when singlet oxygen-generating photosensitizers are used: this hyper-reactive oxygen species has a relatively long natural lifetime, especially in non-polar media, hence it can diffuse to distances of up to almost 1 mm before undergoing deactivation to the ground state. As a consequence, the killing of bacteria physically separated from the immobilized photosensitizers has been observed.[67,68] Therefore, singlet oxygen producers based on photosensitizers bound to solid supports have been described and initial results strongly corroborate the hypothesis that such systems are endowed with antimicrobial photoproperties.[69,70] To implement this research avenue, a variety of inert matrices which can be readily associated with the photosensitizer have been proposed, including cellulose esters,[71] chitosan,[69] agar surfaces,[72] cotton fabrics[73] and silica networks.[74] Finally, Ru(II) coordination complexes immobilized on a variety of supports have been successfully tested[75] for killing of microbial cells and proposed for the disinfection of water specimens which are characterized by the presence of a significant flora of microbial pathogens.

17.4. Photosensitized Processes and Preservation of Biodiversity

Biodiversity loss is, perhaps, the only truly irreversible environmental problem resulting from human activities.[76,77] Of major concern is the conservation future of most amphibian and reptile species.[78] Amphibians, a unique group of vertebrates containing over 6300 known species, are threatened worldwide. A recent assessment of the entire group found that nearly 32% of the world's amphibian species are threatened: amphibians have existed on earth for over 300 million years and in just the last two decades there have been an alarming number of extinctions.[77] Amphibians and reptiles play a major role in ecological food webs, as both predators and prey. As consumers of insects, rodents and other noxious species, reptiles also provide a significant benefit to agriculture and recreational activities. When abundant, amphibians can consume substantial quantities of favoured prey organisms, perhaps serving to limit prey populations.[79,80] For example, salamanders appear to play important roles in organizing many terrestrial and aquatic communities. Amphibians and reptiles are particularly sensitive to habitat disturbance; because of their close contact with air, water and soil, amphibians are considered to be good indicators of environmental health. Indeed, they have highly permeable skin that rapidly absorbs toxic substances in the air, water and soil. Furthermore, many amphibians and reptiles have poor dispersal abilities and are often unable to move to alternate areas when their habitat is disturbed. Habitat loss and fragmentation is a major factor in the decline of amphibian and reptile species. Many habitats are disappearing due to human population growth and development. Of those habitats that do remain, many are not large enough to support populations of a given species. If required habitat components are isolated from each other by agricultural fields, roadways or urban development, the animals may not be able to reach them. Declines of amphibian populations have been extensively documented in Northern and Latin America, to a minor extent in Europe, especially in relation to the anthropogenic activities or their environmental impacts such as habitat destruction, emerging infection diseases and global warming, to the point to consider that amphibians are the group experiencing the most important declines or extinctions from all vertebrates.[81–84] Even if the causes for recent amphibian declines are many, an emerging disease called chytridiomycosis and global climate change are thought to be the biggest threats to amphibians.

Chytridiomycosis is a disease caused by the fungal chytrid pathogen *Batrachochytrium dendrobatidis* (Phylum Chytridiomycota).[85] This pathogen is associated with the global loss of hundreds of species of amphibians and represents a spectacular loss of biodiversity, some say the worst in recorded history. Moreover, the water fungus *Saprolegnia*, common in fish hatcheries, and iridovirus, such as *Ranavirus*, have recently been associated with declines of amphibian populations.[86] In reptiles the decrease of genetic variability and the broad diffusion of freshwater parasitic Hirudinea caused a marked worldwide decimation of many populations. Amphibian and reptile populations are also negatively affected by environmental pollutants, such as heavy metals, pesticides and radioactive waste. Ecological studies concerning amphibians, such as *Rana latastei* (Anura) and *Triturus alpestris* (Urodela), and reptiles, such as *Emys orbicularis*, living in Europe, are relatively scarce, thus limiting the possibility of an efficient coordination of reintroduction plans. To address these problems a novel procedure of phototreatment that combines antimicrobial efficacy with safety and environmental acceptability can be proposed. The basic elements of the strategy are:

1) A porphyrin or chlorin-based compound, which exhibits a multi-band visible absorption spectrum, to enable photoactivation by essentially all wavelengths in the 350–700 nm range. A few porphyrins bearing cationic functional groups protruding from the tetrapyrrolic macrocycle display a preferential affinity for several types of microbial cells.[18]
2) An inert support, namely micrometre-sized particles covalently encapsulating the photosensitizer, can be engineered to provide long-term stability and ready dispersion properties in aqueous media. The size of particles will enable their simple removal from the water using suitable filtering devices.
3) The excitation of porphyrins by means of inexpensive visible light sources; visible light possesses a relatively deep penetration into water, thereby allowing the uniform illumination of large volumes.[12]
4) Monitoring and data collection of the land and freshwater areas where amphibians and reptiles are living are needed. These studies imply a deep historical investigation to determine the diffusion of the species in the past and in the present and knowledge of abiotic and biotic data in the area chosen for a possible repopulation by some species.

Thus a protection of biodiversity can be performed by means of a preventive photosensitization treatment of the waters (especially ponds) in order to control the diffusion of some pathogenic organisms. The success of a preventive treatment will consequently allow an improvement of the animal fitness, which can be verified by means of population studies. Preliminary data obtained in our laboratory indicate also a curative effect by means of porphyrin phototreatment of *Chelus fimbriatus* (mata mata) individuals affected by a dermal mycosis; moreover, the phototreatment of larvae of *Xenopus laevis*, an important model organism, with a tetracationic porphyrin containing an *N*-tetradecyl substituent did not induce particular teratogenic effects and mortality, suggesting the possibility to use the technique based on porphyrin photosensitization to protect at least the larvae of amphibians, while no safe conclusions can be presently drawn as regards the early embryonic stages.

17.5. Conclusions

The photodynamic inactivation of microbial pathogens appears to represent a very flexible technology which has a broad scope and potential for tackling a range of problems connected with the microbiological contamination of waters of different origin and physical or chemical characteristics. Moreover, it represents a potentially very useful tool for the protection of species endangered by infectious diseases, thus contributing to the maintenance of biodiversity. Such a multiplicity of applications is related to the possibility of achieving thorough control of the population of a variety of harmful bacteria, fungi and parasites by using a single protocol. The nature of the photosensitizing agents can be adapted to comply with the properties of the given aqueous medium (*e.g.* fish-farming tanks, natural ponds, industrial wastewater, potable water basins) since the mechanism of action of most porphyrins (as well as of other classes of photosensitizing agents) is essentially independent of the physical and chemical properties of water, namely it is based on the photoinduced generation of one highly reactive and cytotoxic species such as singlet oxygen.

If porphyrins or some analogue thereof (chlorophyll or bacteriochlorophyll, porphycene, *etc.*) are selected as antimicrobial photosensitizers, the process can be considered as based on a natural chemical agent, the porphyrin, and a natural physical agent (sunlight or a solar simulator). Therefore, the impact of the technology on the environment is likely to be very low, especially since the dosages of the chemical agent are generally small and no significant accumulation in the application sites or their surroundings is expected owing to their gradual sunlight-induced bleaching to non-toxic products and the water solubility which promotes their rapid dilution in the ground to (photo)chemically unharmful levels.

Lastly, porphyrins are also being considered as photoactivatable tools for other environmental applications, such as their use as pesticides against noxious insects.[85,86]

References

1. M.R. Hamblin and T. Hasan, Photodynamic therapy: a new antimicrobial approach to infectious disease, *Photochem. Photobiol. Sci.*, 2004, **3**, 436–450.
2. T. Maisch, A new strategy to destroy antibiotic resistant microorganisms: antimicrobial photodynamic treatment, *Mini-review Med. Chem.*, 2009, **9**, 947–983.
3. S.R. Wiegell and H.C. Wulf, Photodynamic therapy of acne vulgaris using methyl aminolevulinate: a blinded, randomized, controlled trial. *Br. J. Dermatol.*, 2006, **154**, 969–976.
4. P.G. Calzavara-Pinton, M. Venturini, R. Capezzera, R. Sala and C. Zane, Photodynamic therapy of interdigital mucose of the feet with topical application of 5-aminolevulinic acid, *Photodermatol. Photoimmun. Photomed.*, 2004, **20**, 144–147.
5. J.W. Harrison and T.A. Svec, The beginning of the end of the antibiotic era, *Quintessence Int.*, 1998, **29**, 223–229.

6. B.A. Cunha, Antibiotic resistance: control strategies, *Crit. Care Clin.*, 1998, **14**, 309–327.
7. G. Jori and O. Coppellotti, Inactivation of pathogenic micro-organisms by photodynamic techniques: mechanistic aspects and perspective applications, *Anti-infective Agents in Medicinal Chemistry*, 2007, **6**, 119–131.
8. M. Ochsner, Photophysical and photobiological processes in the photodynamic therapy of tumours, *J. Photochem. Photobiol., B: Biol.*, 1997, **39**, 1–18.
9. C.M. Carvalho, J.P. Tomé, M.A. Faustino, M.G. Neves, A.C. Tomé, J.A. Cavaleiro, L. Costa, E. Alves, A. Oliveira, A. Cunha and A. Almeida, Antimicrobial photodynamic activity of porphyrin derivatives: potential application on medical and water disinfection, *J. Porph. Phthal.*, 2009, **13**, 574–577.
10. L.F. Paula, R.O. Santos, H.D. Menezes, J.R. Britto, J.B. Vieiram, P.P. Gomtijo and A.C. Oliveira, A comparative study of irradiation systems for photoinactivation of microorganisms, *J. Braz. Chem. Soc.*, 2009, **21**, 1–7.
11. R.F. Donnelly, P.A. McCarron and M.M. Tunney, Antifungal photodynamic therapy, *Microbiol. Res.*, 2008, **163**, 1–12.
12. K.S. Baker and R.C. Smith, Spectral irradiance penetration in natural waters, in *The Role of Solar Ultraviolet Radiation in Marine Ecosystems*, ed. J. Calkins, Plenum Press, New York, 1982, pp. 233–246.
13. T. Maisch, Anti-microbial photodynamic therapy: useful in the future?, *Lasers. Med. Sci.*, 2007, **22**, 83–91.
14. W. Wainwright, Photoantimicrobials – a PACT against resistance and infection, *Drug Future*, 2004, **29**, 85–93.
15. M. Soncin, C. Fabris, A. Busetti, D. Dei, G. Roncucci and G. Jori, Approaches to selectivity in the Zn(II)-phthalocyanine-photosensitised inactivation of wild-type and antibiotic-resistant *Staphylococcus aureus*, *Photochem. Photobiol. Sci.*, 2002, **1**, 815–819.
16. F.M. Lauro, P. Pretto, L. Covolo, G. Jori and G. Bertoloni, Photoinactivation of bacterial strains involved in periodontal diseases sensitised by porphycene-polylysine conjugates, *Photochem. Photobiol. Sci.*, 2002, **1**, 468–470.
17. Z. Luksiene and A. Zukauskas, Prospects of photosensitisation in control of pathogenic and harmful microorganisms, *J. Applied Microbiol.*, 2009, **107**, 1415–1424.
18. G. Jori, C. Fabris, M. Soncin, S. Ferro, O. Coppellotti, D. Dei, L. Fantetti, G. Chiti and G. Roncucci, Photodynamic therapy in the treatment of microbial infections: basic principles and perspective applications, *Lasers Surg. Med.*, 2006, **38**, 468–481.
19. G. Jori and S.B. Brown, Photosensitised inactivation of microorganisms, *Photochem. Photobiol. Sci.*, 2004, **3**, 403–405.
20. M. Merchat, G. Bertoloni, P. Giacomoni, A. Villanueva and G. Jori, Meso-substituted cationic porphyrins as efficient photosensitisers of Gram-positive and Gram-negative bacteria, *J. Photochem. Photobiol., B: Biol.*, 1996, **32**, 153–157.
21. R.W. Redmond and J.N. Gamlin, A compilation of singlet oxygen yields from biologically relevant molecules, *Photochem. Photobiol.*, 1999, **70**, 391–475.
22. L. Polo, A. Segalla, G. Bertoloni, G. Jori, K. Schaffner and E. Reddi, Poly-lysine porphycene conjugates as efficient photosensitisers for the inactivation of microbial pathogens, *J. Photochem. Photobiol., B: Biol.*, 2000, **59**, 152–158.
23. G. Jori and J.D. Spikes, Photobiochemistry of porphyrins, in *Topics in Photomedicine*, ed. K.C. Smith, Plenum Press, New York, 1984, pp.183–319.
24. E. Reddi, M. Ceccon, G. Valduga, G. Jori, J.C. Bommer, F. Elisei, L. Latterini and U. Mazzucato, Photophysical properties and antibacterial activity of meso-substituted cationic porphyrins, *Photochem. Photobiol.*, 2002, **75**, 642–470.

25. D.A. Caminos, M.B. Spesia and E.N. Durantini, Photodynamic inactivation of *Escherichia coli* by novel meso-substituted porphyrins, by 4-(3-*N*,*N*,*N*-trimethylammoniumpropoxy)phenyl and 4-(trifluormethyl)phenyl groups, *Photochem. Photobiol. Sci.*, 2006, **5**, 56–65.
26. J.C. Bommer and G. Jori, Photodynamic porphyrin antimicrobial agents, *US Pat.*, 6,573,258, June 3, 2003.
27. M. Merchat, J.D. Spikes, G. Bertoloni and G. Jori, Studies on the mechanism of bacteria photosensitisation by meso-substituted cationic porphyrins, *J. Photochem. Photobiol., B: Biol.*, 1996, **35**, 149–157.
28. D. Lazzeri, M. Rovera, L. Pascual and E.N. Durantini, Photodynamic studies and photoinactivation of *Escherichia coli* using meso-substituted cationic porphyrin derivatives with asymmetric charge distribution, *Photochem. Photobiol.*, 2004, **80**, 286–291.
29. J.P. Tomé, E.M. Silva, A. Pereira, C.M. Alonso, M.A. Faustino, M.G. Neves, A.C. Tomé, J.A. Cavaleiro, S.A. Tavares, R.R. Duarte, M.F. Caeiro and M.L. Valdeira, Synthesis of neutral and cationic tripyridylporphyrin-D-galactose conjugates and the photoinactivation of HSV-1, *Bioorg. Med. Chem.*, 2007, **15**, 4705–4713.
30. J.P. Tomé, M.G. Neves, A.C. Tomé, J.A. Cavaleiro, A.F. Mendonça, I.N. Pegado, R. Duarte and M.L. Valdeira, Synthesis of glycoporphyrin derivatives and their antiviral activity against herpes simplex virus, *Bioorg. Med. Chem.*, 2005, **13**, 3878–3888.
31. R.L. Droste, *Theory and Practice of Water and Waste Water Treatment*. John Wiley & Sons, New York, 2007.
32. G. Bitton, *Wastewater Microbiology*, John Wiley & Sons, New York, 2005.
33. C.R. Tomas, Wastewater engineering and treatment: problems and solutions, in *Chemical Engineering Techniques in Biotechnology*, ed. M.A. Winkler, Elsevier, Amsterdam, 1990, pp. 23–94.
34. A.C. Twort, F.M. Law and F.W. Croley, *Water Supply*, Arnold Publishing Co., London, 2005.
35. M. Doré, *Chimie des Oxydants et Traitment des Eaux: Techniques et Documentation*, Lavoisier, Paris, 2003.
36. M. Sioi, A. Bolosis, E. Kostopoulou and I. Poulios, Photocatalytic treatment of coloured wastewater from medical laboratories, *J. Photochem. Photobiol., B: Biol.*, 2006, **184**, 18–25.
37. A.C.E. Moore., B. Ortel and T. Hasan, Mechanisms of photodynamic therapy, in *Photodynamic Therapy*, ed. T. Patrice, Royal Society of Chemistry, Cambridge, United Kingdom, 2003, pp. 19–57.
38. D.L. Mitchell and D. Kerentz, The induction and repair of DNA photodamage in the environment, in *Environmental UV Photobiology*, ed. A.R. Young, L.O. Bjorn, J. Moan and W. Nultsch, Plenum Press, New York, 1993, pp. 345–378.
39. R.W. Boyle and D. Dolphin, Structure and biodistribution relationships of photodynamic sensitisers, *Photochem. Photobiol.*, 2006, **14**, 4253–4259.
40. J.P. Percherancier, R. Chapelon and B. Pouyet, Semiconductor-sensitised photodegradation of pesticides in water, *J. Photochem. Photobiol., A: Chem.*, 1995, **87**, 261–266.
41. C. Sichel, J. Blanco, S. Malato and P. Fernandez-Ibanez, Effects of experimental conditions on *Escherichia coli* survival during solar photocatalytic water disinfection, *J. Photochem. Photobiol., A: Chem.*, 2007, **189**, 239–246.
42. L.C. Amalric, C. Guillard and P. Pichat, The photodegradation of 2,3-benzofuran and its intermediates, 2-coumarone and salicylaldehyde, in TiO_2 aqueous suspensions, *J. Photochem. Photobiol., A: Chem.*, 1995, **85**, 239–246.

43. D. Bahnemann, J. Cunningham, M.A. Fox, E. Pelizzetti, P. Pichat and N. Serpone, *Aquatic and Surface Photochemistry*, Lewis Publishing Co., Boca Raton, Florida, 1994.
44. Z. Alouini and M. Jemli, Destruction of helminth eggs by photosensitised porphyrin, *J. Environ. Monit.*, 2002, **3**, 548–551.
45. M. Jemli, Z. Alouini, S. Sabbahi and M. Gueddari, Destruction of fecal bacteria in wastewater by three photosensitisers. *J. Environ. Monit.*, 2002, **4**, 511–516.
46. C.M. Carvalho, A.T. Gomes, S.C. Fernandes, A.C. Prata, M.A. Almeida, M.A. Cunha, J.P. Tomé, M.A. Faustino, M.G.S. Neves, A.C. Tomé, J.A. Cavaleiro, Z. Lin, J.P. Rainho and J. Rocha, Photoinactivation of bacteria in wastewater by porphyrins: bacterial β-galactosidase actiivty and leucine-uptake as methods to monitor the process, *J. Photochem. Photobiol., B: Biology*, 2007, **88**, 112–119.
47. L. Costa, C.M. Carvalho, M.A. Faustino, J.P. Tomé, A.C. Tomé, J.A. Cavaleiro, A. Cunha and A. Almeida, Sewage bacteriophage inactivation by cationic porphyrins: influence of light parameters, *Photochem. Photobiol. Sci.*, 2010, **9**, 1126–1134.
48. M. Magaraggia and G. Jori, Photodynamic approaches to water disinfection, in *CRC Handbook of Organic Photochemistry and Photobiology*, ed. F. Ghetti, CRC Press, Boca Raton, 2010, in press.
49. K.V. Schrader, J.C. Bommer and G. Jori, *In vitro* evaluation of the antimicrobial agent AquaFrin as a bactericide and selective algicide for use in channel catfish aquaculture, *North Amer. J. Aquaculture*, 2010, **72**, 304–308.
50. E. Alves, L. Costa, C. Carvalho, J.P. Tomé, M.A.F. Faustino, M.G. Neves, A.C. Tomé, J.A. Cavaleiro, A. Cunha and A. Almeida, Charge effects on the photoinactivation of Gram-negative and Gram-positive bacteria by cationic meso-substituted porphyrins, *BMC Microbiol.*, 2009, **9**, 70–75.
51. G. Roncucci, D. Dei, M.P. De Filippis, L. Fantetti, I. Masini, B. Cosimelli and G. Jori, Zn(II)phthalocyanine and corresponding conjugates; their preparation and use for photodynamic therapy and as a diagnostic agents, *US Pat.*, 5965598, 1999.
52. O. Coppellotti, J. Bommer, C. Fabris, M. Soncin, M. Magaraggia and L. Guidolin, Porphyrin-photosensitised inactivation of protozoa in the trophozoitic and cystic stage, *8th International Symposium on Photodynamic Therapy and Photodiagnosis in Clinical Practice*, Bressanone, Italy, 2010, Book of Abstracts, Abstract AM-8.
53. C. Fabris, J. Bommer, G. Jori, M. Soncin, M. Magaraggia, L. Guidolin and O. Coppellotti, Photosensitised inactivation of crustacea by a cationic porphyrin, *8th International Symposium on Photodynamic Therapy and Photodiagnosis in Clinical Practice*, Bressanone, Italy, 2010, Book of Abstracts, Abstract FT-6.
54. P.N. Wong, S.K. Mak and M.W. Low, *Vibrio vulnificus* peritonitis after handling of seafood in a patient receiving CAPD, *Am. J. Kidney Dis.*, 2005, **46**, 87–90.
55. T. N Demidova and M.R. Hamblin, Photodynamic inactivation of *Bacillus* endospores mediated by phenothiazinium dyes, *Appl. Environ. Microbiol.*, 2005, **71**, 6918–6925.
56. C. Arrojado, MSc Thesis, University of Aveiro, 2009.
57. J.P. Hawke, R.M. Durborow, R.L. Thune and A.C. Camus, Enteric septicaemia of catfish, *Southern Regional Aquaculture Centre Publication 477*, 1998, Stoneville, Mississippi, USA.
58. R.M. Durborow, R.L. Thune, J.P. Hawke and A.C. Camus, Columnaris disease: a bacterial infection caused by *Flavobacterium columnare*, *Southern Regional Aquaculture Centre Publication 479*, 1998, Stoneville, Mississippi, USA.
59. M. Magaraggia, F. Faccenda, A. Gandolfi and G. Jori, Treatment of microbiologically polluted aquaculture waters by a novel photochemical technique of potentially low environmental impact, *J. Environ. Monit.*, 2006, **8**, 923–931.

60. D.J. Alderman, *Salmon Saprolegniasis*, 1998, Portland, USDOE, Bonneville Power Administration.
61. C.M. Gieseker, S.G. Serfling and R. Reimschuessel, Formalin as a chemotherapeutant for prophylaxis of finfish eggs and fry, *Aquaculture*, 2006, **253**, 120–129.
62. J.C.P. Grancho, M.M. Pereira, M.D.G. Miguel, A.M.R. Gonçalves and H.D. Burrows, Synthesis, spectra and photophysics of some free base tetrafluoroalkyl and tetrafluoroaryl porphyrins, *Photochem. Photobiol.*, 2002, **75**, 249–256.
63. Y. Mikata, Y. Onchi and K. Tabata, Sugar-dependent photocytotoxic properties of tetra- and octa-glycoconjugated tetraphenylporphyrins, *Tetrahedron Lett.*, 1998, **39**, 4505–4508.
64. L. Polo, A. Segalla, G. Bertoloni, G. Jori, K. Schaffner and E. Reddi, Poly-lysine porphycene conjugates as efficient photosensitisers for the inactivation of microbial pathogens, *J. Photochem. Photobiol., B: Biol.*, 2000, **59**, 152–158.
65. J.P.C. Tome, M.G.P.M.S. Neves and A.C. Tome, Synthesis and antibacterial activity of new poly-L-lysine-porphyrin conjugates, *J. Med. Chem.*, 2004, **47**, 6649–6652.
66. J.G. Tomàs, S. Tubby, I.P. Parkin, N. Narband, L. Dekker, S.P. Nair, M. Wilson and C. Street, Lethal photosensitisation of *Staphylococcus aureus* using a toluidine blue O-tiopronin-gold nanoparticles conjugate, *J. Mater. Chem.*, 2007, **17**, 3739–3746.
67. T.A. Dahl, W.R. Midden and P.E. Hartman, Pure singlet oxygen cytotoxicity for bacteria, *Photochem. Photobiol.*, 1987, **46**, 345–352.
68. G. Valduga, G. Bertoloni, E. Reddi and G. Jori, Effect of extra-cellularly generated singlet oxygen on Gram-positive and Gram-negative bacteria, *J. Photochem. Photobiol., B: Biol.*, 1993, **21**, 81–86.
69. R. Bonnett, M.A. Krysteva, I.G. Ialov and S.V. Artarsky, Water disinfection using photosensitisers immobilized on chitosan, *Water Res.*, 2006, **40**, 1269–1275.
70. J. Mosinger, O. Jirsak, P. Kubat, K. Lang and B. Mosinger Jr, Bactericidal nanofabrics based on photoproduction of singlet oxygen, *J. Mater. Chem.*, 2007, **17**, 164–166.
71. M. Krouit, R. Granet, P. Branland, B. Verneuil and P. Krausz, New photoantimicrobial films composed of porphyrinated lipophilic cellulose esters, *Bioorg. Med. Chem. Lett.*, 2006, **16**, 1651–1655.
72. D. Caminos and E. Durantini, Photodynamic inactivation of *Escherichia coli* immobilized on agar surfaces by a tricationic porphyrin, *Bioorg. Med. Chem. Lett.*, 2006, **14**, 4253–4259.
73. C. Ringot, N. Saad, V. Sol, R. Granet, P. Bressolier and P. Krausz, Amino porphyrin-grafted cotton fabric using 1,3,5-triazine link – application to antimicrobial phototherapy, *Abstract ICCP-6*, Albuquerque, USA, 2010, Abstract S26.
74. S. Artrasky, S. Dimitrova, R. Bonnett and M. Krysteva, Immobilization of zinc phthalocyanines in silicate matrices and investigation of their photobactericidal effect on *Escherichia coli*, *Sci. World J.*, 2006, **6**, 374–382.
75. M.E. Jimenez-Hernandez, F. Manjon, D. Garcia-Fresnadillo and G. Orellana, Solar water disinfection by singlet oxygen photogenerated with polymer-supported Ru(II) sensitisers, *Sol. Energy*, 2009, **80**, 1382–1387.
76. G. Ceballos and P.R. Ehrlich, Mammal population losses and extinction crisis, *Science*, 2002, **296**, 904–907.
77. S.N. Stuart, J.S. Chanson, N.A. Cox, B.E. Young, A.S.L. Rodrigues, D.L. Fischman and R.W. Walker, Status and trends of amphibian declines and extinction worldwide, *Science*, 2004, **306**, 1783–1786.

78. G. Santos-Barrera, J. Pacheco, F. Mendoza-Quijano, F. Bolaños, G. Cháves, G.C. Daily, P.R. Ehrlich and G. Ceballos, Diversity, natural history and conservation of amphibians and reptiles from the San Vito Region, southwestern Costa Rica, *Rev. Biol. Trop.*, 2008, **102**, 755–778.
79. M. Capula, *Guide to Reptiles and Amphibians of the World*, Simon & Schuster Inc., New York, NY, USA.
80. J.W. Gibbons, D.E. Scott, T.J. Rayan, K.A. Buhlmann, T.D. Tuberville, B.S. Metts, J.L. Greene, T.M. Mills, Y. Leiden, S. Poppy and C.T. Winne, Reptiles in decline: the global decline of reptiles, déjà vu amphibians, *BioScience*, 2000, **50**, 653–666.
81. R.A. Alford, P.M. Dixon and J.H. Pechman, Ecology. Global amphibian population declines, *Nature*, 2001, **412**, 499–500.
82. B.E. Young, K.R. Lips, J.K. Reaser, R. Ibanez, A.W. Salas, J.R. Cedeno, L.A. Coloma, S. Ron, E. La Marca, J.R. Meyer, A. Munoz, F. Bolaños, G. Cháves and D. Romo, Population declines and priorities for amphibian conservation in Latin America, *Conserv. Biol.*, 2001, **15**, 1212–1223.
83. J.P. Collins and A. Storfer, Global amphibian declines: sorting and hypothesis, *Divers. Distrib.*, 2003, **9**, 89–98.
84. M.B. Araújo, W. Thuiller and R.G. Pearson, Climate warming and the decline of amphibians and reptiles in Europe, *J. Biogeogr.*, 2006, **33**, 1712–1728.
85. L. Berger, R. Speare, P. Daszak, D.E. Green, A.A. Cunningham, C.L. Goggin, R. Slocombe, M.A. Ragan, A.D. Hyatt, K.R. MacDonald, H.B. Hines, K.R. Lips, G. Marantelli and H. Parkes, Chytridiomycosis causes amphibian mortality associated with population declines in the rainforests of Australia and Central America, *Proc. Natl. Acad. Sci. USA*, 1998, **95**, 9031–9036.
86. P. Daszak, L. Berger, A.A. Cunningham, A.D. Hyatt, D.E. Green and R. Speare, Emerging infectious diseases and amphibian population declines, *Emerg. Infect. Dis.*, 1999, **5**, 735–748.
87. T. Ben Amor and G. Jori, Sunlight-activated insecticides: historical background and mechanisms of phototoxic action, *Insect Biochem. Physiol.*, 2000, **30**, 915–925.
88. L. Pujol-Lereis, A. Massaldi, A. Rabossi and L. Quesada-Allué, Photosensitising effect of hematoporphyrin IX on immature stages of *Ceratitis capitata* (Diptera: Tephritidae), *Photochem. Photobiol.*, 2010, in press.

Subject Index

Note: page numbers in *italic* refer to figures and diagrams.